VOLUME SIX HUNDRED AND TWENTY

METHODS IN ENZYMOLOGY

New Approaches for Flavin Catalysis

METHODS IN ENZYMOLOGY

Editors-in-Chief

ANNA MARIE PYLE
Departments of Molecular, Cellular and Developmental Biology and Department of Chemistry
Investigator, Howard Hughes Medical Institute
Yale University

DAVID W. CHRISTIANSON
Roy and Diana Vagelos Laboratories
Department of Chemistry
University of Pennsylvania
Philadelphia, PA

Founding Editors

SIDNEY P. COLOWICK and NATHAN O. KAPLAN

VOLUME SIX HUNDRED AND TWENTY

METHODS IN ENZYMOLOGY

New Approaches for Flavin Catalysis

Edited by

BRUCE A. PALFEY

Associate Professor of Biological Chemistry & Associate Director, Program in Chemical Biology, Department of Biological Chemistry, University of Michigan Medical School, Ann Arbor, United States

ELSEVIER

ACADEMIC PRESS

An imprint of Elsevier

Academic Press is an imprint of Elsevier
50 Hampshire Street, 5th Floor, Cambridge, MA 02139, United States
525 B Street, Suite 1650, San Diego, CA 92101, United States
The Boulevard, Langford Lane, Kidlington, Oxford OX5 1GB, United Kingdom
125 London Wall, London, EC2Y 5AS, United Kingdom

First edition 2019

Notices

Knowledge and best practice in this field are constantly changing. As new research and experience broaden our understanding, changes in research methods, professional practices, or medical treatment may become necessary.

Practitioners and researchers must always rely on their own experience and knowledge in evaluating and using any information, methods, compounds, or experiments described herein. In using such information or methods they should be mindful of their own safety and the safety of others, including parties for whom they have a professional responsibility.

To the fullest extent of the law, neither the Publisher nor the authors, contributors, or editors, assume any liability for any injury and/or damage to persons or property as a matter of products liability, negligence or otherwise, or from any use or operation of any methods, products, instructions, or ideas contained in the material herein.

ISBN: 978-0-12-816829-5
ISSN: 0076-6879

For information on all Academic Press publications
visit our website at https://www.elsevier.com/books-and-journals

Publisher: Zoe Kruze
Acquisition Editor: Ashlie M. Jackman
Editorial Project Manager: Shellie Bryant
Production Project Manager: James Selvam
Cover Designer: Christian J. Bilbow

Typeset by SPi Global, India

Contents

Contributors

Sanjoy Adak
Department of Chemistry, Texas A&M University, College Station, TX, United States

Eyal Akiva
Department of Bioengineering and Therapeutic Sciences and Department of Pharmaceutical Chemistry; Quantitative Biosciences Institute, University of California, San Francisco, CA, United States

Godwin Aleku
School of Chemistry, University of Manchester, Manchester, United Kingdom

Dave W. Anderson
Michael Smith Laboratories, University of British Columbia, Vancouver, BC, Canada

Patricia C. Babbitt
Department of Bioengineering and Therapeutic Sciences and Department of Pharmaceutical Chemistry; Quantitative Biosciences Institute, University of California, San Francisco, CA, United States

Samuel S. Bailey
School of Chemistry, University of Manchester, Manchester, United Kingdom

Arune Balaikaite
School of Chemistry, University of Manchester, Manchester, United Kingdom

Khorcheska A. Batyrova
Department of Chemical Engineering and Applied Chemistry, University of Toronto, Toronto, ON, Canada

Donald F. Becker
Department of Biochemistry, Redox Biology Center, University of Nebraska-Lincoln, Lincoln, NE, United States

Sophia M. Becker
Department of Biochemistry, Redox Biology Center, University of Nebraska-Lincoln, Lincoln, NE, United States

Tadhg P. Begley
Department of Chemistry, Texas A&M University, College Station, TX, United States

Siddarth Chandrasekaran
Department of Chemistry and Chemical Biology, Cornell University, Ithaca, NY, United States

Wei-chen Chang
North Carolina State University, Raleigh, NC, United States

Shelbi L. Christgen
Department of Biochemistry, Redox Biology Center, University of Nebraska-Lincoln, Lincoln, NE, United States

Matthew J. Cliff
Manchester Institute of Biotechnology and School of Chemistry, The University of Manchester, Manchester, United Kingdom

Janine N. Copp
Michael Smith Laboratories, University of British Columbia, Vancouver, BC, Canada

Brian R. Crane
Department of Chemistry and Chemical Biology, Cornell University, Ithaca, NY, United States

H.D. Duan
Department of Chemistry, University of Kentucky, Lexington, KY, United States

Elizabeth A. Edwards
Department of Chemical Engineering and Applied Chemistry, University of Toronto, Toronto, ON, Canada

Holly R. Ellis
Department of Chemistry and Biochemistry, Auburn University, Auburn, AL, United States

Karl Fisher
School of Chemistry, University of Manchester, Manchester, United Kingdom

Robert Flick
Department of Chemical Engineering and Applied Chemistry, University of Toronto, Toronto, ON, Canada

Kevin Francis
Department of Chemistry, Texas A&M University—Kingsville, Kingsville, TX, United States

Peter L. Freddolino
Department of Biological Chemistry and Department of Computational Medicine & Bioinformatics, University of Michigan, Ann Arbor, MI, United States

Jarrod B. French
Department of Chemistry; Department of Biochemistry and Cell Biology, Stony Brook University, Stony Brook, NY, United States

Giovanni Gadda
Department of Chemistry; Department of Biology; Center for Biotechnology and Drug Design; Center for Diagnostics and Therapeutics, Georgia State University, Atlanta, GA, United States

Deepankar Gahloth
School of Chemistry, University of Manchester, Manchester, United Kingdom

George T. Gassner
Department of Chemistry and Biochemistry, San Francisco State University, San Francisco, CA, United States

Irina Gostimskaya
School of Chemistry, University of Manchester, Manchester, United Kingdom

Sam Hay
Manchester Institute of Biotechnology and School of Chemistry, The University of Manchester, Manchester, United Kingdom

Huanming Huang
School of Chemistry, University of Manchester, Manchester, United Kingdom

Andreea I. Iorgu
Manchester Institute of Biotechnology and School of Chemistry, The University of Manchester, Manchester, United Kingdom

James N. Iuliano
Department of Chemistry, Stony Brook University, Stony Brook, NY, United States

Kalani Karunaratne
Department of Chemistry, University of Iowa, Iowa City, IA, United States

Anna N. Khusnutdinova
Department of Chemical Engineering and Applied Chemistry, University of Toronto, Toronto, ON, Canada

David Leys
School of Chemistry, University of Manchester, Manchester, United Kingdom

Changfan Lin
Department of Chemistry and Chemical Biology, Cornell University, Ithaca, NY, United States

Hung-wen Liu
University of Texas, Austin, TX, United States

Stephen A. Marshall
School of Chemistry, University of Manchester, Manchester, United Kingdom

A.-F. Miller
Department of Chemistry, University of Kentucky, Lexington, KY, United States

Tatiana V. Mishanina
Department of Chemistry and Biochemistry, University of California San Diego, La Jolla, CA, United States

N. Mohamed Raseek
Department of Chemistry, University of Kentucky, Lexington, KY, United States

Graham R. Moran
Department of Chemistry and Biochemistry, Loyola University Chicago, Chicago, IL, United States

Daniel Nohr
Institut für Physikalische Chemie, Albert-Ludwigs-Universität Freiburg, Freiburg, Germany

Karl A.P. Payne
School of Chemistry, University of Manchester, Manchester, United Kingdom

David Procter
School of Chemistry, University of Manchester, Manchester, United Kingdom

Stephen E.J. Rigby
School of Chemistry, University of Manchester, Manchester, United Kingdom

John M. Robbins
School of Chemical & Biomolecular Engineering, Georgia Institute of Technology, Atlanta, GA, United States

Annica Saaret
School of Chemistry, University of Manchester, Manchester, United Kingdom

Raspudin Saleem-Batcha
ZBSA, Center for Biological Systems Analysis; Faculty of Biology, University of Freiburg, Freiburg, Germany

Erik Schleicher
Institut für Physikalische Chemie, Albert-Ludwigs-Universität Freiburg, Freiburg, Germany

Nigel S. Scrutton
Manchester Institute of Biotechnology and School of Chemistry, The University of Manchester, Manchester, United Kingdom

Emil Sjulstok
Department of Physics, Chemistry and Pharmacy, University of Southern Denmark, Odense, Denmark

Pablo Sobrado
Department of Biochemistry, Center for Drug Discovery, Virginia Tech, Blacksburg, VA, United States

Ilia A. Solov'yov
Department of Physics, Chemistry and Pharmacy, University of Southern Denmark, Odense, Denmark

Robert J. Stanley
Department of Chemistry, Temple University, Philadelphia, PA, United States

Robin Teufel
ZBSA, Center for Biological Systems Analysis; Faculty of Biology, University of Freiburg, Freiburg, Germany

Christopher J. Thibodeaux
McGill University, Montréal, QC, Canada

Nobuhiko Tokuriki
Michael Smith Laboratories, University of British Columbia, Vancouver, BC, Canada

Peter J. Tonge
Department of Chemistry; Department of Radiology; Center for Advanced Study of Drug Action, Stony Brook University, Stony Brook, NY, United States

Hannah Valentino
Department of Biochemistry, Center for Drug Discovery, Virginia Tech, Blacksburg, VA, United States

Cornelius J. van Galen
Department of Chemistry, Temple University, Philadelphia, PA, United States

T.A. Varner
Department of Chemistry, University of Kentucky, Lexington, KY, United States

Jonathan P. Waltho
Manchester Institute of Biotechnology and School of Chemistry, The University of Manchester, Manchester, United Kingdom

Po-Hsiang Wang
Department of Chemical Engineering and Applied Chemistry, University of Toronto, Toronto, ON, Canada

Stefan Weber
Institut für Physikalische Chemie, Albert-Ludwigs-Universität Freiburg, Freiburg, Germany

Johnny Xiao
Department of Chemical Engineering and Applied Chemistry, University of Toronto, Toronto, ON, Canada

Alexander F. Yakunin
Department of Chemical Engineering and Applied Chemistry, University of Toronto, Toronto, ON, Canada; Centre for Environmental Biotechnology, School of Natural Sciences, Bangor University, Bangor, United Kingdom

Estella F. Yee
Department of Chemistry and Chemical Biology, Cornell University, Ithaca, NY, United States

Preface

The rigorous study of flavin-dependent enzymes started more than eight decades ago. In that time, much has been learned, with perhaps the most important fact being that there is much more yet to be understood. The first purified flavoenzymes were instrumental in elucidating the role of vitamins in biochemistry and prosthetic groups in enzymology. Flavoenzymes were recognized early on for catalyzing a remarkable array of redox reactions, and for bridging the two-electron processes in intermediary metabolism to the one-electron processes of respiration that supply so much biological energy. A vast amount of research has been aimed at understanding the mechanisms of canonical flavoenzymes. These efforts were especially aided by the fact that the flavin prosthetic group is a chromophore which undergoes readily detected changes in the visible/near UV region as events such as ligand binding or intermediate formation occurred in the active site. Thus the direct study of the enzyme by rapid-reaction methods was often rewarded by the detection of intermediates whose plausibility could be matched with models from bioorganic chemistry for virtuous cycles of hypothesis, experiment, and (extremely vigorous) debate. Many of the fundamental models for particular flavoenzyme reaction types were mature well before the first crystal structures trickled—and then flooded—in. With the addition of structures and site-directed mutagenesis to complement the rich information derived from kinetics, spectroscopy, and chemistry, an understanding of the canonical flavoenzymes of unparalleled depth has emerged. A key lesson from this is the humbling realization that we still do not understand the classic paradigms fully—with the embarrassment of riches in experimental approaches and data that come to a flavoenzymologist, so also comes an expanded responsibility in explaining enzyme action in more detail.

Meanwhile, biochemistry, powered by the proliferation of technologies from molecular biology, genomics, and analytical chemistry, continually uncovers new metabolic pathways and metabolites in new organisms, leading, not unsurprisingly, to the discovery of the classic flavoenzyme reaction types applied to novel substrates. More surprisingly, however, is the new flavin chemistry—new mechanisms and new intermediates—beyond the classical paradigms that continues to emerge. Flavins now appear to form catalytic N^5 carbon and oxygen adducts. Reduced flavin has been proposed to act as a strong acid catalyst, not a redox reaction at all, in isomerizing alkenes. Isoprenylated flavin derivatives catalyze decarboxylations in reactions proposed to go through biologically rare cycloadditions. And there are more reported noncanonical activities, detected but barely studied—hydroxylations of alkanes, reductions of alkenes, halogenations and dehalogenations, photodecarboxylations and so on. This expanding repertoire shows that despite decades of study, we have not really grasped the full chemical potential of the conjugated isoalloxazine moiety of flavins.

And so it is hoped that this volume provides a resource to support the additional work needed to understand flavoenzymes. Key fundamental techniques are described that take full advantage of the flavin as a reporter group. There are also updates on classic methods from spectroscopy that should aid in probing reactivity in yet greater detail. Important aspects are also addressed in applying powerful computer methods and in the biochemistry of important emerging systems.

Bruce A. Palfey
Associate Professor of Biological Chemistry &
Associate Director, Program in Chemical Biology
Department of Biological Chemistry
University of Michigan Medical School
Ann Arbor, United States

CHAPTER ONE

Methods for determining the reduction potentials of flavin enzymes

Shelbi L. Christgen, Sophia M. Becker, Donald F. Becker*
Department of Biochemistry, Redox Biology Center, University of Nebraska-Lincoln, Lincoln, NE, United States
*Corresponding author: e-mail address: dbecker3@unl.edu

Contents

Abstract

A key factor for flavoenzyme activity is the reduction potential of the bound flavin. The reduction potentials of protein-bound flavins span approximately a 500-mV range consistent with flavoenzymes having critical roles in metabolism and a variety of biological processes. Redox potentials of flavoenzymes have traditionally been determined using an electrode-based system with either direct or indirect electrochemical coupling between the protein and the working electrode. An electrode independent method, however, is also now commonly used and involves calculating the unknown flavin reduction potential of the protein from the known reduction potential of a reference or indicator dye. Here, the "classic" potentiometric method and the xanthine/xanthine oxidase methods are described. Both methods rely on equilibrium between protein-bound flavin and redox dyes. The potentiometric method measures the equilibrated redox potential of the protein-dye mixture whereas the xanthine/xanthine oxidase technique relies on slow continuous enzymatic reduction to maintain a constant equilibrium between the protein and the dyes. Because electrochemical equipment is not required, the xanthine/xanthine oxidase method is more accessible and convenient for

Methods in Enzymology, Volume 620
ISSN 0076-6879
https://doi.org/10.1016/bs.mie.2019.03.004

researchers seeking to determine reduction potentials of flavoproteins or other biological redox centers such as hemes. The xanthine/xanthine oxidase method has been used to determine flavin reduction potentials from +132 to −417 mV, demonstrating it is suitable for characterizing the redox properties of most flavoproteins.

Abbreviations

***Ec*PutA**	PutA from *Escherichia coli*
E_m	reduction potential
hq	hydroquinone
ox	oxidized
PutA	proline utilization A
red	reduced
sq	semiquinone
XO	xanthine oxidase

1. Introduction

In the study of oxidation-reduction (redox) reactions, it is essential to understand the electrochemical properties of the electron acceptor and donor molecules involved. The reduction potential (aka, midpoint potential, E_m) is a measure of the thermodynamic tendency of a redox center to accept or donate an electron(s). In a reaction involving two redox centers, the center with the lowest E_m value will serve as the electron donor whereas the center with the more positive E_m value will act as the electron acceptor. A useful analogy is proton transfer between two partners with the molecule having the lower p*K*a value donating a proton to the molecule with the higher p*K*a value. Electron transfer that flows from a more negative potential redox center to a more positive potential center is characterized by a positive electromotive force (i.e., $\Delta E > 0$ with $\Delta E = E_{m(acceptor)} - E_{m(donor)}$) and is thus thermodynamically favored ($\Delta G = -nF\Delta E$ where n is the number of electrons transferred and F is the Faraday constant).

Flavoenzymes catalyze a large array of redox reactions in biology, which requires proper orientation and electrochemical tuning of the flavin cofactor for optimal reactivity with the acceptor or donor redox molecule. An important consideration for reactivity is the redox potential of the flavin cofactor when bound to the enzyme. Unbound flavin has a redox potential of −0.219 V at pH 7.0 (Clark, 1960; Clark & Lowe, 1956). If the redox potential of the flavin cofactor did not change upon binding to the enzyme, there would be no need for flavin electrochemistry. This, however, would severely narrow the role of flavin-dependent enzymes in biology. The ability of flavin enzymes to catalyze a wide-range of redox reactions in variety of biological processes relies on the electrochemical tuning of the enzyme bound flavin redox cofactor (Heuts, Scrutton, McIntire, & Fraaije, 2009; Pellet & Stankovich, 2002; Pellett, Becker, Saenger, Fuchs, & Stankovich, 2001). Enzyme active sites are able to modulate the flavin redox potential generally from +100 mV to −400 mV, spanning a 500 mV range that enables flavoenzymes to catalyze a variety of redox reactions.

Flavins are unique in their ability to catalyze one and two electron transfer reactions, equilibrating between quinone (ox), semiquinone (one-electron reduced anionic and neutral forms), and hydroquinone (hq) species that can be readily distinguished in the UV–visible spectrum (Fig. 1). The reduction of flavin occurs across the diazadiene portion of the isoalloxazine ring which includes atoms N(1), C(10a), C(4a), and N(5). Oxidized flavin is yellow with absorption maxima around 370 and 450 nm whereas the hydroquinone form (two-electron reduced) has only weak broad absorption in the UV–visible spectrum. Flavin semiquinone exhibits two distinctive absorption bands depending on the protonation state of the N(5) atom. In the unprotonated form or red anionic semiquinone, a strong absorption peak is found at 370 nm. If the N(5) position is protonated, a strong absorbance band around 570 nm is observed indicative of blue neutral semiquinone. The flavin-binding site in the protein and the pH environment dictate the amount and type of semiquinone formed during electron transfer reactions. If the reduction potential for the first single-electron transfer step (ox/sq or E_1) is significantly more positive than the reduction potential for the second single-electron transfer step (sq/hq, E_2), that is $E_1 \gg E_2$, flavin semiquinone will be stabilized during the reduction process. If $E_2 \gg E_1$, semiquinone is not thermodynamically favored and therefore does not appear. In redox reactions involving semiquinone, the overall reduction potential (E_m) for two-electron reduction of the flavin can be calculated by averaging the reduction potentials of E_1 and E_2.

In this article we will consider two methods for characterizing the redox potential of an enzyme bound flavin cofactor. The first is the spectroelectrochemistry method that involves stepwise manipulation of the redox potential of the flavoenzyme solution (Stankovich, 1980; Wilson, 1978), whereas the second method involves an enzyme system to slowly and continuously

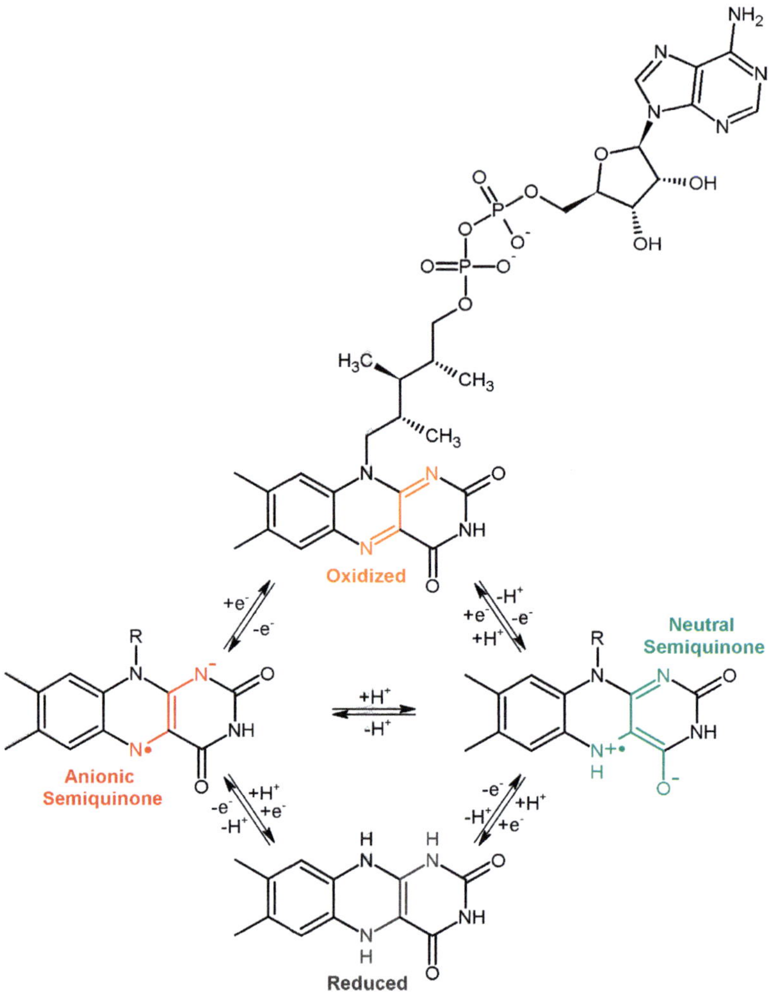

Fig. 1 Redox structures of flavin adenine dinucleotide. Flavins cycle between three redox states of oxidized, semiquinone, and hydroquinone (two-electron reduced). Flavin semiquinone exists in anionic and neutral forms depending on its protonation state.

reduce the flavoenzyme (Massey, 1991). Both methods rely on equilibration between redox indicator dyes and the protein-bound flavin. The main difference is the source of reducing equivalents. In the spectroelectrochemical or potentiometric method, a potentiostat is used to reduce the indicator dyes and then is switched off to allow equilibration between the dyes and the protein-bound flavin (Fig. 2A). In the enzymatic method, xanthine is the source of reducing equivalents via xanthine oxidase (XO) providing a

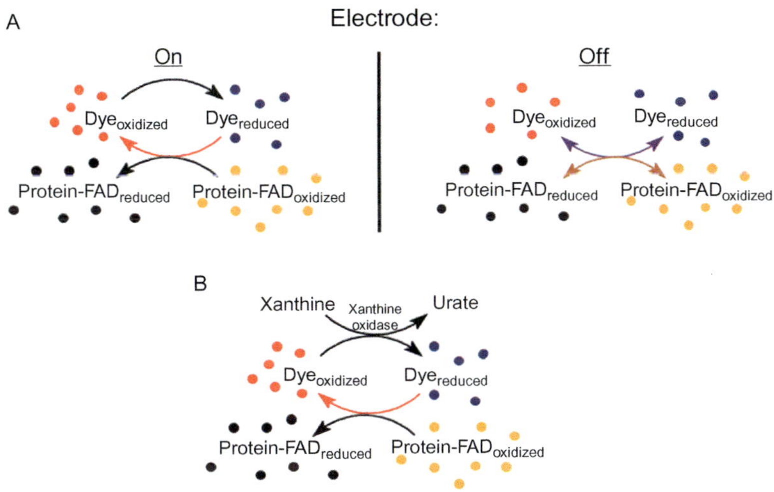

Fig. 2 Methods for determining reduction potentials. (A) The potentiometric method involves repeated cycles of reducing the protein-dye mixture by bulk electrolysis "On" and then turning the applied potential "Off" to allow equilibration of the system and the redox potential of the cell to be measured. (B) The xanthine/XO method is a slow continuous reduction of the protein-dye mixture using small amounts of XO while scanning the UV–visible spectrum at regular time intervals during reduction.

controlled reduction of the indicator dyes and protein-bound flavin (Fig. 2B). Both methods have their advantages and are important tools for studying the electrochemical properties of flavoenzymes.

2. Spectroelectrochemistry method

The reduction potential (E_m) value for a redox center in a protein is determined using the Nernst equation (1), which relates the ratio of oxidized and reduced species with E_{meas}, which is

$$E_{meas} = E_m + (2.3RT/nF) \log [ox]/[red] \quad (1)$$

the redox potential measured at equilibrium during a potentiometric titration. In the Nernst equation E_m is the standard reduction potential of the redox center, R is the gas constant ($8.314 J K^{-1} mol^{-1}$), F is the Faraday constant ($96,485 J V^{-1} mol^{-1}$), and n is the number of electrons transferred. In a potentiometric titration, E_{meas} is recorded at different concentrations of the oxidized [ox] and reduced [red] species (Stankovich, 1980; Wilson, 1978). A plot of E_{meas} versus [ox]/[red], called a Nernst plot, is then used to determine E_m, which is equal to E_{meas} when [ox]/[red] = 1. A Nernst plot also reveals the number of electrons transferred during the titration. A slope of 0.0295 V (25 °C) indicates a two-electron reduction process whereas a slope of 0.059 V (25 °C) reflects one-electron transfer. Observation of a flavin semiquinone during the potentiometric titration would be consistent with one-electron transfer.

The potentiometric method for determining the E_m of redox centers involves measuring the redox potential (E_{meas}) of the protein solution in the cell at different concentrations of [ox] and [red] species. The [ox]/[red] ratio is determined by monitoring a unique feature of the redox center, such as a change in the UV–visible spectrum of the protein-bound flavin. Examples of other techniques used for monitoring [ox] and [red] concentrations are electron paramagnetic resonance (Becker, Leartsakulpanich, Surerus, Ferry, & Ragsdale, 1998; Harder, Feinberg, & Ragsdale, 1989) and limited proteolysis (Zhu & Becker, 2003). Potentiometric cells can thus be modified accordingly for the type of measurement used for monitoring [ox] and [red] species. Regardless of the modifications made, a potentiometric cell houses three electrodes, which includes a working electrode (e.g., gold or platinum), an auxiliary electrode (Ag/AgCl), and a reference electrode (Ag/AgCl). A spectroelectrochemical cell with a three-electrode system was designed by Stankovich (1980) and has been the archetype for potentiometric measurements of flavoproteins (Pellet & Stankovich, 2002). Miniaturized spectroelectrochemical cells can now also be designed to reduce the amount of protein needed for an experiment.

During a potentiometric titration, the contents of the cell are reduced by bulk electrolysis using a potentiostat with an applied negative potential (e.g., −600 mV). Reduction of the cell contents occurs at the working electrode while oxidation reactions occur at the auxiliary electrode. In theory, it would seem that the redox center of a protein could be reduced directly at the working electrode. Electron transfer between a protein and an electrode, however, can be very slow due to the buried nature of the redox center, which is typical of flavoproteins. Often in enzymes, the flavin cofactor is buried and less accessible to a solid electrode surface, thus making direct electron transfer generally difficult with flavoenzymes (Pellet & Stankovich, 2002). Additionally, proteins sometimes adsorb onto the electrode surface causing irreversible electron transfer behavior. To circumvent these problems, a redox mediator dye such as benzyl viologen or methyl viologen is added to the protein solution to facilitate reduction of the flavin. Benzyl viologen and methyl viologen are particularly useful because they have redox potentials that are more negative (−350 mV and −446 mV, respectively) than most protein-bound flavins and they have no visible spectrum in the oxidized state (Clark, 1960; Massey, 1991; Pellet & Stankovich, 2002). Additionally, benzyl viologen and methyl viologen remain positively charged upon one-electron reduction (methyl viologen^{++} + 1e^{-} = methyl viologen$^{\bullet +}$) thereby reducing kinetic barriers for reducing flavins in proteins. The other type of dye used in a potentiometric titration experiment is a reference or indicator dye. These dyes link the equilibrium of the protein-dye mixture with the working electrode when the potential of the cell is recorded. Once equilibrium is achieved, the ratio of the oxidized and reduced dye in solution reports on the ratio of oxidized and reduced flavin in the enzyme, thus the measured potential reflects the redox potential of the protein-bound flavin. The indicator dyes used for a potentiometric titration need to be chosen carefully and should have well defined optical and electrochemical properties and, as a general guideline, should have a redox potential that is ±30 mV (n = 2) from the E_m of the protein-bound flavin (Pellet & Stankovich, 2002).

Potentiometric experiments need to be performed under anaerobic conditions using a sealed anaerobic electrochemical cell or under a nitrogen atmosphere in an anaerobic chamber. To make a cell anaerobic, the cell containing the protein-dye mixture is flushed with nitrogen or argon and residual oxygen is removed by repeated cycles of vacuum and inert gas using a gas/vacuum line. Because oxygen reacts readily with redox mediators and indicators, oxygen must be excluded from potentiometric experiments and maintained at <5 ppm. A potentiometric titration is performed by stepwise reduction of the cell contents by cycling between "On," when a negative potential is applied, and "Off," when no potential is applied (Fig. 2A). In the "Off" state, the potential at the working electrode is monitored as the protein-dye mixture is given time to equilibrate (~30 min–2 h). The protein solution is considered to be at equilibrium when there is minimal drifting (drift is less than 1 mV in 5 min) of the measured redox potential (E_{meas}) and no further change in the UV–visible spectrum. Once the measured potential is stable, the potential between the reference (e.g., Ag/AgCl) and working electrode is recorded along with the UV–visible spectrum of the dye-protein mixture. The titration resumes upon applying a negative potential again to the system. The protein-dye solution is thus reduced in a step-wise fashion from the fully oxidized (quinone) to the fully reduced (hydroquinone) state with the redox potential and UV–visible spectrum being recorded at each step of reduction. The measured potentials are reported in reference to the normal hydrogen electrode by calibrating the reference electrode using a potassium ferricyanide/ferrocyanide standard solution (Stankovich, 1980). The concentration of oxidized and reduced species is determined at each point in the titration from the UV–visible spectra of the flavin. To accurately quantify the oxidized and reduced forms of the protein-bound flavin, the indicator dye spectra are subtracted from the spectra of the dye-protein mixture at each corresponding measured potential. Spectra of the dye alone must thus be acquired at similarly measured redox potentials as for the protein-dye mixture.

The advantage of using a potentiostat is the ability to manipulate the redox environment of the cell in the reductive and oxidative directions. The E_m value determined from a potentiometric titration ideally should not depend on whether equilibrium was approached from the reductive or oxidative direction. To check this, an oxidative potentiometric titration is performed in which the protein is first fully reduced and then a positive potential is applied in a stepwise fashion to oxidize the redox center. For an oxidative titration, ferrocyanide or ferrocene (+422 mV) are good choices as a mediator dye for readily oxidizing flavin proteins upon applying a positive potential to the protein-dye mixture (Pellet & Stankovich, 2002; Vinod, Bellur, & Becker, 2002). In practice, it is best to apply an oxidative potential to the protein-dye mixture periodically throughout the reductive titration to ensure that the protein-dye mixture is truly in equilibrium.

In lieu of a potentiostat, step-wise reduction of a protein can also be performed with chemical reducing agents as outlined by Dutton (1978) using sodium dithionite or by photoreduction using 5-deaza-flavin (Klopprogge & Schmitz, 1999; McIver et al., 1998; Ravasio, Curti, & Vanoni, 2001). Similarly to that described above, mediator and indicator dyes are included in the cell solution and after each addition of reducing equivalent, the redox potential is recorded after the protein-dye mixture has fully equilibrated. To test the reversible behavior of a redox center the reduced protein can be titrated with a chemical oxidizing agent such as potassium ferricyanide.

3. Xanthine oxidase method

Over 50 years ago a method was described by Minnaert (1965) for determining equilibrium constants for the reaction between cytochromes *c* and *a*, and cytochrome *c* with the reference dye dichlorophenol indophenol. It was shown that the reduction potential of a redox center could be determined using a reference or indicator dye of known redox potential by plotting the log of oxidized/reduced concentrations of the redox center versus the indicator dye (Minnaert, 1965). Minnaert used an elaborate titration apparatus that allowed the redox environment to be manipulated in the reductive and oxidative directions using NADH/NADH dehydrogenase and potassium ferricyanide, respectively (Minnaert, 1965). The apparatus allowed drops of NADH and $K_3Fe(CN)_6$ to be titrated into the solution to adjust the redox environment. A variation of the Minnaert method is to reduce the protein-dye mixture with sodium dithionite in a stepwise manner and quantify the concentrations of the oxidized and reduced forms of the protein and dye after each addition of dithionite (Crane, Yeh, Luba, & Claiborne, 2000; Melby, Li, & Mitchell, 2014; Ukaegbu, Henery, & Rosenzweig, 2006). The innovation by Massey was to use continuous reduction of the dye-protein mixture with xanthine using catalytic amounts of XO in a manner that is slow enough to allow full equilibration between the indicator dyes and protein-bound flavin during the reduction process (Massey, 1991). Thus, the equilibration time after each addition of reducing agent was eliminated.

The materials needed and a technical outline for determining the reduction potential of protein-bound flavin using the xanthine/XO method of Massey are provided below.

3.1 Materials

1. Xanthine, XO, benzyl viologen, methyl viologen, and appropriate indicator dye(s) are readily available from commercial sources. Stock solutions of xanthine (10 mM in water, pH 10–11 with sodium hydroxide) and xanthine oxidase (10–100 μM in 10 mM sodium phosphate, pH 7.5–8.0) are made prior to the experiment and stored at −20 °C.
2. A UV–visible cell that can be degassed and flushed with nitrogen or argon (nitrogen/argon) and kept sealed from ambient oxygen is required. Some researchers have used a UV–visible cell that is modified to include a side arm from which XO can be tipped into the solution after the cuvette is made anaerobic. Alternatively, a septum-tight cuvette can be used in which the contents of the experiment are first degassed and flushed with nitrogen/argon. A separate solution of XO that has been flushed with nitrogen/argon is then added through the septum using a gastight syringe. Researchers have also used a stopped-flow apparatus that is housed in an anaerobic chamber (Langer et al., 2013).
3. UV–visible spectrophotometer capable of collecting repetitive scans from 700 to 300 nm.

3.2 Protocol

1. The flavoprotein should be highly purified and be ≥10 μM (final concentration) in an appropriate buffer to ensure an absorbance of >0.1 near 450 nm. The UV–visible spectrum of the oxidized protein is recorded prior to adding any indicator dyes.
2. Benzyl viologen or methyl viologen (2–100 μM final concentration), a suitable indicator dye (5–20 μM final concentration), and xanthine (200–400 μM final concentration) are next added to the flavoprotein solution. The UV–visible spectrum of the flavoprotein-dye mixture is then recorded.
3. The flavoprotein-dye mixture is next made anaerobic by flushing the sealed cuvette cell with nitrogen/argon. If a gas/vacuum line is available, the cell can also be made anaerobic by repeated cycles of vacuum and nitrogen/argon gas.
4. The UV–visible spectrum of the anaerobic flavoprotein-dye mixture is then recorded. Next, XO is tipped into the solution or if using a septum-tight cuvette, XO is added via a gastight syringe as noted above. If using a septum-tight cuvette, some researchers have reported it is convenient to continually flush the headspace of the cuvette with nitrogen/argon during the entire course of the experiment to maintain an anaerobic environment (van den Heuvel, Fraaije, & van Berkel, 2002). Another precaution that is sometimes used to maintain anaerobic conditions is to add glucose, glucose oxidase and catalase to the dye-protein mixture to scrub residual oxygen (Arents, Perez, Hendriks, & Hellingwerf, 2011).
5. After adding XO, the spectra are recorded in regular time intervals (1–5 min) depending on the rate of reduction. The reduction is determined to be complete once the absorbance of the flavoprotein and indicator dye is unchanged over multiple scans. Most reductions are complete within 3 h but some researchers have reported collecting spectra for up to 8 h. A key indicator of complete reduction is the absorbance change of benzyl viologen or methyl viologen, which have no absorbance in the oxidized state but exhibit a sharp absorbance peak at 395 nm in the one-electron reduced forms (e.g., methyl viologen$^{\bullet +}$) (Massey, 1991; Massey & Hemmerich, 1980; Pellet & Stankovich, 2002). If the flavoprotein of interest has a reduction potential near that of benzyl viologen (−350 mV, pH 7), concomitant changes in the absorbance of the flavin and benzyl viologen will be observed.

3.3 Considerations

The xanthine/XO method is a convenient method for characterizing the spectral changes of the flavin during reduction by using only a mediator dye (methyl viologen) and no indicator dye. Changes in the absorbance spectrum and the observation of whether a semiquinone species is formed during reduction can be readily monitored (Hu, Chuenchor, & Rokita, 2015; Ingavat, Kavran, Sun, & Rokita, 2017). Also, the molar absorptivity values for the oxidized, semiquinone, and hydroquinone species can be determined. The molar absorptivity values can be used to quantify the amount of each flavin redox species during the reduction process for calculations of the redox potential.

In preparation for the xanthine/XO method, it is important to note that the choice of which indicator dye to use is mainly dependent on two factors. First, the redox potential of the indicator dye should generally be within 30 mV of the reduction potential for the flavoprotein of interest. Because a priori knowledge of the redox potential of the flavoprotein is often not possible, researchers will need to perform the xanthine/XO assay with different indicator dyes before concluding which dye is most suitable for their flavoprotein. Second, spectral interference between the indicator dye and the flavoprotein needs to be taken into account. Ideally, reduction of both components can be cleanly monitored at wavelengths of minimal interference such as following reduction of the flavin at an isosbestic point of the indicator dye. The redox potential and spectral properties of most indicator dyes are available from different resources, thus helping the researcher choose the appropriate dye(s) (Clark, 1960; Fasman, 1985). If spectral interference of the dye with the protein-bound flavin cannot be avoided, then the dye will need to be reduced by the xanthine/XO method in the absence of protein under identical conditions to those used in the presence of protein. Spectra of the dye alone are then subtracted at similar time points from the spectra of the dye-protein mixture during the xanthine/XO reduction. This results in clean spectra of the flavoprotein that can be used for directly monitoring changes in flavin absorbance during the xanthine/XO reduction.

Another important consideration is the amount of XO to use for the reduction assays. For the determinations of the reduction potentials listed in Table 1, the XO concentration ranged from 10 to 100 nM. The rate at which the components in the assay are reduced needs to be slow enough to allow proper equilibration. The best practice is to perform the xanthine/XO assay with different amounts of XO to test whether the rate

Table 1 Redox potentials of enzyme-bound flavins determined by the xanthine oxidase method.

Flavoenzyme	Redox potential (mV)	Indicator dye	Reference
Berberine bridge enzyme	+132 (pH 7.0)	Toluylene blue	Winkler, Kutchan, and Macheroux (2007)
Pyranose dehydrogenase	+92 (7.0)	Phenazine methosulfate	Graf et al. (2015)
p-Cresol methylhydroxylase	+85 (ox/sq) (pH7) +15 (sq/hq) (pH7)	Thionin	Efimov, Cronin, and McIntire (2001)
Monomeric sarcosine oxidase	+80 (pH8)	Toluylene blue	Zhao, Song, Chen, Mathews, and Jorns (2002)
Vanillyl-alcohol oxidase	+55 (pH7)	Thionin	Fraaije, van den Heuvel, van Berkel, and Mattevi (1999)
Amadoriase	+48 (ox/sq) (pH7) −52 (sq/hq) (pH7)	Thionin Indigo-tetrasulfonate	Wu, Palfey, Mossine, and Monnier (2001)
Cytokinin oxidase/dehydrogenase (*covalent flavin*) *also did CV*	+22 (ox/sq) (pH7.5) −4 (sq/hq) (pH7.5)	Thionin Methylene blue	Frebortova et al. (2004)
Alditol oxidase	−4 (ox/sq) (pH7.5) −213 (sq/hq) (pH7.5)	Methylene blue Anthraquinone-2-sulfonate	Heuts, van Hellemond, Janssen, and Fraaije (2007)
Electron transfer flavoprotein	−21 (pH7)	Pyocyanine	Roberts, Salazar, Fulmer, Frerman, and Kim (1999)
Glycine oxidase $E_m = -179$	−35 (ox/sq) (pH7) −323 (sq/hq) (pH7)	Indigo-tetrasulfonate Benzyl viologen	Pedotti, Ghisla, Motteran, Molla, and Pollegioni (2009)
6-hydroxy-D-nicotine oxidase	−37 (pH7.5)	Pyocyanine/resorufin	Kopacz and Fraaije (2014)
Adenosine 5′-phosphosulfate (APS) reductase	−45 (pH7)	Resorufin	Fritz, Buchert, and Kroneck (2002)
Sdha1 subunit of succinate: ubiquinone oxidoreductase	−55 (pH7.0)	Indigo-tetrasulfonate	Maklashina, Rajagukguk, Iverson, and Cecchini (2018)
L-lactate: cytochrome *c* oxido reductase	−81 mV (pH7)	Resorufin	Cenas, Le, Terrier, and Lederer (2007)

(S)-Mandelate dehydrogenase	−91 mV (pH 7.5)	Indigo-disulfonate	Dewanti, Xu, and Mitra (2004)
L-lactate oxidase	−95 mV (ox/sq) (pH 7)	Indigo-disulfonate	Yorita, Matsuoka, Misaki, and Massey (2000)
D-Aspartate oxidase	−97 mV (pH 7)	Indigo-trisulfonate	Negri, Tedeschi, Ceciliani, and Ronchi (1999)
D-amino acid oxidase	−98 (pH 7.5)	Indigo-disulfonate	Pollegioni, Porrini, Molla, and Pilone (2000)
Pyranose 2-oxidase	−105 mV (pH 7.0)	Indigo-disulfonate	Kujawa et al. (2006)
PutA	−106 (pH 7.5)	Resorufin	Zhu et al. (2013)
Glutaryl-CoA dehydrogenase	−132 (pH 7)	Indigo-disulfonate	Dwyer, Rao, Goodman, and Frerman (2000)
Γ-*N*-methyl-aminobutyrate oxidase	−135 (ox/sq) (pH 7.5)	Indigo-disulfonate	Chiribau, Sandu, Fraaije, Schiltz, and Brandsch (2004)
PqsL monooxygenase	−140 (pH 8.0)	Indigo-disulfonate	Drees et al. (2018)
Azobenzene reductase	−145 (pH 7.0)	Indigo-disulfonate	Langer et al. (2013)
Bacillus cereus flavodoxin 1	−149 (ox/sq) (pH 7.1)	2-Hydroxy-1, 4-naphthoquinone	Gudim, Hammerstad, Lofstad, and Hersleth (2018)
Monoamine oxidase A	−159 (ox/sq) (pH 7.2) −262 (sq/hq) (pH 7.2)	Indigo disulfonate Anthraquinone-2–sulfonate	Sablin and Ramsay (2001)
p-Hydroxybenzoate hydroxylase	−163 (pH 7)	1-Hydroxyphenazine	Palfey, Basu, Frederick, Entsch, and Ballou (2002)
L-α-glycerophosphate oxidase	−167 (pH 7.0)	Cresyl violet	Maenpuen et al. (2015)
3-Hydroxybenzoate 6-hydroxylase	−175 (pH 8)	Indigo-disulfonate	Sucharitakul, Wongnate, Montersino, van Berkel, and Chaiyen (2012)
RebC	−179 (pH 7.0)	1-hydroxyphenazine	Ryan et al. (2007)
Cholesterol oxidase	−179 (ox/sq) (pH 7.5) −255 (sq/hq) (pH 7.5)	Cresyl violet Safranin T	Gadda et al. (1997)

Continued

Table 1 Redox potentials of enzyme-bound flavins determined by the xanthine oxidase method.—cont'd

Flavoenzyme	Redox potential (mV)	Indicator dye	Reference
Anacystis nidulans flavodoxin	−221 (ox/sq) (pH 7)	Anthraquinone-2-sulfonate	Hoover et al. (1999)
Ferredoxin/flavodoxin $NADP^+$ oxidoreductase 1 (FNR1)	−228 (ox/sq) (pH 7.1)	Anthraquinone-2-sulfonate	Gudim et al. (2018)
p-Hydroxyphenylacetate (HPA) hydroxylase	−236 mV (pH 7)	Phenosafranine	Sucharitakul, Chaiyen, Entsch, and Ballou (2005)
Methylenetetrahydrofolate reductase	−237 mV (pH 7.2)	Phenosafranine	Trimmer, Ballou, and Matthews (2001)
Aidb	−258 (sq/hq) (pH 8.0)	Phenosafranine	Hamill et al. (2012)
Appa-BLUF domain	−260 (pH 8)	Phenosafranine	Arents et al. (2011)
Cytochrome b_5 reductase	−273 (pH 7)	Phenosafranine	Roma, Crowley, and Barber (2006)
NifL	−277 (pH 8)	Phenosafranine	Klopprogge and Schmitz (1999)
Thioredoxin reductase	−305 (pH 7)	Safranine O	Hamill, Chobot, Hernandez, Drennan, and Elliott (2008)
Aminoacetone oxidase	−324 (pH 8.0)	Benzyl viologen	Molla et al. (2014)
NADH-dependent ferredoxin-$NADP^+$ oxidoreductase I (Nfn)	−326 (pH 8.0)	Phenosafranine	Lubner et al. (2017)
Apoptosis-inducing factor	−353 (pH 8.0)	Benzyl viologen	Sevrioukova (2016)
Nitroalkane oxidase	−367 (pH 7)	Benzyl viologen	Gadda and Fitzpatrick (1998)
Lipoamide dehydrogenase	−382 (pH 7)	Benzyl viologen	Argyrou, Blanchard, and Palfey (2002)
A. nidulans flavodoxin D100N mutant	−417 (sq/hq) (pH 8.5)	Methyl viologen	Hoover et al. (1999)

of reduction is appropriate for determining the reduction potential (see Section 3.4). If changing the amount of XO does not impact the estimate of the reduction potential, then the equilibration time between the dye and the protein is adequate during the experiment. A question may also arise as to why benzyl viologen or methyl viologen is used in the xanthine/XO method. Similarly to that noted above for potentiometric experiments, benzyl viologen and methyl viologen facilitate the reduction of the protein-dye mixture by shuttling electrons between xanthine oxidase and the flavoprotein. Xanthine oxidase generates one-electron semiquinone species during the reaction with xanthine. Benzyl viologen and methyl viologen are obligate one-electron transfer redox dyes, thus, they act as efficient electron carriers between XO and the indicator dye-protein mixture to help ensure full and facile equilibrium during the reduction.

3.4 Data analysis

The absorbance changes accompanying reduction of the indicator dye and flavoprotein are recorded at each point during the xanthine/XO reduction. From these data, the ratio of (oxidized/reduced) dye and (oxidized/reduced) flavoprotein is calculated from different time points during the reduction (Eq. 2).

$$\text{ox}/\text{red} = (\Delta A_{\text{total}} - \Delta A_{\text{obs}})/\Delta A_{\text{obs}} \tag{2}$$

The log (ox/red) of the flavoprotein is then plotted versus the log (ox/red) of the dye. From this plot the ΔE_m between the flavoprotein and the standard dye can be determined when ox/red = 1 for the protein. Fig. 3 summarizes the equations for relating the equilibrium of the protein and dye and how to calculate the reduction potential of the protein ($E_{m(\text{protein})}$) from the log-log plot of protein versus dye.

In addition to calculating the reduction potential of the flavin, the log-log plot also provides information about whether one or two electrons are transferred between the protein and the dye. A linear slope of 1 indicates that the protein and the dye accept the same number of electrons, such as two electrons being required for complete reduction of each component. A linear slope of 0.5, however, indicates that the flavoprotein accepts only one electron whereas the dye accepts two electrons for complete reduction (see explanation in Fig. 3). In the scenario of a one-electron transfer reaction, a flavin semiquinone species may be observed in the UV–visible spectrum to provide corroborating evidence for one-electron transfer to the flavin. If the

Equations for calculating the unknown reduction potential of the protein

$$E_{(Protein)} = E_{m(Protein)} + \frac{2.3\,RT}{n(p)F}\log\left(\frac{Ox(protein)}{Red(protein)}\right)$$

$$E_{(Dye)} = E_{m(Dye)} + \frac{2.3\,RT}{n(dye)F}\log\left(\frac{Ox(dye)}{Red(dye)}\right)$$

At equilibrium, $E_{(Protein)} = E_{(Dye)}$, thus

$$E_{m(Protein)} + \frac{2.3\,RT}{n(p)F}\log\left(\frac{Ox(protein)}{Red(protein)}\right) = E_{m(Dye)} + \frac{2.3\,RT}{n(dye)F}\log\left(\frac{Ox(dye)}{Red(dye)}\right)$$

and upon rearranging gives

$$\log\left(\frac{Ox(protein)}{Red(protein)}\right) = \frac{n(p)}{n(dye)} * \log\left(\frac{Ox(dye)}{Red(dye)}\right) - \left(E_{m(Protein)} - E_{m(Dye)}\right) * \frac{nF}{2.3\,RT}$$

from which a plot of log (ox/red) protein versus log (ox/red) dye gives a slope of 1 when $n(\mathrm{p}) = n(\mathrm{dye})$, and a slope of 0.5 when $n(\mathrm{p}) = 1$ and $n(\mathrm{dye}) = 2$. When Ox(dye)/Red(dye) = 1, the y-intercept value is equal to

$$y = -\left(E_{m(Protein)} - E_{m(Dye)}\right) * \frac{nF}{2.3\,RT}$$

which upon rearranging allows the reduction potential for the protein $E_{m(Protein)}$ to be calculated from the known value of $E_{m(Dye)}$.

$$E_{m(Protein)} = E_{m(Dye)} - y * \left(\frac{2.3\,RT}{nF}\right)$$

Fig. 3 Equations used for calculating the unknown reduction potential of a protein from data obtained by the xanthine/XO method.

slope of the plot is not near 0.5 or 1, than the protein-dye mixture may not be equilibrating adequately during the reduction process. In this case, the experiment should be repeated with different amounts of XO and it should be verified that absorbance changes in the dye are not interfering at the wavelength used for monitoring flavin absorbance. It should be noted that sometimes in the literature, the log (ox/red) dye is plotted on the *y*-axis and the log (ox/red) protein is on the *x*-axis. In this case, a one-electron transfer process is indicated by a slope of 2 while a slope of 1 still indicates a two-electron reduction process (Gadda et al., 1997).

Fig. 4 shows an example of an xanthine/XO reduction experiment with the flavoprotein proline utilization A (PutA) from *Escherichia coli* (*Ec*PutA). *Ec*PutA (1320 residues) is a bifunctional enzyme that catalyzes the oxidation of proline to glutamate using a flavin-dependent proline dehydrogenase (PRODH) domain and a NAD^+-dependent γ-glutamate-semialdehyde dehydrogenase (GSALDH) domain. *Ec*PutA also contains a N-terminal ribbon-helix-helix DNA-binding domain that enables it to bind the

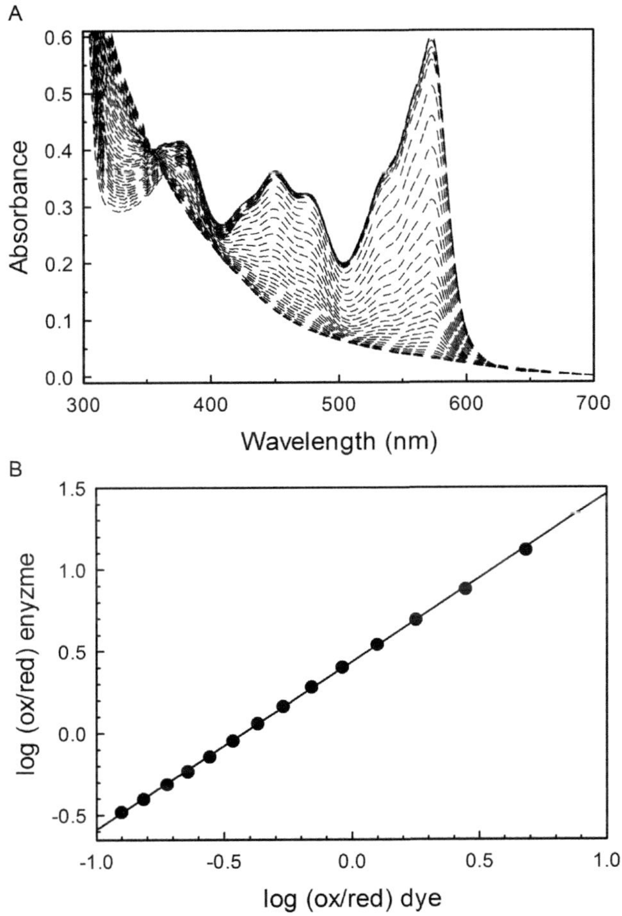

Fig. 4 Reduction potential determination using the xanthine/XO method. (A) *Ec*PutA mutant Arg376Ala (25 μM) was mixed with 200 μM xanthine, 10 μM resorufin (indicator dye), and 100 μM methyl viologen in 50 mM potassium phosphate buffer (10% glycerol, 50 mM NaCl, pH 7.5). XO (20 nM final concentration) was added to the reaction and spectra were recorded every 5 min over 6 h. (B) Plot of log ox/red enzyme versus log ox/red dye from which a slope of 1 unit and a *y*-intercept of 0.436 was determined. The calculated reduction potential ($E_{m(protein)}$) for the protein-bound flavin is −92 mV.

intergenic region between the *putA* and *putP* genes, thereby functioning as a transcriptional regulator as well as a bifunctional enzyme. As reported in Table 1, a reduction potential of −106 mV was previously determined for wild-type *Ec*PutA using the xanthine/XO method (Zhu et al., 2013). Fig. 4 shows the reduction potential determination for the *Ec*PutA mutant Arg376Ala, which was performed under identical conditions as that of the wild-type enzyme using resorufin as the indicator dye. During the reduction, no semiquinone was observed (Fig. 4A). A plot of log (ox/red) flavoprotein versus log (ox/red) dye gave a y-intercept of 0.436 and a slope of 1, consistent with a two-electron transfer between the protein-bound flavin and the dye. Using the y-intercept value, a reduction potential of −92 mV was calculated (see Fig. 3).

4. Effects of ligands

Besides the E_m value of a protein bound flavin, it may also be important to know whether other factors such as ligand binding influence the reduction of the flavin (i.e., redox-linked). Just as the redox potential of protein-bound flavin is linked to the redox potential of free flavin, the influence of ligand binding on the redox potential of protein-bound flavin can be evaluated via a thermodynamic relationship between the oxidized and reduced forms of a protein without ($E_{m(free)}$) and with ligand bound ($E_{m(ligand)}$) using the following relationships.

$$E_{meas} = E_{m(free)} + (2.3RT/nF)\log[E_{ox}]/[E_{red}] \tag{3}$$

$$E_{meas} = E_{m(ligand)} + (2.3RT/nF)\log\left[E_{ox(ligand)}\right]/\left[E_{red(ligand)}\right] \tag{4}$$

$$K_{d(ox)} = [L][E_{ox}]/[L - E_{ox}] \tag{5}$$

$$K_{d(red)} = [L][E_{red}]/[L - E_{red}] \tag{6}$$

Combining these relationships into Eq. (7) shows that differences in ligand binding affinity between the oxidized and reduced forms of the enzyme influence the redox potential of the ligand-bound protein ($E_{m(ligand)}$).

$$E_{m(ligand)} = E_{m(free)} + (2.3RT/nF)\log K_{d(ox)}/K_{d(red)} \tag{7}$$

The greater the difference in redox potential between the free and ligand-bound protein, the more highly redox-regulated are the ligand binding interactions with the protein.

The effects of ligands can be evaluated using either a potentiometric method or the xanthine/XO method (Negri et al., 1999; Sucharitakul et al., 2012). Using the xanthine/XO method, L-(+)-tartrate, which is an inhibitor of D-aspartate acid oxidase, was shown to shift the redox potential of the flavin 57 mV negative indicating that L-tartrate binds more tightly to the oxidized form of the enzyme (Negri et al., 1999). In cytochrome b_5 reductase, the flavin reduction potentials were determined in the presence and absence of NAD^+ (Roma et al., 2006). The reduction potential of the NAD^+-bound enzyme was determined to be −190 mV, which is 83 mV higher than the enzyme without NAD^+ (see Table 1).

5. Conclusion

A fairly extensive list of reduction potentials determined by the xanthine/XO method for various flavoproteins is shown in Table 1 along with the indicator or reference dye used for the experiment. For some protein-bound flavins, the reduction potential is for the two-electron transfer from the oxidized to fully reduced hydroquinone state. In other instances, the reduction potential is determined for each one-electron transfer process in which the semiquinone forms during the xanthine/XO reduction. Because the reduction potentials for ox/sq and sq/hq are significantly different, a separate indicator dye is used for determining the individual one-electron reduction potentials as shown in Table 1 (Murthy & Massey, 1998).

The range of potentials determined by the xanthine/XO method is from +132 mV to −417 mV, which basically covers the biological range of flavin reduction potentials. Covalently bound flavins dominate the positive end of the reported reduction potentials (Heuts et al., 2009) whereas a flavodoxin from *Anacystis nidulans* wins the award for the most negative reduction potential determined using the xanthine/XO method. The redox potential of the xanthine/urate couple is −410 mV, thus allowing xanthine/XO system to be a suitable reductant for most protein-bound flavins. To determine the reduction potential for the sq/hq couple of flavodoxin below −410 mV, however, Ludwig and colleagues cleverly exploited the pH dependence of the xanthine/urate reduction potential, which drops to −440 mV at pH 8.5 (Hoover et al., 1999). Because the reduction potential of flavodoxin is conveniently pH independent between 7 and 9, xanthine/XO was able to fully reduce the flavodoxin mutant at pH 8.5. In this experiment, methyl viologen was dual-purposed as the mediator and indicator dye. In that study, the potentiometric method was used to determine a redox potential of −442 mV for sq/hq in wild-type flavodoxin which was too low for the xanthine/XO method (Hoover et al., 1999).

The main advantage of the xanthine/XO method is the ability for researchers to characterize the redox properties of the flavin protein without the need for sophisticated electrochemistry equipment. Researchers have reported using a closed UV–visible cell, which includes a side arm from which XO can be tipped into the solution to initiate reduction. Other groups have performed the reduction with a septum-sealed cuvette. Another significant advantage is that the xanthine/XO reduction process is typically complete within 4h, which is a much shorter time than that required for a potentiometric titration, which involves $\sim$1 h equilibration periods after each reductive step. The shorter time for conducting a xanthine/XO experiment is particularly beneficial for less stable flavoproteins. Also, because electrodes are not involved, the xanthine/XO technique is easier to miniaturize than a potentiometric experiment, thus reducing the amount of required protein sample. Besides flavoproteins, xanthine/XO method is also used as a convenient method for determining reduction potentials of heme proteins (Efimov et al., 2014).

The advantages of the potentiometric technique include the actual measurement the redox potential of the protein solution and the ability to manipulate the redox potential of the system in the reductive and oxidative directions. Thus, the electrochemical technique can provide a more rigorous characterization of the electrochemical properties such as observing to what extent the reduction process is reversible. Ideally, the redox potential of the protein-flavin should not be dependent on whether the titration is performed in the reductive or oxidative direction. The reversibility of reduction, however, can also be tested in experiments using the xanthine/XO method. For example, after xanthine/XO reduction of the BLUF (sensors of blue light using FAD)-domain photoreceptor protein was complete, the cell was opened and flushed with oxygen and the spectral changes were recorded (Arents et al., 2011). The log ox/red protein versus the log ox/red dye of data obtained in the reductive and oxidative directions was then plotted and shown to overlap, consistent with a reversible reductive reaction (Arents et al., 2011).

In his original study, Massey reported excellent agreement between redox potentials determined by the xanthine/XO method and those reported previously by standard methods. Other groups have also reported consistency in redox potentials determined using xanthine/XO and potentiometric methods. For the NifL flavoprotein, the reduction potentials for the NifL-bound flavin were obtained using xanthine/XO and dithionite/electrode measurements (Klopprogge & Schmitz, 1999). In a study of the thermophilic thioredoxin reductase, the xanthine/XO method resulted in a reduction potential (-305 mV) for the two-electron reduction of the bound-FAD that was near that (-322 mV) determined by protein film voltammetry. For thioredoxin reductase from *Escherichia coli*, a reduction potential of -291 mV was determined by the xanthine/XO method, which was reported to agree reasonably well with the −252 mV value determined using a potentiometric method. For the enzyme PutA, a value of −106 mV (pH 7.5) was reported using the xanthine/XO method whereas −76 mV (pH 7.5) was reported from potentiometric measurements. Thus, the xanthine/XO method has been shown to agree quite well with potentiometric and protein film voltammetry methods.

References

Arents, J. C., Perez, M. A., Hendriks, J., & Hellingwerf, K. J. (2011). On the midpoint potential of the FAD chromophore in a BLUF-domain containing photoreceptor protein. *FEBS Letters*, *585*(1), 167–172. https://doi.org/10.1016/j.febslet.2010.11.035.

Argyrou, A., Blanchard, J. S., & Palfey, B. A. (2002). The lipoamide dehydrogenase from mycobacterium tuberculosis permits the direct observation of flavin intermediates in catalysis. *Biochemistry*, *41*(49), 14580–14590.

Becker, D. F., Leartsakulpanich, U., Surerus, K. K., Ferry, J. G., & Ragsdale, S. W. (1998). Electrochemical and spectroscopic properties of the iron-sulfur flavoprotein from *Methanosarcina thermophila*. *The Journal of Biological Chemistry*, *273*(41), 26462–26469.

Cenas, N., Le, K. H., Terrier, M., & Lederer, F. (2007). Potentiometric and further kinetic characterization of the flavin-binding domain of *Saccharomyces cerevisiae* flavocytochrome b_2. Inhibition by anions binding in the active site. *Biochemistry*, *46*(15), 4661–4670. https://doi.org/10.1021/bi602634y.

Chiribau, C. B., Sandu, C., Fraaije, M., Schiltz, E., & Brandsch, R. (2004). A novel gamma-N-methylaminobutyrate demethylating oxidase involved in catabolism of the tobacco alkaloid nicotine by Arthrobacter nicotinovorans pAO1. *European Journal of Biochemistry*, *271*(23–24), 4677–4684. https://doi.org/10.1111/j.1432-1033.2004.04432.x.

Clark, W. M. (1960). *Oxidation-reduction potentials of organic systems*. Baltimore, MD: Williams and Wilkins.

Clark, W. M., & Lowe, H. J. (1956). Studies on oxidation-reduction. XXIV. Oxidation-reduction potentials of flavin adenine dinucleotide. *The Journal of Biological Chemistry*, *221*(2), 983–992.

Crane, E. J., 3rd, Yeh, J. I., Luba, J., & Claiborne, A. (2000). Analysis of the kinetic and redox properties of the NADH peroxidase R303M mutant: Correlation with the crystal structure. *Biochemistry*, *39*(34), 10353–10364.

Dewanti, A. R., Xu, Y., & Mitra, B. (2004). Role of glycine 81 in (S)-mandelate dehydrogenase from Pseudomonas putida in substrate specificity and oxidase activity. *Biochemistry*, *43*(33), 10692–10700. https://doi.org/10.1021/bi049005p.

Drees, S. L., Ernst, S., Belviso, B. D., Jagmann, N., Hennecke, U., & Fetzner, S. (2018). PqsL uses reduced flavin to produce 2-hydroxylaminobenzoylacetate, a preferred PqsBC substrate in alkyl quinolone biosynthesis in *Pseudomonas aeruginosa*. *The Journal of Biological Chemistry*, *293*(24), 9345–9357. https://doi.org/10.1074/jbc.RA117.000789.

Dutton, P. L. (1978). Redox potentiometry: Determination of midpoint potentials of oxidation-reduction components of biological electron-transfer systems. *Methods in Enzymology*, *54*, 411–435.

Dwyer, T. M., Rao, K. S., Goodman, S. I., & Frerman, F. E. (2000). Proton abstraction reaction, steady-state kinetics, and oxidation-reduction potential of human glutaryl-CoA dehydrogenase. *Biochemistry*, *39*(37), 11488–11499.

Efimov, I., Cronin, C. N., & McIntire, W. S. (2001). Effects of noncovalent and covalent FAD binding on the redox and catalytic properties of p-cresol methylhydroxylase. *Biochemistry*, *40*(7), 2155–2166.

Efimov, I., Parkin, G., Millett, E. S., Glenday, J., Chan, C. K., Weedon, H., et al. (2014). A simple method for the determination of reduction potentials in heme proteins. *FEBS Letters*, *588*(5), 701–704. https://doi.org/10.1016/j.febslet.2013.12.030.

Fasman, G. (1985). Physical and chemical data. In G. Fasman (Ed.), *CRC handbook of biochemistry and molecular biology: Vol. 1.* (3rd ed., pp. 122–130).

Fraaije, M. W., van den Heuvel, R. H., van Berkel, W. J., & Mattevi, A. (1999). Covalent flavinylation is essential for efficient redox catalysis in vanillyl-alcohol oxidase. *The Journal of Biological Chemistry*, *274*(50), 35514–35520.

Frebortova, J., Fraaije, M. W., Galuszka, P., Sebela, M., Pec, P., Hrbac, J., et al. (2004). Catalytic reaction of cytokinin dehydrogenase: Preference for quinones as electron acceptors. *The Biochemical Journal*, *380*(Pt. 1), 121–130. https://doi.org/10.1042/BJ20031813.

Fritz, G., Buchert, T., & Kroneck, P. M. (2002). The function of the [4Fe-4S] clusters and FAD in bacterial and archaeal adenylylsulfate reductases. Evidence for flavin-catalyzed reduction of adenosine 5′-phosphosulfate. *The Journal of Biological Chemistry*, *277*(29), 26066–26073. https://doi.org/10.1074/jbc.M203397200.

Gadda, G., & Fitzpatrick, P. F. (1998). Biochemical and physical characterization of the active FAD-containing form of nitroalkane oxidase from *Fusarium oxysporum*. *Biochemistry*, *37*(17), 6154–6164. https://doi.org/10.1021/bi973085y.

Gadda, G., Wels, G., Pollegioni, L., Zucchelli, S., Ambrosius, D., Pilone, M. S., et al. (1997). Characterization of cholesterol oxidase from Streptomyces hygroscopicus and *Brevibacterium sterolicum*. *European Journal of Biochemistry*, *250*(2), 369–376.

Graf, M. M., Sucharitakul, J., Bren, U., Chu, D. B., Koellensperger, G., Hann, S., et al. (2015). Reaction of pyranose dehydrogenase from *Agaricus meleagris* with its carbohydrate substrates. *The FEBS Journal*, *282*(21), 4218–4241. https://doi.org/10.1111/febs.13417.

Gudim, I., Hammerstad, M., Lofstad, M., & Hersleth, H. P. (2018). The characterization of different flavodoxin reductase-flavodoxin (FNR-Fld) interactions reveals an efficient FNR-Fld redox pair and identifies a novel FNR subclass. *Biochemistry*, *57*(37), 5427–5436. https://doi.org/10.1021/acs.biochem.8b00674.

Hamill, M. J., Chobot, S. E., Hernandez, H. H., Drennan, C. L., & Elliott, S. J. (2008). Direct electrochemical analyses of a thermophilic thioredoxin reductase: Interplay between conformational change and redox chemistry. *Biochemistry*, *47*(37), 9738–9746. https://doi.org/10.1021/bi800676g.

Hamill, M. J., Jost, M., Wong, C., Bene, N. C., Drennan, C. L., & Elliott, S. J. (2012). Electrochemical characterization of *Escherichia coli* adaptive response protein AidB. *International Journal of Molecular Sciences*, *13*(12), 16899–16915. https://doi.org/10.3390/ijms131216899.

Harder, S. R., Feinberg, B. A., & Ragsdale, S. W. (1989). A spectroelectrochemical cell designed for low temperature electron paramagnetic resonance titration of oxygen-sensitive proteins. *Analytical Biochemistry*, *181*(2), 283–287.

Heuts, D. P., Scrutton, N. S., McIntire, W. S., & Fraaije, M. W. (2009). What's in a covalent bond? On the role and formation of covalently bound flavin cofactors. *The FEBS Journal*, *276*(13), 3405–3427. https://doi.org/10.1111/j.1742-4658.2009.07053.x.

Heuts, D. P., van Hellemond, E. W., Janssen, D. B., & Fraaije, M. W. (2007). Discovery, characterization, and kinetic analysis of an alditol oxidase from Streptomyces coelicolor. *The Journal of Biological Chemistry*, *282*(28), 20283–20291. https://doi.org/10.1074/jbc.M610849200.

Hoover, D. M., Drennan, C. L., Metzger, A. L., Osborne, C., Weber, C. H., Pattridge, K. A., et al. (1999). Comparisons of wild-type and mutant flavodoxins from *Anacystis nidulans*. Structural determinants of the redox potentials. *Journal of Molecular Biology*, *294*(3), 725–743. https://doi.org/10.1006/jmbi.1999.3152.

Hu, J., Chuenchor, W., & Rokita, S. E. (2015). A switch between one- and two-electron chemistry of the human flavoprotein iodotyrosine deiodinase is controlled by substrate. *The Journal of Biological Chemistry*, *290*(1), 590–600. https://doi.org/10.1074/jbc.M114.605964.

Ingavat, N., Kavran, J. M., Sun, Z., & Rokita, S. E. (2017). Active site binding is not sufficient for reductive deiodination by iodotyrosine deiodinase. *Biochemistry*, *56*(8), 1130–1139. https://doi.org/10.1021/acs.biochem.6b01308.

Klopprogge, K., & Schmitz, R. A. (1999). NifL of Klebsiella pneumoniae: Redox characterization in relation to the nitrogen source. *Biochimica et Biophysica Acta*, *1431*(2), 462–470.

Kopacz, M. M., & Fraaije, M. W. (2014). Turning a monocovalent flavoprotein into a bicovalent flavoprotein by structure-inspired mutagenesis. *Bioorganic & Medicinal Chemistry*, *22*(20), 5621–5627. https://doi.org/10.1016/j.bmc.2014.05.051.

Kujawa, M., Ebner, H., Leitner, C., Hallberg, B. M., Prongjit, M., Sucharitakul, J., et al. (2006). Structural basis for substrate binding and regioselective oxidation of monosaccharides at C3 by pyranose 2-oxidase. *The Journal of Biological Chemistry*, *281*(46), 35104–35115. https://doi.org/10.1074/jbc.M604718200.

Langer, S., Nakanishi, S., Mathes, T., Knaus, T., Binter, A., Macheroux, P., et al. (2013). The flavoenzyme azobenzene reductase AzoR from *Escherichia coli* binds roseoflavin mononucleotide (RoFMN) with high affinity and is less active in its RoFMN form. *Biochemistry*, *52*(25), 4288–4295. https://doi.org/10.1021/bi400348d.

Lubner, C. E., Jennings, D. P., Mulder, D. W., Schut, G. J., Zadvornyy, O. A., Hoben, J. P., et al. (2017). Mechanistic insights into energy conservation by flavin-based electron bifurcation. *Nature Chemical Biology*, *13*(6), 655–659. https://doi.org/10.1038/nchembio.2348.

Maenpuen, S., Watthaisong, P., Supon, P., Sucharitakul, J., Parsonage, D., Karplus, P. A., et al. (2015). Kinetic mechanism of L-alpha-glycerophosphate oxidase from Mycoplasma pneumoniae. *The FEBS Journal*, *282*(16), 3043–3059. https://doi.org/10.1111/febs.13247.

Maklashina, E., Rajagukguk, S., Iverson, T. M., & Cecchini, G. (2018). The unassembled flavoprotein subunits of human and bacterial complex II have impaired catalytic activity and generate only minor amounts of ROS. *The Journal of Biological Chemistry*, *293*(20), 7754–7765. https://doi.org/10.1074/jbc.RA118.001977.

Massey, V. (1991). *A simple method for the determination of redox potentials flavins and flavoproteins* (pp. 59–66). Berlin: Walter de Gruyter and Company.

Massey, V., & Hemmerich, P. (1980). Active-site probes of flavoproteins. *Biochemical Society Transactions*, *8*, 246–257.

McIver, L., Leadbeater, C., Campopiano, D. J., Baxter, R. L., Daff, S. N., Chapman, S. K., et al. (1998). Characterisation of flavodoxin NADP$^+$ oxidoreductase and flavodoxin; key components of electron transfer in *Escherichia coli*. *European Journal of Biochemistry*, *257*(3), 577–585.

Melby, J. O., Li, X., & Mitchell, D. A. (2014). Orchestration of enzymatic processing by thiazole/oxazole-modified microcin dehydrogenases. *Biochemistry*, *53*(2), 413–422. https://doi.org/10.1021/bi401529y.

Minnaert, K. (1965). Measurement of the equilibrium constant of the reaction between cytochrome *c* and cytochrome *a*. *Biochimica et Biophysica Acta*, *110*(1), 42–56.

Molla, G., Nardini, M., Motta, P., D'Arrigo, P., Panzeri, W., & Pollegioni, L. (2014). Aminoacetone oxidase from Streptococcus oligofermentans belongs to a new three-domain family of bacterial flavoproteins. *The Biochemical Journal*, *464*(3), 387–399. https://doi.org/10.1042/BJ20140972.

Murthy, Y. V., & Massey, V. (1998). Synthesis and properties of 8-CN-flavin nucleotide analogs and studies with flavoproteins. *The Journal of Biological Chemistry*, *273*(15), 8975–8982.

Negri, A., Tedeschi, G., Ceciliani, F., & Ronchi, S. (1999). Purification of beef kidney D-aspartate oxidase overexpressed in *Escherichia coli* and characterization of its redox potentials and oxidative activity towards agonists and antagonists of excitatory amino acid receptors. *Biochimica et Biophysica Acta*, *1431*(1), 212–222.

Palfey, B. A., Basu, R., Frederick, K. K., Entsch, B., & Ballou, D. P. (2002). Role of protein flexibility in the catalytic cycle of p-hydroxybenzoate hydroxylase elucidated by the Pro293Ser mutant. *Biochemistry*, *41*(26), 8438–8446.

Pedotti, M., Ghisla, S., Motteran, L., Molla, G., & Pollegioni, L. (2009). Catalytic and redox properties of glycine oxidase from *Bacillus subtilis*. *Biochimie*, *91*(5), 604–612. https://doi.org/10.1016/j.biochi.2009.02.007.

Pellet, J., & Stankovich, M. T. (2002). Potentiometric measurements of proteins. In G. S. Wilson (Ed.), *Encyclopedia of electrochemistry*: *Vol. 9. Bioelectrochemistry* (pp. 487–509), Weinheim, Germany: Wiley-VCH.

Pellett, J. D., Becker, D. F., Saenger, A. K., Fuchs, J. A., & Stankovich, M. T. (2001). Role of aromatic stacking interactions in the modulation of the two-electron reduction potentials of flavin and substrate/product in *Megasphaera elsdenii* short-chain acyl-coenzyme A dehydrogenase. *Biochemistry*, *40*(25), 7720–7728.

Pollegioni, L., Porrini, D., Molla, G., & Pilone, M. S. (2000). Redox potentials and their pH dependence of D-amino-acid oxidase of *Rhodotorula gracilis* and *Trigonopsis variabilis*. *European Journal of Biochemistry*, *267*(22), 6624–6632.

Ravasio, S., Curti, B., & Vanoni, M. A. (2001). Determination of the midpoint potential of the FAD and FMN flavin cofactors and of the 3Fe-4S cluster of glutamate synthase. *Biochemistry*, *40*(18), 5533–5541.

Roberts, D. L., Salazar, D., Fulmer, J. P., Frerman, F. E., & Kim, J. J. (1999). Crystal structure of *Paracoccus denitrificans* electron transfer flavoprotein: Structural and electrostatic analysis of a conserved flavin binding domain. *Biochemistry*, *38*(7), 1977–1989. https://doi.org/10.1021/bi9820917.

Roma, G. W., Crowley, L. J., & Barber, M. J. (2006). Expression and characterization of a functional canine variant of cytochrome b_5 reductase. *Archives of Biochemistry and Biophysics*, *452*(1), 69–82. https://doi.org/10.1016/j.abb.2006.04.021.

Ryan, K. S., Howard-Jones, A. R., Hamill, M. J., Elliott, S. J., Walsh, C. T., & Drennan, C. L. (2007). Crystallographic trapping in the rebeccamycin biosynthetic enzyme RebC. *Proceedings of the National Academy of Sciences of the United States of America*, *104*(39), 15311–15316. https://doi.org/10.1073/pnas.0707190104.

Sablin, S. O., & Ramsay, R. R. (2001). Substrates but not inhibitors alter the redox potentials of monoamine oxidases. *Antioxidants & Redox Signaling*, *3*(5), 723–729. https://doi.org/10.1089/15230860152664920.

Sevrioukova, I. F. (2016). Structure/function relations in AIFM1 variants associated with neurodegenerative disorders. *Journal of Molecular Biology*, *428*(18), 3650–3665. https://doi.org/10.1016/j.jmb.2016.05.004.

Stankovich, M. T. (1980). An anaerobic spectroelectrochemical cell for studying the spectral and redox properties of flavoproteins. *Analytical Biochemistry*, *109*(2), 295–308.

Sucharitakul, J., Chaiyen, P., Entsch, B., & Ballou, D. P. (2005). The reductase of p-hydroxyphenylacetate 3-hydroxylase from *Acinetobacter baumannii* requires p-hydroxyphenylacetate for effective catalysis. *Biochemistry*, *44*(30), 10434–10442. https://doi.org/10.1021/bi050615e.

Sucharitakul, J., Wongnate, T., Montersino, S., van Berkel, W. J., & Chaiyen, P. (2012). Reduction kinetics of 3-hydroxybenzoate 6-hydroxylase from *Rhodococcus jostii* RHA1. *Biochemistry*, *51*(21), 4309–4321. https://doi.org/10.1021/bi201823c.

Trimmer, E. E., Ballou, D. P., & Matthews, R. G. (2001). Methylenetetrahydrofolate reductase from *Escherichia coli*: Elucidation of the kinetic mechanism by steady-state and rapid-reaction studies. *Biochemistry*, *40*(21), 6205–6215.

Ukaegbu, U. E., Henery, S., & Rosenzweig, A. C. (2006). Biochemical characterization of MmoS, a sensor protein involved in copper-dependent regulation of soluble methane monooxygenase. *Biochemistry, 45*(34), 10191–10198. https://doi.org/10.1021/bi060693h.

van den Heuvel, R. H., Fraaije, M. W., & van Berkel, W. J. (2002). Redox properties of vanillyl-alcohol oxidase. *Methods in Enzymology, 353*, 177–186.

Vinod, M. P., Bellur, P., & Becker, D. F. (2002). Electrochemical and functional characterization of the proline dehydrogenase domain of the PutA flavoprotein from Escherichia coli. *Biochemistry, 41*(20), 6525–6532.

Wilson, G. S. (1978). Determination of oxidation-reduction potentials. *Methods in Enzymology, 54*, 396–410.

Winkler, A., Kutchan, T. M., & Macheroux, P. (2007). 6-S-cysteinylation of bi-covalently attached FAD in berberine bridge enzyme tunes the redox potential for optimal activity. *The Journal of Biological Chemistry, 282*(33), 24437–24443. https://doi.org/10.1074/jbc.M703642200.

Wu, X., Palfey, B. A., Mossine, V. V., & Monnier, V. M. (2001). Kinetic studies, mechanism, and substrate specificity of amadoriase I from Aspergillus sp. *Biochemistry, 40*(43), 12886–12895.

Yorita, K., Matsuoka, T., Misaki, H., & Massey, V. (2000). Interaction of two arginine residues in lactate oxidase with the enzyme flavin: Conversion of FMN to 8-formyl-FMN. *Proceedings of the National Academy of Sciences of the United States of America, 97*(24), 13039–13044. https://doi.org/10.1073/pnas.250472297.

Zhao, G., Song, H., Chen, Z. W., Mathews, F. S., & Jorns, M. S. (2002). Monomeric sarcosine oxidase: Role of histidine 269 in catalysis. *Biochemistry, 41*(31), 9751–9764.

Zhu, W., & Becker, D. F. (2003). Flavin redox state triggers conformational changes in the PutA protein from Escherichia coli. *Biochemistry, 42*(18), 5469–5477. https://doi.org/10.1021/bi0272196.

Zhu, W., Haile, A. M., Singh, R. K., Larson, J. D., Smithen, D., Chan, J. Y., et al. (2013). Involvement of the beta3-alpha3 loop of the proline dehydrogenase domain in allosteric regulation of membrane association of proline utilization A. *Biochemistry, 52*(26), 4482–4491. https://doi.org/10.1021/bi400396g.

CHAPTER TWO

Anaerobic methods for the transient-state study of flavoproteins: The use of specialized glassware to define the concentration of dioxygen

Graham R. Moran*

Department of Chemistry and Biochemistry, Loyola University Chicago, Chicago, IL, United States

*Corresponding author: e-mail address: gmoran3@luc.edu

Contents

Abstract

The methods described here concern the use specialized glassware to achieve accurate sub-micromolar to low-millimolar dioxygen concentrations specifically for the study of flavoproteins. Since the early 1960s, these methods have been passed through the lineage of flavin enzymologists almost exclusively as an oral tradition. Given that

Methods in Enzymology, Volume 620
ISSN 0076-6879
https://doi.org/10.1016/bs.mie.2019.03.005

these methods have been used to systematically reveal a multitude of natural phenomena, it is a curiosity that they have not been comprehensively committed to print prior to this entry. It should be noted that the author is describing these methods with scant added modernity and should not be attributed with their invention. The investigators to whom such ingenuity should be accredited include David P. Ballou, Quentin H. Gibson, Vincent Massey, Charles H. Williams, and Graham Palmer (Foust, Burleigh, & Mayhew, 1969; Gibson, Massey, & Atherton, 1962; Massey, Brumby, Komai, & Palmer, 1969; Massey, Gibson, & Veeger, 1960), who, out of necessity, developed the vessel design and approach over a period of two decades.

1. Introduction

The mole fraction of molecular oxygen in air is 21%, and it is therefore one of nature's more pervasive oxidants. At one atmosphere of pressure and 25 °C, atmospheric dioxygen is dissolved in aqueous media to a concentration of ~250 μM. Reduced flavins are one of a small number of protein-bound organic prosthetic groups that readily react with O_2 molecules. The two-electron reduced isoalloxazine can oxidize in one-electron steps, stabilizing a one-electron reduced, semiquinone radical form (Fig. 1A). This capacity imparts the ability to reduce the dioxygen ground-state triplet that, despite its high reduction potential, is generally exceedingly slow to react with ground-state singlet molecules. As a protein cofactor the propensity of the flavin hydroquinone and semiquinone forms to reduce dioxygen is characteristic (Palfey, Ballou, & Massey, 1995). The protein environment can promote or suppress the transfer of electrons from the flavin hydroquinone to dioxygen but is exceedingly rare for it to abolish this reactivity entirely (Ouedraogo et al., 2017). As such, investigators who seek a full accounting of the catalytic behavior of a redox-active flavoprotein must incorporate methods that control the dissolved molecular oxygen concentration. This entry is concerned with describing flavoenzymes generally and the methods that rely on the use of specialized glassware that were developed by flavoprotein enzymologists to control dioxygen concentration *in vitro*. Obviously, many of these approaches have the utility to be applied to the study of other enzymes that exhibit dioxygen reactivity.

1.1 O_2 as a substrate or a reactive contaminant

Molecular oxygen is one of the most potent and ubiquitous oxidants in nature. Its high reduction potential is the basis for the relative high energy flux and vibrance of aerobic life. The reduction of dioxygen is highly

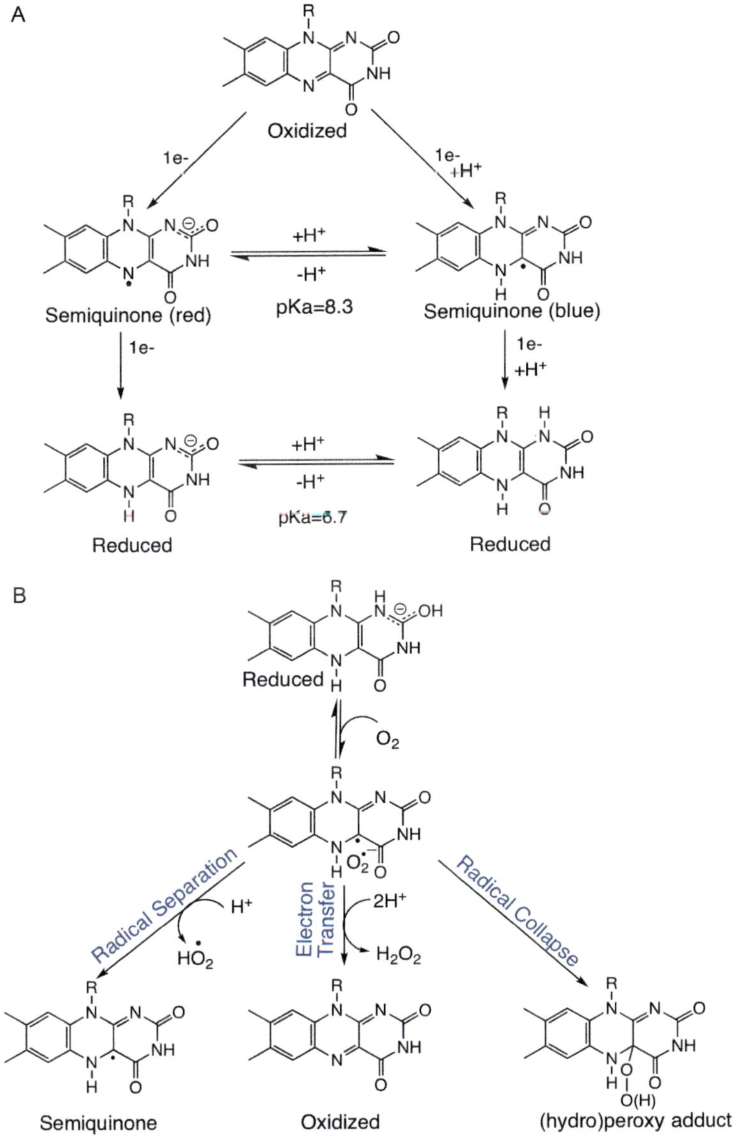

Fig. 1 Oxidation states and dioxygen reactions of flavins. (A) The oxidation and protonation states of flavins. (B) The reactions of dihydroflavin with dioxygen.

exothermic and enzymes that have O_2 as a substrate frequently have ostensibly irreversible catalytic cycles. This irreversibility can simplify the interpretation of kinetic data and this has been a contributing factor to the wealth of unambiguous, detailed mechanistic evidence available for numerous flavoproteins. For flavo-enzymes that utilize dioxygen as a substrate the initial steps of catalysis are concerned with two-electron reduction of the flavin isoalloxazine ring (Fig. 1A), either by a non-specific co-substrate such as NAD(P)H or by oxidation of a more *specific substrate* of the enzyme. The flavin hydroquinone tends to react with molecular oxygen in a bimolecular manner and the second-order rate constant for such reactions can range from 10^1 to 10^5 M^{-1} s^{-1} (Ballou, 1982, 1984) indicating that flavoenzymes have considerable latitude to modulate the dioxygen reduction reaction. The dioxygen triplet has two electrons of the same spin that lie in degenerate non-bonding orbitals and cannot be reduced by two paired electrons from a singlet-state molecule in one step (Bruice, 1984). For flavins the initial step is a single electron transfer to form a superoxide/semiquinone radical pair and it is the fate of this pair that is characteristic for categories of flavoprotein reactions.

In Fig. 1B three fates for the radical pair are depicted. Radical separation as a result of diffusion of the superoxide ion is, with noteworthy exceptions such as NAD(P)H oxidase (Cachat, Deffert, Hugues, & Krause, 2015), a rare unintended bifurcation inefficiency that forms hydroperoxyl radical (Fig. 1B). More often enzymes catalyze further reduction of the superoxide forming hydrogen peroxide or alternatively the two radicals will collapse to form a bond, yielding a flavin-(hydro)peroxy adduct. The formation of hydrogen peroxide is often associated with the latter steps of an enzyme's mechanism, as the two-electron oxidized form of the flavin is regenerated with this product. There are exceptions that retain and utilize the H_2O_2 further such as choline oxidase and lactate monooxygenase (Maeda-Yorita, Aki, Sagai, Misaki, & Massey, 1995; Smitherman, Rungsrisuriyachai, Germann, & Gadda, 2015). By contrast, formation of the flavin-(hydro) peroxy adduct is more often mid-catalytic cycle and poises the enzyme for chemistry that will transform the enzyme's specific substrate by oxygenation/oxidation.

Dioxygen is a significant complicating factor for the study of numerous redox active and non-redox active flavoproteins that do not use it as a substrate (Oppenheimer, Poulin, Lowary, Helm, & Sobrado, 2010; Sobrado, 2012). For these enzymes the radical separation and electron transfer pathways result in futile loss of electrons. The study of non-dioxygen utilizing flavoproteins requires a systematic assessment of dioxygen reactivity for reduced forms of the flavin in order to fully account for observed stoichiometry and kinetic behavior.

1.2 Categories of flavoprotein reactions

Fig. 2 illustrates schematically generalized reaction schemes for the major classes of flavoproteins and the portions of the catalytic cycle that can be studied independently by controlling the dissolved molecular oxygen concentration. These parts of the catalytic cycle are shown in filled rectangles that each represent a single experiment; green for reductive processes, blue for native oxidative and yellow for non-native oxidative (with respect

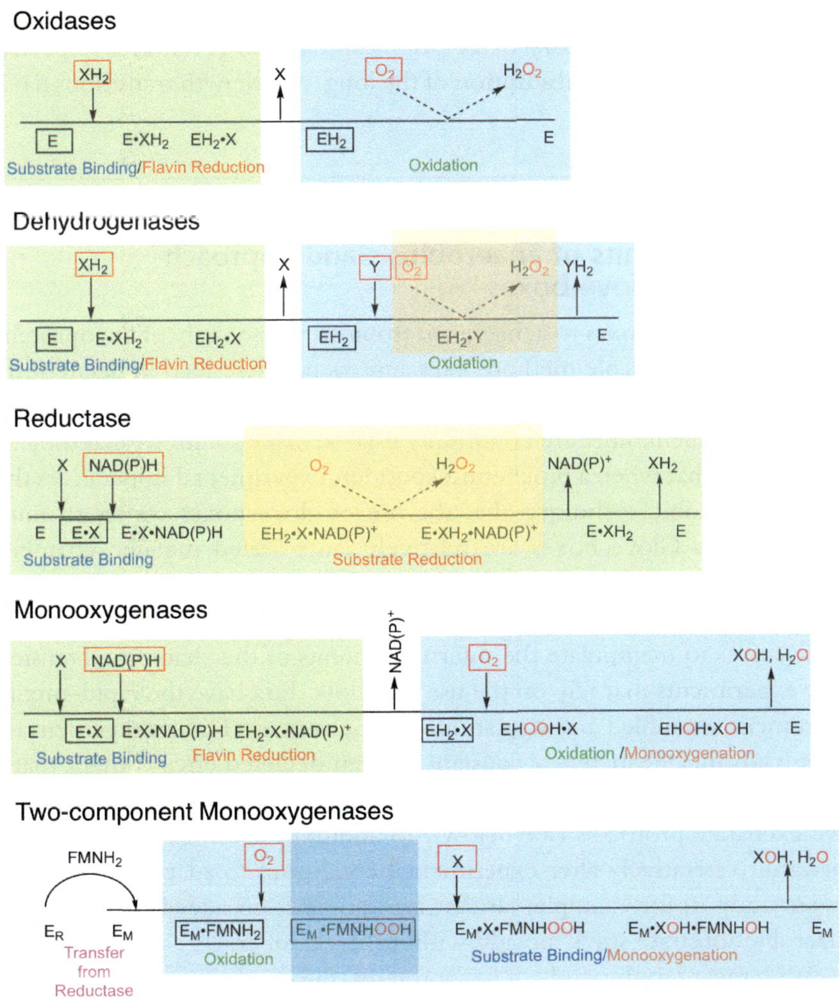

Fig. 2 A schematic representation of the major classes of dioxygen reactive flavoproteins and associated experimental approaches.

to the flavin). Fig. 2 also indicates reactants and reactant complexes using lined boxes. These reactants are those that can be prepared either anaerobically or with a defined dioxygen concentration and mixed to exclusively observe a portion of the catalytic cycle by transient state methods. These are depicted as an enzyme or enzyme-ligand complex by a black line and a substrate by a red line. Traditionally, the substrate concentration would be pseudo-first-order with respect to that of the enzyme and varied during the course of the experiment to establish the associated dependence of the observed rates. For each of these enzyme classes, flavin reduction and reoxidation are the primary observations and are observed as a respective decrease and increase in absorption of the long-wavelength transitions flavin (~350–450 nm).

2. Controlling the concentration of molecular oxygen

2.1 The limits of anaerobiosis and approach with glove-boxes

The use of glove-boxes to achieve anaerobiosis in the study of flavoproteins is in most cases a viable method. This approach is described in detail in the subsequent chapter of this volume. In biochemistry laboratories, both glove-boxes and Schlenk lines are commonly used to make solutions anaerobic. It is likely true that when a biochemist considers experimental imperatives that require anaerobic technique that the use of glove-boxes comes to mind more readily. Glove-boxes are relatively large sealed metallic chambers (though acrylic resin boxes and vinyl bags are also available) that have a large polycarbonate or acrylic window(s) in which rubber gloves are mounted to allow access to manipulate the internal contents of the chamber. Transient state experiments that rely on the use of a glove-box have the rapid-mixing instrument assembled and operating in the chamber. This arrangement has certain advantages such as a constant oxygen depleted environment that is always available for use without the need for dioxygen scrubbing or extensive exchange protocols (see below). In addition, it can be more straightforward to iteratively alter experimental conditions to adapt to the system under study if, for example, all that is required is an added dilution of an anaerobic substrate stock already within the chamber.

With regard to the study of flavoproteins, the disadvantages of the glove-box primarily relate to the extent of anaerobiosis that can be achieved. Glove-boxes are effective instruments for the study of low to mid dioxygen reactive systems. A generic single user glove-box has an internal volume of around 120 L and is comprised of multiple sheets of metal bonded together with bolts and/or welds. Basal dioxygen infiltration at the joints of the box is a continuous and unavoidable complication that is mitigated by continuously passing the internal atmosphere through a canister of dioxygen reactive material to provide a steady rate of dioxygen removal. As previously stated, some flavoproteins exhibit dioxygen reactivity rate constants of $10^5\ M^{-1}\ s^{-1}$. A typical claim for attainable dioxygen contamination in a glove-box is 1 ppm (0.0001%) dictating that the constant dissolved dioxygen concentration in aqueous solutions at room temperature will be around 1–2 nM, providing an observed reaction rate constant for a $10^5\ M^{-1}\ s^{-1}$ reaction of $\sim 0.0002\ s^{-1}$ (or put more explicitly, $t_{1/2} \sim 1\ h$). As such, for a number of flavoproteins the usable window to study reduced complexes of the cofactor is from an experimental perspective relatively short and will tend to diminish the observed amplitude and rates of reaction as the sample ages in the glove-box.

2.2 The advantages of specialized glassware

The primary advantage of the use of specialized glassware over other anaerobic methods is that glass is impermeable to the dominant atmospheric gases, N_2 and O_2. While interfaces to atmosphere exist, they are all sparingly greased high-tolerance ground-glass joints that ostensibly prevent the ingress of external gases. Purpose-made vessels of this type can maintain near complete anaerobiosis for >20 h. However, in order for this to be possible specific types of glassware components must be incorporated to ensure integrity of the barrier formed to prevent intrusion of atmospheric gases. The vessels must be made from high quality joints that have been ground together such that the male and female surfaces have near uniform contact tolerance. Moreover, the glassware must be used in a deliberate manner in accordance with its design that minimizes contact of the solutions within the vessels with atmospheric gases while being made anaerobic and while in use (see below). When constructed to specification and used correctly, glassware of this type offers a convenient method for use of rapid-mixing instruments with anaerobic samples.

The importance of convenience cannot be over-stated as the nature of experiments is that the outcome is unknown. It is more often true that deviations in the behavior of the system under study from what was expected are encountered. These deviations require, for example, that additional concentration ranges be assessed or specific data be re-acquired to test conformity to a model. Glassware is less taxing to the researcher as it allows the instrument to be used in a manner more consistent with the manufacturer's intention.

2.3 Instrument design considerations

The use of the glassware described here to achieve and maintain anaerobic samples for transient state experiments assumes certain design features of the rapid-mixing instruments. The first is the necessity for vertical mounting of the samples. Vertical mounting permits the anaerobic vessel to rest on the glass luer-tip used to connect to the instrument (see Section 2.5). Due to the low shear stress strength of a glass luer-tip, the vessels cannot be safely mounted horizontally. In addition, given that samples introduced from a tonometer (see below) have no mechanical mechanism to deliver the sample to the instrument (such as a syringe plunger), when mounted horizontally all or most of the sample will be inaccessible. The second consideration is the spacing between ports used to mount vessels on the instrument. Glass syringes and tonometers are the primary forms of glassware used to introduce anaerobic samples to rapid-mixing instruments. The diameter of these vessels can be adjusted to some extent, but ultimately adequate spacing between sample ports improves sustained anaerobiosis as the vessels do not crowd one another and rest at odd angles that promote liquid and/or gas leaks. The third consideration is the flow path within the instrument. Instruments that rely heavily on polytetrafluoroethylene (Teflon) tend to harbor greater amounts of diffusible dioxygen dissolved within the polymer. Such internal surfaces must be prepared prior to anaerobic experiments by incubation with a solution that will react and eliminate these internal caches of dioxygen (see below).

Flavins are highly diagnostic spectrophotometric reporters. The oxidation state, adduct form, ionization state or perturbation of transitions for any one of the observable states permits immediate and often sequential mapping of the reaction in which they are involved. Instruments that include a photo-diode array (PDA) or a charge-coupled device (CCD) for simultaneous multiwavelength detection are suited to the study of flavoprotein systems, as the pure spectra of short-lived intermediate states can be deconvoluted from 3-dimensional data sets (wavelength, absorption, time) using decomposition algorithms (SVD). As such, instruments with respective hardware and software to support this type of data acquisition are advantageous.

2.4 The Schlenk line

The modern day Schlenk line consists of a manifold made from a row of double stopcocks connected to an inert gas line and a vacuum line. This arrangement allows the sample connected to one of the stopcocks to be made depleted of dioxygen by exchange, alternating exposure to the inert gas and the vacuum. Though more often an integral part of a synthetic chemistry lab, the basic design is retained in the apparatuses intended for biochemical samples (Fig. 3). The primary adaptations for flavoenzyme anaerobic work are not necessarily unique to these systems and were incorporated initially by Ballou and coworkers. In truth there exists a great deal of latitude in the design of a Schlenk line for the study of oxygen-sensitive biochemical systems. What is described here is a general system in which the use of glass is maximized. The principle adaptations for biochemical samples are the inclusion of a sparger, the use of precision, all-glass stopcocks and double stopcocks in place of those that have Teflon plugs. The purpose of the sparger is to add moisture to the inert gas such that the gas does not cause net evaporation from the surface of the sample during exchange cycles. The sparger also has a secondary role that allows visual assessment of the flow of gases through the line. Slow movement of gas aids the dioxygen scrubbing function of the O_2 reactive cylinder. The stopcock plugs should be retained by either a washer and nut tightened onto a thread at the narrow end or a

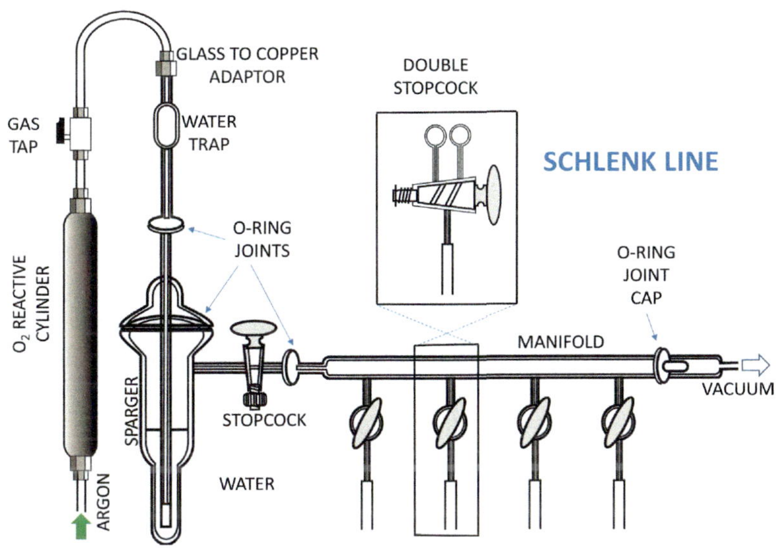

Fig. 3 The Schlenk line with adaptations required for protein samples.

spring, washer and cotter pin. Clips and rubber washers should be avoided as horizontal movement of the stopcocks will greatly undermine the extent of anaerobiosis that can be achieved. Each lower stem/port of the stopcocks should have a 1/4 in. or 6 mm internal diameter butyl rubber or polyvinyl chloride reinforced pressure tubing to connect the line to tonometers or anaerobic cuvettes.

Practical design additions for the Schlenk line include the incorporation of a stopcock on the sparger, such that the manifold can be isolated from the rest of the system for cleaning. It is also advisable to include a gas tap after the dioxygen scrubbing cylinder to ensure that maintenance and partial disassembly do not exhaust the contents of this cylinder. The inert gas used for displacement of the dissolved molecular oxygen can be argon or dinitrogen. Though not rigorously tested, argon is generally regarded as superior of the two as it is more profoundly inert, has higher solubility in water (similar to dioxygen) and has a greater specific gravity (SG) than dinitrogen gas (1.4-fold). The pressure of the gas used within the Schlenk line should be maintained (even when not in use) at ~5 psi. This constant internal pressure minimizes back flow of atmosphere should a leak occur and permits pressurization of tonometers and anaerobic cuvettes for the same reason. It is recommended that small capacity vacuum pumps be used as these build vacuum more slowly providing greater control over the exchange process (see below).

2.5 Anaerobic vessels: Tonometers and cuvettes

Tonometers are the key anaerobic vessel that allows transient state experiments to be pursued without the aid of a glove-box (Fig. 4). Anaerobic cuvettes are also an integral part of the glassware required as these vessels permit simple experiments to be assessed spectrophotometrically by combining the enzyme with ligands or other reactants under inert atmosphere. This is particularly useful with regard to the measurement of binding constants for reduced states of the enzyme. In any case, the protocol to prepare a sample in a tonometer or anaerobic cuvette is the same. This subsection will describe the design of both of these vessels briefly and then move to explain how they are employed. Fig. 4 depicts the basic design of a tonometer and an anaerobic cuvette. Much like the Schlenk line in Fig. 3, the drawings are quite stylized and not meant as a substitute for precise schematics. However, there is no critical need for precision in fabrication of the tonometer from its parts. Departures from the design shown could function as, or more, effectively.

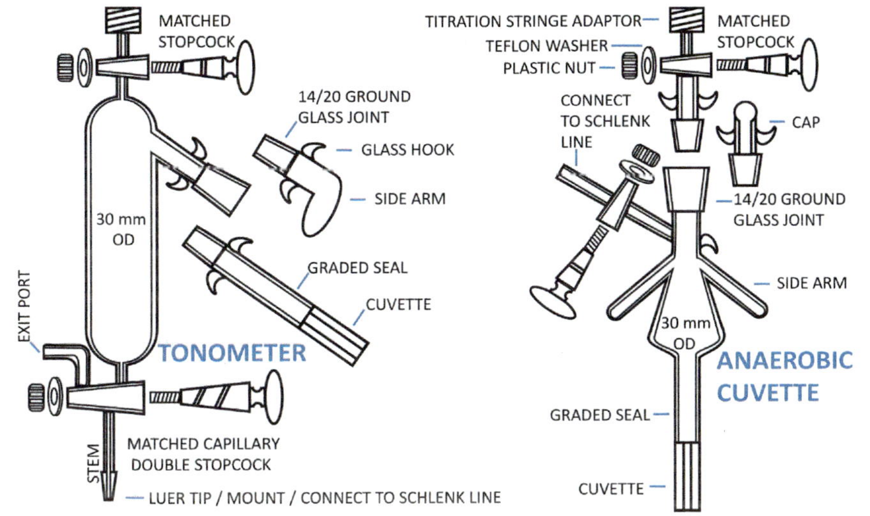

Fig. 4 The design of glass tonometers and anaerobic cuvettes.

The tonometer is essentially a glass cylinder sealed by stopcocks at either end. As mentioned the stopcocks should be all glass and of the highest tolerance available. Major glassware manufactures sell glass stopcocks with matched barrel and plug that each carry the same number to ensure they are used as a pair. The design of the stopcocks should be such that the plugs can be firmly seated in the barrel. Retention mechanisms such as springs or threaded plugs with washers and nuts (the latter is depicted in Fig. 4) ensure no displacement of the two parts of the stopcock during use. The double stopcock is to be bonded to the lower end of the tonometer body. This stopcock will need to be ordered with capillary-sized internal diameter glass tubing above and below. This diameter of tubing greatly facilitates the control of menisci as the sample is shunted into the instrument (see Section 2.8). The lower common port of the double stopcock is fused to a ground glass luer-tip as most rapid-mixing device manufacturers assume that luer-tip syringes will be used to mount samples. This luer interface supports the entire weight of the device when mounted and must be treated with care to avoid snapping the stem below the stopcock. This part of the tonometer must be glass and cannot be adapted to a steel luer for greater strength as the sample meniscus must be observable as it enters the instrument. The second upper port of the stopcock is used for venting liquid and gases at defined stages during mounting. This port simply has a glass capillary tube that directs material contained within the stem beneath the double stopcock out of the vessel (see section below on mounting). The single stopcock at the top of the tonometer body is used for venting if needed but also for mounting of a gas tight syringe for titration. Syringes used for this purpose have threaded plungers that wind through a collar at the top of the syringe barrel. When mounted each turn of the plunger delivers a precise amount of solution to the body of the tonometer (or anaerobic cuvette).

The body of the tonometer may have one or more removable side-arms that are coupled via a ground glass joint. Side-arms are used to make additions to the sample in the body of the tonometer once it has been made anaerobic. This avoids inducing reactions in the presence of dioxygen, but this protocol could also be used to initiate enzymatic oxygen scrubbing whereby the substrate for the scrubbing enzyme is placed in the solution in the body and the enzyme in the side-arm. The side-arm port can also be used to mount a side-arm cuvette, to provide the added functionality of allowing the user to assess the state of the sample spectrophotometrically prior to mounting on the rapid-mixing instrument. This is particularly useful for titrations that form dioxygen reactive complexes or for evaluating the extent of enzyme (flavin) reduction for oxidative half-reactions. The side-arm cuvette is constructed from three pieces, a ground glass joint, a graded seal tube (to bond borosilicate glass at one end to quartz at the other), and a suprasil grade quartz cuvette. Ideally the cuvette is polished on four sides and not masked so that it can also be used for fluorescence measurements. For all side-arm attachments, the orientation of the male and female parts of the ground glass joint is important. The male portion should be part of the removable arm as this facilitates returning solutions to the body of the tonometer.

Anaerobic cuvettes are vessels designed for addition or titration under anaerobic atmosphere. While not directly used in transient state experiments, they are an important part of the glassware required for the study of flavoproteins. The cuvette usually has a rounded body and at least two fused side-arms (though they could be made removable if desired). The side-arms emerge from the body at an approximate angle of 120° with respect to one another and are angled downward (~45°). These are used to make bolus additions to the sample once it is made anaerobic. The cuvette also has a glass stem to connect to the Schlenk line that emerges at 120° with respect to either side-arm that is angled upward (~45°, shown incorrectly as parallel with a side-arm in Fig. 4 for the sake of clarity). As such the side-arms and stem that connects to the Schlenk line emerge at roughly equal spacing about the body of the vessel when viewed from above. The Schlenk line stem must include a single stop cock to seal off the vessel once anaerobiosis is achieved. The central top port of the cuvette can be used for titration using a threaded plunger syringe as described above. The device that couples to the syringe also requires a stopcock so that gas exchanges can be made before the syringe is mounted. This avoids drawing the entire contents of the syringe into the main body of the cuvette while under vacuum. If titrations are not performed, a glass cap is needed to seal the cuvette at this port, though the syringe adaptor could be used for this purpose. The cuvette is bonded to the base of the device using a graded seal tube as described above.

Assembly of both tonometers and anaerobic cuvettes requires greasing of all ground glass contact surfaces. The joints and stopcocks should not be assembled without grease as this will abrade and degrade both surfaces. For this purpose, traditionally high vacuum grease has been used by laboratories that study flavoproteins. Silicon grease is more common and considerably less expensive and will suffice for this purpose (though tends to spread and smear more easily). The amount of grease used is critically important. Small amounts are superior to thick applications as excess grease tends to nullify the advantages of high tolerance surfaces and also plugs and fouls the capillary tubes on either side of the stopcocks. One small smear down opposite faces of the stopcock plug or male portion of a ground glass joint is generally sufficient to form a clear glassy film between the two parts. Look for irregularities in the film and add more grease as required. A uniform clear film without streaks formed by trapped air or excess accumulations are indicative that the desired amount was used.

2.6 Exchanging argon for dissolved dioxygen: Making a sample anaerobic

What follows is a largely empirically established method for achieving near complete anaerobiosis within a tonometer or anaerobic cuvette. It includes the key elements of the protocol and lore of the method as described for the succession of flavoprotein enzymologists. Anaerobic vessels should be cleaned thoroughly prior to use. It is important to remain cognizant of the fact that debris within the vessel can induce protein to precipitate or can be liberated as insoluble or soluble matter that causes experimental artifacts. It is more often true that rapid-mixing experiments involve optical measurements and an utterly clear sample will yield higher quality data. The enzyme sample should be centrifuged at high speed (>10,000*g*) for at least 10 min to sediment colloidal aggregates that present similar complications as an unclean vessel. Once clean the anaerobic vessel should be greased, assembled and placed horizontally in ice. The clarified enzyme can be transferred to the body of the tonometer through a side-arm port. Additives to be combined after anaerobiosis is achieved can be added to side-arms and the side-arms positioned such that minor agitation of the vessel will not result in unintended mixing. At this time the stem of the vessel can be connected to the tubing of the Schlenk line, the vacuum pump of the line having been turned on some time prior.

Exchanging argon for dissolved dioxygen involves alternating exposure of the sample within the anaerobic vessel to vacuum and argon gas. This process is not the same as freeze-pump-thaw methods where high vacuum is held for extended periods. Instead this method requires that the vacuum be used with care to prevent the formation of bubbles. The purpose of the vacuum is to remove the majority of the atmosphere above the liquid sample, not to accelerate the liberation of dissolved dioxygen from the sample.

The vessel is laid flat generally in a bed of ice. Once the vessel is connected to the Schlenk line, the initial step is to pull vacuum and refill the tubing with argon a few times before opening the stopcock to begin exchanges. The stopcock of the anaerobic vessel should be opened while the Schlenk line stopcock is in the argon position. At this stage the exchange process can commence. The stopcocks of the Schlenk line manifold should be turned with care to avoid stressing the glass stems that connect them to the vacuum and argon tubes but also so that vacuum and argon are introduced to the sample in a controlled manner. As vacuum is applied the researcher should watch the sample closely for the formation of bubbles. Extreme environments form at the surface of bubbles that may cause proteins to precipitate, most particularly if they burst. It is likely that it is not possible to completely avoid the formation of bubbles while applying vacuum, but with control of the Schlenk line stopcock the size of the bubbles formed and the number that burst can be minimized. When returning the Schlenk line stopcock to the argon position it is best to bleed the argon gas into the vessel relatively slowly. Observation of the sparger is the best way to perceive the rate of gas flow. As a rough rule of thumb, the vessel should come to pressure equilibrium with the Schlenk line in around 30 s. One application of vacuum and one fill with argon is regarded as one cycle. After the completion of three cycles the vessel is gently agitated by hand for 2–3 min while under argon pressure to encourage the exchange of dissolved gases at the gas-liquid interface. This process of three cycles and dissolved gas equilibration is repeated 10–12 times. It is known that this number of exchanges is sufficient to achieve near complete anaerobiosis for the vessels described here;

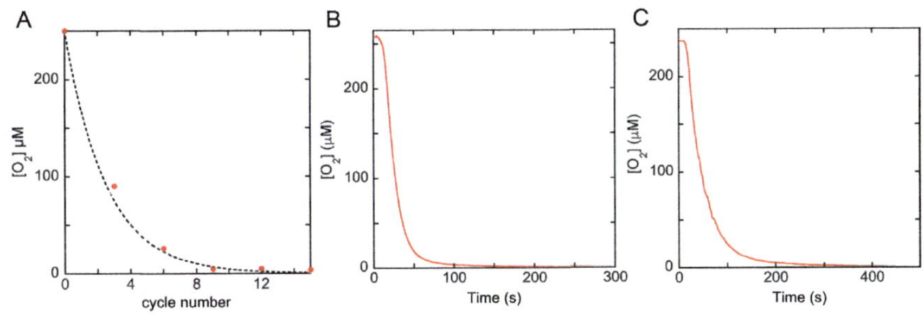

Fig. 5 Evaluation of the methods used to obtain anaerobiosis. (A) Gas exchange with the Schlenk line. (B) Argon sparging. (C) Enzymatic dioxygen depletion.

however, the minimum required number has, to the author's knowledge, not previously been rigorously established and the method may require adaptation for vessel size and/or sample volumes. Fig. 5A depicts the concentration of dissolved dioxygen in a 6 mL sample of buffer in a 60 mL tonometer (total internal volume) as a function of exchange cycles. The data were generated by mixing oxygenated buffer subject to exchange cycles with anaerobic $FADH_2$ and observing the extent of flavin reoxidation. This plot (corrected for 1:1 mixing) indicates, as might be expected, that the depletion of dioxygen from a tonometer by the above methods follows an exponential dependence on the cycle number. The line indicates that the concentration of dissolved dioxygen decreases by half for every 1.7 cycles and that sub-micromolar dioxygen is achieved at around 14 exchange cycles. While added cycles will achieve greater stringency, this example suggests that the 30–36 cycles defined empirically is in excess of what is required.

The above methods have been devised to prepare delicate enzyme samples for mounting onto rapid-mixing instruments. Small molecule substrates, inhibitors, and/or ions are not generally susceptible to degradation or denaturation by mechanical manipulations and so can be prepared anaerobically quickly and easily by sparging. This is advantageous as it is more often these molecules that are varied to measure dependences for the development of kinetic models. Anaerobiosis is accomplished by placing the small molecule solution in a glass syringe. The volume chosen should permit at least 2–3 mL of air or head space to be accommodated in the syringe with the sample. The syringe is then inverted and held in a clamp beneath the Schlenk line manifold such that the 2–3 mL head space is retained. A tube from one of the manifold stopcocks is then coupled to a ~5 cm needle. This is used to sparge the sample with argon gas. The gas flow should be modest to avoid displacing the sample out of the luer-tip of the inverted syringe. Five to fifteen minutes of sparging is generally regarded as sufficient to achieve complete anaerobiosis. The needle is then removed and the argon atmosphere is captured by placing a gloved fingertip over the syringe luer tip. The plunger is then used to carefully displace the gas and the anaerobic sample is captured using the fingertip, inverted and mounted onto the rapid-mixing instrument (see mounting below). Fig. 5B depicts the dioxygen concentration of 2 mL of buffer placed in the glass chamber of a Clark dioxygen electrode and sparged with argon gas. The chamber was capped with an O-ringed polymethylmethacrylate plug. The plug was drilled to form a 2 mm port to accept the sparge needle and positioned such that the chamber had ~2 mL of head space, overall creating a vessel similar to a 5 mL inverted luer-tipped glass syringe. These data indicate that the 2 mL of buffer is anaerobic within 4 min.

2.7 Anaerobic preparation of the rapid-mixing instrument

For stringent control of dioxygen within the stopped flow instrument it is commonly necessary to treat the internal cavities of the instrument with an agent that will consume oxygen molecules prior to the start of the experiment. This is most important when a significant fraction of the internal surface area is composed of Teflon or other oxygen permeable plastics. It is most common to introduce a solution that will react with dioxygen 12 h before the start of the experiment; however the optimal time has in most cases not been determined. Sodium hydrosulfite is a common oxygen reactive solution. However, it is somewhat difficult to displace from the instrument by flushing with anaerobic buffer and will certainly cause artifacts if it is not completely removed. An alternative is to anaerobically prepare enzymes that react with dioxygen along with their organic substrate in a tonometer and then introduce this solution to the rapid-mixing instrument after it has been flushed with anaerobic buffer. Common examples are glucose oxidase (GO) and protocatechuate dioxygenase (PCD) (Patil & Ballou, 2000). The dioxygen scrubbing enzyme solution can be prepared as a 2 U/mL stock and combined with its substrate from a side-arm (5–50 mM final) once anaerobic (see Section 2.6). These oxygen-scrubbing solutions can be kept within the tonometer and used for multiple experiments. GO can be stored in this manner on the bench for 1–3 weeks, whereas PCD is remarkably stable and can be expected to maintain activity for many months. Fig. 5C depicts the rate and extent to which 2 U/mL GO in the presence of 50 mM glucose eliminates dioxygen from a 2 mL solution equilibrated with atmosphere.

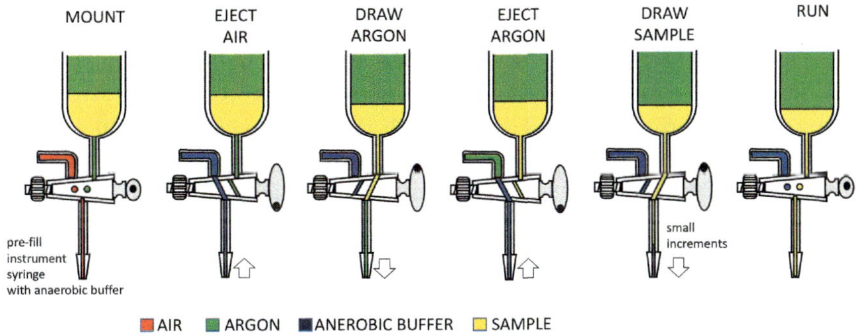

Fig. 6 The sequence required to introduce a sample from a tonometer to a rapid-mixing instrument without compromising anaerobiosis.

2.8 Method for mounting tonometers and syringes on the kinetic instrument

The design of the tonometer described in this entry is in part based on a protocol to mount the vessel onto a rapid-mixing instrument without the contents coming into contact with atmosphere. Fig. 6 depicts the sequence of steps required to mount a tonometer and deliver its contents to the instrument with anaerobiosis maintained. In all two gaseous and two aqueous environments must be kept separate to successfully mount the tonometer and introduce the sample. These are shown as separate colors in Fig. 6. It is incumbent on the researcher to remain aware of each of these environments during mounting. Initially the instrument should be prepared by scrubbing the residual dioxygen (see above). This scrubbing reagent should be displaced by anaerobic buffer that has been prepared by sparging (see above). The instrument is filled with this buffer prior to mounting the tonometer as the mounting procedure requires that the internal sample syringe, to which the tonometer will be connected, be filled with anaerobic buffer. To avoid oxygen infiltration, all instrument ports should be capped in some way at all times. It is common to leave syringes in each port until they are to be replaced by another syringe or a tonometer.

When first mounting the tonometer, it is necessary to remove the syringe that was used to flush the instrument with anaerobic buffer. All vessels mounted to the instrument should be retrieved or seated with a twisting motion. The tonometer can then be seated in place of the syringe. Due to the maximum width of the tonometer occurring at the double stopcock, it is recommended that the vessel be oriented such that the barrel of the stopcock is perpendicular to the face of the instrument.

When initially mounted, the tonometer has its stem, one half of the double stopcock plug and the exit port filled with atmosphere (shown in red). The first step after mounting is to turn the stopcock to the position that connects the stem and the exit port such that the anaerobic buffer within the instrument (shown in blue) can be used to displace the trapped air within the stem and stopcock. Then the double stopcock is rotated through 180° to connect the stem and the tonometer body. In this position argon trapped in the capillary tube above the stopcock (shown in green) can be drawn into the stem and/or instrument syringe. The objective in this step is to clear the lower tonometer of argon and bring the enzyme (shown in yellow) to the base of the stop cock plug. Turning the stopcock through another 180° permits the trapped argon to be ejected out the exit port using the anaerobic buffer within the instrument. Continue to eject the anaerobic buffer until the internal syringe is empty (you may need a paper towel to catch the overflow from the exit port). Turn the stopcock again through 180°. At this stage you will have formed an interface between the remaining anaerobic buffer in the stem and the enzyme in the plug of the stopcock. Bring the sample into the instrument in small increments of around ~100 μL and eject each through the instrument waste line. The objective is to bring the sample to the instrument without significant mixing with the anaerobic buffer from the stem. Flavoproteins have the advantage of color, and so it is possible to discern when the enzyme sample has entered the instrument. However, if the sample is dilute, reduced or both it may be difficult to establish precisely at what point all of the buffer is ejected and only sample is being drawn into the instrument. The number of draw and eject steps required is a function of the length of the stem and the internal volume of the instrument; however in general five to six steps is sufficient to remove all the anaerobic buffer. At this juncture, the internal syringe can be filled with sample and the experiment can commence.

It is more often true that only one tonometer is mounted during the course of a typical rapid-mixing experiment. All other reactant solutions are generally brought to the instrument in syringes. Cleanly mounting a syringe on a rapid-mixing instrument is relatively simple. Syringe samples that have been made anaerobic by sparging can be mounted into the female luer port of the instrument only after the port has been manually filled with the anaerobic buffer or previous sample from instrument syringe. While this does induce some equilibration with atmospheric oxygen within the luer port, it is done to avoid capturing bubbles of atmosphere into the instrument syringes that will cause artifacts and certainly undermine anaerobiosis. In a similar manner to the final stages of mounting tonometers, once mounted, the anaerobic sample should be introduced with small draw and eject steps of ~100 μL each. This process brings in the initial partially oxygenated buffer and ejects it immediately. So long as the movement of the loading is in only one direction (into the instrument) this method removes any molecular oxygen captured during loading.

2.9 Gas blending and sparging to control the dioxygen concentration

For experiments designed to measure the dependence of the observed rate of reaction on the concentration of dissolved dioxygen a range of concentrations of dioxygen are required. As is the case for all such experiments, the greater the range of reactant concentration explored, the more certainty of the mechanistic conclusion. At 4 °C, molecular oxygen has a maximal practical concentration in water of around 1.8 mM (0.9 mM after mixing). As such the available concentration range is somewhat constrained. For this reason, there are advantages to gas blending to achieve numerous dioxygen partial pressures that can be used to equilibrate buffer by sparging for rapid-mixing experiments. Medical gas blenders are highly suitable for this purpose. These devices are intended to blend air (SG = 1.00) and oxygen (SG = 1.10) but work very well to blend nitrogen (SG = 0.97) and oxygen gases as air is 78% nitrogen gas with the bulk of the remainder (21%) being oxygen. Some of these devices have inline O_2 detectors that give a digital reading of the dioxygen percentage partial pressure. Sparging the blend of gases into a buffer solution will yield a predictable concentration of dissolved dioxygen at a specific ionic strength, temperature and atmospheric pressure and tables and web-based calculators exist to make this calculation. However, a more accurate method of assessing the dissolved dioxygen is to sparge the gas mixture into the thermostatically controlled cavity of a Clark electrode that is filled with the buffer used in the kinetic experiment. A correctly calibrated Clark electrode will measure the actual dissolved dioxygen concentration directly to within 1–2 μM. Once this value is known, the buffer can be withdrawn from the electrode chamber into a syringe and used, or another sample of buffer can be equilibrated with the same blend of gases by sparging it in an inverted syringe.

3. Preparing reduced flavoproteins

Observation of oxidative half-reactions for flavoproteins requires the preparation of stable reduced states of the enzyme. This is accomplished by first preparing the enzyme anaerobically as described above and then reducing it in some manner such that it can then be mixed with dioxygen (or some other oxidant). Four methods of preparation of reduced enzyme will be briefly described. A more comprehensive description of these methods is provided elsewhere in this volume.

3.1 Titration with sodium hydrosulfite

Sodium hydrosulfite (dithionite) is a useful relatively potent reductant ($E^o = -660\,mV$) that rapidly reduces enzyme bound flavins that generally have potentials between −200 and +100 mV. This is accomplished by preparing the dithionite solution in a vessel with a narrow neck (such as a volumetric flask). It should be prepared by assuming that ~250 μM dithionite will be consumed by reaction with dissolved dioxygen in the buffer. The resulting anaerobic dithionite solution will be introduced to the enzyme by attaching the threaded plunger syringe to the anaerobic vessel. For tonometers, this process also requires that the cuvette side-arm be attached. Together these two attachments make the tonometer rather top-heavy and precarious when mounted to the rapid-mixing instrument. It is recommended that the device is supported by a stand and clamp to avoid snapping the stem of the tonometer. Coupling the threaded gas tight syringe requires opening the single stopcock at the top of tonometer. The objective is to do this as quickly as is possible and only when attached to the Schlenk line and under argon flow. It is assumed that the flow of argon out of the small orifice of the stopcock impedes the influx of atmosphere to some degree. During this window the syringe can be mounted. It is common to extend the needle by sheathing it with a ~10 cm section of clear Teflon tubing. The syringe is held in place at the top of the tonometer by a plastic grommet and rubber compression washer that is seated by a glass thread. Once mounted, the progress of reduction can be observed spectrophotometrically via the cuvette side-arm. The target concentration of the dithionite solution should be such that six to eight turns of the syringe plunger is sufficient to fully reduce the flavin. It is common to slightly over-reduce by the enzyme (~20%) by adding one to two additions of the dithionite solution after full reduction is achieved. This added reductant provides a buffer against reoxidation during the course of the experiment.

3.2 Double-mixing and reduction using the native reductant

Rapid-mixing instruments that support double-mixing combine three solutions in two steps. Two reactants are combined in the first and aged for a specific time before the aged sample is mixed with a third. For oxidases, dehydrogenases and single component monooxygenases, anaerobic enzyme samples can be stoichiometrically reduced by their native reductant substrates. In the first mix the anaerobic enzyme is mixed with the anaerobic reductant substrate and allowed to age and become fully reduced. In the second mix the reduced enzyme-first product mixture is combined with the oxidant substrate that is either anaerobic for organic oxidants or oxygenated. Ideally this method requires knowledge of the kinetics of the reductive half-reaction, but for spectrophotometric experiments can be accomplished empirically by varying the age time and observing the extent of reduction at the first data point when mixed with anaerobic buffer in the second mixing step.

3.3 Enzymatic reduction using mediators

Flavoenzymes that are not amenable to other methods of reduction can be reduced using electrons supplied from some other enzymatic reaction. An example of such a reaction is the xanthine/xanthine oxidase system (Massey, 1990) that oxidizes xanthine. The electrons liberated are channeled through the protein and can be shuttled to other acceptors (such as the flavoenzyme under study) in the presence of a redox active mediator molecule such as a viologen. The method involves preparation of the enzyme anaerobically in the presence of the substrate for the reducing enzyme with its substrate and the mediator. The reducing enzyme is kept separate in a side-arm of the anaerobic vessel until all the contents are devoid of dioxygen and then combined to initiate the reduction of the enzyme under study.

3.4 Light mediated reduction

Flavins will reduce when irradiated with sufficient light intensity in the presence of amino acids, carboxylic acids and amines (Frisell, Chung, & Mackenzie, 1959). EDTA has been shown to be one of the more effective photoreductants (Massey, Stankovich, & Hemmerich, 1978). This process is thought to be mediated by small amounts of free flavin. Anaerobic preparation of an enzyme sample with added millimolar EDTA and micromolar free flavin precedes irradiation by a relatively intense light source. Much like dithionite reduction, the extent of reduction can be observed spectrophotometrically if a cuvette side-arm is used on the tonometer.

4. Summary and conclusion

The use of specialized glassware to control the concentration of dissolved dioxygen in transient state experiments developed as an approach and discipline beginning in the early 1960s. Refinements and innovations have improved the vessel designs and methods to the extent that it has become a reliable and convenient means of controlling dioxygen concentration. The glassware described here requires a skilled scientific glassblower to fabricate to the specifications shown. However, even if fabricated commercially, the cost of the glassware is considerably less than commercially made anaerobic chambers. Moreover, glassware requires less dedicated laboratory space and incurs very minimal ongoing costs to maintain. Once acquired and used in the manner described, these vessels permit and almost unlimited number of approaches to the transient state study of oxygen reactive enzyme states that yield quality, interpretable data.

References

Ballou, D. P. (1982). *Flavins and flavoproteins*. North Holland, New York: Elsevier.

Ballou, D. P. (1984). Flavoprotein monooxygenases. In R. C. Bray & P. C. Engel (Eds.), *Flavins and flavoproteins* (pp. 605–618). De Gruyter.

Bruice, T. C. (1984). Oxygen-flavin chemistry. *Israel Journal of Chemistry*, *24*, 54–61.

Cachat, J., Deffert, C., Hugues, S., & Krause, K. H. (2015). Phagocyte NADPH oxidase and specific immunity. *Clinical Science (London, England)*, *128*(10), 635–648. https://doi.org/10.1042/CS20140635.

Foust, G. P., Burleigh, B. D., & Mayhew, S. G. (1969). An anaerobic titration assembly for spectrophotometric use. *Analytical Biochemistry*, *27*, 530–535.

Frisell, W. R., Chung, C. W., & Mackenzie, C. G. (1959). Catalysis of oxidation of nitrogen compounds by flavin coenzymes in the presence of light. *The Journal of Biological Chemistry*, *234*(5), 1297–1302.

Gibson, Q. H., Massey, V., & Atherton, N. M. (1962). The nature of compounds present in mixtures of oxidized and reduced flavin mononucleotides. *The Biochemical Journal*, *85*, 369–383.

Maeda-Yorita, K., Aki, K., Sagai, H., Misaki, H., & Massey, V. (1995). L-Lactate oxidase and L-lactate monooxygenase: Mechanistic variations on a common structural theme. *Biochimie*, *77*(7–8), 631–642.

Massey, V. (1990). A simple method for the determination of redox potentials. In B. Curti, S. Ronchi, & G. Zanetti (Eds.), *Flavins and flavoproteins* (pp. 59–66). New York: Walter de Gruyter & Co.

Massey, V., Brumby, P. E., Komai, H., & Palmer, G. (1969). Studies on milk xanthine oxidase. Some spectral and kinetic properties. *The Journal of Biological Chemistry*, *244*(7), 1682–1691.

Massey, V., Gibson, Q. H., & Veeger, C. (1960). Intermediates in the catalytic action of lipoyl dehydrogenase (diaphorase). *The Biochemical Journal*, *77*, 341–351.

Massey, V., Stankovich, M., & Hemmerich, P. (1978). Light-mediated reduction of flavoproteins with flavins as catalysts. *Biochemistry*, *17*(1), 1–8.

Oppenheimer, M., Poulin, M. B., Lowary, T. L., Helm, R. F., & Sobrado, P. (2010). Characterization of recombinant UDP-galactopyranose mutase from *Aspergillus fumigatus*. *Archives of Biochemistry and Biophysics*, *502*(1), 31–38. https://doi.org/10.1016/j.abb.2010.06.035.

Ouedraogo, D., Ball, J., Iyer, A., Reis, R. A. G., Vodovoz, M., & Gadda, G. (2017). Amine oxidation by d-arginine dehydrogenase in *Pseudomonas aeruginosa*. *Archives of Biochemistry and Biophysics*, *632*, 192–201. https://doi.org/10.1016/j.abb.2017.06.013.

Palfey, B. A., Ballou, D., & Massey, V. (1995). Oxygen activation by flavins and pterins. In *Vol. 2. Active oxygen in biochemistry* (pp 37–83). Springer Science.

Patil, P. V., & Ballou, D. P. (2000). The use of protocatechuate dioxygenase for maintaining anaerobic conditions in biochemical experiments. *Analytical Biochemistry*, *286*(2), 187–192. https://doi.org/10.1006/abio.2000.4802.

Smitherman, C., Rungsrisuriyachai, K., Germann, M. W., & Gadda, G. (2015). Identification of the catalytic base for alcohol activation in choline oxidase. *Biochemistry*, *54*(2), 413–421. https://doi.org/10.1021/bi500982y.

Sobrado, P. (2012). Noncanonical reactions of flavoenzymes. *International Journal of Molecular Sciences*, *13*(11), 14219–14242. https://doi.org/10.3390/ijms131114219.

CHAPTER THREE

Performing anaerobic stopped-flow spectrophotometry inside of an anaerobic chamber

Hannah Valentino, Pablo Sobrado*
Department of Biochemistry, Center for Drug Discovery, Virginia Tech, Blacksburg, VA, United States
*Corresponding author: e-mail address: psobrado@vt.edu

Contents

Methods in Enzymology, Volume 620
ISSN 0076-6879
https://doi.org/10.1016/bs.mie.2019.03.006

Abstract

The catalytic cycle of most flavin-dependent enzymes can be divided into oxidative and reductive half-reactions. Although some enzymes are oxidized by electron carrier proteins or organic compounds, many use oxygen as the final electron acceptor. In order to properly study the reductive half-reaction of flavin-dependent enzyme that react with oxygen, as in the case of oxidases and monooxygenases, it is necessary to establish anaerobic conditions that will only allow the reduction process to be monitored. The reduced flavoenzyme can be further studied by exposing it to oxygen to monitor the oxidative half-reaction. Anaerobic chambers provide an ideal environment for performing these experiments as they reliably maintain an anaerobic atmosphere inside a large workspace. A common tool used to study flavin-dependent enzymes is the stopped-flow spectrophotometry. This chapter describes methods for performing stopped-flow experiments in an anaerobic chamber. We include information about the chamber components, setting up a stopped-flow spectrophotometer inside of a chamber, preparing anaerobic solutions, and performing experiments to measure the reductive and oxidative half-reactions of flavin-dependent monooxygenases.

Abbreviations

GO glucose oxidase
ITC isothermal titration calorimetry
NADH reduced nicotinamide adenine dinucleotide
NADPH reduced nicotinamide adenine dinucleotide phosphate
PCD protocatechuate dioxygenase
SHU sample handling unit
UV–vis ultraviolet–visible

1. Introduction

Flavoenzymes commonly perform redox chemistry where the catalytic cycle can be divided into reductive and oxidative half-reactions (Neims & Hellerman, 1970). In the reductive half-reaction, the oxidized flavin is reduced by a hydride equivalent to form the reduced flavin (Fig. 1) or via single electron transfer steps where the semiquinone species is formed (Bruice, 1980; Edmondson & Tollin, 1983). During the oxidative

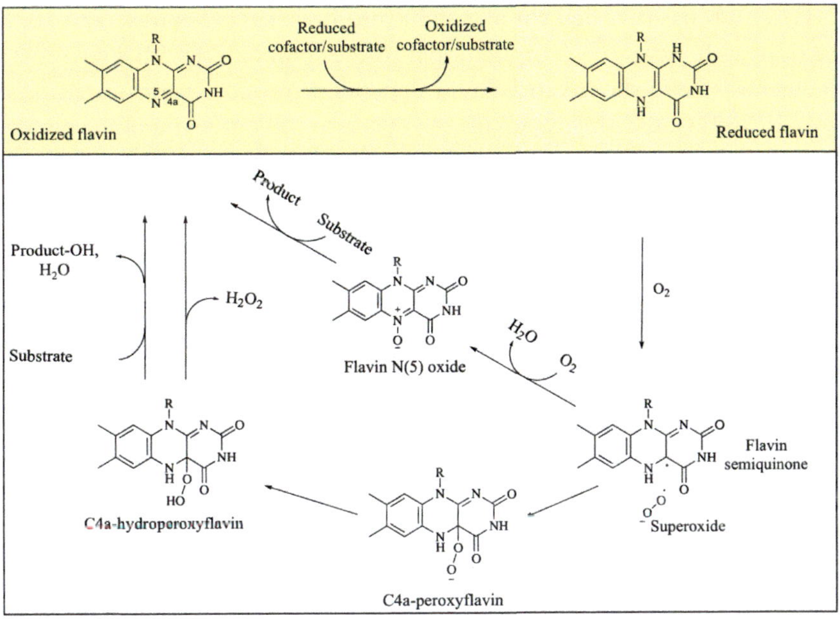

Fig. 1 The general reaction scheme of the redox cycle of flavoenzymes. The reductive half-reaction is highlighted in yellow and the oxidative half-reaction is highlighted in white. During the reductive half-reaction, the oxidized flavin is reduced to its hydroquinone form by NAD(P)H or other substrates. The reduced flavin then reacts with oxygen and forms various covalent intermediates used for oxygenation reactions or to form hydrogen peroxide.

half-reaction, the reduced flavin reacts with oxygen, as in the case of flavin oxidases and monooxygenases, or with electron carrier molecules as in cellular respiration (Dashty, 2013; Romero, Gómez Castellanos, Gadda, Fraaije, & Mattevi, 2018).

Due to the difference in redox potential between oxygen (E_{O_2/H_2O} = + 270 mV) and flavin ($E_{red/ox}$ = − 209 mV), the reaction of these molecules is thermodynamically favorable (Massey, 1994). The reaction begins with a single electron transfer forming superoxide and a flavin semiquinone. This radical pair recombines forming either a C4a-peroxyflavin or C4a-hydroperoxyflavin which are key intermediates in Baeyer-Villiger or hydroxylation reactions, respectively (Fig. 1) (Massey, 2002; Romero et al., 2018). Recently, a flavin-N5 oxide intermediate has also been identified as part of the mechanism of some oxidation reactions (Fig. 1) (Adak & Begley, 2016; Piano, Palfey, & Mattevi, 2017; Teufel et al., 2015).

Research has revealed some insight into how flavoenzymes control the reaction with oxygen. Computational studies combined with site directed mutagenesis have shown that flavin dependent monooxygenases and oxidases use diffusion pathways to direct oxygen to the active site of the enzyme (Baron et al., 2009). In addition, oxygen pressurized X-ray crystallography has shown that the orientation of oxygen in the active site can permit the formation of regiospecific flavin-oxygen adducts (Saleem-Batcha et al., 2018). It has also been shown that key residues regulate access of molecular oxygen to reduced flavin. These have been named "gate keepers" in studies with vanillyl-alcohol oxidases (Leferink et al., 2009).

To effectively study the half-reactions of a flavoenzyme, a completely anaerobic system is required to prevent oxygen interference. Anaerobic chambers establish and maintain an anaerobic workspace that can be used to perform assays to study flavoenzyme half-reactions. This chapter discusses the application of an anaerobic chamber to control the concentration of oxygen when performing these experiments with a focus on using stopped-flow spectrophotometry.

2. Anaerobic chambers

Anaerobic chambers maintain a hypoxic atmosphere inside of a hermetic container permitting strictly anaerobic experiments to be performed without the risk of oxygen contamination. These chambers are regularly used in microbiology to grow anaerobes and in biochemistry to study oxygen sensitive biomolecules (Crack, Green, Thomson, & Le Brun, 2014; Lange & Ahring, 2001). In protein chemistry, their use has been applied for preparation of reagents, purification, storage, crystallization, and to perform enzyme assays (Echavarri-Erasun, Arragain, & Rubio, 2014; Lee, Kwon, Kim, & Rhee, 1998; Tsai & Tainer, 2018). With flavoenzymology in particular, these chambers allow an investigator to work in an environment ideal for studying flavin reduction and oxidation.

2.1 Anaerobic chamber components

All anaerobic chambers share similar elements for their operation. Because these chambers are much larger than other options (e.g., anaerobic jars and glovebags), they require a more permanent set up including the use of an antechamber, catalytic filters, an oxygen and hydrogen monitor, and oxygen free gases. The most common features of an anaerobic chamber are described in Fig. 2 and in the following bullets.

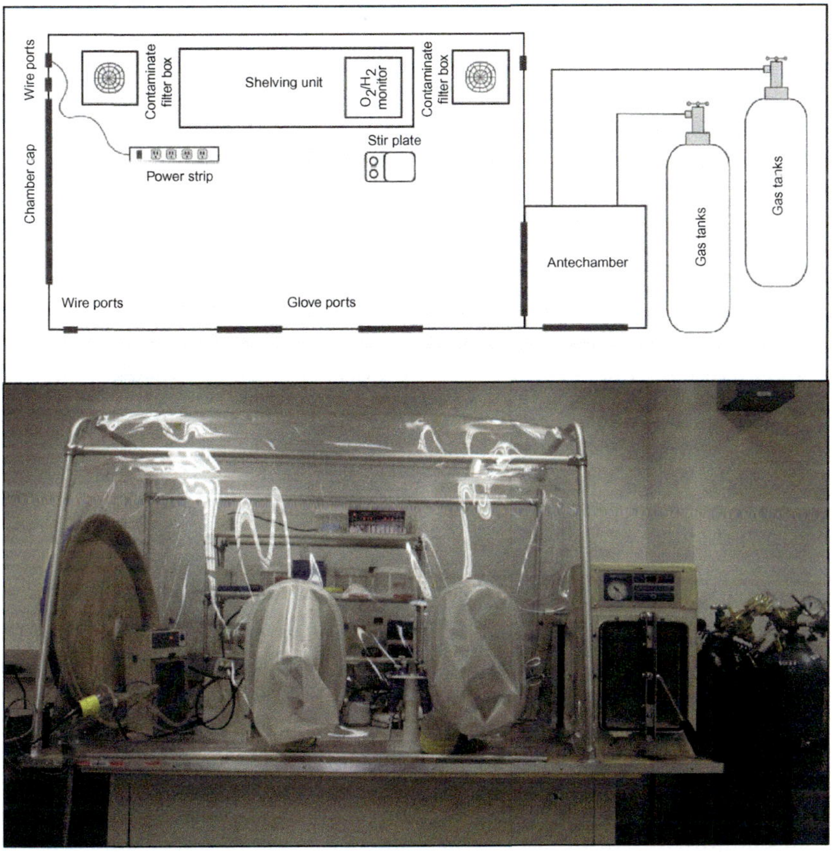

Fig. 2 Top panel: A diagram of an anaerobic chamber displaying the key components. The figure is modeled from a COY Laboratories Type A vinyl anaerobic chamber. Bottom panel: an image of a COY Laboratories Type A vinyl anaerobic chamber. Dimensions are 36″ × 76″ for the full chamber with a 32″ × 59″ workspace.

- *Antechamber*: A chamber attached to the side of the anaerobic chamber with one airlock door inside of the anaerobic chamber and one accessible from the outside. The antechamber is essential for transferring solutions without causing oxygen contamination. Every time material is transferred into or out of the chamber, the antechamber undergoes a cycle of vacuum and gas purges to make the antechamber anaerobic.
- *Glove ports*: Ports built into the chamber shell that allows a user to access material inside the anaerobic chamber. Glove ports can vary depending on the chamber model with a significant difference being between gloved ports compared to gloveless ports. Gloved ports secure interchangeable

gloves of varying sizes to the ends of each sleeve. Gloveless ports use a sleeve-vacuum system allowing an investigator to directly interact with materials inside the chamber. Depending on the size of the chamber, multiple ports can be added.

- *Vacuum pump*: A mechanical pump required to remove air from the antechamber when establishing anaerobiosis. The pump is also used to remove gases from the anaerobic chamber to regulate pressure or to remove oxygen contamination. In the example model shown in Fig. 2, the vacuum pump brings the antechamber to a vacuum pressure of 20 in Hg.
- *Oxygen free gas tanks*: Oxygen free gas is used to fill the chamber and antechamber. The atmospheric conditions used in most chambers consist of a 5% hydrogen and 95% nitrogen mix that can be created in the laboratory with a gas mixer, or purchased premixed.
- *O_2/H_2 monitor*: A device that shares dual purpose in detecting oxygen at ppm and hydrogen as a percentage using different electrochemical sensors.
- *Contaminant filters*: It is required for a chamber to be able to remove oxygen, water, and hydrogen sulfide from the chamber atmosphere. Commonly, a regenerative catalyst is used to convert the contaminants into inert products. Section 2.2 discusses this process in greater detail.
- *Other equipment*: Inside each anaerobic chamber, some basic equipment is often including to improve experimentation. A shelving unit provides extra storage space for chemicals and materials, a power strip is needed so equipment can be used, and a stir plate is commonly added to mix solutions.

Most anaerobic chambers follow a similar process to establish and maintain anaerobiosis. A vacuum pump and oxygen free gasses are used to create a closed circulating system that removes air from the chamber and fills it with a 95% nitrogen/5% hydrogen mix to establish an anaerobic atmosphere. Contaminant filters are then used to maintain these conditions while a monitor detects the levels of oxygen and hydrogen in the chamber. Because the hydrogen concentration decreases due to the reaction in the filters (see Section 2.2.1), the level of hydrogen needs to be maintained between 1.5% and 4% by cycling the chamber with vacuum and mixed gas. Keeping the hydrogen percentage within this range ensures continuous oxygen removal without it being a safety risk as a high percentage of hydrogen can be explosive. Each anaerobic chamber manufacturer provides detailed instructions that should be followed when setting up this equipment. In this chapter, a COY vinyl chamber is used in the methods described.

2.2 Maintaining an anaerobic chamber

Anaerobic chambers operate as a closed system leading to a buildup of unwanted side products that can ruin assays or damage electronics if not properly maintained. For this reason, a filtration system is required to continuously remove contaminants including oxygen, water, and hydrogen sulfide. This section discusses some of the equipment and techniques used to remove contaminants from an anaerobic chamber.

2.2.1 Maintaining chamber anaerobiosis

Oxygen contamination is a serious concern as even low levels can interfere with anaerobic experiments. This problem is addressed using a catalyst that reacts with oxygen forming nonreactive byproducts. COY Laboratories utilize palladium chloride coated alumina beads to react with hydrogen and oxygen to form water (Eq. 1). These beads are placed inside of a metal cassette called a Stak-Pak and inserted into a box that uses a fan to circulates gasses through the Stak-Pak filter. Other metals can also be used to react with oxygen. For example, Belle Technologies use copper/magnesium silicide pellets to react with oxygen forming copper oxide (Eq. 2). This specific reaction does not require hydrogen or form water as a byproduct. The use of copper as a reagent does mean that the copper catalyst is more quickly oxidized than the palladium catalyst used by COY Laboratories.

$$2H_2 + O_2 \xrightarrow{Pd} 2H_2O \quad (1)$$

$$O_2 + 2Cu \rightarrow 2CuO \quad (2)$$

Both reagents described can be regenerated back to their reduced forms using a chemical oven heated to 125–200 °C for 2 h. This allows the catalysts to be reused providing a cheap method for oxygen removal. Both examples maintain atmospheric oxygen concentrations below 2 ppm.

2.2.2 Maintaining humidity and removing hydrogen sulfide

Water vapor is produced from the evaporation of liquids stored in the chamber, exterior moisture entering the chamber, and from oxygen removal depending on the catalyst used. Without any form of regulation, the humidity inside of an anaerobic chamber will reach saturating vapor concentrations causing water condensation that will deactivate metal catalysts used to remove oxygen and potentially damage electronics component of instruments housed in the anaerobic chamber (Selmer, 2005). Different chemicals can been used to maintain humidity including sodium hydroxide, calcium hydroxide, and aluminum oxide. One of the more popular desiccants used is alumina as it readily absorbs water from circulating air and is easily regenerated using a chemical oven. Beside chemical catalysts, a dehumidifier can also be used.

Hydrogen sulfide (H_2S) is a common anaerobic digestion byproduct of sulfate reducing organisms. This chemical is highly explosive and is corrosive, damaging metals, electronic sensors, and the chamber surface with prolonged exposure. If working with anaerobic organisms, it is essential to have an activated carbon and/or iron oxide column present to remove H_2S from the chamber atmosphere.

2.3 Flexible chambers compared to rigid chambers

Anaerobic chambers can be described as either flexible or rigid. Flexible chambers are made of polyvinyl material attached to an aluminum frame, while rigid chambers can be made of clear acrylic or polycarbonate material. Both types of chambers are regularly used for a variety of applications, and cover many of the same requirements including use of an antechamber, catalysts, and oxygen free gases. When purchasing a new chamber, however, there are significant differences that should be considered between these chamber types (Table 1).

3. Introduction to stopped-flow spectrophotometry

Stopped-flow spectrophotometry is used to monitor fast enzymatic reactions as it provides rapid mixing of solutions and records spectra on a millisecond time scale. In flavoenzymology, this instrument is especially

Table 1 A summary of flexible anaerobic chambers compared to rigid anaerobic chambers.

Flexible chambers	Rigid chambers
Flexible chamber minimizes the amount of gas wasted on regulating chamber pressure	Wastes more gas due to pressure regulation
Chamber is clear on all sides	Durable structure
Working area is generally larger	Modular units allow specialization
Optimal ergonomics	Limited user movement
Cheaper to maintain	Can store oxygen sensitive items for long periods of time

useful to determine the rate constants associated with flavin reduction and oxidation. The use of stopped-flow spectrophotometry has contributed to elucidating many flavoenzyme mechanisms and to the detection of intermediates in the catalytic cycle (van Berkel, Benen, Eppink, & Fraaije, 1999).

3.1 Components of a stopped-flow spectrophotometer

Most stopped-flow spectrophotometers contain similar components that rapidly mix solutions and record multiple spectra using different types of detectors. Each stopped-flow spectrophotometer contains a sample handling unit used to rapidly mix samples, a lamp with a power source, a electronics unit, and a computer. When discussing stopped-flow spectrophotometry, this chapter will be referring to the SX20 series from Applied Photophysics. Specifications of this instrument are provided along with a general description of the instrument parts in the following bullets and in Fig. 3.

- *Electronics unit*: This part controls the electornic communications between the various components of the stopped-flow apparatus and processes the information from the detectors connected to the sample handling unit.
- *Lamp power source*: Provides the energy needed to power the stopped-flow spectrophotometer lamp.
- *Lamp housing*: A xenon lamp is commonly used. There is a significant difference between ozone producing and ozone free xenon lamps. Ozone producing lamps can be used in experiments that require wavelength in

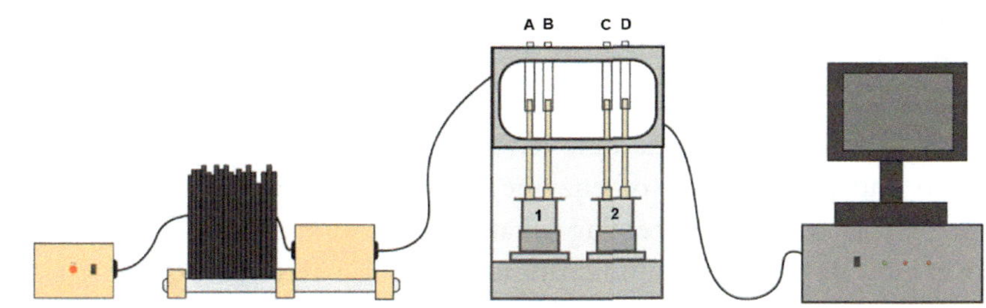

Fig. 3 Diagram of a stopped-flow spectrophotometer modeled from the SX20 unit from Applied Photophysics. The sample handling unit can be configured into a single-mixing mode or a double-mixing mode utilizing the four syringe ports (A–D) and two drives (1 and 2). For the single-mixing mode, only drive 1 is used to mix the solutions from syringe ports (A) and (B). For the double-mixing mode, both drives are used with drive 2 mixing the solutions from syringe ports (C) and (D) in an aging loop before drive 1 mixes the solutions from (A) and (B) with the solution in the aging loop. Detectors with photomultiplier tubes would be attached to the back of the sample handling unit.

the far-UV region (200–1000 nm) while ozone free lamps are limited to 250–725 nm.

- *Monochromator*: This unit is used to select specific wavelengths using a diffraction grating. For Applied Photophysics, a 250 nm holographic diffraction grating is used, which provides a wavelength range of 190–850 nm. The unit is also equipped with adjustable entrance and exit slit widths (0.25–2 mm).
- *Sample handling unit (SHU)*: The SHU performs rapid mixing with one or two drives mixing up to four different solutions. Pneumatic force is used to push the drives and mix solutions. A stop syringe controls the exact volume pushed during each drive by expelling an adjustable volume from its syringe providing only enough room to accommodate the desired volume. The SHU is equipped with a quartz optical cell with a pathlength of 2 or 10 mm.
- *Detector*: The SHU can be equipped with interchangeable fluorescence, absorbance photomultiplier, or photodiode array (PDA) detectors.

3.2 Anaerobic stopped-flow spectrophotometry

When used under anaerobic conditions, a stopped-flow spectrophotometer can be used to study changes to the flavin spectra that are associated with reductive and oxidative half-reactions. Anaerobic chambers serve as an effective way to perform anaerobic stopped-flow spectrophotometry assays as they provide the ability to maintain a stable anaerobic environment for preparing and performing enzyme assays. These chambers also reduce the amount of time and gas spent deoxygenating solutions as it only requires an anaerobic stock solution which can be stored inside of the chamber for days at a time. The anaerobic environment allows a more relaxed preparation and performance of enzyme assays. This setup can also perform anaerobic assays without relying on an oxygen scavenger (glucose oxidase) or a chemical reducer (sodium dithionite) to remove residual oxygen. Using chambers for stopped-flow spectrophotometry is a useful method that is well-established in the flavoenzyme community (Knight & Scrutton, 2002; Robinson, Badieyan, & Sobrado, 2013; Sucharitakul, Wongnate, & Chaiyen, 2011; Yu, Hausinger, Tang, & Xu, 2014). Even with its benefits, anaerobic chambers also take up significant space in a lab and their initial cost and maintenance are expenses that need to be considered when comparing the two methods (Table 2).

Table 2 Summary of the benefits and disadvantages of using an anaerobic chamber for stopped-flow spectrophotometry.

Benefits of using an anaerobic chamber
Stable anaerobic environment
Easy preparation of anaerobic solutions from stock solutions
Lower risk of O_2 contamination, leading to less experimental error
Less time and materials spent making the solution anaerobic
Instrument accessibility is not limited by being inside of the chamber
Disadvantages of an anaerobic chamber
Costs of purchase and maintenance
Requires a large footprint in the lab
The stopped-flow equipment takes up space in the chamber that could be used for other things
Limited mobility
Continuous chamber maintenance required

4. Setting up, preparing, and performing stopped-flow spectrophotometry assays inside of an anaerobic chamber

Performing anaerobic assays with a stopped-flow spectrophotometer inside of an anaerobic chamber requires careful set up of the instrument to allow easy handling once inside the chamber. It is also important to establish a method for preparing deoxygenated solutions that consistently produce anaerobic buffers with limited risk of contamination. This section describes the process of setting up a stopped-flow spectrophotometer inside of an anaerobic chamber, setting up Schlenk line apparatus, deoxygenating buffers, validating solution anaerobiosis, preparing samples for experimentation, and measuring the reductive and oxidative half-reactions of a flavoenzyme using stopped-flow spectrophotometer. While the described methods have proven to be successful in our lab, considerations should always be made to optimize conditions specific to each investigator.

4.1 Setting up a stopped-flow spectrophotometer inside of an anaerobic chamber

When setting up a stopped-flow spectrophotometer inside of an anaerobic chamber, it is important to consider the placement of each component in order to optimize operation and handling. This section provides recommendations for arranging an SX20 Applied Photophysics stopped-flow spectrophotometer inside of a COY Laboratories Type A vinyl chamber. Both the chamber and the stopped-flow model are two of the more popular designs in their respective categories. For this reason, they serve as a good example of how a stopped-flow spectrophotometer can be operated inside of a chamber. It should also be mentioned that hard shell anaerobic chambers are compatible for stopped-flow spectrophotometry studies as well (Knight & Scrutton, 2002).

4.1.1 Materials

- Stopped-flow spectrophotometer (Applied Photophysics).
  - Sample handling unit (SHU).
  - Electronics communication unit.
  - Lamp power source.
  - Lamp holder.
  - Monochromator.
  - Detectors: Photodiode array, absorbance, or fluorescence.
  - Computer.
- Anaerobic chamber (COY Laboratories)
- A large sturdy table or other workspace.
- Water temperature regulator (Fisher Scientific).
- Nitrogen/hydrogen (95%/5%) gas mix.
- Gas regulator.
- Rubber sealant.
- Electrical tape.

4.1.2 Procedure

1. Situate the anaerobic chamber next to a table or bench that is capable of holding stopped-flow spectrophotometer equipment.
2. Remove the chamber cap and unseal all wire ports.
3. Place a sturdy Teflon plastic sheet underneath a disconnected SHU. This is to protect the chamber from damage. Carefully, place the instrument in the chamber, and angle the SHU in a manner so that it is easy to reach from the glove ports.

4. If not already in the chamber, place the catalyst boxes, a shelving unit, and a standard power strip inside of the chamber. Place all detectors that will be used with the SHU into the chamber as well.
5. Place the other stopped-flow equipment and the water temperature regulator in the space next to the chamber. Fig. 4 shows an example of how this equipment can be placed.
6. Secure a mixed gas tank to a wall. Attach the gas regulator to the tank and connect it to the SHU using gas tubing. Make sure to thread the connection through a wire port.
7. Carefully begin to make connections starting with the wires going through the wire ports. A suggested setup is shown in Fig. 4.
8. Finish setting up the other connections of the instrument. Make sure a power strip for the equipment is accessible.

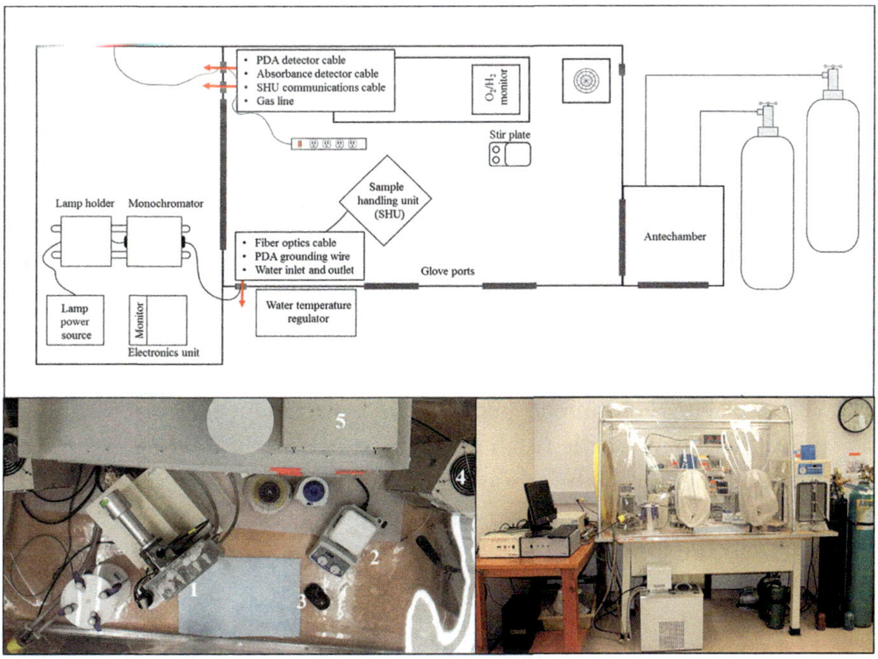

Fig. 4 A stopped-flow spectrophotometer inside an anaerobic chamber. The top panel represents a diagram of the described setup from a top down perspective with the specific wires in each port described. Lower left panel: materials inside of the anaerobic chamber including an Applied Photophysics SHU **(1)**, a magnetic stir plate **(2)**, a wireless computer mouse **(3)**, COY Laboratories catalyst boxes **(4)**, and an O_2/H_2 monitor **(5)**. Lower right panel: the exterior of the chamber with the stopped-flow spectrophotometer setup including electronics unit, lamp power supply, lamp holder, monochromator, computer, water bath temperature regulator, anaerobic chamber, and gas tanks.

9. Test the operation of the equipment by turning it on and performing a dry run. If any component is difficult to reach or awkward to handle, adjust the equipment accordingly.
10. Once the equipment has been optimized, seal the ports with rubber sealant. Let the sealant dry for 24 h.
11. Return the chamber cap to its proper place and seal the entire rim multiple times with electrical tape.
12. Vacuum air out of the chamber as much as possible and refill it with 95%/5% nitrogen/hydrogen mixed gas. Repeat until the O_2/H_2 monitor records a percent hydrogen between 1.5% and 4% and 0 ppm O_2.
13. Inflate the chamber until it is full, and watch if it deflates or if the percentage of hydrogen rapidly decreases over a few hours. This would suggest leak in the system. Remove sealant and chamber cap and repeat steps 9–12 if a leak is suspected.
14. Place equipment needed to perform assays inside of the chamber. This includes a stir plate, Allen wrenches to remove detectors, sample syringes, sample filters, 2 M nitric acid to clean the SHU, and deionized water.

4.1.3 Notes

1. Make sure there are no oxygen sensitive materials inside the anaerobic chamber when opening the chamber to the room atmosphere.
2. If one detector will be used frequently (e.g., absorbance), it is recommended to attach the detector to the SHU before moving the instrument into the chamber.
3. When placing equipment, keep in mind the distance between connections. For example, the reach of the fiber optic light guide is about 1 m, so the orientation of the monochromator and the SHU should be within this range so that an extension is not required.
4. In the setup shown in Fig. 4, there are three wire ports close to the SHU providing access from multiple sides.
5. Be creative with what is tested in the dry run in step 8. Try changing detectors or switching the pathlength while using the glove ports, change the monochromator entrance and exit slit widths, and turn on the water regulator to check for leaks. Operation of the stopped-flow spectrophotometer equipment should not be hindered by being in the chamber. A person should still be able to adjust the settings with little difficulty. The arrangement should also allow plenty of space to put supplies and perform assays without crowding.

6. Make sure there is plenty of slack with the wires and lines inside of the chamber to allow easy removal.
7. Include a wireless mouse inside of the chamber for easier operation of the computer during experimentation.
8. With a careful setup, the SHU can be removed from the chamber for fixing or cleaning without breaking the rubber seal on the wire ports. This saves time and materials as this seal takes up to a day to be fully hardened.

4.2 Setting up a Schlenk line on a cart

An essential part of preparing assays under anaerobic conditions is the process of removing oxygen from buffers, substrate solutions, and enzyme stocks. There are several methods used to deoxygenate buffers, including sonication, sparging, bubbling, boiling with an inert gas headspace, or using a vacuum gas manifold (a Schlenk line) (Butler, Schoonen, & Rickard, 1994; Lowe, Fisher, & Thorneley, 1993; Romero, Robinson, & Sobrado, 2012; Selmer, 2005). This chapter specifically describes using a Schlenk line to obtain anaerobic solutions with this section focusing using a Schlenk line set up on a cart. This method allows the system to be used in various locations without disassembling or dismounting.

A Schlenk line consists of two glass manifolds specially made to withstand changes in pressure and temperature with multiple valves fitted with double-oblique stopcocks. To use a Schlenk line, one manifold is kept under vacuum while the other is filled with a pure inert gas, typically argon. Airtight flasks are attached to the Schlenk line by a tube, and dissolved oxygen is removed through cycles of high vacuum and inert gas purges. These two conditions in combination with rapid stirring exchanges the gasses from a solution with oxygen free inert gas making it anaerobic. A suggested set up of a Schlenk line for degassing buffers is described along with a visual example (Fig. 5).

4.2.1 Equipment

- Large cart.
- Aluminum support frame.
- Glass dual manifold (Schlenk line).
- Norprene tubing.
- Vacuum hosing.
- Stainless steel hose clamps (MSC Industrial Supply Co) and a screwdriver.
- Glass oil bubbler.

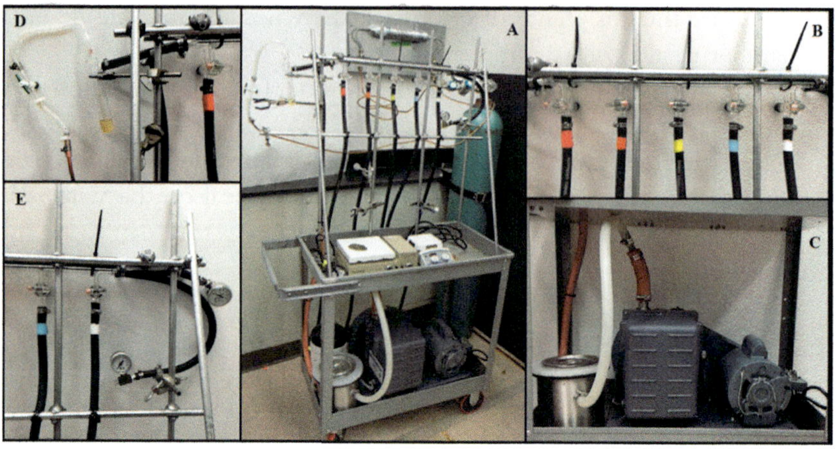

Fig. 5 Setup of a Schlenk line discussed in Section 4.2.2. (A) The overall setup. The bottom of the cart contains a vacuum pump and cold trap. The top of the cart contains two stir plates and an aluminum structure holding a Schlenk line with the attached bubbler and gauges. A high purity argon gas tank connected to an oxygen trap (Agilent) is in the background. (B) The Schlenk line is secured to the aluminum structure using zip-ties. Norprene tubing is attached to the end of the stopcock valve with hose clamps, and each tube is marked with a different colored tape. (C) The bottom of the cart with the vacuum pump (Welch) and cold trap (Labconco). Vacuum hosing attaches the cold trap to the inlet of the vacuum pump. Hosing to the Schlenk is shown in the back going through a hole in the cart top. (D) Close up image of the oil-bubbler and high-vacuum stopcock attached to the Schlenk line. Each attachment is secured with hose clamps. (E) Close up image of the vacuum and gas gauges secured to the Schlenk line with hose clamps.

- Glass high vacuum stopcock.
- Vacuum and gas pressure gauges.
- Vacuum pump (Welch).
- Dry ice cold trap (Labconco).
- An oxygen trap (Agilent).
- Magnetic stir plate.
- Flexible copper tubing.
- Plastic zip ties.
- Vacuum grease.

4.2.2 Procedure

1. Assemble an aluminum frame to hold the Schlenk line. Drill the ends of the frame into the cart to secure it.
2. Secure the Schlenk line to the top of the frame using zip ties (Fig. 5B).

3. Lubricate the stopcocks with vacuum grease and firmly tightened to the manifold using an O-ring and plastic nut.
4. Secure Norprene tubing (~3 ft.) to the stopcocks and secure with hose clamps (Fig. 5B).
5. To the exits of the Schlenk line, secure a gas gauge and a vacuum gauge to their respective manifolds using Norprene tubing with metal hose clamps (Fig. 5E).
6. Place the vacuum pump on the bottom shelf of the cart.
7. From the intake of the vacuum pump, secure vacuum hosing to a cold trap.
8. Attach the cold trap to the inlet of the vacuum manifold using a combination of vacuum hosing and Norprene tubing (Fig. 5C). Secure the tubing with hose clamps.
9. To the entrance of the argon gas manifold attach a glass oil bubbler using Norprene tubing and secure it with metal hose clamps (Fig. 5D).
10. Secure a high purity argon gas tank to a wall and attach a gas regulator.
11. Use the flexible copper tubing to attach the gas tank to an oxygen trap.
12. Attach copper tubing from the trap to rubber tubing with a hose clamp. Connect the tubing to a high vacuum stopcock (Fig. 5D).
13. From the stopcock, connect the rubber tubing to an oil bubbler and secure with a clamp.
14. On the top of the cart, place two stir plates with clamps attached to the aluminum frame.

4.2.3 Notes

1. All oxygen contaminants in the inert gas should be removed by passing the gas through an oxygen trap. Oxygen traps remove oxygen and sulfur from non-oxidizing gases in bulk by using a reactive metal catalyst. In Fig. 5, the oxygen trap can be seen secured to the wall. The gas should also pass through an oil-bubbler containing water for (i) gas hydration and (ii) detection of gas flow.
2. Make sure the vacuum pump has the proper amount of oil. Remember to change the oil every couple months or more often depending on use.

4.3 Preparing anaerobic buffers

This method describes the removal of oxygen from a buffer using a Schlenk line with high purity argon gas. This is useful for kinetic analysis on a stopped-flow spectrophotometer and can be used to prepare anaerobic buffers for other assays as well.

4.3.1 Equipment

- Schlenk line setup (Section 4.2.2).

4.3.2 Buffers and reagents

- Buffer solution, 100 mL.
- High purity argon gas.
- Dry ice for the cold trap.

4.3.3 Procedure

1. Measure 100 mL of a buffer and place into a 250 mL Büchner flask with a stir bar.
2. Cover the flask with a (i) rubber septum or (ii) rubber stopper.
3. Attach the Schlenk line to the side arm of the Büchner flask. If using the designed tubing described in Fig. 6, make sure there is a tight seal by taping the tube to the side of the flask.
4. Place the flask onto a stir plate and begin stirring.
5. Put dry ice into the cold trap and start the vacuum pump.
6. Turn on the argon gas and adjust the pressure to 5 psi.
7. Place the solution under vacuum for 4 min.
8. Purge with argon for 1 min.
9. Repeat for four cycles for a total of 20 min.
10. Turn off vacuum after the last argon purge. Leave the flask under positive pressure from the argon.
11. Detach from the Schlenk line, and transfer flask into the anaerobic chamber.

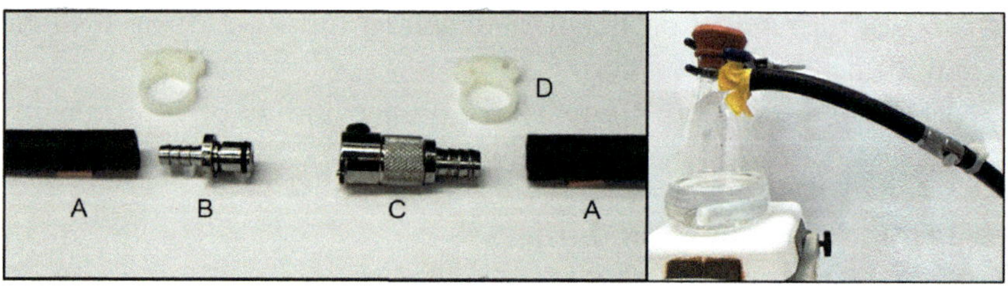

Fig. 6 Left panel: disassembled set up of connectors used to detach and transfer anaerobic solutions to the anaerobic chamber without oxygen contamination. (A) Norprene tubing, (B) straight thru valve, (C) shutoff valve, (D) nylon hose clamps. Right panel: assembled setup attached to a Büchner flask.

4.3.4 *Notes*

1. Be sure to seal the stopper/septum with electrical tape. The rubber stopper risks leaking oxygen into the flask during argon purges, and the septum will pop off in the antechamber when under negative pressure.
2. The buffer loses 4% volume when using this method.
3. The vacuum pressure should be approximately 28 in Hg. Anything higher might not provide enough vacuum to remove dissolve oxygen.
4. Make sure to always have dry ice in the cold trap, and remove all condensed solvent after use. If this is not done, water will collect inside of the vacuum oil and corrode the interior of the pump.

4.4 Transferring anaerobic solutions

Methods to transfer anaerobically prepared solutions to the anaerobic chamber should be considered as even minute exposure to air can cause oxygen contamination. To address this concern, Norprene tubing can be retrofitted to detach from the Schlenk line and move with the flask into the chamber (Fig. 6). This simple design minimizes the risk during transfer from the Schlenk line into the anaerobic chamber.

4.4.1 *Equipment*

- Norprene tubing.
- In-line hose barb LC series chrome plated brass inline body shutoff valve (Colder Products).
- In-line hose barge LC series plates brass insert, straight thru valve (Colder Products).
- Nylon hose clamps.

4.4.2 *Procedure*

1. Cut a short segment of tubing that fits tightly around the arm of a Büchner flask (~6 in.).
2. Insert a shut off valve on the end of the tubing segment.
3. Secure the valve with a nylon hose clamp until it is airtight.
4. On the end of a long tubing attached to a Schlenk line, insert the straight thru valve and secure it with a hose clamp.

4.4.3 *Notes*

1. This system uses a shutoff valve to seal the end of the tube preventing oxygen from entering the flask from the sidearm of the Büchner flask during transfer.

4.5 Addition of enzymes to maintain anaerobic solutions

After a buffer has been made anaerobic, addition of enzymatic oxygen scavengers is sometimes used to maintain anaerobiosis. While reducing agents including sodium dithionite and titanium(III) citrate can be used to remove oxygen, these chemicals will also reduce flavoproteins interfering with the kinetic studies. Oxygen consuming enzymes do not react with the flavin cofactor allowing them to be used in assays with minimal interference. This section describes two enzyme systems that are regularly used to maintain anaerobic solutions.

Glucose oxidase (GO) is a flavoenzyme that converts glucose in the presence of oxygen to gluconolactone and hydrogen peroxide. This enzyme is easy to isolate from *Aspergillus niger* and *Penicillum* species making it commercially available as a lyophilized powder. The stability of this enzyme along with its common substrate has caused GO to be used for a broad range of applications including glucose detection, food preservation, and bio-technical applications (Bankar, Bule, Singhal, & Ananthanarayan, 2009). Among these uses, GO is regularly used as an oxygen scavenger to remove residual oxygen and maintain anaerobiosis. GO is active in most aqueous buffers with a pH range of 3.4–7.5 and is optimal at a temperature between 40 and 45 °C (Selmer, 2005). This enzyme is also used to scrub stopped-flow spectrophotometers of oxygen before experimentation. GO it is not always the ideal choice for oxygen scavenging as it produces hydrogen peroxide. GO also has a high K_m that demands a high concentration of glucose to be added to the assay buffer (Swoboda & Massey, 1965).

Another enzyme system that can be used to maintain anaerobic solutions is protocatechuate dioxygenase (PCD). PCD converts protocatechuate to β-carboxy-*cis*, *cis*-muconic consuming a molecule of oxygen without the creation of hydrogen peroxide or any other reactive species. The PCD system has also been shown to produce solutions with lower dissolved oxygen concentrations than GO (Aitken, Marshall, & Puglisi, 2008). PCD has been shown to be active at a pH 6–9 and a temperature range of 15–25 °C making it very useful for assays performed at room temperature and for scrubbing instruments (Patil & Ballou, 2000). Patil and Ballou in 2000 showed stable reduction of free flavin with sodium dithionite in a solution of 6.5 mg/mL PCD and 0.4 mM protocatechuate. While this enzyme is an excellent substitute, it is the more expensive alternative, with 25 units of PCD costing $298 from Sigma Aldrich compared to 10 kU of GO for $45.

Our lab regularly uses GO for the preparation of enzyme assays with a stopped-flow spectrophotometer. Preparation of this enzyme is a simple procedure that can be adjusted to fit specific lab requirements. The following sections describe the technique for preparing a GO solution for scrubbing a stopped-flow spectrophotometer. For a description of adding GO to assay buffers, refer to note 1 (Section 4.5.4).

4.5.1 Equipment

- Schlenk line setup (Section 4.2.2).
- Stopped-flow spectrophotometer setup (Section 4.1.2).
- 50 mL conical tube.
- Four disposable 5 mL luer lock syringes.

4.5.2 Buffers or reagents

- GO from *Aspergillus niger* (>100,000 U/g, Sigma-Aldrich).
- GO buffer: 100 mM sodium acetate, pH 5.0 (Fisher Scientific).
- Glucose (Fisher Scientific).

4.5.3 Procedure

1. Measure 100 mL of 100 mM Na-acetate, pH 5.0 buffer and pour into a Büchner flask.
2. Deoxygenate buffer using the procedure described in Section 4.3.3.
3. Measure 5 mg of GO and 900 mg of glucose and place components inside of a 50 mL conical tube.
4. Transfer tube into the anaerobic chamber with buffer.
5. Dissolve enzyme and glucose in 50 mL of anaerobic 100 mM Na-acetate, pH 5.0 buffer. Final concentration will be 0.6 μM GO and 100 mM glucose.
6. Draw anaerobic buffer into four luer lock syringes.
7. Replace the sample syringes on the SHU with the syringes filled with GO dissolved in 100 mM Na-acetate, pH 5.0 buffer.
8. Flush the system with 100 mM Na-acetate, pH 5.0 buffer with GO by either using the drive option on the program or manually.
9. Leave the SHU in 100 mM Na-acetate, pH 5.0 buffer with GO overnight.

4.5.4 Notes

1. GO can also be mixed with anaerobic buffers to maintain anaerobiosis during enzyme assays. For this application, 3 mg of GO and 450 mg of

glucose should be dissolved in 50 mL of assay buffer making a final concentration of 0.4 μM GO and 50 mM glucose.

2. For more information, this technique is described in another publication from our lab (Romero et al., 2012).

4.6 Determining oxygen concentration in deoxygenated buffers to validate anaerobiosis

When preparing anaerobic buffers, it is important to quantitate the concentration of oxygen present in a solution to determine how effective the deoxygenating process is. This technique is very helpful especially when troubleshooting for oxygen contamination. Here, the use of the indicator methyl viologen will be discussed. Methyl viologen is a redox-active heterocyclic compound that forms a stable structure with a strong absorbance when reduced with sodium dithionite. This agent is very reactive with oxygen and changes from clear to dark blue between its oxidized and reduced forms (Selmer, 2005). The data shown in Fig. 7 were collected using the method described here.

4.6.1 Equipment

- Anaerobic chamber (COY Laboratories).
- Schlenk line setup (Section 4.2.2).
- Anaerobic cuvette (Section 5.3).
- UV–visible spectrophotometer (Agilent).

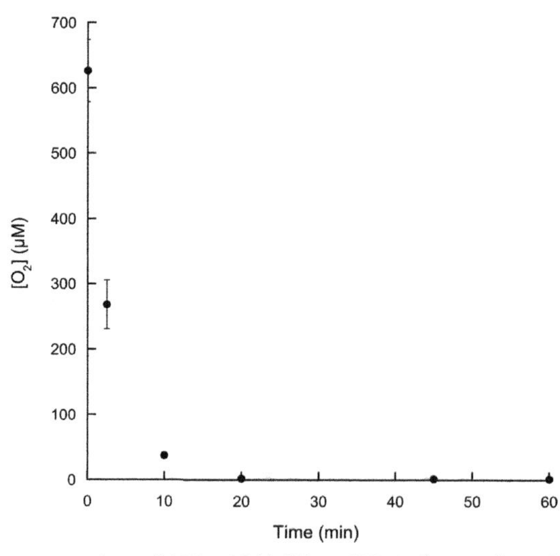

Fig. 7 Oxygen concentration of 100 mM $NaPO_4$, pH 7.5 after cycles of argon/vacuum as a function of time.

4.6.2 Buffers and reagents

- Aqueous buffer, 100 mL.
- Methyl viologen (Sigma-Aldrich). Dissolve 26 mg into 10 mL anaerobic deionized (DI) water for a stock solution of 10 mM.
- Potassium hexocyanoferrate (Sigma-Aldrich). Dissolve 33 mg into 10 mL anaerobic DI water for a stock solution of 10 mM.
- Sodium dithionite (Merck). Dissolve 17.4 mg into 10 mL of anaerobic DI water for a stock solution of 10 mM.
- Oxygen free argon, nitrogen, and nitrogen/hydrogen (95%/5%) gasses.

4.6.3 Procedure

1. Prepare deoxygenated buffer following the protocol in Section 4.3.3.
2. Transfer solutions into the anaerobic chamber.
3. Determine the concentration of sodium dithionite by measuring the reduction of potassium ferricyanide. Use an anaerobic cuvette to measure the concentration of 1 mM potassium ferricyanide on a UV–vis spectrophotometer. Mix 1 mM potassium ferricyanide with 50 μM of the sodium dithionite stock. Transfer the solution to an anaerobic cuvette and record the absorbance with a UV–vis spectrophotometer.
4. From the change in absorbance determined from step 2, calculate the amount of potassium ferricyanide consumed using the Beer-Lambert equation ($\varepsilon_{405} = 1.04\,\text{mM}^{-1}\,\text{cm}^{-1}$). From this, calculate the concentration of sodium dithionite.
5. Prepare a 2.5 mM methyl viologen reduced stock by mixing it with 2.5 mM sodium dithionite and anaerobic buffer.
6. Mix 50 μL of the reduced methyl viologen stock with 1 mL of anaerobic buffer stock and 2 mM sodium dithionite. This solution serves as an anaerobic control. Transfer the solution to an anaerobic cuvette and record the absorbance with an UV–vis spectrophotometer.
7. Mix 50 μL of the reduced methyl viologen stock with 1 mL of anaerobic buffer with no sodium dithionite added. This solution serves as the sample to be compared to the anaerobic control. Transfer the solution to an anaerobic cuvette and record the absorbance with a UV–vis spectrophotometer.
8. Determine the change in absorbance between the anaerobic control and the sample from steps 6 and 7. From this change, calculate the amount of methyl viologen oxidized ($\varepsilon_{605} = 13.7\,\text{mM}^{-1}\,\text{cm}^{-1}$) (Watanabe & Honda, 1982). Calculate the concentration of residual oxygen in the buffer by determining how much methyl viologen was oxidized.

4.6.4 Notes

1. Sodium dithionite will decompose, so it is important to prepare a fresh solution and measure the concentration each time.
2. There are different colorimetric indicators that can be used for testing anaerobic solutions. These chemicals vary depending on stability, extinction coefficients, and the color changes between their reduced and oxidized forms. For a detailed explanation of the types of electron acceptors that can be used to determine oxygen concentrations, refer to table 5.2 from Selmer (2005).
3. If oxygen contamination is high, the methyl viologen will completely oxidize turning the solution clear. To get an exact measurement of oxygen concentration, the methyl viologen concentration can be increased to a point where it is no longer fully oxidized.
4. This method can also quantitate the oxygen concentrations of oxygenated buffers used to study flavin oxidation. To do this, it is recommended to dilute the oxygenated buffer with anaerobic buffer to a point where it does not oxidize all of the reduced methyl viologen.

The method described has been used to determine anaerobiosis of buffers prepared using the method from Section 4.3.3. Reduced methyl viologen stock was mixed with aqueous buffer (100 mM $NaPO_4$, pH 7.5) deoxygenated at different time points from 0 to 60 min. The results from this experiment show that most of the oxygen was removed from the buffer after 10 min and was completely anaerobic after 20 min (Fig. 7).

4.7 Preparing anaerobic enzyme solutions

On the day of experimentation, several samples have to be carefully prepared before analysis can begin. Even with correct buffer preparation and scrubbing of the instrument, using solutions that have not been appropriately deoxygenated will ruin the enzyme assays. As many substrate and cofactors are sold from retailors as solids, these can be dissolved in anaerobic buffer removing any concern of oxygen contamination. However, this typically cannot be accomplished with enzymes as only a few are stable when lyophilized. While degassing methods can be applied, the rigorous method described in Section 4.3.3 would lead to protein denaturation. To prevent this, the Schlenk line can be used in a more gentle manner to deoxygenated enzyme solutions. Since the volume used with protein stocks is much smaller than with buffers, the enzyme solutions can be deoxygenated within a few minutes.

4.7.1 Equipment

- Schlenk line setup (Section 4.1.2).
- A plastic male luer lock with hose barb.
- 5 or 20 mL serum flask with a rubber plug and peel off metal cap (Wheaton).
- Small plastic container (e.g., an empty tip box).
- A small stir bar.
- Disposable needle (BD).

4.7.2 Buffers and reagents

- Ice.
- Dry ice.
- Enzyme stock solution.
- High purity argon gas.

4.7.3 Procedure

1. Thaw the required amounts of enzyme stock solution on ice.
2. Place enzyme into a 5 or 20 mL Wheaton flask depending on volume. Add a small stir bar and cap the flask using a crimper.
3. Fill a small plastic container with ice and water. Place the container onto a stir plate below the Schlenk line with the enzyme solution resting in the ice water slurry.
4. Secure the flask with a clamp.
5. Insert the needle into a male luer-lock attachment secured to the end of the Norprene tubing attached to the Schlenk line. Insert the needle into top of the flask.
6. Start stirring the enzyme solution.
7. Turn on the vacuum pump and argon gas adjusted to 5 psi.
8. Switch the enzyme flask to vacuum for a few seconds until bubbles begin to form. When this is observed, turn the stopcock to argon gas for 15 s.
9. Repeat step 8 for 20 cycles.
10. Remove the needle from the flask.
11. Place the flask into the antechamber.
12. Dilute enzyme solution with anaerobically prepared buffer to the desired concentration needed.

4.7.4 Notes

1. The enzyme solution should be deoxygenated as a concentrated stock and diluted inside of the chamber with anaerobic buffer.
2. When inserted into the flask, the needle should occupy the headspace and not submerge in the solution.

4.8 Preparing anaerobic substrate and cofactor solutions

Substrates and cofactors are commonly used when studying flavoproteins, and require anaerobic preparation for kinetic use. In situations where the ligand is already dissolved, the sample can be diluted with aerobic buffer to the desired concentration and deoxygenated following the protocol in Section 4.7.3. Alternatively, solid substrates and cofactors can be dissolved with anaerobic buffers inside of an anaerobic chamber.

4.8.1 Equipment

- Anaerobic Chamber (COY Laboratories).
- UV–visible spectrophotometer (Agilent).
- Anaerobic cuvette
- Microcentrifuge or conical tubes.

4.8.2 Buffers and reagents

- Pre-prepared anaerobic assay buffer (Section 4.3).
- Cofactor or substrate in a solid form.
- DI water.

4.8.3 Procedure

1. Measure the amount of the cofactor/substrate required for the planned experiments. Place the material into a tube with the appropriate size for the desired dilution.
2. Transfer tube with the compound into the anaerobic chamber.
3. Dissolve cofactor/substrate with anaerobic buffer.

4.8.3.1 Quantifying cofactor/substrate concentration

- When preparing a compound without a chromophore (e.g., L-lysine): determine the concentration of the solution from the exact mass of the compound measured and the volume of buffer it was dissolved in.
- When preparing a compound with a chromophore (e.g., NADH or NADPH): Pipette a 2 μL aliquot of the compound into a 1.5 mL microcentrifuge tube. Remove the tube from the chamber, and dilute the

sample with DI water so the expected absorbance is below one. Measure the diluted sample with a UV–vis spectrophotometer, and calculate for the concentration using the Beer-Lambert equation.

- When preparing an oxygen-sensitive reagents with a chromophore (e.g., methyl viologen): Dilute an aliquot of the sample with anaerobic buffer so the expected absorbance is below one. Transfer the solution into an anaerobic cuvette and remove from the chamber. Measure the diluted sample with a UV–vis spectrophotometer, and calculate for the concentration using the Beer-Lambert equation.

4.9 Preparing oxygenated buffers

In order to study the oxidative half-reactions of some flavoenzymes, it is important to be able to mix concentrations of oxygen into the enzyme assays. This can be accomplished by preparing an oxygen saturated buffer that can be diluted with anaerobic buffer and used to perform the desired assays. Aqueous solutions contain 280 μM of dissolved oxygen at 20 °C, and fully saturates at ~1.2 mM dissolved oxygen at 0 °C (Rice, Baird, & Eaton, 2017).

4.9.1 Equipment

- Pressure regulator.
- Gas lines with a long metal syringe attached.
- A 400 mL beaker.
- A 250 mL volumetric flask with a rubber septum.
- A disposable syringe (BD).
- A stir bar.
- A stir plate.
- A support stand with clamp.

4.9.2 Buffers and reagents

- Oxygen gas.
- Aqueous buffer.
- Ice.

4.9.3 Procedure

1. Measure 100 mL of assay buffer and place into a volumetric round bottom flask with a stir bar.
2. Cap the flask with a rubber septum.
3. Prepare an ice water solution in a beaker and place on top of a stir bar.

4. Secure the flask in the ice bath using a support stand and clamp.
5. Insert a clean disposable needle into the septum and begin stirring.
6. Insert a long needle attached to an oxygen gas tank into the buffer.
7. Turn the tank on and adjust the pressure to 30 psi. Small bubbles should be forming in the solution.
8. Allow the oxygen to bubble for 30 min.
9. Remove the disposable needle for venting and wait 10 s.
10. Remove the oxygen line.

4.9.4 Notes

1. Do not use a septum that has been punctured multiple times.

4.10 Monitoring flavin oxidation and reduction with stopped-flow spectrophotometry

The flavoenzyme SidA can be used as an example of performing stopped-flow experiments inside of an anaerobic chamber. This enzyme catalyzes the hydroxylation of the amino group on the side chain of L-ornithine forming N^5-hydroxy-L-ornithine, which is an important step in siderophore biosynthesis in *Aspergillus fumigatus*. The stopped-flow spectrophotometry setup described in this chapter has been used to extensively study this enzyme (Robinson et al., 2013, 2014, 2015). The following procedure outlines performing oxidative and reductive studies inside of an anaerobic chamber using SidA.

4.10.1 Equipment

- Stopped-flow spectrophotometer setup inside of an anaerobic chamber (Section 4.1.2).
- Schlenk line setup (Section 4.2.2).
- Disposable 5 mL luer lock syringes

4.10.2 Buffers and reagents

- GO buffer: 100 mM sodium acetate, pH 5.0 (Fisher Scientific).
- SidA buffer: 100 mM $NaPO_4$, pH 7.5 (Fisher Scientific).
- L-Ornithine (Sigma-Aldrich).
- SidA enzyme stock.
- Nicotinamide adenine dinucleotide phosphate (NADPH) (Research Products International).

4.10.3 Procedure

4.10.3.1 Day 1

1. Prepare at least 100 mL 100 mM $NaPO_4$, pH 7.5 buffer (SidA) and 100 mM Na-acetate, pH 5.0 (GO) each.
2. Switch the SHU to the desired mixing mode. For this example, the reductive half-reaction uses single-mixing that only uses drive 1 (Fig. 3), while the oxidative half-reaction uses double-mixing that uses drive 1 and 2 (Fig. 3).
3. Prepare anaerobic buffers described in Section 4.3.3.
4. Transfer anaerobic solutions into the anaerobic chamber.
5. Perform oxygen-scrubbing to the SHU as described in Section 4.5.3.
6. Leave 100 mM $NaPO_4$, pH 7.5 buffer stirring inside of the chamber overnight.

4.10.3.2 Day 2

7. Turn on the stopped-flow spectrophotometer lamp and water temperature regulator. Let the lamp warm up for 30 min.
8. Thaw the required amount of enzyme stock and deoxygenated it using a Schlenk line (Section 4.7.3).
9. Prepare NADPH stock (Section 4.8.3).

4.10.3.3 Reductive half-reaction

10. Prepare the desired concentrations of NADPH for the assay from the stock solution prepared in step 9.
11. Remove the GO solution by flushing the SHU ports A and B with anaerobic 100 mM $NaPO_4$, pH 7.5 (Fig. 3).
12. Switch the detection mode to PDA. Collect background using anaerobic 100 mM $NaPO_4$, pH 7.5 buffer.
13. Dilute SidA enzyme stock to 30 μM and draw up into a luer lock syringe.
14. Insert syringe into the SHU port A (Fig. 3). Pull a small volume (~50 μL) into the SHU sample syringe and flush completely out of the SHU by emptying the stop syringe and driving the syringe. Repeat this two times.
15. Obtain spectra of the oxidized flavin for 500 s logarithmically.
16. Draw NADPH assay solution into a luer-lock syringe and insert into the SHU port B (Fig. 3). Repeat the process described in step 15.
17. Acquire data for 500 s on a logarithmic scale.

Date collected of SidA reduction will show the flavin peak at 450 nm disappear as it is reduced (Fig. 8A).

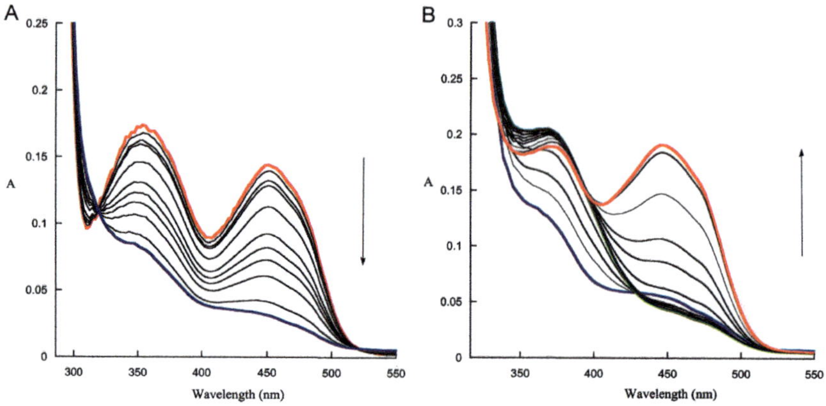

Fig. 8 Experimental spectra showing SidA reduction (A) and oxidation (B). Reduced enzyme is shown in blue (dashed), oxidized enzyme is shown in red (solid), and enzyme when it forms the C4a-hydroperoxyflavin intermediate is shown in green (small dashed). The direction of peak changes at 450 nm is represented with the gray arrows.

4.10.3.4 Oxidative half-reaction (perform after completing Sections 4.10.3.1 and 4.10.3.2)

10. Remove the GO solution by flushing the SHU ports A, B, C, and D with anaerobic 100 mM $NaPO_4$, pH 7.5 (Fig. 3).

11. Switch the detection mode to PDA. Collect background using anaerobic 100 mM $NaPO_4$, pH 7.5 buffer.

12. Dilute the SidA enzyme stock to 60 μM with anaerobic 100 mM $NaPO_4$, pH 7.5 buffer. Insert the sample syringe into the SHU port C (Fig. 3). Pull a small volume (~50 μL) into the SHU sample syringe and flush completely out of the SHU by emptying the stop syringe and driving the syringe. Repeat this two times.

13. Prepare a solution of NADPH stoichiometric to the concentration of SidA (64–88 μM). Draw the solution into a luer-lock syringe. Insert syringe into SHU port D (Fig. 3) and repeat the process described in step 12.

14. Set the delay for the second mixing at 60 s and acquire spectra of the reduced protein for 500 s on a logarithmic scale. During detection, the flavin spectra should be those of a fully reduced enzyme. Repeat two more times to verify that SidA is reduced and remains reduced for the period of 500 s.

15. Prepare oxygenated 100 mM $NaPO_4$, pH 7.5 buffer as described in Section 4.9.3. The concentration of oxygen should be ~912 μM.

16. Dilute oxygenated buffer with anaerobic buffer to get a range of oxygen concentrations. Replace the syringes containing anaerobic buffer in SHU ports A and B (Fig. 3) with syringes containing the desired oxygen concentration. Repeat the process described in step 12.
17. Acquire data. For SidA, the spectra will show a peak formation and decay at 360 nm representing the formation of the C4a-hydroperoxyflavin intermediate.

Date collected of SidA oxidation will show the formation of a C4a-hydroperoxyflavin at 360 nm followed by appearance of the oxidized flavin peak at 450 nm (Fig. 8B).

4.10.3.5 Clean up

18. Flush the system with water to remove the majority of enzyme solution. Wash the SHU with 2 M nitric acid to oxidize the residual enzyme for at a minimum of 2 h and a maximum of overnight.
19. Thoroughly rinse the instrument with water. If the instrument will not be used for a long period of time, leave the system in 20% ethanol.

4.10.4 Notes

1. For oxidation studies, it is usually desired that there are stoichiometric amounts of NADPH to SidA as excess NADPH will react with the oxidized flavin interfering with the initial measurements of flavin oxidation.
2. The varied reactants (e.g., NADPH and O_2) should always be studied from low to high concentration.
3. During sample preparation, be sure to move carefully inside of the chamber and not to act in haste.
4. When changing syringes on the SHU, enough solution should be inside of the instrument to form a positive meniscus on the syringe port. This should be the same with the syringe containing the sample before placing it onto the port. This process helps limit the risk of bubbles being introduced into the instrument which will disrupt the spectra readings.
5. A small volume of the sample in the syringes should be pulled into the SHU and flushed three times before conducting a reading to make sure that the sample in the cell is not diluted.
6. When transferring samples into a COY Laboratories vinyl chamber, go through two automatic cycles to establish anaerobiosis to limit the amount of oxygen that is entering the chamber.

7. Plan ahead so all required materials are transferred into the chamber at one time to avoid multiple transfers.
8. Keep track of the oxygen concentration recorded by the anaerobic detector during experimentation. Even small levels can be absorbed into anaerobic buffers and interfere with kinetic studies. Extending the time between transfers will also limit oxygen contamination from building up as the catalysts will remove the contaminant before introducing more into the chamber.
9. If stopped-flow experiments are planned for the next day with the same enzyme, the same GO used to scrub the stopped-flow spectrophotometer before can be used to scrub the system again.

4.11 Data analysis

The spectra collected from stopped-flow spectrophotometry can be analyzed to determine the observed rate (k_{obs}) under different reactant concentrations. The most basic studies of flavoenzyme chemistry observe the changes to the flavin at 450 nm to determine the observed rate constants during reduction and oxidation. Before analysis, traces collected at 700 nm, where no species absorb, can be subtracted from traces collected at 450 nm in order to compensate for baseline fluctuation. These data can then be plotted against time to show an exponential decay or rise. These data can be fitted to an exponential for decay (Eq. 3) and rise (Eq. 4) equations, where k_{obs1} is the observed rate constant of the fast reaction and k_{obs2} is the observed rate constant of the slow reaction, B_1 or B_2 are the amplitudes of the flavin change, and C is the final absorbance of the flavin during reduction (Eq. 3) or oxidation (Eq. 4).

$$A_{450} = B_1 e^{-k_{obs1} t} + B_2 e^{-k_{obs2} t} + C \tag{3}$$

$$A_{450} = B_1 \left(1 - e^{-k_{obs1} t}\right) + B_2 \left(1 - e^{-k_{obs2} t}\right) + C \tag{4}$$

5. Other uses for anaerobic chambers

Beyond stopped-flow spectrophotometry, anaerobic chambers offer an opportunity to use different techniques to study enzymes. These chambers offer a large anaerobic work space that, with a little creativity, can facilitate any type of experiment a researcher might be planning. An anaerobic chamber can be optimized to perform many of these applications at the same time. This section briefly describes some of other uses of an anaerobic chamber applied to enzymology.

5.1 Protein purification

One common application for an anaerobic chamber is its use in protein purification. Anaerobic protein purification is typically applied to metal containing enzymes where reactions with oxygen will inactivate the metal core (Echavarri-Erasun et al., 2014). An example setup involves placing an FPLC fraction collector and a sonicator inside of a chamber (Tsai & Tainer, 2018). Reducing agents can be added to the buffer and sample unless the enzyme or the column is sensitive to reduction.

Because anaerobic protein purification is commonly used to isolate metalloenzymes, multiple review papers discussing this method have been published (Echavarri-Erasun et al., 2014; Selmer, 2005; Tsai & Tainer, 2018).

5.2 Isothermal titration calorimetry (ITC)

Isothermal titration calorimetry is a common technique used to determine the binding of two molecules. The instrument measures the heat released or absorbed by the interaction of molecules. For the purpose of protein assays, ITC is regularly used to study the binding of substrates, inhibitors, peptides, and cofactors to a protein providing quantitative results within a few hours. With flavoenzymes in particular, it is sometimes also expected that specific ligands bind more effectively to the reduced form of the enzyme over the oxidized form. ITC experiments can be very useful in determining these changes in affinity; however, maintaining a reduced flavin is difficult without an anaerobic environment.

Most modern ITC instruments are small enough to comfortably fit inside an anaerobic chamber with the accompanying solution bottles for washing and waste collection. Using an ITC inside of a chamber allows anaerobic detection that is commonly used to determine interactions of reduced metals and enzymes (Kondapalli, Kok, Dancis, & Stemmler, 2008). During preparation, the enzyme sample and buffer stock are dialyzed aerobically and deoxygenated with high purity argon or degassed with only a vacuum (Henderson, Müller, Hausinger, & Emerson, 2015; Pop, Kolarik, & Ragsdale, 2004). Researchers have also added sodium dithionite or GO to the solutions to maintain anaerobiosis (Bou-Abdallah et al., 2002; Terpstra et al., 2014). With this setup, it is still possible to perform aerobic ITC assays inside of the anaerobic chamber by using samples that have not been deoxygenated.

5.3 Ultraviolet-visible (UV–vis) spectrophotometry

UV–vis spectrophotometry offers other techniques for studying anaerobic enzyme reactions beyond that discussed with stopped-flow spectrophotometry. One of the simplest applications of UV–vis spectrophotometry using an anaerobic chamber is by placing the spectrophotometer inside of the chamber. This setup lends itself to easy preparation and detection of anaerobic enzyme assays (Lopez de Felipe & Hugenholtz, 2001; Rau & Stolz, 2003; Schulte et al., 1998). Most modern spectrophotometers are small enough to fit inside of a chamber leaving plenty of space for preparing enzyme assays. Setting up a spectrophotometer inside of a chamber will require a computer. Integrating the connections required for communications through the wire ports allows the computer to be outside of the chamber providing more space inside of the chamber to operate the instrument.

Another method that is used to perform anaerobic UV–vis spectrophotometry is using an anaerobic cuvette (Kantz, Chin, Nallamothu, Nguyen, & Gassner, 2005; Rothman, Helm, & Poulter, 2007; Zeghouf, Fontecave, Macherel, & Covès, 1998). This technique allows an investigator to study anaerobic reactions without the need of a spectrophotometer inside of a chamber. The solutions can be prepared inside of an anaerobic chamber and placed inside of an anaerobic cuvette with an airtight seal before being transferred out of the chamber for measurement on a UV–vis spectrophotometer. Anaerobic cuvettes can also be customized for specific assays, and some offer the ability to deoxygenate solutions directly inside of the cuvette using a Schlenk line.

Non-continuous assays can also be used to perform anaerobic experiments inside of a chamber. These assays can be used to obtain data quickly because multiple assays can be performed at once. Termination of the enzyme reaction is facilitated with either organic solvent, acid, or heat. This method is useful as oxygen deactivates the enzyme without precipitation and does not affect other factors in the assay. Discontinuous anaerobic assays can be used beyond UV–visible spectrophotometry, and have been used to prepare samples for high performance liquid chromatography (HPLC) studies (Gorny & Schink, 1994; Thiele et al., 2008).

6. Summary and conclusions

In conclusion, the methods outlined in this chapter provide a practical approach to preparing and performing flavoenzyme assays studying the reductive and oxidative half-reactions with a stopped-flow spectrophotometer. There are no significant limitations from placing a stopped-flow spectrophotometer inside of an anaerobic chamber, and this setup decreases the amount of time spent deoxygenating solutions while also reducing the risk of oxygen contamination. Anaerobic chambers also offer a lot of space for other enzymology applications. The chamber cost and maintenance should be considered if a stopped flow spectrophotometer is to be setup inside of an anaerobic chamber.

Acknowledgments

Research reported in this publication was supported by grants from the National Institute of General Medical Sciences award R01GM094469, the National Science Foundation award CHE-1506206.

References

Adak, S., & Begley, T. P. (2016). Dibenzothiophene catabolism proceeds via a flavin-N5-oxide intermediate. *Journal of the American Chemical Society*, *138*(20), 6424–6426.

Aitken, C. E., Marshall, R. A., & Puglisi, J. D. (2008). An oxygen scavenging system for improvement of dye stability in single-molecule fluorescence experiments. *Biophysical Journal*, *94*(5), 1826–1835.

Bankar, S. B., Bule, M. V., Singhal, R. S., & Ananthanarayan, L. (2009). Glucose oxidase—An overview. *Biotechnology Advances*, *27*(4), 489–501.

Baron, R., Riley, C., Chenprakhon, P., Thotsaporn, K., Winter, R. T., Alfieri, A., et al. (2009). Multiple pathways guide oxygen diffusion into flavoenzyme active sites. *Proceedings of the National Academy of Sciences of the United States of America*, *106*(26), 10603–10608.

Bou-Abdallah, F., Arosio, P., Santambrogio, P., Yang, X., Janus-Chandler, C., & Chasteen, N. D. (2002). Ferrous ion binding to recombinant human H-chain ferritin. An isothermal titration calorimetry study. *Biochemistry*, *41*(37), 11184–11191.

Bruice, T. C. (1980). Mechanisms of flavin catalysis. *Accounts of Chemical Research*, *13*(8), 256–262.

Butler, I. B., Schoonen, M. A. A., & Rickard, D. T. (1994). Removal of dissolved oxygen from water: A comparison of four common techniques. *Talanta*, *41*(2), 211–215.

Crack, J. C., Green, J., Thomson, A. J., & Le Brun, N. E. (2014). Techniques for the production, isolation, and analysis of iron–sulfur proteins. In J. C. Fontecilla-Camps & Y. Nicolet (Eds.), *Metalloproteins: Methods and protocols* (pp. 33–48). Totowa, NJ: Humana Press.

Dashty, M. (2013). A quick look at biochemistry: Carbohydrate metabolism. *Clinical Biochemistry*, *46*(15), 1339–1352.

Echavarri-Erasun, C., Arragain, S., & Rubio, L. M. (2014). Purification of O_2-sensitive metalloproteins. In J. C. Fontecilla-Camps & Y. Nicolet (Eds.), *Metalloproteins: Methods and protocols* (pp. 5–18). Totowa, NJ: Humana Press.

Edmondson, D. E., & Tollin, G. (1983). Semiquinone formation in flavoproteins and metalloflavoproteins. *Topics in Current Chemistry*, *108*, 109–138.

Gorny, N., & Schink, B. (1994). Hydroquinone degradation via reductive dehydroxylation of gentisyl-CoA by a strictly anaerobic fermenting bacterium. *Archives of Microbiology*, *161*(1), 25–32.

Henderson, K. L., Müller, T. A., Hausinger, R. P., & Emerson, J. P. (2015). Calorimetric assessment of Fe^{2+} binding to α-ketoglutarate/taurine dioxygenase: Ironing out the energetics of metal coordination by the 2-His-1-carboxylate facial triad. *Inorganic Chemistry*, *54*(5), 2278–2283.

Kantz, A., Chin, F., Nallamothu, N., Nguyen, T., & Gassner, G. T. (2005). Mechanism of flavin transfer and oxygen activation by the two-component flavoenzyme styrene monooxygenase. *Archives of Biochemistry and Biophysics*, *442*(1), 102–116.

Knight, K., & Scrutton, N. S. (2002). Stopped-flow kinetic studies of electron transfer in the reductase domain of neuronal nitric oxide synthase: Re-evaluation of the kinetic mechanism reveals new enzyme intermediates and variation with cytochrome P450 reductase. *Biochemical Journal*, *367*(1), 19–30.

Kondapalli, K. C., Kok, N. M., Dancis, A., & Stemmler, T. L. (2008). *Drosophila* Frataxin: An iron chaperone during cellular Fe–S cluster bioassembly. *Biochemistry*, *47*(26), 6917–6927.

Lange, M., & Ahring, B. K. (2001). A comprehensive study into the molecular methodology and molecular biology of methanogenic Archaea. *FEMS Microbiology Reviews*, *25*(5), 553–571.

Lee, S. R., Kwon, K. S., Kim, S. R., & Rhee, S. G. (1998). Reversible inactivation of protein-tyrosine phosphatase 1B in A431 cells stimulated with epidermal growth factor. *Journal of Biological Chemistry*, *273*(25), 15366–15372.

Leferink, N. G. H., Fraaije, M. W., Joosten, H.-J., Schaap, P. J., Mattevi, A., & van Berkel, W. J. H. (2009). Identification of a gatekeeper residue that prevents dehydrogenases from acting as oxidases. *Journal of Biological Chemistry*, *284*(7), 4392–4397.

Lopez de Felipe, F., & Hugenholtz, J. (2001). Purification and characterization of the water forming NADH-oxidase from *Lactococcus lactis*. *International Dairy Journal*, *11*(1), 37–44.

Lowe, D. J., Fisher, K., & Thorneley, R. N. F. (1993). Klebsiella pneumoniae nitrogenase: Pre-steady-state absorbance changes show that redox changes occur in the MoFe protein that depend on substrate and component protein ratio; a role for P-centres in reducing dinitrogen. *Biochemical Journal*, *292*(1), 93–98.

Massey, V. (1994). Activation of molecular oxygen by flavins and flavoproteins. *Journal of Biological Chemistry*, *269*, 22459–22462.

Massey, V. (2002). The reactivity of oxygen with flavoproteins. *International Congress Series*, *1233*, 3–11.

Neims, A. H., & Hellerman, L. (1970). Flavoenzyme catalysis. *Annual Review of Biochemistry*, *39*(1), 867–888.

Patil, P. V., & Ballou, D. P. (2000). The use of protocatechuate dioxygenase for maintaining anaerobic conditions in biochemical experiments. *Analytical Biochemistry*, *286*(2), 187–192.

Piano, V., Palfey, B. A., & Mattevi, A. (2017). Flavins as covalent catalysts: New mechanisms emerge. *Trends in Biochemical Sciences*, *42*(6), 457–469.

Pop, S. M., Kolarik, R. J., & Ragsdale, S. W. (2004). Regulation of anaerobic dehalorespiration by the transcriptional activator CprK. *Journal of Biological Chemistry*, *279*(48), 49910–49918.

Rau, J., & Stolz, A. (2003). Oxygen-insensitive nitroreductases NfsA and NfsB of *Escherichia coli* function under anaerobic conditions as Lawsone-dependent Azo reductases. *Applied and Environmental Microbiology*, *69*(6), 3448–3455.

Rice, E. W., Baird, R. B., & Eaton, A. D. (2017). Standard methods for the examination of water and wastewater. *Vol. 23*. American Public Health Association.

Robinson, R., Badieyan, S., & Sobrado, P. (2013). C4a-Hydroperoxyflavin formation in N-hydroxylating flavin monooxygenases is mediated by the 2′-OH of the nicotinamide ribose of $NADP^+$. *Biochemistry*, *52*(51), 9089–9091.

Robinson, R., Franceschini, S., Fedkenheuer, M., Rodriguez, P. J., Ellerbrock, J., Romero, E., et al. (2014). Arg279 is the key regulator of coenzyme selectivity in the flavin-dependent ornithine monooxygenase SidA. *Biochimica et Biophysica Acta (BBA)*, *1844*(4), 778–784.

Robinson, R., Qureshi, I. A., Klancher, C. A., Rodriguez, P. J., Tanner, J. J., & Sobrado, P. (2015). Contribution to catalysis of ornithine binding residues in ornithine N5-monooxygenase. *Archives of Biochemistry and Biophysics*, *585*, 25–31.

Romero, E., Gómez Castellanos, J. R., Gadda, G., Fraaije, M. W., & Mattevi, A. (2018). Same substrate, many reactions: Oxygen activation in flavoenzymes. *Chemical Reviews*, *118*(4), 1742–1769.

Romero, E., Robinson, R., & Sobrado, P. (2012). Monitoring the reductive and oxidative half-reactions of a flavin-dependent monooxygenase using stopped-flow spectrophotometry. *Journal of Visualized Experiments: JoVE*, (61), 3803.

Rothman, S. C., Helm, T. R., & Poulter, C. D. (2007). Kinetic and spectroscopic characterization of type II isopentenyl diphosphate isomerase from *Thermus thermophilus*: Evidence for formation of substrate-induced Flavin species. *Biochemistry*, *46*(18), 5437–5445.

Saleem-Batcha, R., Stull, F., Sanders, J. N., Moore, B. S., Palfey, B. A., Houk, K. N., et al. (2018). Enzymatic control of dioxygen binding and functionalization of the flavin cofactor. *Proceedings of the National Academy of Sciences of the United States of America*, *115*(19), 4909–4914.

Schulte, U., Abelmann, A., Amling, N., Brors, B., Friedrich, T., Kintscher, L., et al. (1998). Search for novel redox groups in mitochondrial NADH: Ubiquinone oxidoreductase (complex I) by diode array UV/VIS spectroscopy. *BioFactors*, *8*, 177–186.

Selmer, T. (2005). Anoxic testing and purification of enzymes. In P. Durre (Ed.), *Handbook on Clostridia* (pp. 71–98). Boca Raton: Taylor and Francis.

Sucharitakul, J., Wongnate, T., & Chaiyen, P. (2011). Hydrogen peroxide elimination from C4a-hydroperoxyflavin in a flavoprotein oxidase occurs through a single proton transfer from flavin N5 to a peroxide leaving group. *Journal of Biological Chemistry*, *286*(19), 16900–16909.

Swoboda, B., & Massey, V. (1965). Purification and properties of the glucose oxidase from *Aspergillus niger*. *Journal of Biological Chemistry*, *240*(5), 2209–2215.

Terpstra, T., McNally, J., Han, T.-H.-L., Ha-Duong, N.-T., El-Hage-Chahine, J.-M., & Bou-Abdallah, F. (2014). Direct thermodynamic and kinetic measurements of Fe^{2+} and Zn^{2+} binding to human serum transferrin. *Journal of Inorganic Biochemistry*, *136*, 24–32.

Teufel, R., Stull, F., Meehan, M. J., Michaudel, Q., Dorrestein, P. C., Palfey, B., et al. (2015). Biochemical establishment and characterization of EncM's flavin-N5-oxide cofactor. *Journal of the American Chemical Society*, *137*(25), 8078–8085.

Thiele, B., Rieder, O., Jehmlich, N., von Bergen, M., Müller, M., & Boll, M. (2008). Aromatizing cyclohexa-1,5-diene-1-carbonyl-coenzyme a oxidase: Characterization and its role in anaerobic aromatic metabolism. *Journal of Biological Chemistry*, *283*(30), 20713–20721.

Tsai, C.-L., & Tainer, J. A. (2018). Robust production, crystallization, structure determination, and analysis of [Fe–S] proteins: Uncovering control of electron shuttling and gating in the respiratory metabolism of molybdopterin guanine dinucleotide enzymes. *Methods in Enzymology*, *599*, 157–196.

van Berkel, W. J. H., Benen, J. A. E., Eppink, M. H. M., & Fraaije, M. W. (1999). Flavoprotein kinetics. In S. K. Chapman & G. A. Reid (Eds.), *Flavoprotein protocols* (pp. 61–85). Totowa: Humana Press.

Watanabe, T., & Honda, K. (1982). Measurement of the extinction coefficient of the methyl viologen cation radical and the efficiency of its formation by semiconductor photocatalysis. *Journal of Physical Chemistry*, *86*, 2617–2619.

Yu, H., Hausinger, R. P., Tang, H.-Z., & Xu, P. (2014). Mechanism of the 6-hydroxy-3-succinoyl-pyridine 3-monooxygenase flavoprotein from *Pseudomonas putida* S16. *Journal of Biological Chemistry*, *289*(42), 29158–29170.

Zeghouf, M., Fontecave, M., Macherel, D., & Covès, J. (1998). The flavoprotein component of the *Escherichia coli* sulfite reductase: Expression, purification, and spectral and catalytic properties of a monomeric form containing both the flavin adenine dinucleotide and the flavin mononucleotide cofactors. *Biochemistry*, *37*(17), 6114–6123.

CHAPTER FOUR

Chemical quenching and identification of intermediates in flavoenzyme-catalyzed reactions

Kalani Karunaratne[a], **Tatiana V. Mishanina**[b,*]
[a]Department of Chemistry, University of Iowa, Iowa City, IA, United States
[b]Department of Chemistry and Biochemistry, University of California San Diego, La Jolla, CA, United States
*Corresponding author: e-mail address: tmishanina@ucsd.edu

Contents

Abstract

Chemical quenching offers a complementary approach to studying the mechanism of a flavoenzyme, supplementing the information learned from spectroscopic, structural, and computational methods. Generally, in a chemical quench experiment, an enzymatic turnover is quickly stopped at various stages with a chemical agent, and the individual reaction mixtures at each time point are analyzed for the reactants, products and any intermediates. The order by which bonds are made and broken in the reaction is indicated by the identities of the captured intermediates, and the rates of individual steps in the mechanism are determined from the amounts of various chemical species at different time points. This chapter outlines general considerations in selecting a chemical quencher of a particular enzyme-catalyzed reaction and methods for analyzing captured reaction intermediates, with a focus on flavoenzymes. The investigation of flavin-dependent thymidylate synthase is used as a case study to illustrate the concepts and workflow of quenching, isolating, and characterizing quencher-modified reaction intermediates and drawing mechanistic conclusions from the identities of these molecules.

Methods in Enzymology, Volume 620
ISSN 0076-6879
https://doi.org/10.1016/bs.mie.2019.03.007

Abbreviations

CH_2H_4folate	(6R)-N^5,N^{10}-methylenetetrahydrofolate
dTDP	2′-deoxythymidine-5′-diphosphate
dTMP	2′-deoxythymidine-5′-monophosphate
dUMP	2′-deoxyuridine-5′-monophosphate
EDTA	ethylenediaminetetraacetic acid
EPSP	5-enolpyruvoyl-shikimate-3-phosphate
FAD	flavin adenine dinucleotide
FDTS	flavin-dependent thymidylate synthase
H_4folate	tetrahydrofolate
HPLC	high-pressure liquid chromatography
LC-MS	liquid chromatography-mass spectrometry
NADPH	nicotinamide adenine dinucleotide phosphate
NMR	nuclear magnetic resonance
QF	quench flow
SDS	sodium dodecyl sulfate
Tris	tris(hydroxymethyl)aminomethane
UDP	uridine 5′-diphosphate

1. Introduction

Flavoenzymes catalyze an impressive variety of chemical transformations, ranging from oxidation-reduction reactions to substitutions of hydrogen atoms with oxygens or halogens to carbon-carbon bond formations, and operate on different substrates, including small molecules, nucleic acids and proteins. To accomplish these chemistries, flavoenzymes have developed an arsenal of mechanisms involving non-covalent reaction intermediates, those covalently bound to the flavin cofactor or the protein, or a combination of thereof. When faced with an uncharacterized flavoenzyme, the question is: where does one begin to dissect the enzyme's chemical mechanism? Significant insight into the mechanism can be gained by taking advantage of the spectral features unique to a particular flavin intermediate. For example, covalent flavin C4a adducts, such as C4a-peroxy and C4a-hydroxy intermediates in flavin monooxygenases or a C4a-thiol adduct in thiol-disulfide oxidoreductases, are identified by their UV-visible absorbance maxima between 370 and 390 nm (Pimviriyakul, Thotsaporn, Sucharitakul, & Chaiyen, 2017; Ruangchan, Tongsook, Sucharitakul, & Chaiyen, 2011; Sahlman, Lambeir, & Lindskog, 1986; Thorpe & Williams, 1976; Yeh et al., 2006). Likewise, charge-transfer (CT) intermediates involving the isoalloxazine of the flavin display unique spectroscopic properties: electronic interactions between reduced flavin and oxidized substrates give rise to a blue CT band with absorbance in the 550–650 nm range, while interactions between oxidized flavin and reduced ligands produce a green CT band at longer wavelengths (Ghisla & Thorpe, 2004; Mishanina, Yadav, Ballou, & Banerjee, 2015; Williamson, Engel, Mizzer, Thorpe, & Massey, 1982). If such an intermediate exists, one can use stopped-flow techniques to observe it and characterize the kinetics of its accumulation and decay during the course of the reaction, from which the timing of chemical events involving the flavin can be deduced. What if, however, none of the reaction intermediates have a unique optical signal, or a specific optical signal cannot be readily assigned to a particular intermediate? And how does one probe chemical steps that do not involve the flavin? In these cases, chemical quenching of the reaction can be informative.

In a chemical quench setup (i.e., a quench flow), the reaction of interest is initiated by forcing the reactants together into a reaction loop using a rapid mixer, similarly to a stopped-flow instrument. Unlike in a stopped-flow experiment, however, the reaction is then quickly stopped (quenched) at various time points by adding a chemical agent that disrupts the integrity of the enzyme as a whole or of its active site, thereby preventing any further catalysis (Fig. 1). Examples of chemical quenchers include acids or bases; protein denaturing agents; metal chelators, suited for metalloenzyme-catalyzed reactions; thiol alkylating agents, useful for quenching reactions relying on active-site thiols, and others listed in Table 1 (Barman, Bellamy, Gutfreund, Halford, & Lionne, 2006). Once quenched, the mixture in various stages of reaction is collected and analyzed at each time point for the substrates, products and (hopefully) intermediates that are chemically different from either the starting material or the final product and therefore can be purified from the rest of the mixture and characterized. The order by which bonds are made and broken in the reaction is deduced from the identities of the captured intermediates, and the rates of interconversion between various chemical species are determined from their relative amounts at different time points.

An important caveat of chemical quenching is that most commonly used quenchers, such as strong acid or base, are so reactive that they chemically alter true intermediates to produce their modified forms observed in the analysis of the quenched reaction. This is not to say that the quencher-modified intermediates are not informative. On the contrary,

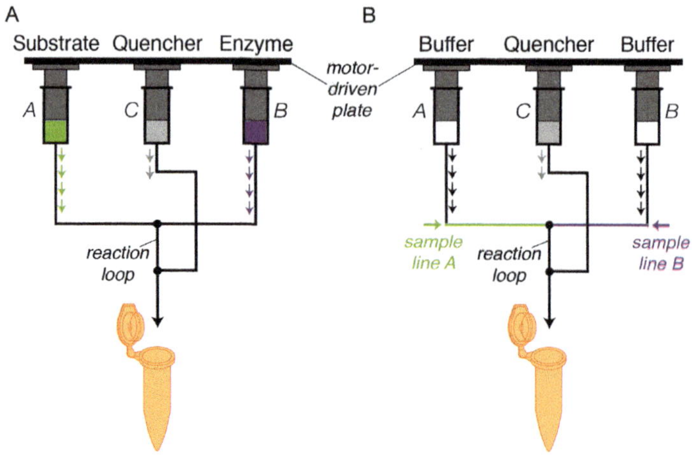

Fig. 1 Schematic of a time delay rapid quench-flow instrument. Two possible setups can be employed: Drive syringes A, B and C can be filled with the substrate, enzyme and chemical quencher solutions, respectively (A), e.g., as in a Hi-Tech (TgK) quench-flow instrument. This setup requires a one-time loading of the syringes with 3–5 mL of each solution. If one of the samples is precious and/or available in a limited amount, a setup in (B) might be more appropriate. Here, a small volume (~20 μL each) of the substrate and enzyme solutions is loaded into the sample lines A and B, respectively, prior to initiating a reaction at each time point (e.g., as in a KinTek quench-flow instrument). In both setups, the substrate and enzyme solutions are rapidly pushed together into the reaction loop by the motor-driven plate (downward arrows). After a defined time delay, the reaction mixture is pushed out of the reaction loop and quenched by the second drive of the plate.

as demonstrated with the case study in this chapter, these modified intermediates serve as valuable reporters on the timing and nature of chemical steps in the mechanism. Milder quenchers, such as organic solvents or denaturing agents, may be tested to avoid chemical modification of the intermediates by the quencher (Table 1). Alternatively, one can abandon chemical quenchers altogether and stop the reaction physically instead, e.g., by flash freezing, filtering, or rapidly evaporating the mixture into a mass spectrometer (Table 1). In the end, the choice of quenching method depends on the enzyme, stability of the predicted intermediates and the specific question at hand, e.g., whether or not an intermediate is covalently bound to the enzyme. Ideally, a variety of approaches should be used to obtain complementary information about the mechanism of a particular flavoenzyme.

Table 1 Examples of chemical and physical methods to stop (quench) an enzyme-catalyzed reaction.

Reaction quencher	When to use	Quencher examples	Applications[a]
Chemical			
Acid	To study kinetics of product formation	Hydrochloric acid; trichloroacetic acid; perchloric acid	Flavin-dependent thymidylate synthase[¶,*] (Mishanina et al., 2012); ThyA thymidylate synthase* (Moore, Ahmed, & Dunlap, 1986); flavin-dependent halogenase RebH[¶] (Yeh et al., 2006); EPSP synthase* (Anderson, Sikorski, & Johnson, 1988); dopamine β-monooxygenase (Brenner & Klinman, 1989); Ca^{2+}-ATPase (Kubo, Suzuki, & Kanazawa, 1990)
Base	Intermediates are likely unstable in acid	Strong base (sodium hydroxide); mild base (trimethylamine, sodium bicarbonate)	Flavin-dependent thymidylate synthase[¶,*] (Mishanina et al., 2016); EPSP synthase* (Anderson, Sikorski, Benesi, & Johnson, 1988); enolpyruvoyl transferase MurZ* (Brown, Marquardt, Lee, Walsh, & Anderson, 1994)
Denaturing agent	Intermediates are labile in either acid or base	SDS; urea; thiourea; guanidinium chloride	dTDP-glucose 4,6-dehydratase* (Gross, Hegeman, Vestling, & Frey, 2000); interaction of myosin with actin (Van Dijk, Céline, Barman, & Chaussepied, 2000)
Reducing agent	Oxidized/electrophilic intermediates (e.g., iminium, keto species)	Sodium borohydride or cyanoborohydride; lithium aluminum hydride	UDP-galactopyranose mutase[¶,*] (Soltero-Higgin, Carlson, Gruber, & Kiessling, 2004); dTDP-glucose 4,6-dehydratase* (Gross et al., 2000)
Metal chelator	Metal-dependent reactions	EDTA	DNA polymerases (Bryant, Johnson, & Benkovic, 1983); ATP sulphurylase (Liu, Martin, & Leyh, 1994; Sukal & Leyh, 2001)

Continued

Table 1 Examples of chemical and physical methods to stop (quench) an enzyme-catalyzed reaction.—cont'd

Reaction quencher	When to use	Quencher examples	Applications[a]
Organic solvent	Intermediates are labile	Methanol; acetonitrile; acetone, or a mixture of organic solvents	β-1,4-Galactosyltransferase (Wu, Takayama, Wong, & Siuzdak, 1997); glutathione *S*-transferase (Ge, Sirich, Beyer, Desaire, & Leary, 2001)
Alkylating agent	Active-site thiol-catalyzed reactions	*N*-ethylmaleimide; iodoacetamide	Ribonucleoside diphosphate reductase (Erickson, 2001)
Competitive inhibitor	Potent competitive inhibitor is known	Methotrexate; 5-fluoro-dUMP	Dihydrofolate reductase (methotrexate) (Williams, Morrison, & Duggleby, 1979); ThyA thymidylate synthase* (5-fluoro-dUMP) (Santi, McHenry, & Sommer, 1974)
Physical			
Filtration	Membrane proteins, immobilized enzymes	0.45 μm Millipore filter	Ca^{2+}-ATPase (Dupont, 1984)
Flash freezing	Radical intermediates are hypothesized	Isopentane at −140 °C	Xanthine oxidase[¶,*] (Palmer, Bray, & Beinert, 1964); dopamine β-monooxygenase (Mitchell C Brenner, Murray, & Klinman, 1989); M-ferritin ferroxidase* (Krebs, Edmondson, & Huynh, 2002)
Evaporation into a mass spectrometer	Intermediates are covalently bound to the enzyme	ESI or MALDI ionization methods	Xylanase* (Zechel, Konermann, Withers, & Douglas, 1998); EPSP synthase* (Paiva, Tilton, Crooks, Huang, & Anderson, 1997); dTDP-glucose 4,6-dehydratase* (Gross et al., 2000)

[a]Note that the list of applications is not meant to be exhaustive. A few examples of enzymatic systems that benefited from the specific quencher are given as a starting point for the reader. In some cases denoted with a "*", a reaction intermediate was isolated as a result of the quench. In others, the quencher was used to follow the kinetics of product formation. Flavoprotein examples are marked with a "¶" symbol.

2. Practical considerations

2.1 Selection of the chemical quencher and reaction conditions

As a part of planning a chemical quench-flow experiment, it is helpful to jot down a few mechanistic possibilities for the reaction in question. This will give an idea of predicted stability of the hypothesized intermediates under different solvent conditions, thus suggesting the most appropriate chemical quencher. A few guidelines for the selection of a quenching agent, with examples, are discussed in the following paragraphs.

Acid and base rapidly halt an enzymatic reaction by drastically changing the pH of the solution. This extreme pH change causes the enzyme to denature and often results in the formation of chemical species that are not part of the normal reaction pathway but are instead quencher-modified versions of the true reaction intermediates. Nonetheless, acid is the most commonly used chemical quencher in rapid-quenching experiments and usually the go-to choice. Several phosphoprotein intermediates have been trapped in acid or base and the choice of acid vs. base depends on the enzymatic residue involved in the covalent bond formation. If the residue is histidine, then base quenching should be employed due to the stability of phosphoryl histidine bond in base and not in acid (Van Etten & Hickey, 1977). Likewise, if the covalent bond to the enzyme is formed via a cysteine residue, then this intermediate can be trapped in base due to the stability of phosphothioketals at high pH conditions (Wanke & Amrhein, 1993). We refer the reader to the extensive compilation of covalent intermediates in enzyme-catalyzed reactions, many of which have been captured by chemical quenching (Allison & Purich, 2002).

If both acid and base are expected to decompose reaction components, then denaturing agents such as urea, guanidine hydrochloride and sodium dodecyl sulfate (SDS) can be employed as quenching reagents. As an example, a reaction catalyzed by dTDP-glucose 4,6-dehydratase was successfully quenched in both acid and base but the analytes were decomposed. However, 6 M guanidine hydrochloride not only quenched the enzyme but also preserved the integrity of analytes (Gross et al., 2000).

Reducing agents, such as sodium borohydride or sodium cyanoborohydride, are known to reduce highly labile iminium ions to form stable amines. Therefore, these reducing agents can be used as quenching agents when iminium ions are proposed as intermediates. Addition of a reducing agent will stop the enzyme turnover by reducing the reactive iminium ions to form non-reactive amines. Sodium cyanoborohydride quenching allowed the trapping of an N5-alkylated flavin intermediate in uridine 5′-diphosphate (UDP)-galactopyranose mutase by reducing the iminium ion formed on N5 of flavin to form a non-reactive N5-alkylated flavin derivative (Soltero-Higgin et al., 2004).

When applicable, employing different quenching agents to study the same system will allow trapping of different intermediates, as demonstrated with 5-enolpyruvoylshikimate-3-phosphate synthase (EPSP synthase) quenching studies. Even though acid quenching suggested the formation and decay of an intermediate, the structure of the intermediate was not resolved because it was labile in acid. This tetrahedral intermediate was eventually trapped in neat triethylamine and identified as a true, unmodified reaction intermediate (Anderson, Sikorski, Benesi, & Johnson, 1988). The trapping experiments with flavin-dependent thymidylate synthase (FDTS) discussed later in this chapter also showcase the importance of using different quenchers to trap intermediates that are stable under different chemical conditions.

A quench-flow experiment can be time-intensive, particular if anaerobiosis is required. Before heading to a quench-flow apparatus, therefore, it is useful to confirm that the selected quencher indeed quickly (relative to the observed rate constant) stops the enzymatic reaction. This can be done in a couple of ways. First, if the quencher successfully compromises enzyme integrity, then pre-mixing the enzyme with the quencher prior to addition of the required reaction components should not generate any product. This sample is the most appropriate "blank" measurement for product formation. Second, a reaction can be initiated under steady-state conditions and the quencher added to the reaction after a few seconds, after which aliquots are analyzed for product at various time points. If the quencher works, the amount of product should remain constant over time. Once these preliminary controls confirm that the quencher rapidly stops enzyme turnover and the reactants and products are stable in the quenching solution, then the selected quencher can be used in a quench-flow setup.

Intermediates in an enzymatic reaction are generally short-lived and do not accumulate in large quantities. Rapid quench experiments performed under single-turnover conditions (excess of enzyme over substrates) increase the chances of trapping reaction intermediates by ensuring that all of the substrate is bound by the enzyme and thus will react. To further increase the accumulation of intermediates, quench-flow experiments can be carried out at a reduced temperature by taking advantage of the temperature-dependence of reaction rates. This will allow an intermediate with a lifetime of microseconds-milliseconds at physiological temperature to accumulate due to its reduced reactivity. However, it is important to note that the reaction pathway at low temperatures might be different from that at the physiological temperature under which the enzyme normally operates. Inhibitors that hinder the subsequent chemical step, thus resulting in an accumulation of covalent intermediates, can also increase the probability of trapping an enzyme-bound intermediate. As an example, a variety of 2-deoxy-2-fluoroglycosides (substrate analogues) were used to successfully trap enzyme bound α-glycosyl intermediates in retaining β-glycosidases (Dan et al., 2000; Wicki, Rose, & Withers, 2002). Finally, mutagenesis of the enzymatic residues predicted to play a role in the chemical mechanism is another tool to trap fleeting reaction intermediates, by either slowing the overall rate of the reaction or blocking chemistry that consumes the intermediate (Lairson et al., 2004; Miller et al., 1990; Recksiek, Selmer, Dierks, Schmidt, & von Figura, 1998).

2.2 Detection and analysis of quencher-modified reaction intermediates

Samples collected during a chemical quenching experiment might require additional workup before the analysis for reaction components. With acid (e.g., trichloroacetic acid, HCl) as the quenching agent, the enzyme along with any enzyme-bound intermediates precipitates and can be easily separated from the soluble reaction components by either centrifugation or filtration of the quenched sample. Soluble components can then be analyzed for substrates, products and non-covalently bound intermediates, while the precipitated enzyme may be studied for the presence of covalent intermediates. Base-quenched samples should be neutralized before injection onto an HPLC column to avoid any damage to the column resin. Additionally, unlike in acid, denatured enzymes do not precipitate in base and should be removed from the sample via ultrafiltration. Finally, if the quenched sample contains excessive amounts of salts that may interfere with downstream analysis, e.g., mass spectrometry, desalting or buffer exchange might be required before analysis.

Quenched reaction mixtures typically contain multiple chemical components, not all of which participate in the enzyme-catalyzed reaction. One can separate these components using various chromatographic techniques, such as normal- or reverse-phase liquid chromatography, gas or thin layer chromatography. To ease tracking of the species directly involved in the enzymatic reaction, substrates and/or cofactors labeled with either radioactive or stable heavy isotopes are valuable. These labeled reaction components not only make the reacting compounds stand out of the background, but also offer the advantage of requiring small amounts of material for detection and analysis, which becomes crucial when the intermediates do not accumulate in large quantities. If such labeled substrates are used, one may keep track of the total "counts" (either radioactive counts or ion counts in a mass spectrum) between different chemical species throughout the time course of the reaction. "Missing" counts at any time point suggest existence of an intermediate bound to the enzyme and may be looked for by either gel electrophoresis or mass spectrometry on the protein fraction of the quenched samples. Beyond its value in detecting and isolating trapped intermediates, isotopic labeling is useful for testing what components of the substrate or cofactor structure are parts of the intermediate. For instance, as discussed in the case study, labeling individual rings of the flavin's isoalloxazine one at a time showed that only two of the rings remain intact in the base-modified intermediate.

Ideally, mass spectrometry and radiolabeling experiments would provide enough information for determination of the structure of quencher-modified intermediates. If not or if sufficient quantity of purified intermediate is easily generated, various NMR experiments can be performed to deduce the chemical structure of the trapped intermediate. Obtaining enough material for an NMR spectrum can be tedious and time consuming but it is possible by collecting multiple quenched samples around the time point of the maximum accumulation of the intermediate and purifying the intermediate out of the crude reaction mixtures.

3. Case study: Flavin-dependent thymidylate synthase

Thymidylate synthase encoded by the *thyA* gene (*TYMS* in humans) catalyzes the reductive methylation of uridylate (dUMP, 2′-deoxyuridine-5′-monophosphate) to form thymidylate (dTMP, 2′-deoxythymidine-5′-monophosphate), an essential building block of DNA (Carreras & Santi, 1995). Genomes of several pathogenic bacteria lack *thyA* (e.g., *Helicobacter pylori, Rickettsia species and Bacillus anthracis*) and instead rely on the flavin-dependent thymidylate synthase (FDTS) encoded by the *thyX* gene for thymidylate biosynthesis (Myllykallio et al., 2002). These two thymidylate synthases are different in structure, chemical mechanism and co-factor requirement. With FDTS, (6R)-N^5,N^{10}-methylenetetrahydrofolate (CH_2H_4folate) serves as the carbon source, in the form of a methylene, and the non-covalently bound flavin adenine dinucleotide (FAD) prosthetic group catalyzes the oxidation-reduction chemistry to form the methyl group of dTMP (Fig. 2A). The chemical mechanism of thyA thymidylate synthase is

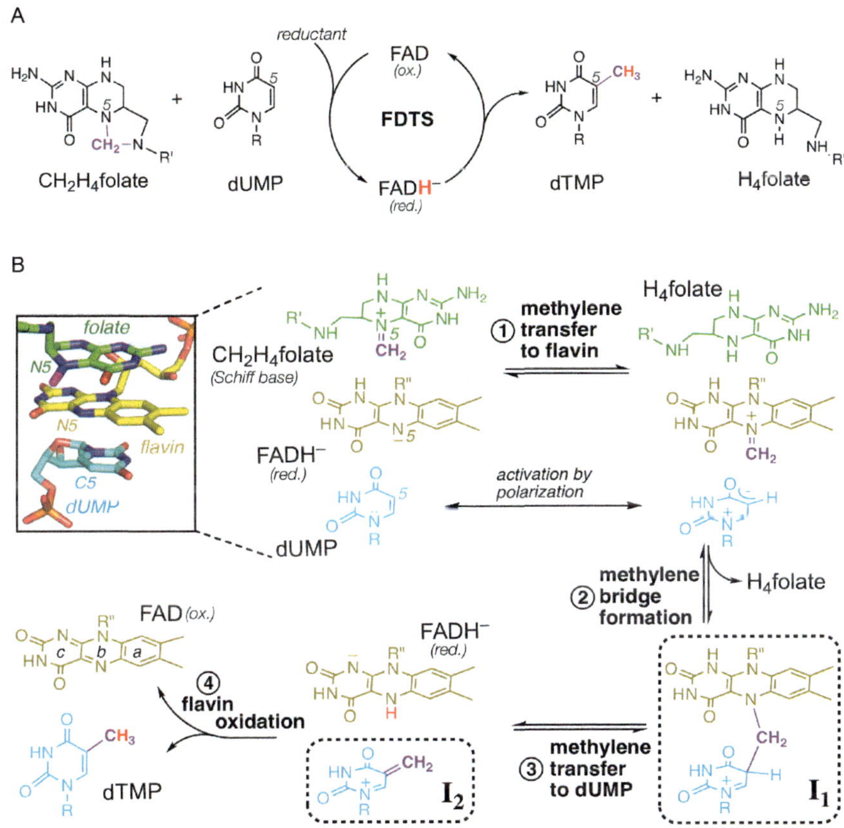

Fig. 2 Reaction and the latest mechanistic proposal for flavin-dependent thymidylate synthase (FDTS). (A) Overall chemical conversion catalyzed by FDTS. In the reductive half-reaction (downward arrow), the oxidized FAD cofactor is reduced to an anionic hydroquinone form of the flavin. The physiological reductant is likely nicotinamide adenine dinucleotide phosphate (NADPH); however, for the purposes of the chemical quench experiment, FAD can be reduced with sodium dithionite. In the oxidative half-reaction (upward arrow), the methylene provided by (6R)-N^5,N^{10}-methylenetetrahydrofolate (CH_2H_4folate) is reduced by the flavin into a methyl group of the final dTMP product. (B) Proposed chemical mechanism for FDTS, where flavin relays the methylene from the folate donor to the nucleotide acceptor. The mechanism was based on the chemical trapping of intermediates I_1 and I_2, described in the main text. Inset: stacking of the folate's pterin, flavin's isoalloxazine and dUMP in the crystal structure of a ternary FDTS complex, consistent with the proposed path of methylene transfer (PDB ID: 4gta; an unreactive CH_2H_4folate analogue, folinic acid, was used in the structure) (Koehn et al., 2012). Note that the chemical structures in the mechanism have been slightly tilted out of the plane of the page to mimic the view in the crystal structure. The rings of the isoalloxazine core are labeled as "a/b/c" for reference. R = 2′-deoxyribose-5′-phosphate; R′ = (p-aminobenzoly)glutamate; R" = adenosine-5′-pyrophosphate-ribityl.

extensively studied and well understood, which led to the development of mechanism-based inhibitors used as clinical drugs, such as 5-fluorouracil and raltitrexed (Carreras & Santi, 1995; Finer-Moore, Santi, & Stroud, 2003). The mechanism of FDTS, on the other hand, is not clearly established. This lack of mechanistic knowledge on FDTS hampers the development of specific antibiotics with minimal toxicity to humans.

Chemical trapping of intermediates along the reaction pathway with both acid and base as reaction quenchers resulted in a revised chemical mechanism of FDTS (Fig. 2B). Specifically, the identity of an intermediate trapped in base suggested that the flavin prosthetic group in FDTS transfers the carbon from the folate to the nucleotide substrate, in addition to providing the reducing hydride in the end. Here, we discuss the use of two chemical quenchers, acid and base, in a rapid quench-flow experiment and characterization of trapped intermediates to solve the mechanistic puzzle of FDTS. Chemical quench experiments were carried out at room temperature with FDTS from a hyperthermophilic organism *Thermotoga maritima*, which normally grows optimally at the physiological temperature 80 °C. Sub-physiological temperature allowed a slower overall reaction and thus increased the magnitude of intermediate accumulation.

3.1 Chemical quenching protocol

3.1.1 *Equipment*

- KinTek Chemical Quench-Flow (QF) instrument model RQF-3
- Anaerobic glass tonometer for deoxygenating and loading the enzyme onto the QF (Fig. 3A)
- Hamilton gastight syringes
- Hamilton threaded plunger 1700
- Schlenk line (a.k.a. vacuum manifold)
- Round-bottom Schlenk flasks, 200 and 50-mL
- Precision seal rubber septa

3.1.2 *Buffers and reagents*

- Reaction buffer for quenching experiment: 200 mM Tris-HCl pH 8.0
- Glucose oxidase powder from *Aspergillus niger* and D-glucose
- [2-^{14}C]dUMP (Moravek Biochemicals)
- CH_2H_4folate solution: 100 mM stock in the reaction buffer
- Formaldehyde stock: 36.5% by weight

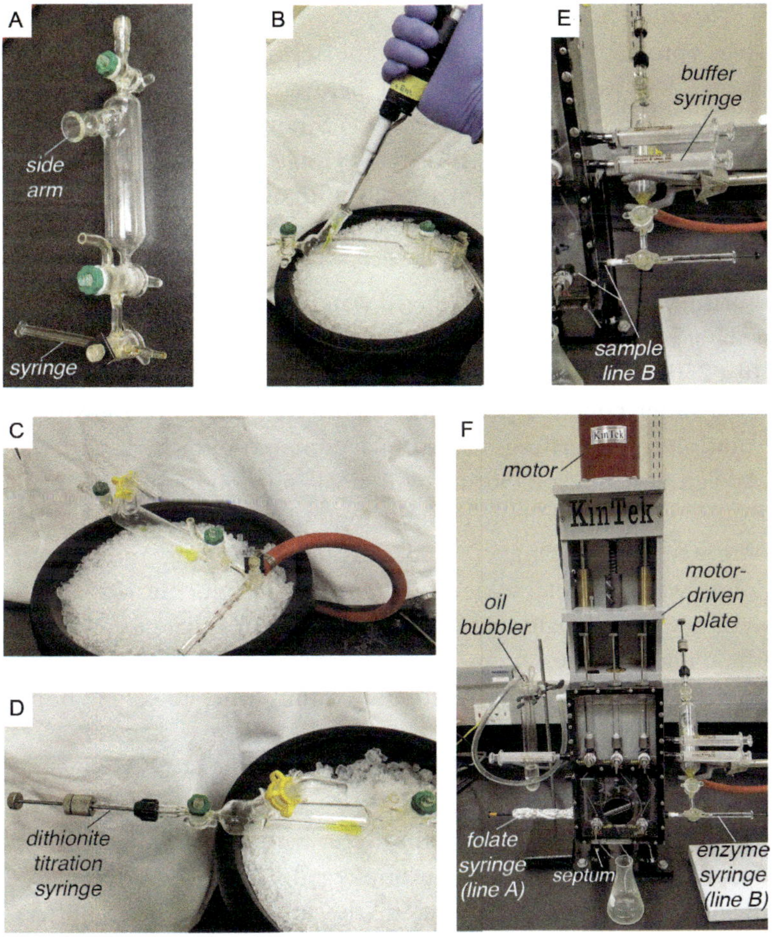

Fig. 3 Overview of the quench-flow instrument setup. (A) Anaerobic tonometer for enzyme reduction, with a gastight syringe attachment for loading the enzyme into the instrument's sample line. (B) Loading of the enzyme (or an enzyme-dUMP complex) into the tonometer through the side-arm. (C) Tonometer connected to the vacuum manifold (a.k.a. the Schlenk line) through a rubber hose for degassing. (D) Titration syringe with dithionite inserted into the anaerobic tonometer. (E) Tonometer connected to the sample line B and secured with a clamp. (F) Complete setup ready for sample collection.

- Purified *T. maritima* FDTS: 1 mL of 100 μM (active-site FAD concentration, $\varepsilon_{450} = 11{,}300\,M^{-1}\,cm^{-1}$) (Mishanina, Corcoran, & Kohen, 2014)
- Sodium dithionite
- Ultra-high purity argon
- Quenching agent: 1 M HCl or 1 M NaOH

3.1.3 Procedure

Note: Fig. 3 provides pictures of a few of the steps described below, and Fig. 4A summarizes the workflow described below. The amount of dithionite required to stoichiometrically reduce FDTS-bound FAD can be determined beforehand by following the FAD UV-visible absorbance spectrum upon gradual addition of dithionite in a spectrophotometer (Fig. 4B).

1. Clean the drive syringes and the reaction loops of the quench-flow system with deionized water.
2. Fill the water bath with deionized water and seal the bottom inlet with a rubber septum (Fig. 3F). Secure the septum with electrical tape.
3. In a 250-mL round-bottom Schlenk flask, deoxygenate 75 mL of reaction buffer containing 10 mM glucose for ~30 mins by exposing the solution to alternating cycles of vacuum and ultra-high purity argon.
4. Add glucose oxidase (50 units/mL final concentration) to the anaerobic buffer with a gastight syringe through the septum of the flask.
5. Rinse the drive syringes (except the quencher syringe) and the reaction loops with the anaerobic buffer containing glucose oxidase, fill the syringes with remaining buffer and allow to incubate for ~6 h to scrub the oxygen in the system.
6. Insert a needle through the rubber septum from step 2 and bubble the water bath with argon.
7. Attach an oil bubbler to the outlet of the water bath to maintain the positive argon pressure (Fig. 3F).
8. Let the argon flow through the water bath while the glucose oxidase from step 5 scrubs the system of oxygen.
9. Repeat step 3 and 4 and set aside the anaerobic reaction buffer.
10. Deoxygenate 3 mL of glucose buffer for ~15 mins in a 50-mL Schlenk flask by exposing the solution to alternating cycles of vacuum and argon.
11. Add formaldehyde (final concentration 30 mM) and glucose oxidase (50 units/mL final concentration) with a gastight syringe through the septum of the flask from step 10 and set aside, kept under the positive pressure of argon to maintain anaerobiosis. This solution is later used for CH_2H_4folate dilution. Formaldehyde stabilizes the labile methylene on the folate.
12. Assemble the tonometer and load the enzyme (1 mL of 100 μM FDTS) mixed with [2-^{14}C]dUMP (30 μM final concentration) through the side arm (Fig. 3A and B).

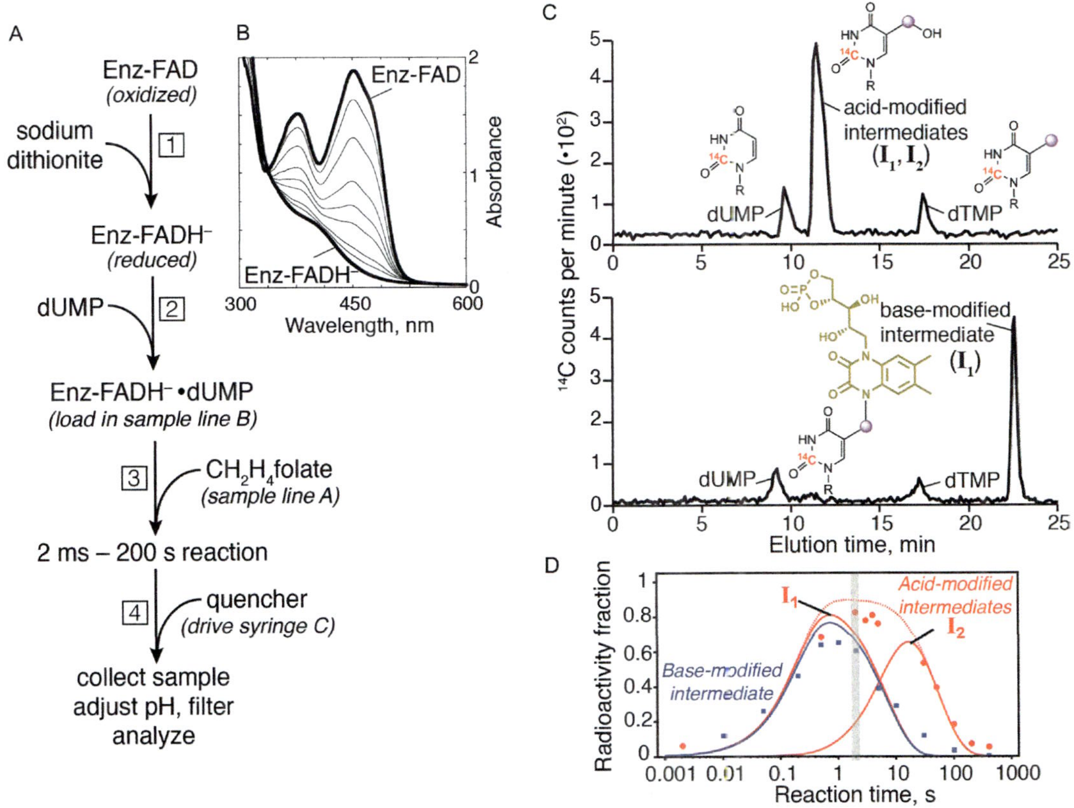

Fig. 4 Chemical quench of a single-turnover FDTS reaction. (A) Flow diagram of the experimental steps. Refer to Fig. 3 for the photos of steps 1–3. (B) Stoichiometric reduction of the oxidized FDTS-bound FAD (Enz-FAD) with sodium dithionite (step 1 in A), followed by the UV-visible absorbance of the flavin. (C) Reverse-phase HPLC radiograms of a 2 s reaction with [2-^{14}C]dUMP as a substrate, quenched with either strong acid (top) or strong base (bottom). The chemical structures of the quencher-modified intermediates are shown (refer to Figs. 5 and 6 for structure determination experiments). The sphere represents the methylene supplied by the folate substrate. (D) Base-modified intermediate kinetics (square data points) overlaid with acid-modified intermediate data (round data points), globally fitted to a two-intermediate model (solid curves). The sum of the two intermediates is shown as a dotted curve. Each data point results from a radiogram like in (C) (shaded slice represents the 2 s time point).

13. Attach the tonometer to the Schenk line and deoxygenate for ~30 mins (Fig. 3C)
14. Bubble 10 mL of reaction buffer in a volumetric flask with ultra-high purity argon for 10 mins.
15. Dissolve dithionite powder (final concentration 5 mM) in the deoxygenated buffer from step 14.
16. Load dithionite into the syringe with the threaded plunger.
17. Insert the dithionite syringe into the tonometer and gently secure it with the screw cap on top of the tonometer (Fig. 3D).
18. Reduce the enzyme-bound FAD by adding the appropriate amount of dithionite and check for the color change of the enzyme from yellow to colorless.
19. Remove the needle connected to water bath from step 6.
20. Empty the drive syringes from step 5 and load the syringes on either side with fresh degassed buffer from step 4 and the middle syringe with the quenching reagent (aerobic).
21. Attach the tonometer to the quench-flow instrument, secure it with a clamp (Fig. 3E) and prime the sample line B with enzyme solution.
22. Add CH_2H_4folate (800 μM final concentration) to the anaerobic buffer from step 11.
23. Load CH_2H_4folate solution into a gastight syringe and connect it to the sample line A of the quench-flow instrument (Fig. 3F).
24. Prime the sample line A with the CH_2H_4folate solution.
25. Turn on the quench-flow instrument and select the appropriate sample loop according to the time point of interest.
26. Load enzyme and the CH_2H_4folate solutions into the sample lines (20 μL each) and start the reaction.
27. Collect the quenched samples in 1.5-mL Eppendorf tubes and freeze them immediately on dry ice till analysis.

3.1.4 Notes

1. All the glass stopcocks on the tonometer, Schlenk flasks and vacuum manifold need to be well greased before each experiment and care should be taken not to over grease to prevent clogs in the bore. When choosing the grease, select a cryogenic high-vacuum grease (Apiezon N-00025).
2. High vacuum ground glass stopcocks and plugs are numbered to avoid accidental mix-ups. When assembling the tonometer, match the numbers of glass joints for a tight seal.

3. When degassing enzyme solution, do not expose it to vacuum for more than 2 s or so, as this will cause the enzyme to bubble up and denature.
4. When inserting the syringe containing dithionite, keep the argon flowing through the tonometer to maintain the anaerobicity.
5. Once the enzyme is stoichiometrically reduced with dithionite, set the tonometer aside for ~5 min to see whether the flavin remains reduced, i.e., colorless. This will ensure the anaerobicity is maintained inside the tonometer.
6. Store CH_2H_4folate in an amber vial or keep it covered to avoid exposure to light and dilute the CH_2H_4folate stock solution toward the end of the setup to avoid decomposition. CH_2H_4folate is known to be stable at a high concentration, hence the choice of 100 mM for the stock concentration.
7. Sample loops should be cleaned at least once by doing a blank run with buffer before first use and should be cleaned between time points in the same manner to avoid contamination with the mixture from previous reactions. *Note*: this is in contrast to the vacuum-water-methanol cleaning protocol typically performed between reactions on the KinTek RQF, which would compromise anaerobicity of the system.
8. Keep the quenched samples frozen until ready to analyze to avoid potential decomposition of the trapped species.
9. This experiment is performed under anaerobic conditions. However, the chemical quenching step is not.
10. If the quench-flow instrument is placed in a glove box, then steps 5 through 8 of the Procedure can be skipped.

3.2 Analyses of the quencher-modified intermediates

3.2.1 Equipment

- HPLC instrument
- Reverse-phase analytical column (Supelco, Discovery series 250 mm × 4.6 mm)
- Table-top centrifuge
- Radioactivity flow-scintillation counter
- LC-MS instrument

3.3 Buffers and reagents

- HPLC separation buffer: 50 mM potassium phosphate (KH_2PO_4) pH 6.0 and methanol
- LC-MS eluent: water and acetonitrile, both containing 0.1% formic acid

3.3.1 Notes

1. Centrifuge or filter the thawed quenched samples to remove any denatured enzyme before injection onto the HPLC column, to prevent clogging of the column.
2. Base-quenched samples should be neutralized to pH 7–8 before injecting them onto the HPLC, to avoid damage to the column resin.
3. Mobile phase for HPLC separations is a gradient of 50 mM KH_2PO_4 pH 6.0 and methanol. LC-MS mobile phase is a gradient of water and acetonitrile, both containing 0.1% formic acid.
4. For NMR experiments, base-modified intermediate was purified out of multiple quenched 1 s reactions by HPLC in phosphate buffer and freeze-dried in a 50-mL Falcon tube (tube 1). The compound was desalted via trituration into methanol: (i) the powder was dissolved in methanol; (ii) the supernatant was transferred into a fresh Falcon tube (tube 2), and (iii) methanol was evaporated by blowing argon over the solution in tube 2. Steps i–iii were then repeated with the remaining salt precipitate in the original tube 1. Desalted compound in tube 2 was dissolved in D_2O for NMR analyses.

Initial indirect evidence for accumulation of an intermediate(s) in the FDTS-catalyzed reaction came from the LC-MS analysis of nucleotide substrate (dUMP) and product (dTMP) in acid-quenched reactions with unlabeled reactants (Mishanina et al., 2012). In the analysis, the sum of ion counts between dUMP and dTMP was not conserved during the course of the reaction: the total counts decreased in the 0.5–10 s time window, pointing to a "missing" nucleotide material different from either dUMP or dTMP. Use of radiolabeled reactants afforded direct detection, isolation, and identification of the reaction intermediates, as detailed below.

When acid-quenched samples of the FDTS reaction with [2-^{14}C]dUMP were analyzed by HPLC, a radioactive peak different from both the dUMP substrate and the dTMP product was indeed identified (Fig. 4C, top panel), suggesting that this new species contained the pyrimidine ring of the nucleotide substrate. The fraction of the total radioactivity in this new radioactive peak first increased and then decayed with time, indicative of an intermediate. The maximum accumulation of this new radioactive peak occurred around 2 s and corresponded to ~80% of the total radioactivity (Fig. 4D, round data points). In addition, when [11-^{14}C]CH_2H_4folate, i.e., radiolabeled at the methylene, was used as the only radioactive species in the reaction, the HPLC chromatogram showed a new peak with the same retention time as the intermediate formed when [2-^{14}C]dUMP was used (data not shown (Mishanina et al., 2012)). This observation suggested that the trapped intermediate is comprised of both the pyrimidine ring of the nucleotide substrate and the methylene group of CH_2H_4folate (sphere on the intermediate structure in Fig. 4C), giving insight into the order of events during FDTS catalysis, i.e., that the methylene transfer from CH_2H_4folate to dUMP occurs prior to the formation of the intermediate trapped in acid.

In order to identify the acid-trapped species, a separate quench-flow experiment was carried out with nonradioactive substrates, and several 2s reactions were collected where the intermediate shows maximum accumulation (Fig. 4D). Trapped intermediate was purified by HPLC and analyzed by electrospray ionization mass spectrometry (ESI-MS). The exact mass and atomic composition of the acid-modified species corresponded to the product dTMP plus a hydroxyl group (Fig. 5A), i.e., 5-hydromethyl-dUMP. An LC standard of 5-hydroxymethyl-dUMP was synthesized to confirm the identity of the trapped intermediate as such. Retention time on the LC for both the synthetic standard and the trapped species overlapped (data not shown (Mishanina et al., 2012)) and the MS spectra showed the same mass for the two molecules. In addition, the MS-MS spectra of the 5-hydroxymethyl-dUMP and trapped intermediate displayed the same fragmentation pattern (Fig. 5B) confirming the identity of the trapped species as 5-hydroxymethyl-dUMP.

Base-quenched FDTS reactions with both [2-^{14}C]dUMP and [11-^{14}C] CH_2H_4folate showed a radioactive peak with same retention time (Fig. 4C; data with the labeled folate not shown (Mishanina et al., 2016)).

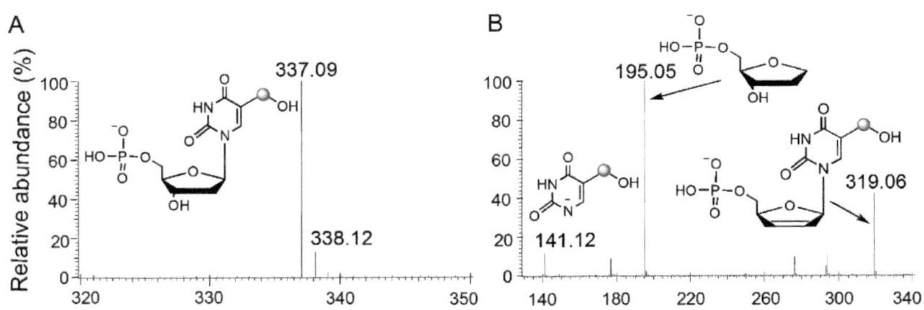

Fig. 5 Mass spectra of the product of acid modification of intermediates I_1 and I_2. (A) Negative-ion ESI-MS spectrum. (B) Negative-ion ESI-MS-MS spectrum, with the structures corresponding to each fragment shown. These data are identical to the MS of an independently synthesized 5-hydroxymethyl-dUMP standard (data not shown (Mishanina et al., 2012)), confirming the identity of the product of acid modification of intermediates I_1 and I_2 as 5-hydroxymethyl-dUMP. The sphere represents the methylene supplied by the folate substrate.

This new peak elutes much later than the acid-modified intermediate under the same LC conditions, suggesting that these two intermediates are chemically dissimilar. Nevertheless, both contain the pyrimidine moiety of the nucleotide and the methylene carbon of the CH_2H_4folate (sphere in the structures in Fig. 4D). The mass of the trapped intermediate was 705 Da (Fig. 6A) and tandem mass spectrum of the trapped intermediate showed a mass fragment of 385 Da (Fig. 6B), which did not account for any folate derivative, buffer or protein components. When [7α,8α-^{3}H] FAD-FDTS was used along with [2-^{14}C]dUMP, the base-modified intermediate contained both ^{14}C and ^{3}H radioactive labels indicating that the trapped species contains both the pyrimidine ring and the dimethylbenzene portion of FAD (Fig. 6C). A trapping experiment with [dioxopyrimidine-^{13}C,^{15}N]FAD reconstituted FDTS showed a 2 Da increase in the mass of the trapped species suggesting only a part of the dioxopyrimidine ring is connected to the trapped intermediate (Fig. 6D). The UV-visible absorbance spectrum of the purified and desalted base-modified intermediate suggested an N5-alkylated flavin (Fig. 6E). The final step in characterization of the base-modified intermediate was obtaining NMR spectra of the intermediate, which corroborated the substrate labeling experiments carried out by using labeled FAD and confirmed the chemical structure in Fig. 6E (data not shown (Mishanina et al., 2016)).

3.4 Mechanistic insights from the chemical quench

Once the structure of a quencher-modified intermediate is fully characterized, one can seek to identify its implications for the chemical mechanism of the enzyme. In the case of FDTS, the unexpected covalent link between the flavin cofactor and the methylenated nucleotide substrate in the base-modified intermediate suggested that the flavin serves as a shuttle, accepting the methylene from the folate (step 1 in Fig. 2B) and passing it on to the pyrimidine substrate (steps 2–3 in Fig. 2B), thus calling for a drastic revision of the previously proposed mechanisms (Conrad, Ortiz-Maldonado, Hoppe, & Palfey, 2014; Koehn et al., 2009, 2012; Mishanina et al., 2014). Here, chemical quench revealed a mechanistic detail other experimental methods missed. The use of different quenchers of the same reaction provided complementary data about the intermediates. Specifically, while the acid converted both intermediates I_1 and I_2 into a single chemical species (5-hydroxymethyl-dUMP, Fig. 4C), alkaline quench resolved these by

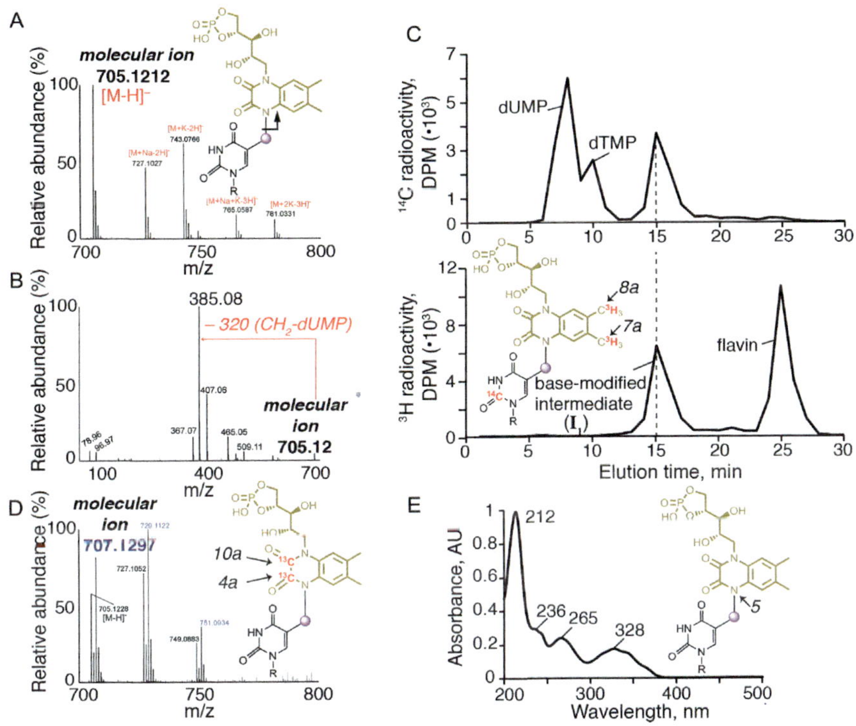

Fig. 6 Characterization of the product of base modification of intermediate I_1. (A) High-resolution negative-ion ESI-MS and (B) ESI-MS-MS of the base-modified intermediate. The major site of fragmentation is marked with an arrow. The mass of the molecular ion is consistent with the chemical formula of the shown structure. (C) Reverse-phase HPLC radiogram of a 1s reaction of [7α,8α-^{3}H]FAD-FDTS with [2-^{14}C]dUMP as a substrate, with the two panels showing signal from ^{14}C-(top) and ^{3}H-(bottom) containing species in the reaction mixture. Importantly, the base-modified intermediate has both ^{14}C and ^{3}H signals associated with it, suggesting that dimethylbenzene of the flavin's isoalloxazine (ring *a* in Fig. 2B) is a component of the base-modified intermediate. Carbons 7α and 8α carrying tritiums are labeled. The sphere represents the methylene supplied by the folate substrate. (D) High-resolution negative-ion ESI-MS of the base-modified intermediate isolated from the reaction of [dioxopyrimidine-^{13}C,^{15}N]FAD-FDTS with unlabeled substrates (dioxopyrimidine is the ring *c* of the flavin's isoalloxazine core, Fig. 2B). The mass of the molecular ion and its sodium adducts is increased by 2 Da, consistent with retention of carbons 4a and 10a in the base-modified intermediates but not the rest of dioxopyrimidine. The final chemical structure of the base-modified intermediate was determined by a suite of NMR experiments (data not shown (Mishanina et al., 2016)). (E) UV-visible absorbance spectrum of the purified base-modified intermediate. The 328-nm peak is characteristic of an N5-monoalkylated flavin (Soltero-Higgin et al., 2004).

capturing exclusively the earlier intermediate I_1 (Fig. 4D). Additionally, the presence of the methylene on the trapped intermediates at early time points suggested that the methylene transfer (steps 1–3 in Fig. 2B) occurs quickly (a few seconds at room temperature), while the remaining time of the turnover (tens-hundreds of seconds) is spent on the reduction of the methylene into a methyl by the reduced flavin (step 4 in Fig. 2B).

The identities and timing of the chemically trapped intermediates I_1 and I_2 agree with the stopped-flow trace of the enzyme-bound FAD at 420nm (Conrad et al., 2014). First, a decrease in 420-nm absorbance observed with I_1 accumulation is consistent with the formation of an N5-monoalkylated flavin adduct, which lacks absorbance at 420nm (Fig. 6E). Second, an increase in 420-nm absorbance corresponding to flavin oxidation accompanies disappearance of the exocyclic methylene intermediate I_2 and appearance of the dTMP product (Conrad et al., 2014; Mishanina et al., 2014). Together, chemical quenching and stopped-flow experiments with FDTS provided information on the role of the flavin prosthetic group in catalysis, as well as the sequence and timing of chemical steps.

4. Concluding remarks

Chemical quenching has been used to dissect a wide range of enzyme mechanisms for over 50 years (Barman & Gutfreund, 1964). As we hope the reader took away from this chapter, chemical quenching will continue to provide valuable insights into the mechanisms of flavoenzymes that other methods cannot access. In particular, chemical quench allows determination of the kinetics for substrate consumption, product formation and accumulation and decay of intermediates, along with effects of enzymatic mutations, and flavin and substrate analogues on these kinetics. Additionally, from the identities of the trapped, spectroscopically silent, intermediates one can learn the order by which bonds are broken and new bonds are formed in the reaction, with sometimes-unexpected connectivities appearing in the isolated intermediates. As with any other experimental tool, however, chemical quenching does come with its own set of limitations that should be kept in mind, such as (i) high likelihood that the quencher chemically modifies the intermediates; (ii) information about the surrounding environment of the intermediate in the enzyme's active site (e.g., ion-pair or charge-transfer interactions) is lost upon the quench, and (iii) analysis of the quenched samples is quite labor intensive as reactions from individual time points have to be analyzed one at a time.

Acknowledgments

The authors dedicate this chapter to Prof. Amnon Kohen (University of Iowa) as thanks for his mentoring, encouragement and support over the years. The authors would like to thank Ilya Gurevic and Nicholas Luedtke (University of Iowa) for providing the images of the quench-flow instrument, and Nicholas Luedtke and Daniel Roston (University of Wisconsin–Madison) for critical reading of the manuscript and valuable input. The work discussed in this chapter was supported by NIH R01 GM110775 to Amnon Kohen (University of Iowa). T.V.M. is supported by the Career Award at the Scientific Interface from Burroughs Wellcome Fund (ID#1016945.01) and Yinan Wang Memorial Chancellor's Endowed Junior Faculty Fellowship from the University of California San Diego.

References

Allison, R. D., & Purich, D. L. (2002). A survey of covalent, ionic, and radical intermediates in enzyme-catalyzed reactions. In D. L. Purich (Ed.), *Methods in Enzymology* (pp. 455–470): *Vol. 354. Academic Press.*

Anderson, K. S., Sikorski, J. A., Benesi, A. J., & Johnson, K. A. (1988). Isolation and structural elucidation of the tetrahedral intermediate in the EPSP synthase enzymic pathway. *Journal of the American Chemical Society, 110*(19), 6577–6579. https://doi.org/10.1021/ja00227a056.

Anderson, K. S., Sikorski, J. A., & Johnson, K. A. (1988). A tetrahedral intermediate in the EPSP synthase reaction observed by rapid quench kinetics. *Biochemistry, 27*(19), 7395–7406.

Barman, T. E., Bellamy, S. R. W., Gutfreund, H., Halford, S. E., & Lionne, C. (2006). The identification of chemical intermediates in enzyme catalysis by the rapid quench-flow technique. *Cellular and Molecular Life Sciences: CMLS, 63*(22), 2571–2583. https://doi.org/10.1007/s00018-006-6243-z.

Barman, T. E., & Gutfreund, H. (1964). A comparison of the resolution of chemical and optical sampling. In B. Chance, Q. H. Gibson, R. H. Eisenhardt, & K. K. Lonberg-Holm (Eds.), *Rapid mixing and sampling techniques in biochemistry* (pp. 339–344). Academic Press.

Brenner, M. C., & Klinman, J. P. (1989). Correlation of copper valency with product formation in single turnovers of dopamine beta-monooxygenase. *Biochemistry, 28*(11), 4664–4670. https://doi.org/10.1021/bi00437a023.

Brenner, M. C., Murray, C. J., & Klinman, J. P. (1989). Rapid freeze and chemical-quench studies of dopamine beta-monooxygenase: Comparison of pre-steady-state and steady-state parameters. *Biochemistry, 28*(11), 4656–4664.

Brown, E. D., Marquardt, J. L., Lee, J. P., Walsh, C. T., & Anderson, K. S. (1994). Detection and characterization of a phospholactoyl-enzyme adduct in the reaction catalyzed by UDP-*N*-acetylglucosamine enolpyruvoyl transferase, MurZ. *Biochemistry, 33*(35), 10638–10645.

Bryant, F., Johnson, K., & Benkovic, S. (1983). Elementary steps in the DNA polymerase I reaction pathway. *Biochemistry, 22*(15), 3537–3546.

Carreras, C. W., & Santi, D. V. (1995). The catalytic mechanism and structure of thymidylate synthase. *Annual Review of Biochemistry, 64*(1), 721–762. https://doi.org/10.1146/annurev.bi.64.070195.003445.

Conrad, J. A., Ortiz-Maldonado, M., Hoppe, S. W., & Palfey, B. A. (2014). Detection of intermediates in the oxidative half-reaction of the FAD-dependent thymidylate synthase from *Thermotoga maritima*: Carbon transfer without covalent pyrimidine activation. *Biochemistry, 53*(32), 5199–5207. https://doi.org/10.1021/bi500648n.

Dan, S., Marton, I., Dekel, M., Bravdo, B.-A., He, S., Withers, S. G., et al. (2000). Cloning, expression, characterization, and nucleophile identification of family 3, *Aspergillus niger* β-glucosidase. *Journal of Biological Chemistry, 275*(7), 4973–4980. https://doi.org/10.1074/jbc.275.7.4973.

Dupont, Y. (1984). A rapid-filtration technique for membrane fragments or immobilized enzymes: Measurements of substrate binding or ion fluxes with a few-millisecond time resolution. *Analytical Biochemistry, 142*(2), 504–510. https://doi.org/10.1016/0003-2697(84)90496-2.

Erickson, H. K. (2001). Kinetics in the pre-steady state of the formation of cystines in ribonucleoside diphosphate reductase: Evidence for an asymmetric complex. *Biochemistry, 40*(32), 9631–9637.

Finer-Moore, J. S., Santi, D. V., & Stroud, R. M. (2003). Lessons and conclusions from dissecting the mechanism of a bisubstrate enzyme: Thymidylate synthase mutagenesis, function, and structure. *Biochemistry, 42*(2), 248–256. https://doi.org/10.1021/bi020599a.

Ge, X., Sirich, T. L., Beyer, M. K., Desaire, H., & Leary, J. A. (2001). A strategy for the determination of enzyme kinetics using electrospray ionization with an ion trap mass spectrometer. *Analytical Chemistry, 73*(21), 5078–5082. https://doi.org/10.1021/ac0105890.

Ghisla, S., & Thorpe, C. (2004). Acyl-CoA dehydrogenases. A mechanistic overview. *European Journal of Biochemistry, 271*(3), 494–508.

Gross, J. W., Hegeman, A. D., Vestling, M. M., & Frey, P. A. (2000). Characterization of enzymatic processes by rapid mix-quench mass spectrometry: The case of dTDP-glucose 4,6-dehydratase. *Biochemistry, 39*(45), 13633–13640.

Koehn, E. M., Fleischmann, T., Conrad, J. A., Palfey, B. A., Lesley, S. A., Mathews, I. I., et al. (2009). An unusual mechanism of thymidylate biosynthesis in organisms containing the *thyX* gene. *Nature, 458*(7240), 919–923. https://doi.org/10.1038/nature07973.

Koehn, E. M., Perissinotti, L. L., Moghram, S., Prabhakar, A., Lesley, S. A., Mathews, I. I., et al. (2012). Folate binding site of flavin-dependent thymidylate synthase. *Proceedings of the National Academy of Sciences of the United States of America, 109*(39), 15722–15727. https://doi.org/10.1073/pnas.1206077109.

Krebs, C., Edmondson, D. E., & Huynh, B. H. (2002). Demonstration of peroxodiferric intermediate in M-ferritin ferroxidase reaction using rapid freeze-quench Mössbauer, resonance Raman, and XAS spectroscopies. In D. L. Purich (Ed.), *Vol. 354. Methods in Enzymology* (pp. 436–454): Academic Press.

Kubo, K., Suzuki, H., & Kanazawa, T. (1990). Characterization of the substrate-induced conformational change of N-iodoacetyl-N′-(5-sulfo-1-naphthyl)ethylenediamine-labeled sarcoplasmic reticulum Ca^{2+}+-ATPase by using different kinds of substrate. *Biochimica et Biophysica Acta (BBA)—Protein Structure and Molecular Enzymology, 1040*(2), 251–259. https://doi.org/10.1016/0167-4838(90)90084-S.

Lairson, L. L., Chiu, C. P. C., Ly, H. D., He, S., Wakarchuk, W. W., Strynadka, N. C. J., et al. (2004). Intermediate trapping on a mutant retaining α-galactosyltransferase identifies an unexpected aspartate residue. *Journal of Biological Chemistry, 279*(27), 28339–28344. https://doi.org/10.1074/jbc.M400451200.

Liu, C., Martin, E., & Leyh, T. S. (1994). GTPase activation of ATP sulfurylase: The mechanism. *Biochemistry, 33*(8), 2042–2047.

Miller, S. M., Massey, V., Ballou, D., Williams, C. H., Distefano, M. D., Moore, M. J., et al. (1990). Use of a site-directed triple mutant to trap intermediates: Demonstration that the flavin C(4a)-thiol adduct and reduced flavin are kinetically competent intermediates in mercuric ion reductase. *Biochemistry, 29*(11), 2831–2841. https://doi.org/10.1021/bi00463a028.

Mishanina, T. V., Corcoran, J. M., & Kohen, A. (2014). Substrate activation in flavin-dependent thymidylate synthase. *Journal of the American Chemical Society, 136*(30), 10597–10600. https://doi.org/10.1021/ja506108b.

Mishanina, T. V., Koehn, E. M., Conrad, J. A., Palfey, B. A., Lesley, S. A., & Kohen, A. (2012). Trapping of an intermediate in the reaction catalyzed by flavin-dependent thymidylate synthase. *Journal of the American Chemical Society*, *134*(9), 4442–4448. https://doi.org/10.1021/ja2120822.

Mishanina, T. V., Yadav, P. K., Ballou, D. P., & Banerjee, R. (2015). Transient kinetic analysis of hydrogen sulfide oxidation catalyzed by human sulfide quinone oxidoreductase. *Journal of Biological Chemistry*, *290*(41), 25072–25080. https://doi.org/10.1074/jbc.M115.682369.

Mishanina, T. V., Yu, L., Karunaratne, K., Mondal, D., Corcoran, J. M., Choi, M. A., et al. (2016). An unprecedented mechanism of nucleotide methylation in organisms containing *thyX*. *Science*, *351*(6272), 507–510. https://doi.org/10.1126/science.aad0300.

Moore, M. A., Ahmed, F., & Dunlap, R. B. (1986). Trapping and partial characterization of an adduct postulated to be the covalent catalytic ternary complex of thymidylate synthase. *Biochemistry*, *25*(11), 3311–3317.

Myllykallio, H., Lipowski, G., Leduc, D., Filee, J., Forterre, P., & Liebl, U. (2002). An alternative flavin-dependent mechanism for thymidylate synthesis. *Science*, *297*(5578), 105.

Paiva, A. A., Tilton, R. F., Crooks, G. P., Huang, L. Q., & Anderson, K. S. (1997). Detection and identification of transient enzyme intermediates using rapid mixing, pulsed-flow electrospray mass spectrometry. *Biochemistry*, *36*(49), 15472–15476. https://doi.org/10.1021/bi971883i.

Palmer, G., Bray, R. C., & Beinert, H. (1964). Direct studies on the electron transfer sequence in xanthine oxidase by electron paramagnetic resonance spectroscopy: I. Techniques and description of spectra. *Journal of Biological Chemistry*, *239*(8), 2657–2666.

Pimviriyakul, P., Thotsaporn, K., Sucharitakul, J., & Chaiyen, P. (2017). Kinetic mechanism of the dechlorinating flavin-dependent monooxygenase HadA. *Journal of Biological Chemistry*, *292*, 4818–4832. https://doi.org/10.1074/jbc.M116.774448.

Recksiek, M., Selmer, T., Dierks, T., Schmidt, B., & von Figura, K. (1998). Sulfatases, trapping of the sulfated enzyme intermediate by substituting the active site formylglycine. *Journal of Biological Chemistry*, *273*(11), 6096–6103. https://doi.org/10.1074/jbc.273.11.6096.

Ruangchan, N., Tongsook, C., Sucharitakul, J., & Chaiyen, P. (2011). pH-dependent studies reveal an efficient hydroxylation mechanism of the oxygenase component of *p*-hydroxyphenylacetate 3-hydroxylase. *Journal of Biological Chemistry*, *286*(1), 223–233. https://doi.org/10.1074/jbc.M110.163881.

Sahlman, L., Lambeir, A.-M., & Lindskog, S. (1986). Rapid-scan stopped-flow studies of the pH dependence of the reaction between mercuric reductase and NADPH. *European Journal of Biochemistry*, *156*(3), 479–488.

Santi, D. V., McHenry, C. S., & Sommer, H. (1974). Mechanism of interaction of thymidylate synthetase with 5-fluorodeoxyuridylate. *Biochemistry*, *13*(3), 471–481.

Soltero-Higgin, M., Carlson, E. E., Gruber, T. D., & Kiessling, L. L. (2004). A unique catalytic mechanism for UDP-galactopyranose mutase. *Nature Structural & Molecular Biology*, *11*(6), 539–543. https://doi.org/10.1038/nsmb772.

Sukal, S., & Leyh, T. S. (2001). Product release during the first turnover of the ATP sulfurylase – GTPase. *Biochemistry*, *40*(49), 15009–15016. https://doi.org/10.1021/bi015735a.

Thorpe, C., & Williams, C. H. (1976). Spectral evidence for a flavin adduct in a monoalkylated derivative of pig heart lipoamide dehydrogenase. *Journal of Biological Chemistry*, *251*(23), 7726–7728.

Van Dijk, J., Céline, F., Barman, T., & Chaussepied, P. (2000). Interaction of myosin with F-actin: Time-dependent changes at the interface are not slow. *Biophysical Journal*, *78*(6), 3093–3102. https://doi.org/10.1016/S0006-3495(00)76846-9.

Van Etten, R. L., & Hickey, M. E. (1977). Phosphohistidine as a stoichiometric intermediate in reactions catalyzed by isoenzymes of wheat germ acid phosphatase. *Archives of Biochemistry and Biophysics*, *183*(1), 250–259. https://doi.org/10.1016/0003-9861(77)90438-6.

Wanke, C., & Amrhein, N. (1993). Evidence that the reaction of the UDP-*N*-acetylglucosamine 1-carboxyvinyltransferase proceeds through the *O*-phosphothioketal of pyruvic acid bound to Cys115 of the enzyme. *European Journal of Biochemistry*, *218*(3), 861–870. https://doi.org/10.1111/j.1432-1033.1993.tb18442.x.

Wicki, J., Rose, D. R., & Withers, S. G. (2002). Trapping covalent intermediates on beta-glycosidases. *Methods in Enzymology*, *354*, 84–105.

Williams, J. W., Morrison, J. F., & Duggleby, R. G. (1979). Methotrexate, a high-affinity pseudosubstrate of dihydrofolate reductase. *Biochemistry*, *18*(12), 2567–2573.

Williamson, G., Engel, P. C., Mizzer, J. P., Thorpe, C., & Massey, V. (1982). Evidence that the greening ligand in native butyryl-CoA dehydrogenase is a CoA persulfide. *Journal of Biological Chemistry*, *257*(8), 4314–4320.

Wu, J., Takayama, S., Wong, C.-H., & Siuzdak, G. (1997). Quantitative electrospray mass spectrometry for the rapid assay of enzyme inhibitors. *Chemistry & Biology*, *4*(9), 653–657. https://doi.org/10.1016/S1074-5521(97)90220-9.

Yeh, E., Cole, L. J., Barr, E. W., Bollinger, J. M., Ballou, D. P., & Walsh, C. T. (2006). Flavin redox chemistry precedes substrate chlorination during the reaction of the flavin-dependent halogenase RebH. *Biochemistry*, *45*(25), 7904–7912. https://doi.org/10.1021/bi060607d.

Zechel, D. L., Konermann, L., Withers, S. G., & Douglas, D. J. (1998). Pre-steady state kinetic analysis of an enzymatic reaction monitored by time-resolved electrospray ionization mass spectrometry. *Biochemistry*, *37*(21), 7664–7669. https://doi.org/10.1021/bi980445o.

CHAPTER FIVE

On the use of noncompetitive kinetic isotope effects to investigate flavoenzyme mechanism

Kevin Francis[a], Giovanni Gadda[b,c,d,e,*]

[a]Department of Chemistry, Texas A&M University—Kingsville, Kingsville, TX, United States
[b]Department of Chemistry, Georgia State University, Atlanta, GA, United States
[c]Department of Biology, Georgia State University, Atlanta, GA, United States
[d]Center for Biotechnology and Drug Design, Georgia State University, Atlanta, GA, United States
[e]Center for Diagnostics and Therapeutics, Georgia State University, Atlanta, GA, United States
*Corresponding author: e-mail address: ggadda@gsu.edu

Contents

Abstract

This account describes the application of kinetic isotope effects (KIEs) to investigate the mechanistic properties of flavin dependent enzymes. Assays can be conducted during steady-state catalytic turnover of the flavoenzyme with its substrate or by using rapid-kinetic techniques to measure either the reductive or oxidative half-reactions of the enzyme. Great care should be taken to ensure that the observed effects are due to isotopic substitution and not other factors such as pH effects or changes in the solvent viscosity of the reaction mixture. Different types of KIEs are described along with a physical description of their origins and the unique information each can provide about the mechanism of an enzyme. Detailed experimental techniques are outlined with special emphasis on the proper controls and data analysis that must be carried out to avoid erroneous conclusions. Examples are provided for each type of KIE measurement from

Methods in Enzymology, Volume 620
ISSN 0076-6879
https://doi.org/10.1016/bs.mie.2019.03.008

references in the literature. It is our hope that this article will clarify any confusion concerning the utility of KIEs in the study of flavoprotein mechanism and encourage their use by the community.

1. Introduction

Kinetic isotope effects (KIEs) have been utilized in biological, organic and physical chemistry for almost a century as a probe of chemical mechanism and catalysis (Allison & Purich, 1979; Cleland, 1982; Klinman & Kohen, 2013; Northrop, 1982). Detailed accounts of the types, physical origins and utility of KIEs can be found elsewhere (Cook & Cleland, 2007; Kohen & Limbach, 2005) and only a brief review will be provided here. The main focus of this chapter is the application of KIE measurements to the study of flavin dependent enzymes. KIEs have been used to elucidate the mechanisms of many classes of reactions ranging from alcohol and amine oxidations (Fitzpatrick, 2010, Fitzpatrick, 2017; Gadda, 2008; Gadda & Francis, 2010) to the detoxification of xenobiotics. Since KIEs arise largely from the differences in the reduced masses of the reacting bonds (Cleland, 1982; Kohen & Limbach, 2005) (Paneth, 1995) the largest effects are observed when protium in the substrate or solvent is replaced with deuterium. Thus, deuterium KIEs are most easily measured and will be the focus of this chapter, although the terminology and theory can be easily extended to heavy atom KIEs as reviewed in Hengge (2015) and Paneth (1995).

We begin the chapter with a brief theoretical explanation and definitions of the most common types of KIEs. We will then explain how KIEs can be used in both steady-state and rapid kinetic measurements of flavoproteins and emphasize that both techniques should be employed to fully understand the catalytic mechanism of enzymes. Finally, we will outline the proper controls and data analysis that must be carried out to avoid erroneous conclusions. It is hoped that this account will clarify common pitfalls encountered with KIE studies and highlight their remarkable utility in the study of flavoprotein catalysis.

1.1 Physical origin of kinetic isotope effects

1.1.1 Primary kinetic isotope effects

A primary KIE (1° KIE) arises from the isotopic substitution of an atom that is directly involved in the bond that is being cleaved in a reaction. It is the ratio of rate constants of the reactant containing a natural abundance of isotopes (k_{light}) to that of a reactant that was enriched with a heavy isotope at the bond being cleaved (k_{heavy}; Eq. 1):

$$\mathrm{KIE} - \frac{k_{light}}{k_{heavy}} \tag{1}$$

The earliest theories to explain KIEs relied heavily on semi-classical approximations involving changes in force constants between the ground state and transition state (F) and the reduced mass of the bond being cleaved in the reaction (μ) on the frequency (ν) of the reacting bond, Eq. (2) (Jacob, 2006):

$$\nu = 1/2\pi\sqrt{F/\mu} \tag{2}$$

Over time an increasing number of reports of isotope effects in both enzymatic and non-enzymatic reactions involving transfers of hydrogen (H^+, H or H^-) demonstrated that this treatment could not fully explain the experimental observations found in many systems. This led to more sophisticated models that invoke the role of quantum mechanical hydrogen tunneling (Hay, Pudney, & Scrutton, 2009; Klinman & Kohen, 2013). Hydrogen tunneling involves the crossing of a particle through a potential barrier between the reactants and products due to its wave-like nature. While several models have been developed to explain such a behavior they all share similar features and are referred to here as activated tunneling models. These models are based on three principles. (1) The hydrogen being transferred should be treated quantum mechanically throughout the reaction coordinate. (2) The rate of transfer is determined by fluctuations of the electronic potential surface of the reaction. (3) Those fluctuations comprise a rearrangement of the coordinate that adjusts the energies of the reactant and product and an orthogonal "gating" coordinate that represents the hydrogen donor and acceptor distance at the transition state (also known as the tunneling-ready state) (Klinman & Kohen, 2013). The rate constant for hydrogen transfer in such models is given by Eq. (3).

The terms before the integral are from Marcus theory (Marcus & Sutin, 1985) and contain the electronic coupling (V), the reorganization energy (λ), the reaction driving force ($\Delta G°$) and the absolute temperature (T). C(T) is the fraction of reactive complexes that can proceed to product. The F.C. term is the Frank-Condon nuclear overlap and is given by Eq. (4), where m_H, ω_H and r are the mass, frequency, and distance, respectively, transferred for the tunneling particle, and 0,0 refers to tunneling from ground state vibrational modes. E(r_x) is the barrier for the fluctuation of the donor-acceptor distance (DAD) and depends on the frequency (ω_x) and the collective mass (m_x) of the heavy atoms controlling distance sampling. The probability of tunneling of all possible donor-acceptor distances once the system has reached the transition (tunneling ready) state is what determines the magnitude of the KIE. As will be discussed below many factors influence the magnitude of the KIE, but those $\leq$500 have been measured under standard atmosphere and pressure in enzymatic systems (Hu, Soudackov, Hammes-Schiffer, & Klinman, 2017). Great care should be taken when drawing mechanistic conclusions from KIE data, but with proper controls, KIEs can be a powerful tool to elucidate the biological roles of enzymes.

$$k = C(T)\frac{[V]^2}{\hbar}\sqrt{\frac{\pi}{\lambda k_B T}}e^{-(\Delta G^\circ + \lambda)/4\lambda k_B T}\int_0^\infty F.C.term_{0,0}e^{-E(r_x)/k_B T}dr_x \quad (3)$$

$$F.C.term_{0,0} = e^{\left(-{}^{m}H^{\omega}H^{r^2}/2\hbar\right)} \quad (4)$$

1.1.2 Secondary kinetic isotope effects

Secondary KIEs (2° KIE) arise from isotopic substitution at sites other than the bond that is directly being cleaved in the reaction. They result from changes in the force constants describing the vibrational energies of the bond being cleaved in the reaction. As a chemical bond is being broken the force constant is weakened and the vibrational frequency of the bond is decreased. Since the frequency of the bond vibration is inversely proportional to the mass of the atoms involved (Eq. 5) the change in force constants is different for the light and heavy reactants:

$$\nu = \frac{1}{2\pi}\sqrt{\frac{x}{m}} \quad (5)$$

In Eq. (5) x is a constant. If the hybridization of the carbon atom in the bond being cleaved changes from sp^2 to sp^3 the bending motion results in a lower frequency and a normal 2° KIE (>1). On the other hand, if bond cleavage results in the change in the hybridization of a carbon atom from sp^3 to sp^2 an inverse (<1) 2° KIE is observed. The effects of 2° KIEs are significantly smaller than 1° KIEs with maximal normal effects typically in the 1.15–1.41 range and inverse values commonly between 0.86 and 0.70 at standard temperature and pressure.

1.1.3 Substrate versus solvent kinetic isotope effects

Isotopic labeling can be either directly on the substrate involved in the enzymatic reaction or the solvent in which the reaction is carried out. Labeling of the substrate results in the isotope effects discussed above and is referred to as substrate KIEs. Running the reaction in deuterium oxide as opposed to water results in the replacement of all solvent-exchangeable protons in the enzyme and the substrate with deuterium. Isotope effects that result from these substitutions are referred to as solvent KIEs and are described further in Section 3. Finally, a combination of substrate and solvent KIEs can be used to study the relative timing of bond cleavages in complex enzymatic reactions where multiple bonds are cleaved. These are discussed in Section 4 and are referred to as multiple KIEs. This chapter will focus mainly on noncompetitive KIE measurements in which separate assays are done with light and heavy reagents. Competitive isotope effects in which KIE values are extracted based on the fraction conversion of mixtures of isotopologues and the use of the Northrop equation are extensively discussed in the following references: Kholodar, Ghosh, and Kohen (2017) and Northrop (1982). These experiments are also a powerful probe of mechanism, but require remote labeling and cumbersome experimental protocols that often make them unfeasible for the study of flavoprotein catalysis.

1.1.4 Nomenclature used in reports of kinetic isotope effects

The nomenclature developed by Cook and Cleland (Cleland, 1982; Cook & Cleland, 2007) will be used in this account. We will focus primarily on hydrogen isotope effects, though the basic nomenclature is easily extended to heavy atom isotope effects. 1° KIEs are denoted by a superscript preceding the kinetic parameter being considered (${}^{D}k$). Solvent isotope effects are denoted by a superscript preceding the kinetic parameter being considered (${}^{D_2O}k$). Finally, multiple KIEs are denoted by both superscripts before the kinetic parameter being described (${}^{D,\,D_2O}k$).

1.2 Observed versus intrinsic kinetic isotope effects

As pointed out by Northrop, Cleland, and coworkers the multi-step nature of enzymatic reactions often results in a masking of the KIE observed for a reaction (KIE_{obs}) from the intrinsic value arising from cleavage of the labeled bond (KIE_{int}) (Cleland, 1982; Northrop, 1982). Even for the simplest enzymatic reaction, catalytic turnover of the enzyme involves the reversible binding of the substrate (S) to the enzyme (E) to form a competent enzyme-substrate complex (ES), the chemical step of catalysis to generate

$$E + S \underset{k_2}{\overset{k_1}{\rightleftharpoons}} ES \xrightarrow{k_3} EP \xrightarrow{k_5} E + P$$

Fig. 1 The kinetic mechanism of an irreversible enzymatic reaction involving a single substrate (S) being converted to a single product (P). The reversible binding of the substrate is governed by the kinetic rate constants k_1 and k_2. The chemical step is governed by k_3 and is isotope sensitive and the product release step is shown as irreversible (k_5) as is the case when initial rates of reaction are measured in the absence of products.

a product bound to the enzyme (EP) and the release of the product (P) from the enzyme. Thus, even for a reaction involving a single substrate and product there are a minimum of three mechanistically distinct kinetic steps, as illustrated in Fig. 1.

The kinetic steps involving chemical bond cleavage are the only ones that are significantly affected by isotopic substitution and therefore only k_3 should give rise to KIEs.[a] As can be seen through inspection of Eqs. (6) and (7) a steady-state kinetic assay cannot isolate the chemical step(s) in which bonds are being broken or formed, and one measures a rate that is a summation of several different events along the catalytic pathway. Eqs. (6) and (7) show the steady-state kinetic parameters determined for the mechanism of Fig. 1. Quantitative expressions of the KIEs on these parameters are found by dividing the rate constant for the unlabeled substrate by that of the labeled substrate (Eqs. 6–8):

$$k_{cat} = \frac{k_3 k_5}{k_3 + k_5} \tag{6}$$

$$\frac{k_{cat}}{K_m} = \frac{k_1 k_3}{k_2 + k_3} \tag{7}$$

$${}^{D}\left(\frac{k_{cat}}{K_m}\right) = \frac{{}^{D}k_3 + \frac{k_3}{k_2}}{1 + \frac{k_3}{k_2}} \tag{8}$$

As illustrated in Eqs. (9) and (10), the multistep nature of enzymatic catalysis makes it difficult to isolate the chemical step of the reaction even for the simplest possible mechanism of Fig. 1. This was eloquently pointed

[a] As pointed out in Stratton, Poulin, and Schramm (2017) isotopic labeling of a substrate can sometimes give rise to binding isotope effects in which the relative stabilities of the Michaelis complexes formed between the light and heavy isotope differ. These effects are usually negligible and below deduction with non-competitive methods using routine laboratory equipment and are therefore not discussed further in this account.

out in a landmark paper by Northrop (1975) and expanded in numerous reviews and book chapters on the application of KIEs measurements to elucidate enzyme mechanisms (Cleland, 1982; Cook & Cleland, 2007; Northrop, 1982). The vast majority of enzyme reactions involve more than one substrate and product and thus multiple chemical steps. Eqs. (9) and (10) give the general expressions for the observed KIE in steady-state assays. In these expressions C_f and C_r are the forward and reverse commitment to catalysis, R_f is the ratio of catalysis, E_f is the equilibration proceeding catalysis and EIE is the equilibrium isotope effect. Expressions for C_f, C_r, R_f and E_f are mechanism-dependent and can be derived as described in Northrop (1982). They are naturally complex expression of individual rate constants. The EIE arises for cases where the KIEs are different in the forward and reverse directions and are commonly negligible in routine non-competitive measurements:

$$\frac{{}^{D}k_{cat}}{k_m} = \frac{{}^{D}k + C_f + C_r EIE}{1 + C_f + C_r} \tag{9}$$

$${}^{D}k_{cat} = \frac{{}^{D}k + \frac{R_f}{E_f} + C_r EIE}{1 + \frac{R_f}{E_f} + C_r} \tag{10}$$

When reporting experimental measurements of KIEs it is critical to report whether the value represents an intrinsic KIE or an observed value that may be masked by the kinetic complexity of the reaction (Francis & Kohen, 2014). This is especially true if the data will be used to extract mechanistic insights on the enzymatic reaction or for computation simulations of the reaction to elucidate the physical nature of catalysis. As will be noted in Section 4, the kinetic complexity of KIE_{obs} can be exploited in measurements of multiple KIEs to deduce the relative timing of chemical bond cleavages in complex enzyme reactions.

Finally, it is important to note that equations analogous to (6)–(10) can be derived when using rapid-kinetic techniques as is commonly carried out for flavin dependent enzymes. Even though measurements of oxidative or reductive half-reactions (see Section 2.1) significantly reduce the kinetic complexity of the measured KIE they still involve the binding of the substrate and possible isomerization of the enzyme-substrate complex to form a catalytically competent complex that proceeds to reaction, which can affect the value of the measured KIE. Examples in which the kinetic complexity of

Table 1 Statistical equations for isotope effect measurements.

calculation	Example	Equation
Addition or subtraction[a]	x=a+b+c	$s_x = \sqrt{s_a^2 + s_b^2 + s_c^2}$
multiplication or division[a]	x=a * b/c	$\frac{s_x}{x} = \sqrt{\left(\frac{s_a}{a}\right)^2 + \left(\frac{s_b}{b}\right)^2 + \left(\frac{s_c}{c}\right)^2}$
Standard deviation[b]		$s_{dev} = \sqrt{\frac{\sum((x_i)^2/N)}{N-1}}$

[a]a, b and c are the individual measurements with their standard deviations of s_a, s_b and s_c, respectively, and s_x is the standard deviation for x.
[b]x_i is the value of the ith measurement; N is the total number of measurements.

a rapid-kinetic measurement affects the observed KIE can be found in choline oxidase (Yuan & Gadda, 2011) and dihydroorotate dehydrogenase (Fagan, Jensen, Bjornberg, & Palfey, 2007) to cite two examples.

1.3 Experimental considerations of KIE measurements

Detailed descriptions of experimental techniques employed to study flavoproteins can be found elsewhere in this volume. In this section we wish to note the critical controls that must be generally carried out when conducting non-competitive KIE studies of enzyme mechanism. Since multiple measurements are employed using light and heavy substrates, error propagation is critical. While it is customary to conduct individual assays multiple times to improve precision, all data points should be plotted and fit globally to the kinetic equations for the model being used. The errors on each of the kinetic parameters must be propagated using the equations in Table 1. When reporting the data, one must always report the determined values with the associated standard deviations. In addition the buffer concentration and composition, pH and temperature should be reported as each can affect the magnitude of the KIE measured for the reaction. More details on the reporting of KIE data in enzymology can be found in Francis and Kohen (2014).

2. Steady-state versus rapid kinetic investigations of flavoenzymes

2.1 Steady-state KIE analysis of flavoproteins

The reactivity of some flavoproteins with molecular oxygen provides a useful probe of their steady-state kinetic parameters using a Clarke-type oxygen electrode (Clark, Wolf, Granger, & Taylor, 1953). The method of initial rates (Allison & Purich, 1979) is used to measure the consumption of oxygen during steady-state turnover of the enzyme with oxygen and the data are globally fit to the kinetic equations pertaining the model that best describes the enzyme being investigated (see Cook & Cleland, 2007 for detailed descriptions of the types of kinetic models). It is critical to perform global fits of the data and to use the kinetic parameters obtained from those fits to judge the quality of the results obtained from the experiment. This avoids bias and the drawing of erroneous conclusions. It is best practice to run assays by alternating the labeled and unlabeled substrates to ensure that the differences in the observed rates are due solely to isotopic substitution and not the degradation of the enzymes or substrates during the time course of the experiment. While conclusions should be drawn from global fits of the data, it is convenient to show secondary Lineweaver-Burk plots of the data that are refit with the parameters obtained from the global fit. This allows a quick visual inspection of the data by the reader to judge the fitness of the model being proposed. An illustration of a non-competitive steady-state KIE experiment is shown in Fig. 2 for a hypothetical enzyme displaying a sequential steady-state kinetic mechanism. In a non-competitive KIE experiment individual kinetic assays would be measured by alternating isotopically unlabeled and labeled substrates. This would be done using at least four concentrations of oxygen, which can be easily controlled by bubbling gas into the reaction chamber. A variety of global fitting software is commercially available to analyze the data using the kinetic equations of multi-substrate enzymes. Some of these software packages include EnzFitter (Biosoft; Cambridge, UK), Sigma Plot (Sysat Software; San Jose, CA) and KinTek Explorer (KinTek; Snow Shoe, PA). Non-competitive KIE measurements have been carried out for a multitude of flavin-dependent enzymes including lysine monooxygenase (Robinson, Rodriguez, & Sobrado, 2014), alkanesulfonate monooxygenase (Robbins & Ellis, 2014) and D-amino acid oxidase (Trimmer, Wanninayake, & Fitzpatrick, 2017) and others (Adachi, Juarez, & Fitzpatrick, 2010; Adachi, Taylor, Hart, & Fitzpatrick, 2012a, 2012b; Adachi, Torres, & Fitzpatrick, 2010; Ball, Reis, Agniswamy, Weber, & Gadda, 2018; Daubner, Gadda, Valley, & Fitzpatrick, 2002; Emanuele & Fitzpatrick, 1995a, 1995b; Emanuele, Heasley, & Fitzpatrick, 1995; Fan & Gadda, 2005a, 2005b, 2007; Fan, Germann, & Gadda, 2006; Finnegan & Gadda, 2008; Fitzpatrick, 2001, 2004, 2007, 2015, 2017; Fitzpatrick et al., 2007; Fitzpatrick, Chadegani, Zhang, & Dougherty, 2017; Fitzpatrick, Chadegani, Zhang,

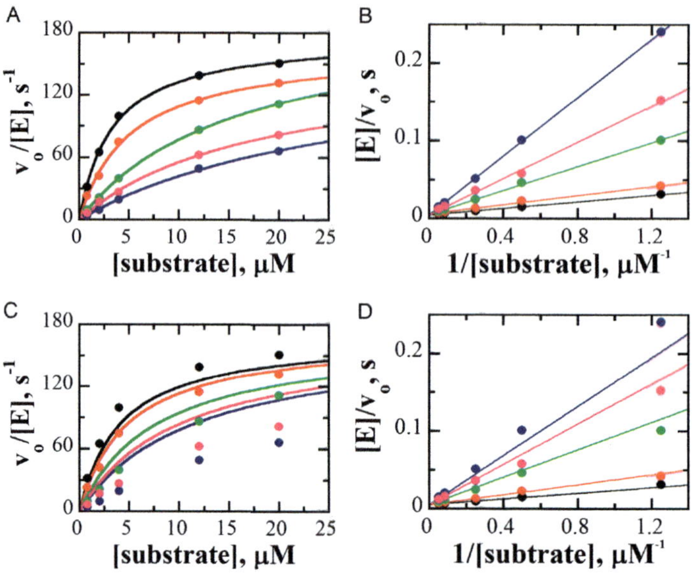

Fig. 2 Global fitting of steady-state kinetic data. The hypothetical data show an experiment in which the steady-state kinetics of a flavoprotein was fit globally to a model describing a sequential steady-state kinetic mechanism. (A and B) The data fit individually to the Michaelis-Menten equation and Lineweaver-Burke equations, respectively. (C and D) The same data that were replotted using the kinetic parameters obtained from a global fit of all data to a sequential steady state kinetic mechanism. As can easily be observed while the individual fits can give the impression that they are fit well to the model, it either incorrectly describes the system or different substrate concentrations that better cover the range of the kinetic parameters of the enzyme should be chosen.

Roberts, & Hinck, 2016; Fitzpatrick, Orville, Nagpal, & Valley, 2005; Flores & Gadda, 2018; Francis, Russell, & Gadda, 2005; Gadda, Banerjee, Dangott, & Fitzpatrick, 2000; Gadda, Banerjee, & Fitzpatrick, 2000; Gadda, Banerjee, Fleming, & Fitzpatrick, 2001; Gadda, Choe, & Fitzpatrick, 2000; Gadda, Dangott, Johnson, Whitman, & Fitzpatrick, 1999; Gadda, Edmondson, Russell, & Fitzpatrick, 1997; Gadda, Fan, & Hoang, 2006; Gadda & Fitzpatrick, 1998, 1999, 2000a, 2000b; Gadda, Powell, & Menon, 2004; Gadda et al., 1997; Gadda & Yuan, 2017; Gaweska & Fitzpatrick, 2011; Gaweska, Henderson Pozzi, Schmidt, McCafferty, & Fitzpatrick, 2009; Gaweska, Roberts, & Fitzpatrick, 2012; Gaweska, Taylor, Hart, & Fitzpatrick, 2013; Ghanem, Fan, Francis, & Gadda, 2003; Ghanem & Gadda, 2005, 2006; Henderson Pozzi & Fitzpatrick, 2010; Henderson Pozzi, Gawandi, & Fitzpatrick, 2009; Hoang & Gadda, 2007; Jonsson,

Edmondson, & Klinman, 1994; Mijatovic & Gadda, 2008; Nagpal, Valley, Fitzpatrick, & Orville, 2004, 2006; Ouedraogo et al., 2017; Ouedraogo, Souffrant, Vasquez, Hamelberg, & Gadda, 2017; Quaye, Lountos, Fan, Orville, & Gadda, 2008; Ralph, Anderson, Cleland, & Fitzpatrick, 2006; Ralph et al., 2007; Robbins, Geng, Barry, Gadda, & Bommarius, 2018; Robbins, Souffrant, Hamelberg, Gadda, & Bommarius, 2017; Roberts & Fitzpatrick, 2017; Roberts, Tormos, & Fitzpatrick, 2014; Roth & Klinman, 2003; Royo & Fitzpatrick, 2005; Rungsrisuriyachai & Gadda, 2008; Sobrado & Fitzpatrick, 2002, 2003a, 2003b, 2003c; Sobrado & Gadda, 2017; Su, Yuan, & Gadda, 2017; Su & Klinman, 1999; Tormos, Henderson Pozzi, & Fitzpatrick, 2012; Tormos, Suarez, & Fitzpatrick, 2016; Tormos, Taylor, Daubner, Hart, & Fitzpatrick, 2010; Trimmer et al., 2017; Uluisik, Romero, & Gadda, 2017; Valley, Tichy, & Fitzpatrick, 2005). While the multi-step nature of enzymatic reactions may make it difficult to isolate the chemical step giving rise to the KIE, a steady-state kinetic analysis can reveal much about an enzyme function and can expose features that are not as easily observed using rapid kinetics (see, for example, Section 4). The technique could also be coupled with site directed mutagenesis to elucidate the roles of specific amino acid residues within the enzyme (Gadda & Yuan, 2017; Henderson Pozzi & Fitzpatrick, 2010; Ralph et al., 2006; Tormos et al., 2012). As discussed extensively in Cleland (1982), Gadda and Sobrado (2018), and Northrop (1982) measuring KIEs as a function of pH or solvent viscosity can significantly reduce or even abolish the kinetic complexity inherent during steady-state turnover of flavoenzymes.

2.2 Rapid kinetic measurements of flavoenzyme KIE

The chemical versatility of the flavin gives flavoenzymes another useful property to exploit for KIE measurements, namely their UV-visible absorption spectra, which can be monitored using stopped-flow absorption spectroscopy. The technique, pioneered by Massey (Gibson, Swoboda, & Massey, 1964; Strickland, Palmer, & Massey, 1975), takes advantage of the distinct UV-visible absorption spectra of the various redox states of the flavin cofactor. The redox states and absorption maxima of the most common sates of flavin cofactors are shown in Fig. 3. In addition to these, the C4a (hydro)peroxy form of flavins commonly observed in monooxygenases (Huijbers, Montersino, Westphal, Tischler, & van Berkel, 2014) and the recently discovered N5-oxide (Teufel et al., 2013) also have

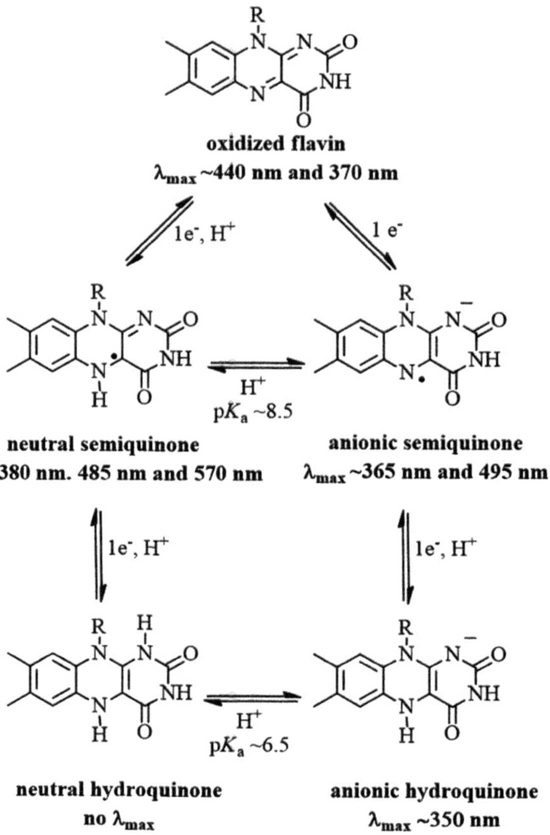

Fig. 3 Common redox states and absorption maxima of flavin cofactors.

distinct spectral characteristics that can be followed to measure KIEs of flavoenzymes. Detailed stopped-flow methodologies are provided elsewhere in this volume (refer to Chapters "Anaerobic methods for the transient-state study of flavoproteins: The use of specialized glassware to define the concentration of dioxygen" by Moran and "Performing anaerobic stopped-flow spectrophotometry inside of an anaerobic chamber" by Valentino and Sobrado). This chapter will introduce the half-reactions of flavoprotein catalysis and the ways that they can be probed using noncompetitive KIE measurements. Using a stopped-flow spectrophotometer, substrate and enzyme solutions are rapidly mixed in front of the detector and absorbance changes of the cofactor are monitored. Data are fit to exponential functions to determine limiting rate constants for the half-reaction being studied.

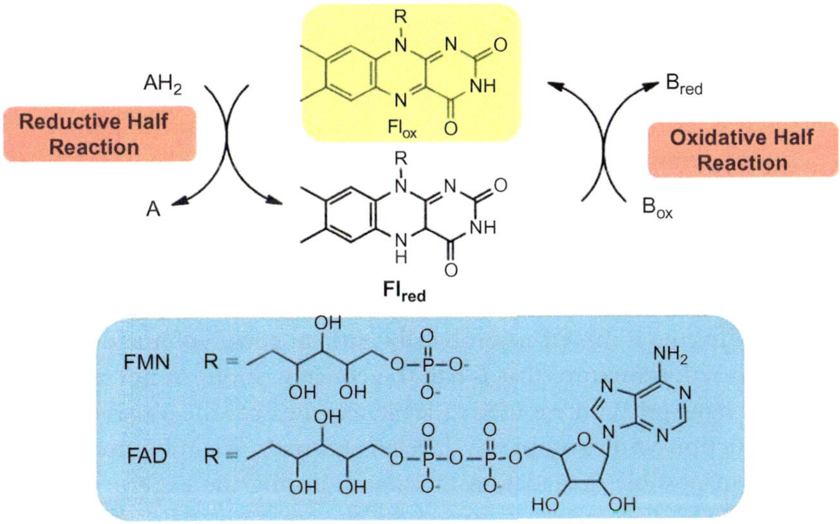

Fig. 4 The half-reactions of flavoenzymes.

Fig. 4 shows the reductive and oxidative half-reactions typically observed for flavoenzymes. In the reductive half-reaction an organic molecule binds to the free enzyme and electrons are transferred to the flavin. The oxidized flavin is then reduced with the concomitant oxidation of the organic substrate of the reaction. The opposite occurs in the oxidative half-reaction in which electrons are transferred from a reduced flavin to an oxidized substrate. The nature of the oxidation products depends on the class of flavoenzyme and the remarkable variety of reactions are the subject of many review articles (Chaiyen, Fraaije, & Mattevi, 2012; Leys & Scrutton, 2016; Massey, 2000). This section will outline how the spectral changes of the flavin cofactor can be monitored to measure non-completive KIEs of flavoenzyme reaction.

2.2.1 Non-competitive kinetic isotope effects on the reductive half reaction of flavoenzymes

Reductive half-reactions are measured through anaerobic mixing of the enzyme with substrate in a stopped-flow spectrophotometer and monitoring absorbance changes of the flavin cofactor. Maximal absorbance changes are typically observed at ~450 nm, which decreases to limiting values as the reaction occurs and the flavin is reduced. Enzymes at concentrations ≥20 μM in buffered solutions are loaded into a tonometer and made anaerobic by a 25-cycle treatment of applying vacuum and flushing with oxygen-free argon (pretreated with an oxygen-scrubbing cartridge). The anaerobic enzyme solution is then mounted onto the stopped-flow instrument, which is pre-treated overnight with an oxygen-scrubbing system composed of 5 mM glucose and 450 U of glucose oxidase. Alternatively, the protocatechuate/protocatechuate dioxygenase oxygen scrubbing system has been described to work equally well (0.2 mM substrate, 2.6 mg/mL enzyme) (Patil & Ballou, 2000). Unlabeled substrate is prepared in aqueous buffered solution and made anaerobic by flushing with oxygen-free argon for at least 20 min before mounting onto the stopped-flow spectrophotometer. The enzyme is mixed anaerobically with an equal volume of substrate to give a reaction mixture that is half the concentrations of the stock solutions. The same procedure is used to separately measure the observed rates of flavin reduction with isotopically labeled substrate. To minimize artifacts arising from possible degradation of samples during the experiment assays of light and heavy isotopes are carried out concurrently. The observed rate constants of flavin reduction are then plotted as a function of substrate concentrations and fit to Eq. (11). This yields the limiting rate constants of flavin reduction (k_{red}) and the apparent dissociation constants for the binding of the substrate to the enzyme (${}^{\mathrm{app}}K_{\mathrm{D}}$):

$$k_{obs} = \frac{k_{red}\, S}{{}^{app}K_D + S} \tag{11}$$

Fig. 5 illustrates a measurement of the KIE on the reductive half reaction of nitronate monooxygenase from *Neurospora crassa* (Francis & Gadda, 2009). In this case, the reductive half-reaction results in the formation of an anionic flavosemiquinone which was monitored using either ethylnitronate or [1-^{2}H]ethylnitronate as a substrate. Isotopic substitution of the substrate had no effect on the observed rate constants of flavin reduction, which ruled out changes of the hybridization states of the substrate α-carbon during the course of the reaction as the mechanistic cause if the inflated steady state KIEs observed with the enzyme.

The reductive-half reactions of many flavoproteins give rise to multiphasic kinetic traces and non-competitive KIEs can be used to identify the mechanistic origins of these results. Conformational sampling of the Michaelis-complex prior to the reductive half is common and is emerging as a key feature of many enzymatic reactions (Jones, Levy, Hay, & Scrutton, 2013; Klinman, 2015). The microenvironment around the flavin cofactor can have dramatic effects on its UV-visible absorption properties (Massey, 2000) and results in shifts or modulations of extinction coefficients

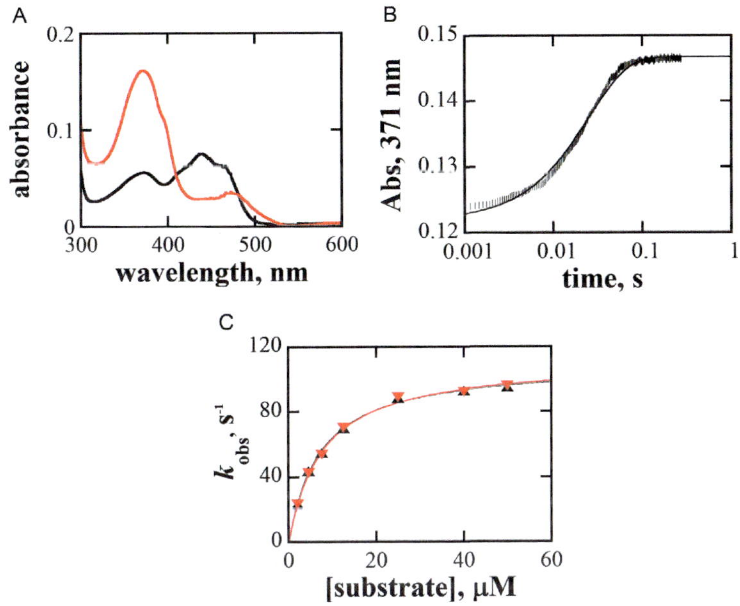

Fig. 5 Kinetic isotope effects on the reductive half-reaction of flavoenzymes. Enzyme and substrate were anaerobically mixed in a stopped-flow spectrophotometer, and the absorbance changes of the flavin cofactor were monitored. (A) Absorption spectra of the enzyme before (black) and after (red) mixing. (B) The absorbance at maximal signal change is plotted as a function of time and fit to an exponential function to obtain observed rates of flavin reduction. The time indicated is at the end of flow (~2 ms). (C) The k_{obs} is plotted against the concentration of either the light (black) or heavy (red) substrate concentration and fit to Eq. (11) to obtain isotope effects on the reductive half-reaction of the enzyme.

that will appear in stopped-flow traces. However, since these conformational changes do not involve the cleavage or formation of chemical bonds in the substrate they are isotope insensitive and do not give rise to KIEs. Measurements of the KIEs on 6-hydroxynicotine oxidase (Fitzpatrick, 2017) and morphinone reductase (Basran, Harris, Sutcliffe, & Scrutton, 2003) have been used to reveal the catalytic mechanisms of those enzymes.

2.2.2 Non-competitive kinetic isotope effects on the oxidative half-reaction of flavoenzymes

Non-competitive KIEs can also be used to probe the mechanism of the oxidative half-reaction with O_2 in pyranose 2-oxidase (Sucharitakul, Wongnate, & Chaiyen, 2011), aryl-alcohol oxidase (Carro, Ferreira, Martinez, & Gadda, 2018), choline oxidase (Gannavaram & Gadda, 2013) and glycolate oxidase (Pennati & Gadda, 2009, 2011), detecting the deprotonation of N5 during oxidation. The experiments involve measuring the oxidative half-reaction of the enzyme in water and deuterium oxide (D_2O) and comparing the observed rate constants. The critical controls that must be carried out in general when performing experiments in D_2O are detailed in Section 3.

Double-mixing stopped-flow was used to study aryl-alcohol oxidase (Carro et al., 2018). Buffered enzyme solutions are loaded into the body of the tonometer and an oxygen scrubbing system is loaded into the side arm. The sample is made anaerobic by a 25-round cycle of applying vacuum and flushing with oxygen-free argon before mounting onto the stopped-flow. The instrument is made anaerobic through a ~12 h treatment with glucose/glucose oxidase or protocatechuate/protocatechuate dioxygenase. The reducing substrates (isotopically labeled and unlabeled) are dissolved in buffer and made anaerobic by bubbling oxygen-free argon for at least 20 min before mounting onto the stopped-flow.

The anaerobic enzyme and organic substrate solutions are mixed with a slight excess of enzyme (1.0:0.9) to avoid catalytic turnover. The samples are aged to allow flavin reduction to occur and then mixed with an equal volume of buffer containing known amounts of dissolved oxygen. The concentration of dissolved oxygen is easily controlled and measured as discussed in Section 2.1. A range of oxygen concentrations are used. Traces are fit to exponential functions in the same manner as in studies of the reductive half-reactions and the observed rate constants of flavin oxidation are plotted as a function of oxygen concentration. Data are fit to Eq. (12) to obtain the second order rate constant for the oxidative half reaction (k_{ox}) and the rate constant for the reverse step of the reaction (k_{rev}):

$$k_{obs} = k_{ox}[O_2] + k_{rev} \tag{12}$$

Fig. 6 shows the results of non-competitive KIE measurements on the oxidative half-reaction of aryl-alcohol oxidase (Carro et al., 2018). In that case a pL-independent primary KIE of ~1.5 was observed with no significant solvent KIEs which was taken as evidence for a stepwise mechanism for the oxidative half reaction that involves a slow hydrogen transfer from the flavin N(5) atom to dioxygen followed by a relatively fast proton transfer from the solvent.

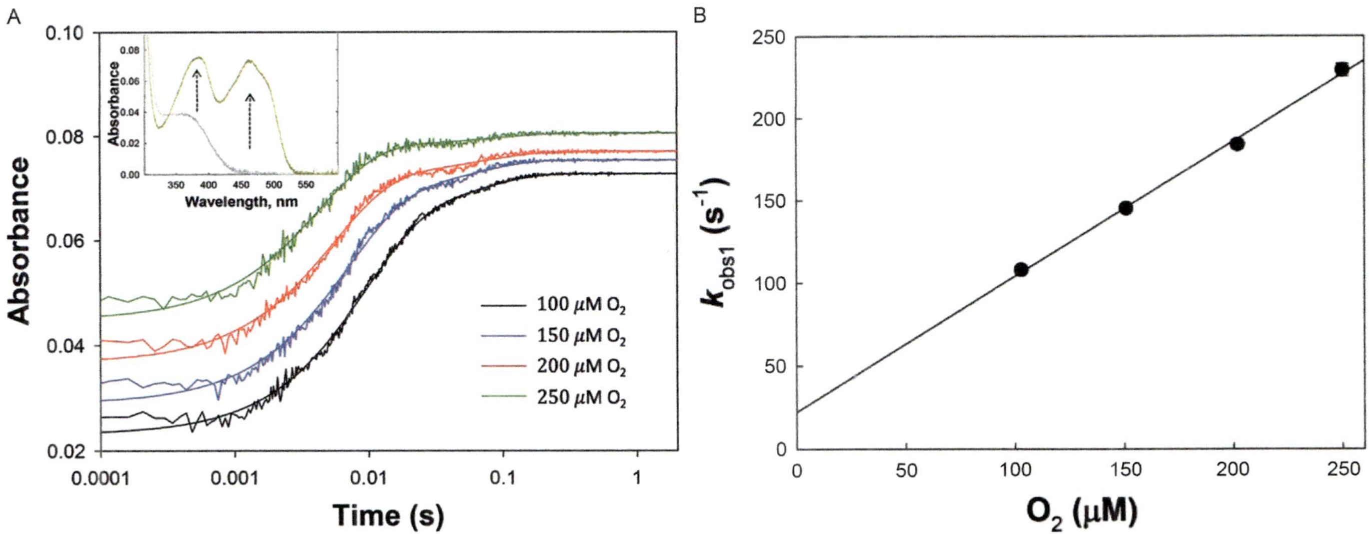

Fig. 6 (A) Time-resolved absorbance spectroscopy of the oxidation of reduced aryl amine oxidase at varying O_2 concentrations. The anaerobic enzyme was premixed with [α-$^1H_2$]-*p*-methoxybenzyl alcohol (1:0.9-fold the enzyme concentration) and allowed to age for 1 s and the reduced enzyme was finally mixed with buffers containing various O_2 concentrations (shown in the legend). Traces at 463 nm show the two phases of the reaction, fast and slow. The solid line indicates the fit of the traces to a double-exponential process. The inset displays spectra of the initial (reduced, gray) and final (oxidized, dark yellow) flavin species, with arrows indicating the direction of the spectral changes. Conditions: enzyme (5 μM after double mixing) in 50 mM sodium phosphate (pH 6.0) at 12 °C. (B) Plot of the observed rate constants (k_{obs}) for the fast phases of AAO oxidation as a function of O_2 concentration. *Reprinted from Carro, J., Ferreira, P., Martinez, A. T., & Gadda, G. (2018). Stepwise hydrogen atom and proton transfers in dioxygen reduction by aryl-alcohol oxidase.* Biochemistry, *57(11), 1790–1797. doi:10.1021/acs.biochem.8b00106, with permission from the American Chemical Society.*

3. Solvent kinetic isotope effects to probe flavoenzyme mechanism

Solvent KIEs (SKIEs) are measured by conducting kinetic assays in both water and D_2O and taking the ratio of the resulting rate constants. They can be measured using steady-state or rapid-kinetic methods. In addition to the precautions noted above, one must also ensure that the effects observed are due to isotopic substitution in the solvent and not due to other effects. In principle, substitution of H_2O with D_2O might lead to solvent effects due to changes in the isotopic composition of solvent-exchangeable sites participating in the reaction (SKIEs), increases in the relative viscosity of the solution, perturbations of the observed pK_a values (pH effects) or a combination of these factors (Karsten, Lai, & Cook, 1995; Purich, 1982; Quinn, Sutton, & Schowen, 1991). A control can easily be carried out by determining the kinetic parameters of the enzyme at increasing concentrations of a viscosigen. Typical viscosigens used in the study of flavin dependent enzymes include glycerol, glucose or sucrose. Solutions of increasing solvent viscosity are prepared using different concentrations of viscosigen. Tables showing the relative viscosities as a function of viscosigens can be found in Lide (2005). If the kinetic parameter of interest is not limited by the diffusion of substrates or products into or away from the active site, and not by slow conformational changes of the protein, no effect will be observed. If any of these factors influence the rate constants governing the enzymatic reaction, on the other hand, the observed SKIE on the kinetic parameter interest must be corrected for the solvent viscosity effect. Not only is this control critical to avoid drawing incorrect conclusion about the enzyme, the observation of solvent viscosity effects can reveal valuable information about the mechanism being studied as outlined in detail in Gadda and Sobrado (2018).

After establishing the possible effects of solvent viscosity on the reaction, SKIEs can be measured to determine if any solvent-exchangeable protons are in flight during the formation of the transition state of the reaction. All substrates and the enzyme solution should be lyophilized and dissolved in D_2O to ensure any solvent-exchangeable protons are replaced by deuterium. The alternative procedure of diluting the enzyme solution in D_2O has been extensively used and works out for most of applications, although one should be cautious when unexpected results are obtained. Kinetic assays are then conducted in both water and D_2O using either steady-state or rapid-kinetic approaches. If no protons are in flight during the rate-limiting step of the kinetic step(s) reflected in the rate constant, no SKIE will be observed. However if protons are in flight there will be a SKIE. Proton inventories can then be measured to determine the number of protons that are in flight as described extensively in Kohen and Limbach (2005). This is done by conducting the assays in buffers containing different ratios of H_2O/D_2O and plotting the ratios of the kinetic parameter determined in light/heavy solvent as a function of the mole fraction of D_2O of the solvent (n). If a linear function with a slope of one is seen then one proton is in flight. Non-linear proton inventories indicate that more than one proton is in flight during the solvent-sensitive step of the reaction. Approaches to elucidate the patterns and the information it could reveal about the enzyme system under study are the subject of the following reviews: Kohen and Limbach (2005) and Quinn (1985). SKIEs have been successfully used to elucidate the mechanism of amine and alcohol oxidation (Fitzpatrick, 2010; Sucharitakul et al., 2011), alkanesulfonate monooxgenation (Robbins & Ellis, 2014) and others (Basran et al., 2003; Fitzpatrick, 2007, 2015; Fitzpatrick et al., 2017; Gaweska et al., 2012; Henderson Pozzi & Fitzpatrick, 2010; Quinn, 1985; Sobrado & Fitzpatrick, 2003c; Su & Klinman, 1999; Tormos et al., 2016; Trimmer et al., 2017).

4. Multiple KIE studies of flavin dependent enzyme

A combination of isotope effects can be measured to determine the relative timing of bond-cleavage during flavoenzyme catalysis. Such an approach has been successfully used in many systems including alkane sulfonate monooxygenase (Robbins & Ellis, 2014), choline oxidase (Yuan & Gadda, 2011) and D-amino acid oxidase (Fitzpatrick, 2010). Experiments involve kinetic assays of both light and heavy substrates conducted in water and D_2O. Thus, all combinations of isotope effects are measured; 1° and 2° KIEs arise from using labeled substrate and SKIEs arise from using D_2O. Experiments can be conducted using either steady-state or rapid-kinetic approaches and all the critical controls outlined above must be used to avoid drawing erroneous conclusions based on artifacts in the experiments. If there is a discernable reaction intermediate the mechanism is defined as being stepwise. If no discernable intermediates are formed the reaction is defined as being concerted. Concerted mechanisms can occur in an asynchronous manner in which the progress of one bond-cleavage is greater than that of the second bond-cleavage. Alternatively, concerted mechanisms can be synchronous in which the two bonds are cleave concurrently. Different patterns are observed for each type of mechanism as described below. The structures of the transition states are obviously different for each type of mechanism, so these types of experiments can prove to be invaluable for rational inhibitor design.

To illustrate the approach, a hypothetical reaction involving a flavin dependent oxidation of an alcohol will be considered. As shown in Fig. 7 a probable mechanism for alcohol oxidation could involve the base-catalyzed proton abstraction from the hydroxyl group and a hydride ion transfer to the N(5) position of the flavin cofactor. Such a mechanism would involve two bonds being cleaved and would give a variety of non-competitive KIEs as illustrated in Fig. 8. Substitution of the hydride on the substrate α-carbon with deuterium would result in a 1° substrate KIE, but a 2° KIE on the proton transfer reaction from the hydroxyl proton and the enzymatic base of the reaction. When the assay is carried out in D_2O the hydroxyl proton would be replaced with a deuteride and a 1° KIE for the proton transfer reaction would be observed that would also be a 2° KIE on the proton transfer reaction.

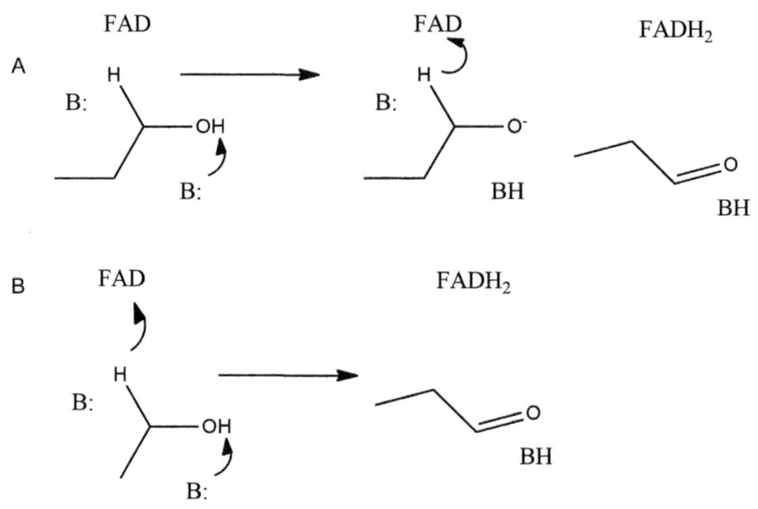

Fig. 7 Probable mechanisms for the oxidation of an alcohol by a flavoenzyme. Path A illustrates a stepwise mechanism involving a proton transfer between an enzymatic base and the hydroxyl group and a separate hydride transfer from the substrate α-carbon and the enzyme bound flavin in a second kinetic step. Path B shows the same oxidation occurring in a concerted fashion where the proton and hydride transfers occur in the same kinetic step.

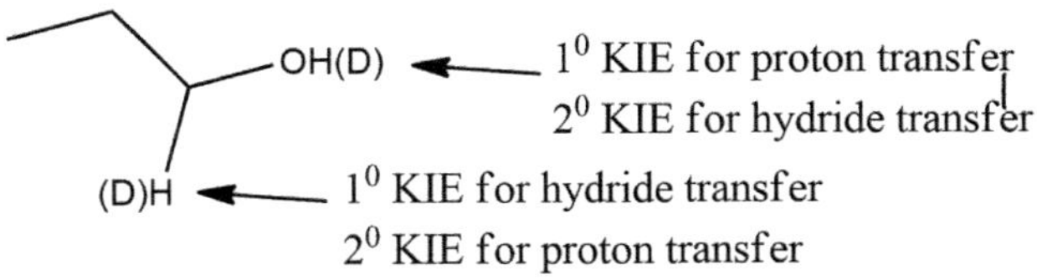

Fig. 8 Isotopic substitution of an alcohol for multiple KIE studies. The bond cleavages would give rise to different patterns of observed KIEs. If proton and hydride transfer occur in a stepwise mechanism and hydride transfer is fully rate-limiting, then slowing the rate of proton transfer with D_2O would have no effect on the 1° KIE observed when the hydride on the α-carbon is replaced with a deuteride. If the bonds are cleaved in a concerned fashion then the observed 1° KIE would be larger in D_2O than in H_2O. Measuring all possible combinations of KIEs would elucidate the timing of the bond cleavages events.

5. Conclusions

The measurements of non-competitive KIEs can reveal much about the mechanism of a flavin dependent enzyme. The reactivity of the isoalloxazine ring with molecular oxygen allows the enzyme to be studied during steady-state turnover with a substrate using a Clarke-type oxygen electrode. Rapid-kinetic techniques using a stopped-flow spectrophotometer can also be employed to measure non-competitive KIEs by taking advantage of the distinct UV-visible absorption spectra of flavin cofactors. This chapter outlines advantages provided by each method and urges the investigator to utilize both to obtain a comprehensive picture of catalysis by a flavoenzyme. With the proper data analysis and controls KIE data can not only strengthen conclusions drawn from crystal structures of the enzyme, but can reveal information static snapshots might not be able to reveal. We hope this chapter will encourage the use of isotope to probe flavoprotein mechanism.

Acknowledgments

This work was supported by the University Research Award from Texas A&M University-Kingsville Texas Compressive Research Fund, Grant AC-00006 from the Welch Foundation to KF and by the National Science Foundation Grant CHE-1506518.

References

Adachi, M. S., Juarez, P. R., & Fitzpatrick, P. F. (2010). Mechanistic studies of human spermine oxidase: Kinetic mechanism and pH effects. *Biochemistry*, *49*(2), 386–392. https://doi.org/10.1021/bi9017945.

Adachi, M. S., Taylor, A. B., Hart, P. J., & Fitzpatrick, P. F. (2012a). Mechanistic and structural analyses of the role of His67 in the yeast polyamine oxidase Fms1. *Biochemistry*, *51*(24), 4888–4897. https://doi.org/10.1021/bi300517s.

Adachi, M. S., Taylor, A. B., Hart, P. J., & Fitzpatrick, P. F. (2012b). Mechanistic and structural analyses of the roles of active site residues in yeast polyamine oxidase Fms1: Characterization of the N195A and D94N enzymes. *Biochemistry*, *51*(43), 8690–8697. https://doi.org/10.1021/bi3011434.

Adachi, M. S., Torres, J. M., & Fitzpatrick, P. F. (2010). Mechanistic studies of the yeast polyamine oxidase Fms1: Kinetic mechanism, substrate specificity, and pH dependence. *Biochemistry*, *49*(49), 10440–10448. https://doi.org/10.1021/bi1016099.

Allison, R. D., & Purich, D. L. (1979). Practical considerations in the design of initial velocity enzyme rate assays. *Methods in Enzymology*, *63*, 3–22.

Ball, J., Reis, R. A. G., Agniswamy, J., Weber, I. T., & Gadda, G. (2018). Steric hindrance controls pyridine nucleotide specificity of a flavin-dependent NADH:quinone oxidoreductase. *Protein Science*, *28*, 167–175. https://doi.org/10.1002/pro.3514.

Basran, J., Harris, R. J., Sutcliffe, M. J., & Scrutton, N. S. (2003). H-tunneling in the multiple H-transfers of the catalytic cycle of morphinone reductase and in the reductive half-reaction of the homologous pentaerythritol tetranitrate reductase. *The Journal of Biological Chemistry*, *278*(45), 43973–43982. https://doi.org/10.1074/jbc.M305983200.

Carro, J., Ferreira, P., Martinez, A. T., & Gadda, G. (2018). Stepwise hydrogen atom and proton transfers in dioxygen reduction by aryl-alcohol oxidase. *Biochemistry*, *57*(11), 1790–1797. https://doi.org/10.1021/acs.biochem.8b00106.

Chaiyen, P., Fraaije, M. W., & Mattevi, A. (2012). The enigmatic reaction of flavins with oxygen. *Trends in Biochemical Sciences*, *37*(9), 373–380. https://doi.org/10.1016/j.tibs.2012.06.005.

Clark, L. C., Jr., Wolf, R., Granger, D., & Taylor, Z. (1953). Continuous recording of blood oxygen tensions by polarography. *Journal of Applied Physiology*, *6*(3), 189–193. https://doi.org/10.1152/jappl.1953.6.3.189.

Cleland, W. W. (1982). The use of isotope effects to determine transition-state structure for enzymic reactions. *Methods in Enzymology*, *87*, 625–641.

Cook, P. F., & Cleland, W. W. (2007). *Enzyme kinetics and mechanism*. New York, NY: Taylor and Francis.

Daubner, S. C., Gadda, G., Valley, M. P., & Fitzpatrick, P. F. (2002). Cloning of nitroalkane oxidase from Fusarium oxysporum identifies a new member of the acyl-CoA dehydrogenase superfamily. *Proceedings of the National Academy of Sciences of the United States of America*, *99*(5), 2702–2707. https://doi.org/10.1073/pnas.052527799.

Emanuele, J. J., & Fitzpatrick, P. F. (1995a). Mechanistic studies of the flavoprotein tryptophan 2-monooxygenase. 1. Kinetic mechanism. *Biochemistry*, *34*(11), 3710–3715.

Emanuele, J. J., & Fitzpatrick, P. F. (1995b). Mechanistic studies of the flavoprotein tryptophan 2-monooxygenase. 2. pH and kinetic isotope effects. *Biochemistry*, *34*(11), 3716–3723.

Emanuele, J. J., Heasley, C. J., & Fitzpatrick, P. F. (1995). Purification and characterization of the flavoprotein tryptophan 2-monooxygenase expressed at high levels in Escherichia coli. *Archives of Biochemistry and Biophysics*, *316*(1), 241–248. https://doi.org/10.1006/abbi.1995.1034.

Fagan, R. L., Jensen, K. F., Bjornberg, O., & Palfey, B. A. (2007). Mechanism of flavin reduction in the class 1A dihydroorotate dehydrogenase from Lactococcus lactis. *Biochemistry*, *46*(13), 4028–4036. https://doi.org/10.1021/bi602460n.

Fan, F., & Gadda, G. (2005a). On the catalytic mechanism of choline oxidase. *Journal of the American Chemical Society*, *127*(7), 2067–2074. https://doi.org/10.1021/ja044541q.

Fan, F., & Gadda, G. (2005b). Oxygen- and temperature-dependent kinetic isotope effects in choline oxidase: Correlating reversible hydride transfer with environmentally enhanced tunneling. *Journal of the American Chemical Society*, *127*(50), 17954–17961. https://doi.org/10.1021/ja0560377.

Fan, F., & Gadda, G. (2007). An internal equilibrium preorganizes the enzyme-substrate complex for hydride tunneling in choline oxidase. *Biochemistry*, *46*(21), 6402–6408. https://doi.org/10.1021/bi700255v.

Fan, F., Germann, M. W., & Gadda, G. (2006). Mechanistic studies of choline oxidase with betaine aldehyde and its isosteric analogue 3,3-dimethylbutyraldehyde. *Biochemistry*, *45*(6), 1979–1986. https://doi.org/10.1021/bi0517537.

Finnegan, S., & Gadda, G. (2008). Substitution of an active site valine uncovers a kinetically slow equilibrium between competent and incompetent forms of choline oxidase. *Biochemistry*, *47*(52), 13850–13861. https://doi.org/10.1021/bi801424p.

Fitzpatrick, P. F. (2001). Substrate dehydrogenation by flavoproteins. *Accounts of Chemical Research*, *34*(4), 299–307.

Fitzpatrick, P. F. (2004). Carbanion versus hydride transfer mechanisms in flavoprotein-catalyzed dehydrogenations. *Bioorganic Chemistry*, *32*(3), 125–139. https://doi.org/10.1016/j.bioorg.2003.02.001.

Fitzpatrick, P. F. (2007). Insights into the mechanisms of flavoprotein oxidases from kinetic isotope effects. *Journal of Labelled Compounds and Radiopharmaceuticals*, *50*(11 – 12), 1016–1025. https://doi.org/10.1002/jlcr.1400.

Fitzpatrick, P. F. (2010). Oxidation of amines by flavoproteins. *Archives of Biochemistry and Biophysics*, *493*(1), 13–25. https://doi.org/10.1016/j.abb.2009.07.019.

Fitzpatrick, P. F. (2015). Combining solvent isotope effects with substrate isotope effects in mechanistic studies of alcohol and amine oxidation by enzymes. *Biochimica et Biophysica Acta—Proteins and Proteomics*, *1854*(11), 1746–1755. https://doi.org/10.1016/j.bbapap.2014.10.020.

Fitzpatrick, P. F. (2017). Nitroalkane oxidase: Structure and mechanism. *Archives of Biochemistry and Biophysics*, *632*, 41–46. https://doi.org/10.1016/j.abb.2017.05.012.

Fitzpatrick, P. F., Bozinovski, D. M., Heroux, A., Shaw, P. G., Valley, M. P., & Orville, A. M. (2007). Mechanistic and structural analyses of the roles of Arg409 and Asp402 in the reaction of the flavoprotein nitroalkane oxidase. *Biochemistry*, *46*(48), 13800–13808. https://doi.org/10.1021/bi701557k.

Fitzpatrick, P. F., Chadegani, F., Zhang, S., & Dougherty, V. (2017). Mechanism of flavoprotein l-6-hydroxynicotine oxidase: pH and solvent isotope effects and identification of key active site residues. *Biochemistry*, *56*(6), 869–875. https://doi.org/10.1021/acs.biochem.6b01160.

Fitzpatrick, P. F., Chadegani, F., Zhang, S., Roberts, K. M., & Hinck, C. S. (2016). Mechanism of the flavoprotein L-hydroxynicotine oxidase: Kinetic mechanism, substrate specificity, reaction product, and roles of active-site residues. *Biochemistry*, *55*(4), 697–703. https://doi.org/10.1021/acs.biochem.5b01325.

Fitzpatrick, P. F., Orville, A. M., Nagpal, A., & Valley, M. P. (2005). Nitroalkane oxidase, a carbanion-forming flavoprotein homologous to acyl-CoA dehydrogenase. *Archives of Biochemistry and Biophysics*, *433*(1), 157–165. https://doi.org/10.1016/j.abb.2004.08.021.

Flores, E., & Gadda, G. (2018). Kinetic characterization of PA1225 from Pseudomonas aeruginosa PAO1 reveals a new NADPH:quinone reductase. *Biochemistry*, *57*(21), 3050–3058. https://doi.org/10.1021/acs.biochem.8b00090.

Francis, K., & Gadda, G. (2009). Inflated kinetic isotope effects in the branched mechanism of Neurospora crassa 2-nitropropane dioxygenase. *Biochemistry*, *48*(11), 2403–2410. https://doi.org/10.1021/bi802238j.

Francis, K., & Kohen, A. (2014). Standards for the reporting of kinetic isotope effects in enzymology. *Perspectives in Science*, *1*(1), 110–120. https://doi.org/10.1016/j.pisc.2014.02.009.

Francis, K., Russell, B., & Gadda, G. (2005). Involvement of a flavosemiquinone in the enzymatic oxidation of nitroalkanes catalyzed by 2-nitropropane dioxygenase. *The Journal of Biological Chemistry*, *280*(7), 5195–5204. https://doi.org/10.1074/jbc.M411249200.

Gadda, G. (2008). Hydride transfer made easy in the reaction of alcohol oxidation catalyzed by flavin-dependent oxidases. *Biochemistry, 47*(52), 13745–13753. https://doi.org/10.1021/bi801994c.

Gadda, G., Banerjee, A., Dangott, L. J., & Fitzpatrick, P. F. (2000). Identification of a cysteine residue in the active site of nitroalkane oxidase by modification with N-ethylmaleimide. *The Journal of Biological Chemistry, 275*(41), 31891–31895. https://doi.org/10.1074/jbc.M003679200.

Gadda, G., Banerjee, A., & Fitzpatrick, P. F. (2000). Identification of an essential tyrosine residue in nitroalkane oxidase by modification with tetranitromethane. *Biochemistry, 39*(5), 1162–1168.

Gadda, G., Banerjee, A., Fleming, G. S., & Fitzpatrick, P. F. (2001). Evidence for an essential arginine in the flavoprotein nitroalkane oxidase. *Journal of Enzyme Inhibition, 16*(2), 157–163.

Gadda, G., Choe, D. Y., & Fitzpatrick, P. F. (2000). Use of pH and kinetic isotope effects to dissect the effects of substrate size on binding and catalysis by nitroalkane oxidase. *Archives of Biochemistry and Biophysics, 382*(1), 138–144. https://doi.org/10.1006/abbi.2000.2009.

Gadda, G., Dangott, L. J., Johnson, W. H., Jr., Whitman, C. P., & Fitzpatrick, P. F. (1999). Characterization of 2-oxo-3-pentynoate as an active-site-directed inactivator of flavoprotein oxidases: Identification of active-site peptides in tryptophan 2-monooxygenase. *Biochemistry, 38*(18), 5822–5828. https://doi.org/10.1021/bi982777z.

Gadda, G., Edmondson, R. D., Russell, D. H., & Fitzpatrick, P. F. (1997). Identification of the naturally occurring flavin of nitroalkane oxidase from Fusarium oxysporum as a 5-nitrobutyl-FAD and conversion of the enzyme to the active FAD-containing form. *The Journal of Biological Chemistry, 272*(9), 5563–5570.

Gadda, G., Fan, F., & Hoang, J. V. (2006). On the contribution of the positively charged headgroup of choline to substrate binding and catalysis in the reaction catalyzed by choline oxidase. *Archives of Biochemistry and Biophysics, 451*(2), 182–187. https://doi.org/10.1016/j.abb.2006.04.005.

Gadda, G., & Fitzpatrick, P. F. (1998). Biochemical and physical characterization of the active FAD-containing form of nitroalkane oxidase from Fusarium oxysporum. *Biochemistry, 37*(17), 6154–6164. https://doi.org/10.1021/bi973085y.

Gadda, G., & Fitzpatrick, P. F. (1999). Substrate specificity of a nitroalkane-oxidizing enzyme. *Archives of Biochemistry and Biophysics, 363*(2), 309–313. https://doi.org/10.1006/abbi.1998.1081.

Gadda, G., & Fitzpatrick, P. F. (2000a). Iso-mechanism of nitroalkane oxidase: 1. Inhibition studies and activation by imidazole. *Biochemistry, 39*(6), 1400–1405.

Gadda, G., & Fitzpatrick, P. F. (2000b). Mechanism of nitroalkane oxidase: 2. pH and kinetic isotope effects. *Biochemistry, 39*(6), 1406–1410.

Gadda, G., & Francis, K. (2010). Nitronate monooxygenase, a model for anionic flavin semiquinone intermediates in oxidative catalysis. *Archives of Biochemistry and Biophysics, 493*(1), 53–61. https://doi.org/10.1016/j.abb.2009.06.018.

Gadda, G., Powell, N. L., & Menon, P. (2004). The trimethylammonium headgroup of choline is a major determinant for substrate binding and specificity in choline oxidase. *Archives of Biochemistry and Biophysics, 430*(2), 264–273. https://doi.org/10.1016/j.abb.2004.07.011.

Gadda, G., & Sobrado, P. (2018). Kinetic solvent viscosity effects as probes for studying the mechanisms of enzyme action. *Biochemistry, 57*(25), 3445–3453. https://doi.org/10.1021/acs.biochem.8b00232.

Gadda, G., Wels, G., Pollegioni, L., Zucchelli, S., Ambrosius, D., Pilone, M. S., et al. (1997). Characterization of cholesterol oxidase from Streptomyces hygroscopicus and Brevibacterium sterolicum. *European Journal of Biochemistry, 250*(2), 369–376.

Gadda, G., & Yuan, H. (2017). Substitutions of S101 decrease proton and hydride transfers in the oxidation of betaine aldehyde by choline oxidase. *Archives of Biochemistry and Biophysics*, *634*, 76–82. https://doi.org/10.1016/j.abb.2017.10.004.

Gannavaram, S., & Gadda, G. (2013). Relative timing of hydrogen and proton transfers in the reaction of flavin oxidation catalyzed by choline oxidase. *Biochemistry*, *52*(7), 1221–1226. https://doi.org/10.1021/bi3016235.

Gaweska, H., & Fitzpatrick, P. F. (2011). Structures and mechanism of the monoamine oxidase family. *Biomolecular Concepts*, *2*(5), 365–377. https://doi.org/10.1515/bmc.2011.030.

Gaweska, H., Henderson Pozzi, M., Schmidt, D. M., McCafferty, D. G., & Fitzpatrick, P. F. (2009). Use of pH and kinetic isotope effects to establish chemistry as rate-limiting in oxidation of a peptide substrate by LSD1. *Biochemistry*, *48*(23), 5440–5445. https://doi.org/10.1021/bi900499w.

Gaweska, H. M., Roberts, K. M., & Fitzpatrick, P. F. (2012). Isotope effects suggest a stepwise mechanism for berberine bridge enzyme. *Biochemistry*, *51*(37), 7342–7347. https://doi.org/10.1021/bi300887m.

Gaweska, H. M., Taylor, A. B., Hart, P. J., & Fitzpatrick, P. F. (2013). Structure of the flavoprotein tryptophan 2-monooxygenase, a key enzyme in the formation of galls in plants. *Biochemistry*, *52*(15), 2620–2626. https://doi.org/10.1021/bi4001563.

Ghanem, M., Fan, F., Francis, K., & Gadda, G. (2003). Spectroscopic and kinetic properties of recombinant choline oxidase from Arthrobacter globiformis. *Biochemistry*, *42*(51), 15179–15188. https://doi.org/10.1021/bi035435o.

Ghanem, M., & Gadda, G. (2005). On the catalytic role of the conserved active site residue His466 of choline oxidase. *Biochemistry*, *44*(3), 893–904. https://doi.org/10.1021/bi048056j.

Ghanem, M., & Gadda, G. (2006). Effects of reversing the protein positive charge in the proximity of the flavin N(1) locus of choline oxidase. *Biochemistry*, *45*(10), 3437–3447. https://doi.org/10.1021/bi052514m.

Gibson, Q. H., Swoboda, B. E., & Massey, V. (1964). Kinetics and mechanism of action of glucose oxidase. *The Journal of Biological Chemistry*, *239*, 3927–3934.

Hay, S., Pudney, C. R., & Scrutton, N. S. (2009). Structural and mechanistic aspects of flavoproteins: Probes of hydrogen tunnelling. *The FEBS Journal*, *276*(15), 3930–3941. https://doi.org/10.1111/j.1742-4658.2009.07121.x.

Henderson Pozzi, M., & Fitzpatrick, P. F. (2010). A lysine conserved in the monoamine oxidase family is involved in oxidation of the reduced flavin in mouse polyamine oxidase. *Archives of Biochemistry and Biophysics*, *498*(2), 83–88. https://doi.org/10.1016/j.abb.2010.04.015.

Henderson Pozzi, M., Gawandi, V., & Fitzpatrick, P. F. (2009). pH dependence of a mammalian polyamine oxidase: Insights into substrate specificity and the role of lysine 315. *Biochemistry*, *48*(7), 1508–1516. https://doi.org/10.1021/bi802227m.

Hengge, A. C. (2015). Kinetic isotope effects in the characterization of catalysis by protein tyrosine phosphatases. *Biochimica et Biophysica Acta—Proteins and Proteomics*, *1854*(11), 1768–1775. https://doi.org/10.1016/j.bbapap.2015.03.010.

Hoang, J. V., & Gadda, G. (2007). Trapping choline oxidase in a nonfunctional conformation by freezing at low pH. *Proteins*, *66*(3), 611–620. https://doi.org/10.1002/prot.21238.

Hu, S., Soudackov, A. V., Hammes-Schiffer, S., & Klinman, J. P. (2017). Enhanced rigidification within a double mutant of soybean lipoxygenase provides experimental support for vibronically nonadiabatic proton-coupled electron transfer models. *ACS Catalysis*, 7(5), 3569–3574. https://doi.org/10.1021/acscatal.7b00688.

Huijbers, M. M., Montersino, S., Westphal, A. H., Tischler, D., & van Berkel, W. J. (2014). Flavin dependent monooxygenases. *Archives of Biochemistry and Biophysics*, *544*, 2–17. https://doi.org/10.1016/j.abb.2013.12.005.

Jacob, B. (2006). Theoretical basis for isotope effects from an autobiographical perspective. In A. Kohen & H.-H. Limbach (Eds.), *Isotope effects in chemistry and biology* (pp. 1–40). Boca Raton, FL: CRC Press.

Jones, A. R., Levy, C., Hay, S., & Scrutton, N. S. (2013). Relating localized protein motions to the reaction coordinate in coenzyme B(1)(2)-dependent enzymes. *The FEBS Journal, 280*(13), 2997–3008. https://doi.org/10.1111/febs.12223.

Jonsson, T., Edmondson, D. E., & Klinman, J. P. (1994). Hydrogen tunneling in the flavoenzyme monoamine oxidase B. *Biochemistry, 33*(49), 14871–14878.

Karsten, W. E., Lai, C.-J., & Cook, P. F. (1995). Inverse solvent isotope effects in the NAD-malic enzyme reaction are the result of the viscosity difference between D_2O and H_2O: Implications for solvent isotope effect studies. *Journal of the American Chemical Society, 117*(22), 5914–5918. https://doi.org/10.1021/ja00127a002.

Kholodar, S. A., Ghosh, A. K., & Kohen, A. (2017). Measurement of enzyme isotope effects. *Methods in Enzymology, 596*, 43–83. https://doi.org/10.1016/bs.mie.2017.06.033.

Klinman, J. P. (2015). Dynamically achieved active site precision in enzyme catalysis. *Accounts of Chemical Research, 48*(2), 449–456. https://doi.org/10.1021/ar5003347.

Klinman, J. P., & Kohen, A. (2013). Hydrogen tunneling links protein dynamics to enzyme catalysis. *Annual Review of Biochemistry, 82*, 471–496. https://doi.org/10.1146/annurev-biochem-051710-133623.

Kohen, A., & Limbach, H.-H. (2005). *Isotope effects in chemistry and biology*. Boca Raton, FL: CRC Press.

Leys, D., & Scrutton, N. S. (2016). Sweating the assets of flavin cofactors: New insight of chemical versatility from knowledge of structure and mechanism. *Current Opinion in Structural Biology, 41*, 19–26. https://doi.org/10.1016/j.sbi.2016.05.014.

Lide, D. R. (2005). *CRC handbook of chemistry and physics*. CRC Press.

Marcus, R. A., & Sutin, n. (1985). Electron transfers in chemistry and biology. *Biochimica et Biophysica Acta—Bioenergetics, 811*, 265–322.

Massey, V. (2000). The chemical and biological versatility of riboflavin. *Biochemical Society Transactions, 28*(4), 283–296.

Mijatovic, S., & Gadda, G. (2008). Oxidation of alkyl nitronates catalyzed by 2-nitropropane dioxygenase from Hansenula mrakii. *Archives of Biochemistry and Biophysics, 473*(1), 61–68. https://doi.org/10.1016/j.abb.2008.02.029.

Nagpal, A., Valley, M. P., Fitzpatrick, P. F., & Orville, A. M. (2004). Crystallization and preliminary analysis of active nitroalkane oxidase in three crystal forms. *Acta Crystallographica Section D, Biological Crystallography, 60*, 1456–1460. https://doi.org/10.1107/s0907444904013289. Pt 8.

Nagpal, A., Valley, M. P., Fitzpatrick, P. F., & Orville, A. M. (2006). Crystal structures of nitroalkane oxidase: Insights into the reaction mechanism from a covalent complex of the flavoenzyme trapped during turnover. *Biochemistry, 45*(4), 1138–1150. https://doi.org/10.1021/bi051966w.

Northrop, D. B. (1975). Steady-state analysis of kinetic isotope effects in enzymic reactions. *Biochemistry, 14*(12), 2644–2651. https://doi.org/10.1021/bi00683a013.

Northrop, D. B. (1982). Deuterium and tritium kinetic isotope effects on initial rates. *Methods in Enzymology, 87*, 607–625.

Ouedraogo, D., Ball, J., Iyer, A., Reis, R. A. G., Vodovoz, M., & Gadda, G. (2017). Amine oxidation by D-arginine dehydrogenase in Pseudomonas aeruginosa. *Archives of Biochemistry and Biophysics, 632*, 192–201. https://doi.org/10.1016/j.abb.2017.06.013.

Ouedraogo, D., Souffrant, M., Vasquez, S., Hamelberg, D., & Gadda, G. (2017). Importance of loop L1 dynamics for substrate capture and catalysis in Pseudomonas aeruginosa D-arginine dehydrogenase. *Biochemistry, 56*(19), 2477–2487. https://doi.org/10.1021/acs.biochem.7b00098.

Paneth, P. (1995). Theoretical calculations of heavy-atom isotope effects. *Computers & Chemistry, 19*(3), 231–240.

Patil, P. V., & Ballou, D. P. (2000). The use of protocatechuate dioxygenase for maintaining anaerobic conditions in biochemical experiments. *Analytical Biochemistry, 286*(2), 187–192. https://doi.org/10.1006/abio.2000.4802.

Pennati, A., & Gadda, G. (2009). Involvement of ionizable groups in catalysis of human liver glycolate oxidase. *The Journal of Biological Chemistry, 284*(45), 31214 31222. https://doi.org/10.1074/jbc.M109.040063.

Pennati, A., & Gadda, G. (2011). Stabilization of an intermediate in the oxidative half-reaction of human liver glycolate oxidase. *Biochemistry, 50*(1), 1–3. https://doi.org/10.1021/bi101387w.

Purich, D. L. (1982). [1] Criteria for evaluating the catalytic competence of enzyme-substrate covalent compounds. In D. L. Purich (Ed.), *Vol. 87. Methods in enzymology.* (pp. 3–20). Academic Press.

Quaye, O., Lountos, G. T., Fan, F., Orville, A. M., & Gadda, G. (2008). Role of Glu312 in binding and positioning of the substrate for the hydride transfer reaction in choline oxidase. *Biochemistry, 47*(1), 243–256. https://doi.org/10.1021/bi7017943.

Quinn, D. M. (1985). Solvent isotope effects for lipoprotein lipase catalyzed hydrolysis of water-soluble p-nitrophenyl esters. *Biochemistry, 24*(13), 3144–3149.

Quinn, D. M., Sutton, L. D., & Schowen, K. B. (1991). In P. F. Cook (Ed.), *Enzyme mechanism from isotope effects* (pp. 73–126). Boca Raton, FL: CRC Press.

Ralph, E. C., Anderson, M. A., Cleland, W. W., & Fitzpatrick, P. F. (2006). Mechanistic studies of the flavoenzyme tryptophan 2-monooxygenase: Deuterium and 15N kinetic isotope effects on alanine oxidation by an L-amino acid oxidase. *Biochemistry, 45*(51), 15844–15852. https://doi.org/10.1021/bi061894o.

Ralph, E. C., Hirschi, J. S., Anderson, M. A., Cleland, W. W., Singleton, D. A., & Fitzpatrick, P. F. (2007). Insights into the mechanism of flavoprotein-catalyzed amine oxidation from nitrogen isotope effects on the reaction of N-methyltryptophan oxidase. *Biochemistry, 46*(25), 7655–7664. https://doi.org/10.1021/bi700482h.

Robbins, J. M., & Ellis, H. R. (2014). Steady-state kinetic isotope effects support a complex role of Arg226 in the proposed desulfonation mechanism of alkanesulfonate monooxygenase. *Biochemistry, 53*(1), 161–168. https://doi.org/10.1021/bi401234e.

Robbins, J. M., Geng, J., Barry, B. A., Gadda, G., & Bommarius, A. S. (2018). Photoirradiation generates an ultrastable 8-formyl FAD semiquinone radical with unusual properties in formate oxidase. *Biochemistry, 57*(40), 5818–5826. https://doi.org/10.1021/acs.biochem.8b00571.

Robbins, J. M., Souffrant, M. G., Hamelberg, D., Gadda, G., & Bommarius, A. S. (2017). Enzyme-mediated conversion of flavin adenine dinucleotide (FAD) to 8-formyl FAD in formate oxidase results in a modified cofactor with enhanced catalytic properties. *Biochemistry, 56*(29), 3800–3807. https://doi.org/10.1021/acs.biochem.7b00335.

Roberts, K. M., & Fitzpatrick, P. F. (2017). Measurement of kinetic isotope effects in an enzyme-catalyzed reaction by continuous-flow mass spectrometry. *Methods in Enzymology, 596*, 149–161. https://doi.org/10.1016/bs.mie.2017.07.001.

Roberts, K. M., Tormos, J. R., & Fitzpatrick, P. F. (2014). Characterization of unstable products of flavin- and pterin-dependent enzymes by continuous-flow mass spectrometry. *Biochemistry, 53*(16), 2672–2679. https://doi.org/10.1021/bi500267c.

Robinson, R. M., Rodriguez, P. J., & Sobrado, P. (2014). Mechanistic studies on the flavin-dependent N(6)-lysine monooxygenase MbsG reveal an unusual control for catalysis. *Archives of Biochemistry and Biophysics, 550-551*, 58–66. https://doi.org/10.1016/j.abb.2014.04.006.

Roth, J. P., & Klinman, J. P. (2003). Catalysis of electron transfer during activation of O2 by the flavoprotein glucose oxidase. *Proceedings of the National Academy of Sciences of the United States of America, 100*(1), 62–67. https://doi.org/10.1073/pnas.252644599.

Royo, M., & Fitzpatrick, P. F. (2005). Mechanistic studies of mouse polyamine oxidase with N1,N12-bisethylspermine as a substrate. *Biochemistry, 44*(18), 7079–7084. https://doi.org/10.1021/bi050347k.

Rungsrisuriyachai, K., & Gadda, G. (2008). On the role of histidine 351 in the reaction of alcohol oxidation catalyzed by choline oxidase. *Biochemistry, 47*(26), 6762–6769. https://doi.org/10.1021/bi800650w.

Sobrado, P., & Fitzpatrick, P. F. (2002). Analysis of the roles of amino acid residues in the flavoprotein tryptophan 2-monooxygenase modified by 2-oxo-3-pentynoate: Characterization of His338, Cys339, and Cys511 mutant enzymes. *Archives of Biochemistry and Biophysics, 402*(1), 24–30. https://doi.org/10.1016/s0003-9861(02)00063-2.

Sobrado, P., & Fitzpatrick, P. F. (2003a). Analysis of the role of the active site residue Arg98 in the flavoprotein tryptophan 2-monooxygenase, a member of the L-amino oxidase family. *Biochemistry, 42*(47), 13826–13832. https://doi.org/10.1021/bi035299n.

Sobrado, P., & Fitzpatrick, P. F. (2003b). Identification of Tyr413 as an active site residue in the flavoprotein tryptophan 2-monooxygenase and analysis of its contribution to catalysis. *Biochemistry, 42*(47), 13833–13838. https://doi.org/10.1021/bi035300i.

Sobrado, P., & Fitzpatrick, P. F. (2003c). Solvent and primary deuterium isotope effects show that lactate CH and OH bond cleavages are concerted in Y254F flavocytochrome b2, consistent with a hydride transfer mechanism. *Biochemistry, 42*(51), 15208–15214. https://doi.org/10.1021/bi035546n.

Sobrado, P., & Gadda, G. (2017). Introduction to flavoproteins: Beyond the classical paradigms. *Archives of Biochemistry and Biophysics, 632*, 1–3. https://doi.org/10.1016/j.abb.2017.08.016.

Stratton, C. F., Poulin, M. B., & Schramm, V. L. (2017). Binding isotope effects for interrogating enzyme-substrate interactions. *Methods in Enzymology, 596*, 1–21. https://doi.org/10.1016/bs.mie.2017.07.019. Epub 2017 Aug 31. PubMed PMID: 28911767.

Strickland, S., Palmer, G., & Massey, V. (1975). Determination of dissociation constants and specific rate constants of enzyme-substrate (or protein-ligand) interactions from rapid reaction kinetic data. *The Journal of Biological Chemistry, 250*(11), 4048–4052.

Su, Q., & Klinman, J. P. (1999). Nature of oxygen activation in glucose oxidase from Aspergillus niger: The importance of electrostatic stabilization in superoxide formation. *Biochemistry, 38*(26), 8572–8581. https://doi.org/10.1021/bi990044o.

Su, D., Yuan, H., & Gadda, G. (2017). A reversible, charge-induced intramolecular C4a-S-cysteinyl-flavin in choline oxidase variant S101C. *Biochemistry, 56*(51), 6677–6690. https://doi.org/10.1021/acs.biochem.7b00958.

Sucharitakul, J., Wongnate, T., & Chaiyen, P. (2011). Hydrogen peroxide elimination from C4a-hydroperoxyflavin in a flavoprotein oxidase occurs through a single proton transfer from flavin N5 to a peroxide leaving group. *The Journal of Biological Chemistry, 286*(19), 16900–16909. https://doi.org/10.1074/jbc.M111.222976.

Teufel, R., Miyanaga, A., Michaudel, Q., Stull, F., Louie, G., Noel, J. P., et al. (2013). Flavin-mediated dual oxidation controls an enzymatic Favorskii-type rearrangement. *Nature, 503*, 552. https://doi.org/10.1038/nature12643. https://www.nature.com/articles/nature12643#supplementary-information.

Tormos, J. R., Henderson Pozzi, M., & Fitzpatrick, P. F. (2012). Mechanistic studies of the role of a conserved histidine in a mammalian polyamine oxidase. *Archives of Biochemistry and Biophysics, 528*(1), 45–49. https://doi.org/10.1016/j.abb.2012.08.007.

Tormos, J. R., Suarez, M. B., & Fitzpatrick, P. F. (2016). (13)C kinetic isotope effects on the reaction of a flavin amine oxidase determined from whole molecule isotope effects. *Archives of Biochemistry and Biophysics, 612*, 115–119. https://doi.org/10.1016/j.abb.2016.10.018.

Tormos, J. R., Taylor, A. B., Daubner, S. C., Hart, P. J., & Fitzpatrick, P. F. (2010). Identification of a hypothetical protein from Podospora anserina as a nitroalkane oxidase. *Biochemistry*, *49*(24), 5035–5041. https://doi.org/10.1021/bi100610e.

Trimmer, E. E., Wanninayake, U. S., & Fitzpatrick, P. F. (2017). Mechanistic studies of an amine oxidase derived from D-amino acid oxidase. *Biochemistry*, *56*(14), 2024–2030. https://doi.org/10.1021/acs.biochem.7b00161.

Uluisik, R., Romero, E., & Gadda, G. (2017). Evidence for proton tunneling and a transient covalent flavin-substrate adduct in choline oxidase S101A. *Biochimica et Biophysica Acta—Proteins and Proteomics*, *1865*(11 Pt A), 1470–1478. https://doi.org/10.1016/j.bbapap.2017.08.004.

Valley, M. P., Tichy, S. E., & Fitzpatrick, P. F. (2005). Establishing the kinetic competency of the cationic imine intermediate in nitroalkane oxidase. *Journal of the American Chemical Society*, *127*(7), 2062–2066. https://doi.org/10.1021/ja043542f.

Yuan, H., & Gadda, G. (2011). Importance of a serine proximal to the C(4a) and N(5) flavin atoms for hydride transfer in choline oxidase. *Biochemistry*, *50*(5), 770–779. https://doi.org/10.1021/bi101837u.

CHAPTER SIX

Isotopically labeled flavoenzymes and their uses in probing reaction mechanisms

Andreea I. Iorgu, Matthew J. Cliff, Jonathan P. Waltho, Nigel S. Scrutton, Sam Hay*

Manchester Institute of Biotechnology and School of Chemistry, The University of Manchester, Manchester, United Kingdom

*Corresponding author: e-mail address: sam.hay@manchester.ac.uk

Contents

Abstract

The incorporation of stable isotopes into proteins is beneficial or essential for a range of experiments, including NMR, neutron scattering and reflectometry, proteomic mass spectrometry, vibrational spectroscopy and "heavy" enzyme kinetic isotope effect (KIE) measurements. Here, we present detailed protocols for the stable isotopic labeling of pentaerythritol tetranitrate reductase (PETNR) via recombinant expression in *E. coli*. PETNR is an ene-reductase belonging to the Old Yellow Enzyme (OYE) family of flavoenzymes, and is regarded as a model system for studying hydride transfer reactions. Included is a discussion of how efficient back-exchange of amide protons in the protein core can be achieved and how the intrinsic flavin mononucleotide (FMN) cofactor can be exchanged, allowing the production of isotopologues with differentially labeled protein and cofactor. In addition to a thorough description of labeling strategies, we briefly exemplify how data analysis and interpretation of "heavy" enzyme KIEs can be performed.

Methods in Enzymology, Volume 620
ISSN 0076-6879
https://doi.org/10.1016/bs.mie.2019.03.009

1. Introduction

The substitution of natural abundance isotopes within a biomacromolecule with their heavier counterparts (such as ^{2}H, ^{3}H, ^{13}C, ^{14}C, ^{15}N, ^{18}O, etc.) has enabled a deeper understanding of the structure and function of a wealth of biological systems. The use of isotopically labeled proteins (and, in particular, of deuterated proteins) originated in the 1960s from the need to reduce complexity and increase resolution of nuclear magnetic resonance (NMR) spectra (Crespi & Katz, 1969; Farmer & Venters, 2002; Katz & Crespi, 1966; Markley, Putter, & Jardetzky, 1968). Starting as a necessity to access structural information from NMR studies, stable isotope labeling of biomacromolecules, along with developments in NMR methodologies, enabled an exponential expansion in the number of studies dedicated to the structure and function of proteins in solution. Concomitant with the advances in NMR techniques, developments in molecular biology resulted in *E. coli* becoming a preferred organism of choice for the production of recombinant proteins (or at least, for those proteins that do not require post-translational modifications). As one of the most characterized protein synthesis machineries, *E. coli* is also the most utilized system for expressing isotopically labeled proteins, given its relatively rapid growth in minimal media, including $^{2}H_2O$ (D_2O) solutions (Rosano & Ceccarelli, 2014).

The necessity to study eukaryotic proteins that are not readily expressed in *E. coli* or other bacterial strains has recently driven developments in heavy atom labeling using insect cells or mammalian expression systems (such as adenoviral vector-based systems) as expression hosts (Sastry, Bewley, & Kwong, 2015; Skora, Shrestha, & Gossert, 2015). Furthermore, a growing number of publications feature cell-free expression systems (Staunton, Schlinkert, Zanetti, Colebrook, & Campbell, 2006; Takeda & Kainosho, 2012). Apart from the now vast choice of expression hosts available that can incorporate heavy atoms, there are many variations in which proteins can be labeled with stable isotopes: uniform or partial isotopic labeling (Duff, Wilde, Rekas, Lake, & Holden, 2015), site-specific labeling (Peuker et al., 2016), selective labeling (Lin et al., 2015; Suarez & Schramm, 2015) and segmental labeling (Michel & Allain, 2015). It seems that there is a lot of choice available when it comes to selecting methods of producing isotopically labeled proteins, and also types of labeling schemes, as required for particular techniques. Nevertheless, peculiarities still exist for each class of proteins, and it is not uncommon to notice that a highly established protocol for incorporating heavy atoms into one protein could not be readily replicated for another related protein.

While a wealth of developments has stemmed from the need to facilitate structural and dynamic NMR studies, the use of isotopic labeling schemes for proteins has expanded well beyond this primary scope. Isotopically labeled biomacromolecules are now commonly used in studies involving a large number of techniques, including small-angle neutron scattering, proteomic mass spectrometry, neutron reflectometry and site-directed vibrational spectroscopy (Chahrour, Cobice, & Malone, 2015; Haertlein et al., 2016; Heinrich, 2016; Peuker et al., 2016). Recently, the scope of isotopically labeled enzymes has extended even further, with "heavy" (primarily ^{2}H, ^{13}C and ^{15}N labeled) enzymes being now used to probe dynamical effects on the chemical step. This approach was popularized by Silva et al. in 2011, with the introduction of the concept of a Born-Oppenheimer enzyme (Silva, Murkin, & Schramm, 2011). A Born-Oppenheimer enzyme can be defined as an enzyme that is labeled with heavy atoms (^{2}H, ^{13}C and ^{15}N), and the isotopic substitution is postulated to reduce femtosecond bond vibration frequencies, while not affecting electrostatic properties of the enzyme. These experiments showed that, as computationally predicted, the increase in the mass of the purine nucleoside phosphorylase (PNP) enzyme by isotopic substitution led to a decrease in the rate of the chemical step as a result of the decreased probability of transition-state barrier crossing (Silva et al., 2011). The observation is consistent with coupling of the vibrational modes of the protein with the transition state, which is distorted in the "heavy" enzyme (compared to the "light," natural abundance unlabeled enzyme). Since this first study, more studies have been dedicated to probing the role of dynamics in the phosphorolysis step catalyzed by PNP (Antoniou, Ge, Schramm, & Schwartz, 2012; Harijan, Zoi, Antoniou, Schwartz, & Schramm, 2017; Suarez & Schramm, 2015; Zoi et al., 2016). Moreover, several other enzyme systems catalyzing diverse catalytic steps, such as proton transfer (Toney, Castro, & Addington, 2013), hydride transfer (Francis, Sapienza, Lee, & Kohen, 2016; Longbotham, Hardman, Görlich, Scrutton, & Hay, 2016; Luk et al., 2013; Pudney et al., 2013; Wang, Chang, & Schramm, 2016), and hydrolysis (Kipp, Silva, & Schramm, 2011; Krzemińska, Moliner, & Świderek, 2016) have been investigated using isotopically labeled enzymes, and measurements of the enzyme isotope effect have shown that catalytic rates are typically affected by isotopic substitution. The approaches used and the findings of these studies have been recently reviewed by Kholodar, Ghosh, and Kohen (2017) and Świderek, Javier Ruiz-Pernía, Moliner, and Tuñón (2014).

Herein, we will focus on discussing methodologies we have employed for the isotopic labeling of pentaerythritol tetranitrate reductase (PETNR), an ene-reductase from the Old Yellow Enzyme (OYE) family of flavoenzymes. PETNR is regarded as a model system for studying hydride transfer reactions, and is a flavin mononucleotide (FMN)-containing enzyme, which catalyzes hydride transfer (H-transfer) from NAD(P)H to a wide range of activated α,β-unsaturated substrates via a bi-bi ping-pong mechanism that uses the FMN cofactor (Basran, Harris, Sutcliffe, & Scrutton, 2003; Khan et al., 2002; Pudney, Hay, & Scrutton, 2009). The H-transfer mechanism from NAD(P)H coenzyme to the FMN cofactor (known as the reductive half-reaction of the catalytic cycle, RHR) has been extensively studied using temperature and pressure dependence studies of coenzyme deuterium kinetic isotope effects (KIEs) (Pudney, Hay, Levy, et al., 2009; Pudney et al., 2013). More recently, KIEs between "heavy" and "light" PETNR have been studied (Pudney et al., 2013) and differential labeling of the enzyme was used to study the different contribution of protein and cofactor vibrational modes to catalysis (Longbotham et al., 2016). Moreover, PETNR is the first enzyme belonging to the OYE family of ene-reductases for which near-complete ^{1}H, ^{15}N and ^{13}C backbone resonance assignments have been reported, which opens new avenues for investigating functional dynamics through structural NMR studies (Iorgu et al., 2018a). In this chapter, we will describe experimental methods used for isotopic labeling of PETNR, from production of the labeled enzyme to efficient methods used for FMN exchange, which enable kinetic studies of PETNR isotopologues (such as "light" apoenzyme-"heavy" FMN). We will also cover aspects regarding the distribution of "heavy" atoms in PETNR, and present methods for efficient deuterium-hydrogen (D-H) exchange of backbone amides, which are essential for structural NMR studies. These methods can be readily applied to other enzymes from the Old Yellow Enzyme (OYE) family of oxidoreductases, given their highly similar architecture (Scholtissek, Tischler, Westphal, van Berkel, & Paul, 2017; Toogood & Scrutton, 2018) and we expect that the protocols presented herein can be readily adapted for efficient labeling/cofactor exchange of other flavoenzymes. The last section of this chapter is dedicated to the recent applications of "heavy" PETNR in studying dynamical contributions to enzyme catalysis and will focus on data analysis and interpretation of KIEs when using "heavy" enzymes.

2. PETNR as a model flavoenzyme from the OYE family

PETNR is a monomeric 40 kDa enzyme that possesses broad substrate specificity, and it can catalyze the reduction of a wide variety of α,β-unsaturated compounds, often with high enantioselectivity toward desired products (Fryszkowska et al., 2012, 2009; Toogood et al., 2011). This promiscuity gives PETNR high potential to be used in biocatalytic processes. Moreover, PETNR is regarded as a model system for studying biological hydride transfer mechanisms. Several kinetic studies have indicated that quantum mechanical tunneling contributes to the enzymatic hydride transfer step from NAD(P)H to the FMN cofactor, and fast (ps–ns) protein dynamics have been inferred to contribute to catalysis from the temperature and pressure dependence of the deuterium KIEs on the RHR (Basran et al., 2003; Pudney, Hay, Levy, et al., 2009; Pudney, Hay, Sutcliffe, & Scrutton, 2006). Recently, we developed an approach to generate isotopologues of PETNR in which the apoprotein and/or the FMN cofactor have been labeled with different isotopes (Longbotham et al., 2016). These were studied using temperature dependence KIEs along with fluorescence lifetime measurements in order to dissect whether the dynamical contributions coupled with the H-transfer chemistry catalyzed by PETNR arise from the apoenzyme alone and/or the FMN cofactor. These studies suggested that vibrations local to the active site (which appear to be on the 10–100 ps timescale) play a role in the H-transfer chemistry, with observed isotope effects arising from both the protein and FMN cofactor (Longbotham et al., 2016). Moreover, having assigned the ^{1}H, $^{15}N_{H}$, $^{13}C_{\alpha}$, $^{13}C_{\beta}$ and $^{13}C'$ backbone resonances in the NMR spectra of PETNR to a great extent (Iorgu et al., 2018a), we were able to unravel more information regarding the role of functional dynamics in the H-transfer chemistry catalyzed by PETNR through more detailed NMR studies (Iorgu et al., 2018b). In the next sections, we will present details regarding the production of different PETNR isotopologues and strategies for determination of efficient conditions for D-H backbone amide exchange.

3. Expression and purification of isotopically labeled PETNR in *E. coli*

Throughout our studies, we used two different constructs for over-expression in *E. coli*. One expression system involves production of PETNR using the pBlueScript II SK(+) plasmid encoding the pONR1 gene. pBlueScript II SK(+) is a legacy plasmid, used since the first isolation of

PETNR from soil contaminated with explosives in the mid-1990s (French, Nicklin, & Bruce, 1996). Many of the early kinetic studies on this enzyme, along with the more recent NMR studies, were performed with recombinant PETNR expressed from this plasmid. More recently, we moved to use a more convenient expression plasmid (pET21a), from which a C-terminal His_8-tagged (or His_6-tagged) version of PETNR can be produced and easily purified via nickel-affinity chromatography. As we are now aware that there are no major differences in the kinetics and structure between PETNR and His-tagged PETNR (Hulley et al., 2010; Longbotham et al., 2016), we would suggest production (and, in particular, purification) of PETNR using the pET21a vector is more straightforward (and therefore, preferable) than using the pBlueScript II SK(+) plasmid. However, it is likely that, for other enzymes, a His tag is not desirable to be used, so we will describe herein both protocols. We will focus on the preparation and use of $^{2}H,^{13}C,^{15}N$-labeled PETNR for NMR studies, which require a uniformly deuterated sample in which all backbone amides are back-exchanged (at least partially) to protium atoms (D-H exchange). As we have previously encountered difficulties in developing efficient methods for this D-H exchange, we aim to provide as many details as possible to perform backbone amide exchange on the PETNR system and, by extension, on other flavoenzymes.

Labeling of flavoenzymes with heavy atoms (such as ^{2}H, ^{13}C and ^{15}N) in *E. coli* can be done using well-established techniques that are applicable to a vast majority of soluble proteins. The common route for expression of isotopically labeled flavoenzymes in *E. coli* involves the use of M9 minimal media (M9mm), which contains salts and trace elements required for bacteria growth. This basic M9mm is then supplemented with carbon and nitrogen sources, which can be either of natural abundance (leading to expression of unlabeled proteins) or isotopically labeled with specific heavy atoms. Expression of ^{2}H-labelled proteins is achieved using media prepared in D_2O (cf. H_2O). In the cases where uniform ^{2}H incorporation (perdeuteration) is required for the target protein, a ^{2}H-labeled carbon source is also required. The most common carbon source used for isotopic labeling in *E. coli* is $^{13}C_6$-D-glucose, which is replaced with $^{2}H_7,^{13}C_6$-D-glucose when expressing uniformly $^{2}H,^{13}C$-labeled proteins.

3.1 Assembly of M9 minimal media for isotopic labeling of PETNR

The expression in *E. coli* of isotopically labeled PETNR involves the use of a modified M9mm, which is prepared following a combination of protocols developed for growth in supplemented minimal growth medium (Sambrook, Fritsch, & Maniatis, 1989; Studier, 2005). Strategies for the preparation of various isotopically labeled M9mm that we use for overexpression of PETNR are presented below.

^{15}N-labeled M9mm. First, 6 g Na_2HPO_4, 3 g KH_2PO_4 and 0.5 g NaCl are dissolved in ~900 mL double-distilled (dd) H_2O. The pH of the solution is then adjusted to 7.4 (using concentrated KOH), followed by dilution to 980 mL and sterilization (autoclave at 121 °C for 15 min). Under sterile conditions, the following components are further added to the 980 mL of M9mm salts solution: 1 mL ampicillin solution (0.1 g/mL), 1 mL 1 M $MgSO_4$ solution, 1 mL thiamine solution (1 mg/mL), 100 μL 1 M $CaCl_2$ solution, 200 μL trace metals solution, 3 g glucose (15 mL 20% (*w/v*) solution) and 1 g $^{15}NH_4Cl$ (5 mL 20% (w/v) solution). The overall composition of ^{15}N M9mm is detailed in Table 1. The trace metals solution (5000 × stock solution) is prepared using a combination of metal salts, as presented in Table 2. The stock solutions used above are prepared individually and filter-sterilized prior to use in M9mm assembly. The final concentration of antibiotic used in the medium is 0.1 mg/mL.

^{2}H, ^{15}N-labeled M9mm. The preparation of ^{2}H, ^{15}N-labeled M9mm is the same as for ^{15}N labeled M9mm, except $^{2}H_7$-D-glucose is used as a carbon source and all reagents are dissolved in D_2O (except for the 5000 × trace metals solution). The solution is sterilized by vacuum filtration using a vacuum filter system with a 0.2 μM pore size membrane, and is not autoclaved. Also, given the medium consists of D_2O solution, the pH is adjusted to $pH_{read} = 6.99$, as the actual pD of the solution is $pD = pH_{read} + 0.41$.

Table 1 Components for 1 L of modified M9 minimal medium.

Component	**Amount**
Na_2HPO_4	6 g
KH_2PO_4	3 g
NaCl	0.5 g
$MgSO_4$	1 mL of 1 M solution
Thiamine	1 mL of 1 mg/mL solution
$CaCl_2$	100 μL of 1 M solution
Trace metals solution	200 μL
D-glucose	15 mL of 20% solution (3 g)
$^{15}NH_4Cl$	5 mL of 20% solution (1 g)

Table 2 Recipe for the preparation of 100 mL 5000 × trace metals solution.

Trace metal	Stock solution	Volume (mL)
Iron	0.1 M $FeCl_3$ in 0.1 M HCl	50
Manganese	1 M $MnCl_2$	1
Zinc	1 M $ZnSO_4$	1
Cobalt	0.2 M $CoCl_2$	1
Copper	0.1 M $CuCl_2$	2
Nickel	0.2 M $NiCl_2$	1
Molybdenum	0.1 M Na_2MoO_4	2
Selenium	0.1 M Na_2SeO_3	2
Boron	0.1 M H_3BO_3	2
–	dd H_2O	38

$^2H,^{13}C,^{15}N$-labeled M9mm. Preparation is the same as for ^{2}H,^{15}N-labeled M9mm, with the only difference being the use of $^2H_7,^{13}C_6$-D-glucose as a carbon source instead of $^2H_7$-D-glucose.

3.2 Overexpression and purification of isotopically labeled PETNR

Our procedure for production of ^{2}H,^{13}C,^{15}N-labeled PETNR, which is similar to the production of ^{2}H and ^{2}H,^{15}N-labeled PETNR, is presented below.

1. The *E. coli* strain JM109 is transformed with 1–100 ng of pBlueScript II SK(+) plasmid encoding the pONR1 gene used for PETNR overexpression expression, following the protocols provided by the manufacturer. Cells are plated on LB agar containing ampicillin (0.1 mg/mL) and incubated at 37 °C for ~16 h.
2. A single cell colony is selected from the plate and used to inoculate 5 mL of ^{15}N M9mm containing ampicillin (0.1 mg/mL). Cells are grown at 37 °C with shaking at 200 rpm, until an OD_{600} of >0.5 is reached. Following growth of the starting culture, 10–100 mL of ^{2}H,^{13}C,^{15}N M9mm (containing ampicillin) are inoculated in a 1:100 dilution with cells from the starting culture. Cells are further grown at 37 °C with shaking at 200 rpm until the OD_{600} reaches 0.8–1 (or overnight).

3. Final cultures, containing 500 mL of ^2H, ^{13}C, ^{15}N M9mm and ampicillin, are inoculated in a 1:50–1:100 dilution with cells from the intermediate culture, and grown at 37 °C with shaking at 200 rpm until the OD_{600} reaches 0.8–1. Overexpression of PETNR is then induced by addition of 0.5 mM isopropyl β-D-1-thiogalactopyranoside (IPTG, final concentration, added as 500 μL of freshly prepared and filter-sterilized 0.5 M IPTG stock in D_2O). At this point, the temperature is lowered to 25 °C. Cells are harvested after a further 24–36 h growth by centrifugation at 3000 × *g* for 10 min at 4 °C.

Purification of PETNR enriched with heavy atoms (^2H, ^{13}C and/or ^{15}N) is performed in the same manner as for unlabeled PETNR samples and is performed with buffers made in H_2O, not D_2O. The methodology for separating PETNR produced in JM109 cells has been previously described in detail (Barna et al., 2002; French et al., 1996; Iorgu et al., 2018a), and involves purification by affinity exchange chromatography, using a Mimetic Orange 2 column, followed by Source 15Q anion exchange chromatography. In the case of C-His_8 tagged PETNR expressed in NiCo(DE3) cells, purification of the target enzyme is performed using a HisTrap HP nickel-charged IMAC column. Usually, given that the yields are relatively high (>200 mg purified protein/L of culture), the C-His_8 tagged PETNR purification is performed with only one chromatographic step, and PETNR is found to be >95% pure following purification, as judged by SDS-PAGE analysis. However, in line with the NiCo(DE3) cell user manual, if contaminants are found after nickel affinity purification, an extra step using chitin affinity chromatography can be added to remove the three well expressed *E. coli* proteins that contain sequences that bind to the chitin column. If purified proteins are being used for crystallography, enzymes from both purification methods (His-tagged or not) can be further purified using a gel filtration step. As all the purification steps are done in protiated buffers, the backbone amide groups that are solvent exposed in the ^2H labeled PETNR samples will undergo D-H exchange. If a completely ^2H labeled sample is required for further experiments, we would suggest that, rather than using expensive perdeuterated buffer solutions for purification, it would be more economical to purify the protein in protiated buffer and, following purification, exchange the solvent-exchangeable backbone amides back to ^2H by using a perdeuterated buffer in the last step.

Notes

- For the production of C-His$_8$ PETNR, *E. coli* strain NiCo21(DE3) is transformed with 1–100 ng of pET21a plasmid encoding the C-His$_8$ PETNR gene, following the protocols provided by the manufacturer. All subsequent steps are identical for when the JM109 cells are used.
- We usually observe that expression yields are higher when the temperature is lowered to 20–25 °C following induction with IPTG (or, ideally, 1–2 h before induction).
- Extra glucose (1–3 g/L culture) can be added at the IPTG induction stage to afford higher cell density and, therefore, higher yields of target protein.
- The main difference between using the NiCo21(DE3) and JM109 cells is related to the doubling time of the cells: JM109 cells grow about 2 × slower than NiCO21(DE3). Also, although it is well known cells grow slower in deuterated medium, the time necessary to reach induction point, as compared to growth in rich supplemented medium, is specific to each cell line. JM109 *E. coli* cells grow 6 × slower in deuterated M9mm than in protiated M9mm: 36–40 h for $^{2}H,^{15}N/^{2}H,^{13}C,^{15}N$-labeled M9mm to OD_{600} of 0.8–1.0, as compared to 5–6 h for ^{15}N-labeled M9mm/LB Broth (growth from 1:100 dilution in final flasks). It was observed that deuteration influences both the timescale of growth, but also the maximal cell density (2–2.5 g of cell pellet/L of $^{2}H,^{15}N$ or $^{2}H,^{13}C,^{15}N$ M9mm, as compared to 4.8 g/L ^{15}N media). Typical yields for PETNR production in JM109 and NiCo21(DE3) cells are commonly in the range of 70–100 and 200–250 mg of purified PETNR/L of culture, respectively.
- As yields of protein expression can be very high, especially when using NiCo21(DE3) cells for C-His$_8$ PETNR production, we often observe that a significant fraction of purified PETNR does not contain bound FMN. We suspect this arises through an inability of the host to produce sufficient quantities of FMN. Thus, FMN incorporation is required at the purification stage and can be done at the lysis stage: the lysis buffer is supplemented with 0.5–1 mM FMN, which will enable rapid incorporation of FMN into the apoenzyme following cell lysis, and the excess FMN will be washed off when loading the protein sample onto the column. It is worth nothing that this step should be done in this manner if the labeling of the FMN cofactor is not relevant for further experiments, as unlabeled FMN will be introduced (in some proportion) in the samples.

4. Spectral characteristics and quality control of isotopically labeled PETNR

An interesting observation for ^{2}H-labeled PETNR is the presence of a blue-shift of the FMN absorption (Fig. 1). While the origin of this shift is uncertain, it indicates that there is a perturbation to the environment of the FMN cofactor in the deuterated samples. This could be an effect of deuteration on the hydrogen bonding interaction between the FMN cofactor and the enzyme, and must involve exchangeable protons, as the effect is lost following D-H exchange of backbone amides.

The assessment of isotopic composition of a protein that was grown in media supplemented with heavy atoms is most commonly checked by mass spectrometry and/or ^{1}H-^{15}N TROSY NMR spectroscopy. While the first technique allows one to confirm the quantitative incorporation of the stable isotopes, the NMR spectra of a ^{2}H,^{13}C,^{15}N-labeled protein allow qualitative visualization of the absence (or presence) of deuterium atoms in exchangeable backbone amide groups. Moreover, inspection of the ^{1}H NMR spectrum of the protein can indicate if a high degree of sample deuteration was achieved. As it can be observed in Fig. 2, the region between −1 and 4 ppm in the 1D ^{1}H NMR spectrum of ^{15}N-labeled PETNR is highly populated with signals arising from the aliphatic region of the enzyme (e.g., protium atoms from side chain methyl groups), while in the case of ^{2}H,^{15}N and ^{2}H,^{13}C,^{15}N-labeled PETNR, this region has no visible signals present, suggesting effective incorporation of deuterium atoms into the enzyme.

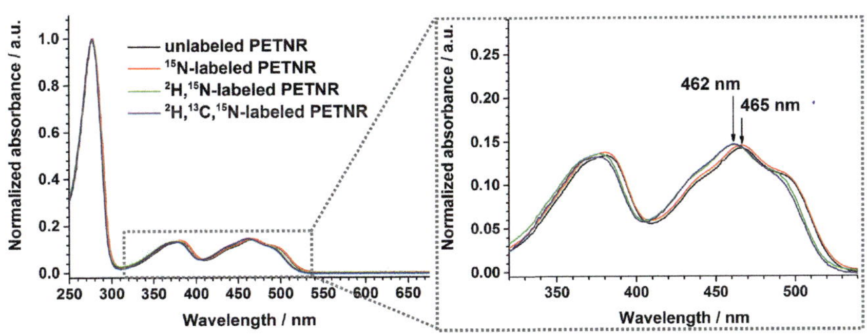

Fig. 1 UV-vis absorbance spectra of unlabeled, ^{15}N-labeled, ^{2}H,^{15}N-labeled and ^{2}H,^{13}C,^{15}N-labeled PETNR. The spectra are normalized to a maxima corresponding to the 277 nm peak. The insert shows the blue-shifted FMN absorbance in the deuterated samples.

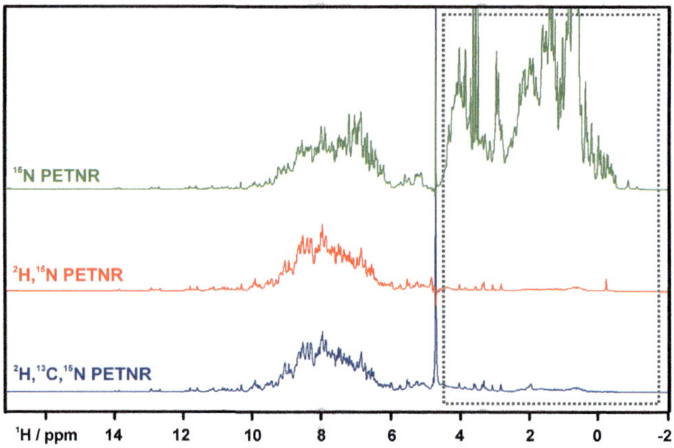

Fig. 2 ^{1}H 1D NMR spectra of PETNR samples of various PETNR isotopologues. Spectra were recorded at 298 K on an 800 MHz Bruker Avance III spectrometer.

Isotopic labeling of PETNR using the expression protocol discussed in Section 3.2 was also previously investigated using QTOF mass spectrometry, with mass increases of 9.8–10.3%, 5.3% and 1.3% observed for ^{2}H,^{13}C,^{15}N-labeled, ^{2}H,^{15}N-labeled and ^{15}N-labeled PETNR, respectively, when compared to "light" unlabeled PETNR (Longbotham et al., 2016; Pudney et al., 2013). These results are consistent with expected/calculated values of protein mass increase upon isotopic substitution, and similar values have been observed for several other systems (Kholodar et al., 2017).

Apart from quantifying the degree of isotopic enrichment through the use of mass spectrometry, the conservation of the structural features and thermal stability of isotopically labeled enzymes, when compared to their corresponding "light" isotopologues, can be interrogated through the use of a range of methods, including circular dichroism spectroscopy, differential scanning calorimetry, differential scanning fluorimetry and NMR spectroscopy. More details regarding the use of these techniques for assessing structural similarity (or otherwise) between "heavy" and "light" enzyme are summarized by Kholodar et al. (2017)). Comparison of the NMR temperature coefficients (the temperature dependencies of the backbone NH chemical shifts) is a powerful probe for structure perturbation. In the case of PETNR, we have previously shown evidence for the isostructural nature of ^{2}H,^{13}C,^{15}N-labeled PETNR and ^{15}N-labeled PETNR by correlating the NMR temperature coefficients for the backbone residues of both isotopologues (Pudney et al., 2013).

5. Analysis of deuterium-hydrogen exchange of backbone amides in PETNR

Large proteins (>25 kDa) require uniform deuteration in order to be structurally studied by NMR, as ^{2}H side-chain labeling increases significantly the sensitivity of 3D NMR experiments designed for backbone assignments (Nietlispach et al., 1996). However, maximum sensitivity can only be achieved if complete re-protonation of all amide groups can be performed. With a molecular weight of 40 kDa, PETNR belongs to the class of relatively large proteins that require ^{2}H labeling for such measurements. The PETNR structure comprises of an $(\alpha,\beta)_8$-barrel structural scaffold, a tertiary fold consisting of an 8-fold repeat of linked α-helix and β-sheet structures, typical to all enzymes belonging to the OYE family (Scholtissek et al., 2017; Williams & Bruce, 2002). This structural feature means the enzyme contains a highly hydrophobic, solvent-inaccessible core, with the *re* side of the flavin prosthetic group completely hidden from the active site, as it is located at the C-terminal end of this hydrophobic cavity. This structural feature makes exchange of amide backbone protons very slow. In order to anticipate and avoid problems in the sequential backbone assignment of the enzyme caused by lack of proton signals arising from deuteration, we initially analyzed the ^{1}H-^{15}N TROSY spectrum of a protonated ^{15}N-labeled PETNR sample and studied its H-D exchange behavior. We observed that >70 backbone amides (peaks) are essentially non-exchangeable over a period of 3 days (Fig. 3). These peaks were later mapped, almost exclusively, to residues located in the β-barrel of the protein. The results indicated a necessity for the use of a denaturant that would enhance flexibility in the core of the enzyme. We thus studied the equilibrium unfolding curve of the enzyme in presence of guanidine hydrochloride in order to determine what conditions could be used that would enable efficient D-H exchange, while avoiding irreversible unfolding of the protein.

6. Spectroscopic characterization of the equilibrium unfolding behavior of PETNR

Initial trials to determine the unfolding-refolding behavior of PETNR showed that after adding moderate concentrations of guanidine hydrochloride (GuHCl) solutions (2–4 M), irreversible precipitation of the protein

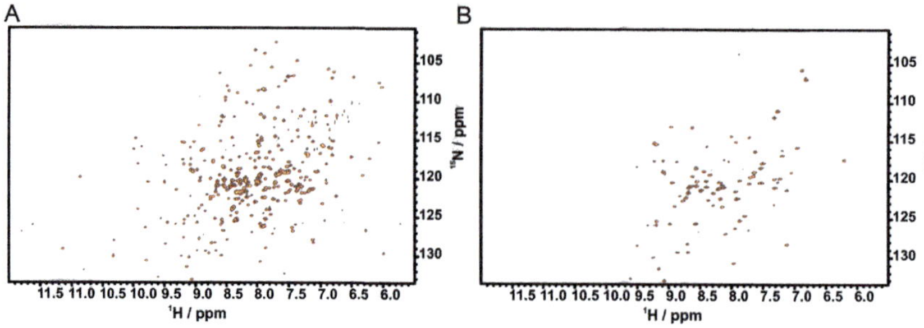

Fig. 3 Comparison between ^{1}H-^{15}N TROSY spectra of ~1 mM ^{15}N-labeled PETNR recorded in (A) standard buffer conditions (50 mM potassium phosphate buffer, pH 7.0, supplemented with 1 mM NaN_3 and 10% D_2O) and (B) 50 mM potassium phosphate buffer, pD 7.0, supplemented with 1 mM NaN_3, constituted in 100% D_2O. The spectrum in D_2O was recorded on an enzyme samples that was exchanged by repeated steps of concentrating by centrifugation into the desired deuterated buffer, starting from standard buffer conditions. Before recording the spectrum, the sample was left at 298 K for 72 h to promote D-H exchange. The ^{15}N-labeled samples were expressed and purified as described for ^{2}H,^{13}C,^{15}N-labeled PETNR, except that unlabeled nitrogen and carbon sources were used during cell overexpression. Both spectra were recorded at 298 K on an 800 MHz Bruker Avance III spectrometer.

occurs during the refolding step. Therefore, we decided to perform a thorough analysis of the unfolding behavior of PETNR, monitoring the changes in UV-vis, fluorescence emission and circular dichroism spectral features when the enzyme is subjected to gradually increasing concentrations of GuHCl denaturant (Fig. 4).

UV-vis absorption shows that FMN gradually loses the vibrational features characteristic of the bound flavin (the peak with a maxima at 465 nm with two shoulders on both sides of the peak) with increasing concentrations of denaturant. At 1 M GuHCl concentration, the spectral features of the sample are suggestive of complete dissociation of the FMN into solution. The same pattern is observed in the fluorescence emission from the FMN, where a steep increase in fluorescence is observed when the denaturant concentration is above 0.75 M. This is in contrast to the fluorescence emission observed in the enzyme in the absence of any denaturant, in which case efficient quenching of the FMN is observed (Hardman, Pudney, Hay, & Scrutton, 2013). The circular dichroism data show a gradual loss in the secondary structure of the enzyme as the GuHCl concentration is increased, and suggests that some secondary structure is present in the apoenzyme in 1 M GuHCl.

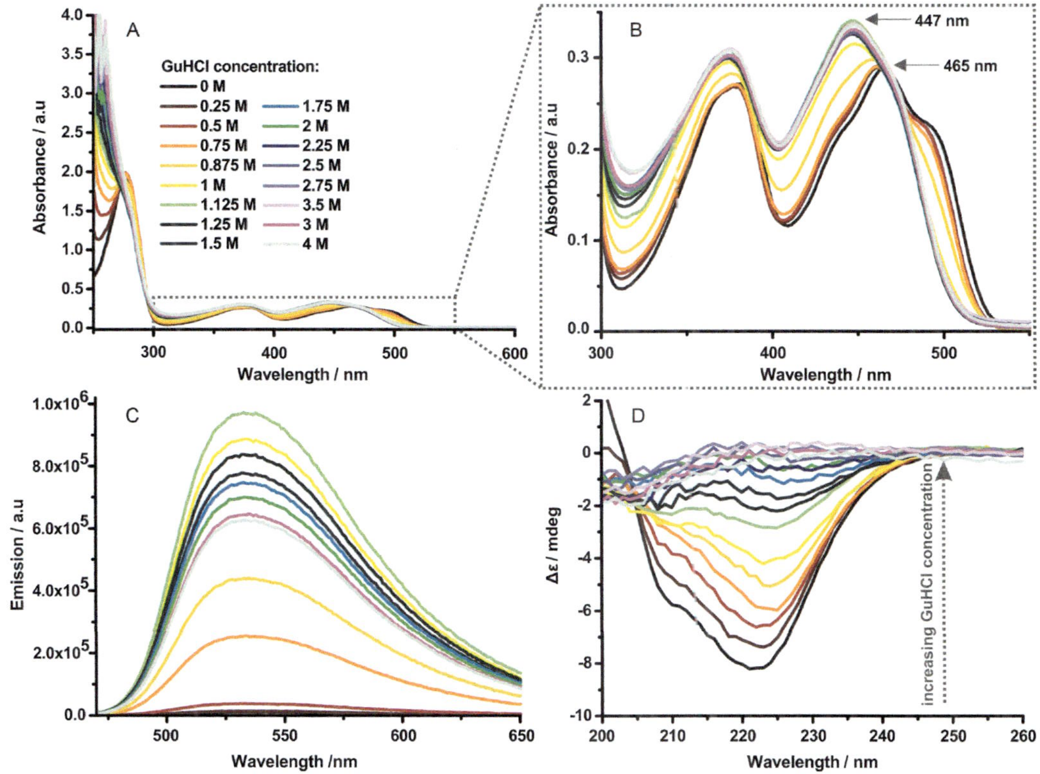

Fig. 4 Characterization of the behavior of PETNR in the presence of gradually increasing concentrations (0–4 M) of guanidine hydrochloride. The unfolding equilibrium was monitored using UV-vis absorption (A and B), fluorescence emission upon excitation at 465 nm (C) and circular dichroism spectroscopy (D). The samples for UV-vis and CD spectroscopy contained 25 μM PETNR in a solution comprising 50 mM potassium phosphate buffer, pH 7.0 and various concentrations of GuHCl. A concentration of 8.3 μM PETNR was used for fluorescence experiments to prevent inner filter effects. All spectra were recorded after 60 min of sample incubation (upon mixing the enzyme with GuHCl). UV-vis spectra were collected on a Cary 50 UV-vis spectrophotometer (Agilent Tech., UK), using a 1-cm two-window quartz cell. Fluorescence emission spectra of PETNR were recorded at 298 K using an FLS920 fluorescence spectrometer (Edinburgh Instruments, UK), with 1 nm excitation and 1 nm emission slit-widths in 1 mL quartz cells (1 cm excitation path length). Circular dichroism measurements were performed on an Applied Photophysics (Leatherhead, UK) Chirascan qCD spectrometer using a sealed cuvette with a 0.1 mm path length.

We also investigated how the samples behaved when subjected to the refolding procedure (fast dilution into native buffer) and found that the samples with concentrations of GuHCl higher than 0.875 M precipitated completely upon dilution. Together, the data indicated that complete unfolding of the enzyme is unadvisable, as rapid refolding fails when the FMN cofactor is unbound. Therefore, we made a compromise, using 0.75 M GuHCl to promote back-exchange, which leads to a partial D-H exchange (~30 backbone amides groups, mainly located in the β-sheets making up the hydrophobic core of the enzyme, could be only partially exchanged; this exchange, although not complete, affords enough NMR signal to enable complete sequential backbone assignments of PETNR). Overall, the main observation is that the loss of much of the secondary structure is concomitant with the loss of the bound FMN, which makes refolding difficult. Therefore, a successful unfolding-refolding procedure can only be performed by partially unfolding of PETNR under conditions where the FMN remains predominantly bound to the protein. Fortunately, this method enabled partial protonation of all amide groups in the enzyme (all peaks were essentially visible in the ^{1}H-^{15}N TROSY spectra of back-exchanged ^{2}H,^{13}C,^{15}N-labeled PETNR when compared to the spectrum of protonated ^{15}N-labeled PETNR; data not shown), and ultimately enabled the near-complete sequential backbone assignment of PETNR, with 333 out of 344 non-proline residues successfully assigned in the ^{1}H-^{15}N TROSY experiment (Iorgu et al., 2018a).

7. Differential labeling of PETNR: Exchanging the FMN cofactor

As mentioned in Section 2, we have recently developed a strategy to generate isotopologues of PETNR in which the protein and/or the intrinsic FMN cofactor is isotopically labeled (Longbotham et al., 2016). This was achieved by binding the His_8-tagged PETNR to a HisTrap HP nickel-charged IMAC column, previously equilibrated with 50 mM sodium acetate, pH 5.5. Next, the FMN cofactor was removed by passing over the column a solution consisting of 2 M KBr, 50 mM sodium acetate, pH 5.5 until no yellow color can be observed on the column and in the eluent. The apoenzyme (which was retained on the column) is then washed with 50 mM potassium phosphate, pH 7.0. PETNR is reconstituted with FMN by recirculating a solution of 0.1 mM FMN in 50 mM potassium phosphate, pH 7.0 through the column for several hours (until complete binding is observed). Next, the column is washed with 50 mM potassium phosphate, pH 7.0, in order to remove any unbound FMN prior to eluting the protein with 200 mM imidazole, 50 mM potassium phosphate, pH 7.0. This method enables almost complete incorporation of, e.g., an unlabeled FMN cofactor into an isotopically labeled PETNR. We also prepared samples of unlabeled PETNR where the intrinsic FMN was removed and replaced with commercially-available (unlabeled) FMN. We compared the RHR kinetics of these reconstituted samples with original samples containing the intrinsic FMN. No significant differences in coenzyme binding or hydride transfer kinetics were observed, suggesting that this approach is non-disruptive and any differences observed between protein and/or FMN isotopologues are a result of isotopic effects (Longbotham et al., 2016).

It is possible to incorporate "heavy" labeled FMN into PETNR, but first the "heavy" FMN must be obtained by removing it from a "heavy" labeled PETNR sample as described above. This "heavy" FMN can then be incorporated into another sample by performing a 3-fold dilution of the FMN-containing solution (which has 2 M KBr) with non-denaturing buffer (50 mM potassium phosphate, pH 7.0). Thus, a lower concentration of KBr will enable subsequent rebinding of FMN.

We have also recently described how this method can be used to replace the FMN cofactor with a 5-deazaflavin-mononucleotide analogue in nitric oxide synthase, a complex multi-domain enzyme containing several different cofactors (Hedison, Leferink, Hay, & Scrutton, 2016), which suggests the approach can be used on a wide variety of flavin-containing proteins. Moreover, we observed that this FMN exchange procedure also enables D-H exchange of the backbone amide groups that are normally solvent inaccessible (unpublished), providing a good alternative to the method described in Section 6 for His-tagged flavoenzymes.

8. Uses of isotopically labeled PETNR in probing biological hydride transfer mechanisms

"Heavy" isotopically labeled enzymes can be used to measure "heavy" enzyme KIEs. In this case, the KIE is simply the ratio of rate constants measured with two protein isotopologues and is expressed as KIE $= k_{\text{"light"}}/k_{\text{"heavy"}}$. "Heavy" enzyme KIEs have been measured in several enzymatic systems, including PNP (Antoniou et al., 2012; Harijan et al., 2017; Suarez & Schramm, 2015; Zoi et al., 2016), HIV-1 protease (Kipp et al., 2011;

Krzemińska et al., 2016), alanine racemase (Toney et al., 2013), dihydrofolate reductase (Francis et al., 2016; Luk et al., 2013), and PETNR. The "heavy" enzyme KIEs are relatively small and it can be challenging to measure significant KIEs. The RHR of OYEs such as PETNR are particularly amenable to characterization using anaerobic stopped-flow spectroscopy, which monitors the absorbance associated with the redox state of the bound FMN and/or the charge-transfer complex formed between the FMN and NAD(P)H in the Michaelis complex (Basran et al., 2003; Hay, Pudney, & Scrutton, 2009; Khan et al., 2002; Pudney et al., 2013, 2006; Pudney, Hay, Levy, et al., 2009; Pudney, Hay, & Scrutton, 2009). In the case of PETNR, primary and secondary deuterium KIEs can be measured using this approach with a precision of 2–3% (Pudney et al., 2006). To date, the magnitude of the "heavy" enzyme KIEs measured on PETNR (Longbotham et al., 2016; Pudney et al., 2013) and the homolog morphinone reductase (unpublished) has been typically no >5% (0.95–1.05), which makes their individual significance questionable. However, by measuring the temperature dependence of these KIEs and by measuring primary KIEs (with deuterated NAD(P)H) with "heavy" enzymes, we can also consider the "heavy" enzyme KIEs on the apparent activation enthalpy and entropy ($\Delta H^{\ddagger}$ and $\Delta S^{\ddagger}$; Fig. 5). By combining these data, we have made the case that isotopic labeling of the protein and/or FMN cofactor of PETNR leads to a measurable alteration in RHR

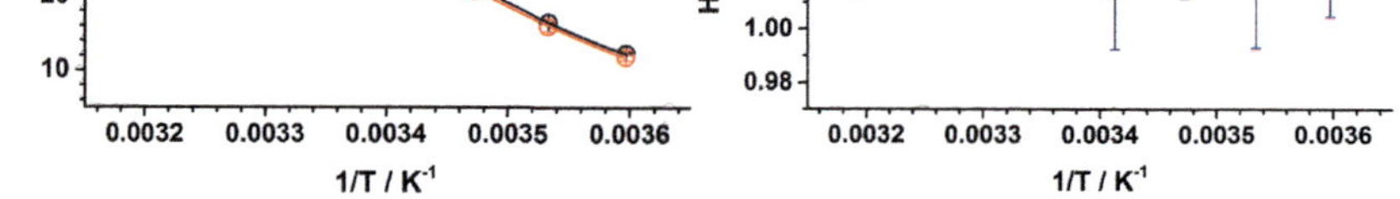

Fig. 5 "Heavy" enzyme KIEs on the RHR of PETNR with NADPH. Left, the temperature dependence of the reaction of "light" and "heavy" $^{2}H,^{13}C,^{15}N$-labeled PETNR with 5 mM NADPH fitted to the nonlinear Eyring-based equation (Arcus et al., 2016; Arcus & Pudney, 2015), yielding $\Delta H^{\ddagger}$ and $\Delta S^{\ddagger} = 6.0 \pm 0.6\,kJ\,mol^{-1}$ and $4.1 \pm 1.9\,J\,mol^{-1}\,K^{-1}$ for the "light" enzyme and $6.9 \pm 0.9\,kJ\,mol^{-1}$ and $7.8 \pm 3.2\,J\,mol^{-1}\,K^{-1}$ for the "heavy" enzyme. The heavy enzyme KIE calculated with these data is shown in the right panel. *Data are taken from Longbotham, J. E., Hardman, S. J. O., Görlich, S., Scrutton, N. S., & Hay, S. (2016). Untangling heavy protein and cofactor isotope effects on enzyme-catalyzed hydride transfer.* Journal of the American Chemical Society, 138, *13693–13699.*

kinetics, consistent with some vibrational coupling between the protein, the FMN cofactor and the reaction coordinate (Longbotham et al., 2016).

Future work could include the specific uniform labeling of specific amino acid type(s) by adding labeled amino acid(s) to the expression medium, as recently employed by Schramm et al. (Harijan, Zoi, Antoniou, Schwartz, & Schramm, 2018), or in a cell-free expression system. More sophisticated experiments could employ specific incorporation of individual isotopically labeled non-natural amino acid(s) using an orthogonal tRNA/aminoacyl-tRNA synthase pair encoded by an amber codon suppression (Venditti, Fawzi, & Clore, 2012). In both cases, there is no apparent reason why such protocols could not be carried out using flavoenzymes such as PETNR.

9. Conclusions

In summary, we have described some uses of stable isotopically labeled flavoenzymes, focusing on NMR and "heavy" enzyme KIE measurements with PETNR. Other experiments that require or benefit from such samples that are not discussed include neutron diffraction, proteomic mass spectrometry, and site-directed vibrational spectroscopy (Chahrour et al., 2015; Haertlein et al., 2016; Heinrich, 2016; Peuker et al., 2016). Detailed protocols for the preparation of isotopically labeled PETNR via recombinant expression in *E. coli* are provided with the hope they will be simply adapted to use with other flavoenzymes.

Acknowledgments

A.I.I. was an Early Stage Researcher supported by the EU's Seventh Framework Programme through the Marie Curie Initial Training Network MAGIC (Grant Agreement No. 606831). This work was also supported by the Biotechnology and Biological Sciences Research Council (BBSRC; BB/M007065/1).

References

Antoniou, D., Ge, X., Schramm, V. L., & Schwartz, S. D. (2012). Mass modulation of protein dynamics associated with barrier crossing in purine nucleoside phosphorylase. *The Journal of Physical Chemistry Letters*, *3*, 3538–3544.

Arcus, V. L., Prentice, E. J., Hobbs, J. K., Mulholland, A. J., Van Der Kamp, M. W., Pudney, C. R., et al. (2016). On the temperature dependence of enzyme-catalyzed rates. *Biochemistry*, *55*, 1681–1688.

Arcus, V. L., & Pudney, C. R. (2015). Change in heat capacity accurately predicts vibrational coupling in enzyme catalyzed reactions. *FEBS Letters*, *589*, 2200–2206.

Barna, T., Messiha, H. L., Petosa, C., Bruce, N. C., Scrutton, N. S., & Moody, P. C. E. (2002). Crystal structure of bacterial morphinone reductase and properties of the C191A mutant enzyme. *Journal of Biological Chemistry*, *277*, 30976–30983.

Basran, J., Harris, R. J., Sutcliffe, M. J., & Scrutton, N. S. (2003). H-tunneling in the multiple H-transfers of the catalytic cycle of morphinone reductase and in the reductive half-reaction of the homologous pentaerythritol tetranitrate reductase. *The Journal of Biological Chemistry*, *278*, 43973–43982.

Chahrour, O., Cobice, D., & Malone, J. (2015). Stable isotope labelling methods in mass spectrometry-based quantitative proteomics. *Journal of Pharmaceutical and Biomedical Analysis*, *113*, 2–20.

Crespi, H. L., & Katz, J. J. (1969). High resolution proton magnetic resonance studies of fully deuterated and isotope hybrid proteins. *Nature*, *224*, 560–562.

Duff, A. P., Wilde, K. L., Rekas, A., Lake, V., & Holden, P. J. (2015). Robust high-yield methodologies for2H and2H/15N/13C labeling of proteins for structural investigations using neutron scattering and NMR. *Methods in Enzymology*, *565*, 3–25.

Farmer, B. T., & Venters, R. A. (2002). NMR of perdeuterated large proteins. *Modern Techniques in Protein NMR*, *16*, 75–120.

Francis, K., Sapienza, P. J., Lee, A. L., & Kohen, A. (2016). The effect of protein mass modulation on human dihydrofolate reductase. *Biochemistry*, *55*, 1100–1106.

French, C. E., Nicklin, S., & Bruce, N. C. (1996). Sequence and properties of pentaerythritol tetranitrate reductase from Enterobacter cloacae PB2. *Journal of Bacteriology*, *178*, 6623–6627.

Fryszkowska, A., Toogood, H. S., Mansell, D., Stephens, G., Gardiner, J. M., & Scrutton, N. S. (2012). A surprising observation that oxygen can affect the product enantiopurity of an enzyme-catalysed reaction. *The FEBS Journal*, *279*, 4160–4171.

Fryszkowska, A., Toogood, H., Sakuma, M., Gardiner, J. M., Stephens, G. M., Scrutton, N. S., et al. (2009). Asymmetric reduction of activated alkenes by pentaerythritol tetranitrate reductase: Specificity and control of stereochemical outcome by reaction optimisation. *Advanced Synthesis & Catalysis*, *351*, 2976–2990.

Haertlein, M., Moulin, M., Devos, J. M., Laux, V., Dunne, O., & Trevor Forsyth, V. (2016). Biomolecular deuteration for neutron structural biology and dynamics. *Methods in Enzymology*, *566*, 113–157.

Hardman, S. J. O., Pudney, C. R., Hay, S., & Scrutton, N. S. (2013). Excited state dynamics can be used to probe donor-acceptor distances for H-tunneling reactions catalyzed by flavoproteins. *Biophysical Journal*, *105*, 2549–2558.

Harijan, R. K., Zoi, I., Antoniou, D., Schwartz, S. D., & Schramm, V. L. (2017). Catalytic-site design for inverse heavy-enzyme isotope effects in human purine nucleoside phosphorylase. *Proceedings of the National Academy of Sciences of the United States of America*, *114*, 6456–6461.

Harijan, R. K., Zoi, I., Antoniou, D., Schwartz, S. D., & Schramm, V. L. (2018). Inverse enzyme isotope effects in human purine nucleoside phosphorylase with heavy asparagine labels. *Proceedings of the National Academy of Sciences of the United States of America*, *115*, E6209–E6216.

Hay, S., Pudney, C. R., & Scrutton, N. S. (2009). Structural and mechanistic aspects of flavoproteins: Probes of hydrogen tunnelling. *The FEBS Journal*, *276*, 3930–3941.

Hedison, T. M., Leferink, N. G. H., Hay, S., & Scrutton, N. S. (2016). Correlating calmodulin landscapes with chemical catalysis in neuronal nitric oxide synthase using time-resolved FRET and a 5-deazaflavin thermodynamic trap. *ACS Catalysis*, *6*, 5170–5180.

Heinrich, F. (2016). Deuteration in biological neutron reflectometry. *Methods in Enzymology*, *566*, 211–230.

Hulley, M. E., Toogood, H. S., Fryszkowska, A., Mansell, D., Stephens, G. M., Gardiner, J. M., et al. (2010). Focused directed evolution of pentaerythritol tetranitrate reductase by using automated anaerobic kinetic screening of site-saturated libraries. *Chembiochem: A European Journal of Chemical Biology*, *11*, 2433–2447.

Iorgu, A. I., Baxter, N. J., Cliff, M. J., Waltho, J. P., Hay, S., & Scrutton, N. S. (2018a). 1H,15N and13C backbone resonance assignments of pentaerythritol tetranitrate reductase from *Enterobacter cloacae* PB2. *Biomolecular NMR Assignments*, *12*, 79–83.

Iorgu, A. I., Baxter, N. J., Cliff, M. J., Levy, C., Waltho, J. P., Hay, S., & Scrutton, N. S. (2018b). Nonequivalence of second sphere "noncatalytic" residues in pentaerythritol tetranitrate reductase in relation to local dynamics linked to H-transfer in reactions with NADH and NADPH coenzymes. *ACS Catalalysis*, *8*, 11589–11599.

Katz, J. J., & Crespi, H. L. (1966). Deuterated organisms: Cultivation and uses. *Science (New York, N.Y.)*, *151*, 1187–1194.

Khan, H., Harris, R. J., Barna, T., Craig, D. H., Bruce, N. C., Munro, A. W., et al. (2002). Kinetic and structural basis of reactivity of pentaerythritol tetranitrate reductase with NADPH, 2-cyclohexenone, nitroesters, and nitroaromatic explosives. *The Journal of Biological Chemistry*, *277*, 21906–21912.

Kholodar, S. A., Ghosh, A. K., & Kohen, A. (2017). Measurement of enzyme isotope effects. *Methods in Enzymology*, *596*, 43–83.

Kipp, D. R., Silva, R. G., & Schramm, V. L. (2011). Mass-dependent bond vibrational dynamics influence catalysis by HIV-1 protease. *Journal of the American Chemical Society*, *133*, 19358–19361.

Krzemińska, A., Moliner, V., & Świderek, K. (2016). Dynamic and electrostatic effects on the reaction catalyzed by HIV-1 protease. *Journal of the American Chemical Society*, *138*, 16283–16298.

Lin, M. T., Fukazawa, R., Miyajima-Nakano, Y., Matsushita, S., Choi, S. K., Iwasaki, T., et al. (2015). Escherichia coli auxotroph host strains for amino acid-selective isotope labeling of recombinant proteins. *Methods in Enzymology*, *565*, 45–66.

Longbotham, J. E., Hardman, S. J. O., Görlich, S., Scrutton, N. S., & Hay, S. (2016). Untangling heavy protein and cofactor isotope effects on enzyme-catalyzed hydride transfer. *Journal of the American Chemical Society*, *138*, 13693–13699.

Luk, L. Y. P., Javier Ruiz-Pernia, J., Dawson, W. M., Roca, M., Loveridge, E. J., Glowacki, D. R., et al. (2013). Unraveling the role of protein dynamics in dihydrofolate reductase catalysis. *Proceedings of the National Academy of Sciences of the United States of America*, *110*, 16344–16349.

Markley, J. L., Putter, I., & Jardetzky, O. (1968). High-resolution nuclear magnetic resonance spectra of selectively deuterated staphylococcal nuclease. *Science (New York, N.Y.)*, *161*, 1249–1251.

Michel, E., & Allain, F. H. T. (2015). Selective amino acid segmental labeling of multidomain proteins. *Methods in Enzymology*, *565*, 389–422.

Nietlispach, D., Clowes, R. T., Broadhurst, R. W., Ito, Y., Keeler, J., Kelly, M., et al. (1996). An approach to the structure determination of larger proteins using triple resonance NMR experiments in conjunction with random fractional deuteration. *Journal of the American Chemical Society*, *118*, 407–415.

Peuker, S., Andersson, H., Gustavsson, E., Maiti, K. S., Kania, R., Karim, A., et al. (2016). Efficient isotope editing of proteins for site-directed vibrational spectroscopy. *Journal of the American Chemical Society*, *138*, 2312–2318.

Pudney, C. R., Guerriero, A., Baxter, N. J., Johannissen, L. O., Waltho, J. P., Hay, S., et al. (2013). Fast protein motions are coupled to enzyme H-transfer reactions. *Journal of the American Chemical Society*, *135*, 2512–2517.

Pudney, C. R., Hay, S., Levy, C., Pang, J., Sutcliffe, M. J., Leys, D., et al. (2009). Evidence to support the hypothesis that promoting vibrations enhance the rate of an enzyme catalyzed H-tunneling reaction. *Journal of the American Chemical Society*, *131*, 17072–17073.

Pudney, C. R., Hay, S., & Scrutton, N. S. (2009). Bipartite recognition and conformational sampling mechanisms for hydride transfer from nicotinamide coenzyme to FMN in pentaerythritol tetranitrate reductase. *The FEBS Journal*, *276*, 4780–4789.

Pudney, C. R., Hay, S., Sutcliffe, M. J., & Scrutton, N. S. (2006). α-secondary isotope effects as probes of "tunneling-ready" configurations in enzymatic H-tunneling: Insight from environmentally coupled tunneling models. *Journal of the American Chemical Society*, *128*, 14053–14058.

Rosano, G. L., & Ceccarelli, E. A. (2014). Recombinant protein expression in Escherichia coli: Advances and challenges. *Frontiers in Microbiology*, *5*, 172.

Sambrook, J., Fritsch, E. F., & Maniatis, T. (1989). *Molecular cloning: A Laboratory Manual*. New York: Cold Spring Harbor Laboratory Press.

Sastry, M., Bewley, C. A., & Kwong, P. D. (2015). Effective isotope labeling of proteins in a mammalian expression system. *Methods in Enzymology*, *565*, 289–307.

Scholtissek, A., Tischler, D., Westphal, A., van Berkel, W., & Paul, C. (2017). Old yellow enzyme-catalysed asymmetric hydrogenation: Linking family roots with improved catalysis. *Catalysts*, *7*, 130.

Silva, R. G., Murkin, A. S., & Schramm, V. L. (2011). Femtosecond dynamics coupled to chemical barrier crossing in a Born-Oppenheimer enzyme. *Proceedings of the National Academy of Sciences of the United States of America*, *108*, 18661–18665.

Skora, L., Shrestha, B., & Gossert, A. D. (2015). Isotope labeling of proteins in insect cells. *Methods in Enzymology*, *565*, 245–288.

Staunton, D., Schlinkert, R., Zanetti, G., Colebrook, S. A., & Campbell, I. D. (2006). Cell-free expression and selective isotope labelling in protein NMR. *Magnetic Resonance in Chemistry*, *44*, S2–S9.

Studier, F. W. (2005). Protein production by auto-induction in high density shaking cultures. *Protein Expression and Purification*, *41*, 207–234.

Suarez, J., & Schramm, V. L. (2015). Isotope-specific and amino acid-specific heavy atom substitutions alter barrier crossing in human purine nucleoside phosphorylase. *Proceedings of the National Academy of Sciences of the United States of America*, *112*, 11247–11251.

Świderek, K., Javier Ruiz-Pernía, J., Moliner, V., & Tuñón, I. (2014). Heavy enzymes-experimental and computational insights in enzyme dynamics. *Current Opinion in Chemical Biology*, *21*, 11–18.

Takeda, M., & Kainosho, M. (2012). Cell-free protein production for NMR. *Methods in Molecular Biology*, *831*, 71–84.

Toney, M. D., Castro, J. N., & Addington, T. A. (2013). Heavy-enzyme kinetic isotope effects on proton transfer in alanine racemase. *Journal of the American Chemical Society*, *135*, 2509–2511.

Toogood, H. S., Fryszkowska, A., Hulley, M., Sakuma, M., Mansell, D., Stephens, G. M., et al. (2011). A site-saturated mutagenesis study of pentaerythritol tetranitrate reductase reveals that residues 181 and 184 influence ligand binding, stereochemistry and reactivity. *Chembiochem: A European Journal of Chemical Biology*, *12*, 738–749.

Toogood, H. S., & Scrutton, N. S. (2018). Discovery, characterization, engineering, and applications of Ene-reductases for industrial biocatalysis. *ACS Catalysis*, *8*, 3532–3549.

Venditti, V., Fawzi, N. L., & Clore, G. M. (2012). An efficient protocol for incorporation of an unnatural amino acid in perdeuterated recombinant proteins using glucose-based media. *Journal of Biomolecular NMR*, *52*, 191–195.

Wang, Z., Chang, E. P., & Schramm, V. L. (2016). Triple isotope effects support concerted hydride and proton transfer and promoting vibrations in human heart lactate dehydrogenase. *Journal of the American Chemical Society*, *138*, 15004–15010.

Williams, R. E., & Bruce, N. C. (2002). "New uses for an old enzyme"—The old yellow enzyme family of flavoenzymes. *Microbiology*, *148*, 1607–1614.

Zoi, I., Suarez, J., Antoniou, D., Cameron, S. A., Schramm, V. L., & Schwartz, S. D. (2016). Modulating enzyme catalysis through mutations designed to alter rapid protein dynamics. *Journal of the American Chemical Society*, *138*, 3403–3409.

CHAPTER SEVEN

Unraveling flavoenzyme reaction mechanisms using flavin analogues and linear free energy relationships

Christopher J. Thibodeaux[a,*], Wei-chen Chang[b], Hung-wen Liu[c]
[a]McGill University, Montréal, QC, Canada
[b]North Carolina State University, Raleigh, NC, United States
[c]University of Texas, Austin, TX, United States
*Corresponding author: e-mail address: christopher.thibodeaux@mcgill.ca

Contents

Abstract

Flavoenzymes mediate a large number of different chemical transformations that employ the flavin coenzyme for many different purposes. Flavins are commonly utilized in a variety of both 1- and 2-electron transfers and reactions involving oxygen activation. In addition, flavins have also been shown to function as acid/base catalysts or nucleophilic catalysts, to electrostatically stabilize transition states, and to serve simply as structural components in some enzymes. In all of these functions, the electronic properties of the flavin play a critical role. Studies carried out over a number of years have shown that these electronic properties (and subsequently, the catalytic properties of the flavoenzyme) can be manipulated by altering the substituents on the isoalloxazine ring system of the flavin. Here, we detail methods for the chemoenzymatic preparation and purification of flavin analogues, the reconstitution of apo-flavoenzymes with these

Methods in Enzymology, Volume 620
ISSN 0076-6879
https://doi.org/10.1016/bs.mie.2019.03.010

analogues, and the use of linear free energy relationships (LFERs) to correlate the perturbations induced by the altered substituents. Using examples from the literature, we highlight how the use of flavin analogues and LFERs can provide mechanistic insight into the transition state structures that define flavoenyzme chemical mechanisms.

Abbreviations

3,4-DOHB 3,4-dihydroxybenzoate
ATP adenosine-5′-triphosphate
CD circular dichroism
DEAE diethylaminoethyl
DMAPP dimethylallyl pyrophosphate
FAD flavin adenine dinucleotide
FMN flavin mononucleotide
HPLC high performance liquid chromatography
IDI-2 type II isopentenyl diphosphate:dimethylallyl diphosphate isomerase
IPP isopentenyl pyrophosphate
LFER linear free energy relationship
MW molecular weight
PHBH *p*-hydroxybenzoate hydroxylase
***p*-OHB** *p*-hydroxybenzoate
UV ultraviolet

1. Introduction

Understanding the catalytic versatility of flavoenzymes (Mansoorabadi, Thibodeaux, & Liu, 2007) requires an equally versatile arsenal of experimental tools. In contrast to many other coenzymes, flavins are usually bound non-covalently by flavoenzymes, often permitting the relatively facile substitution of the naturally occurring flavins (flavin adenine dinucleotide, FAD, and flavin mononucleotide, FMN) with unnatural flavin analogues. This approach has enabled researchers to probe the variable chemistry of the flavin coenzyme in great mechanistic detail. While a number of modified flavins have been chemically synthesized, those analogues substituted at the 7- and 8-position of the isoalloxazine moiety (Scheme 1) have proven to be especially useful for elucidating flavoenzyme reaction mechanisms. Through inductive effects (Edmondson & Ghisla, 1999; Hasford & Rizzo, 1998; Legrand, Gray, Cooke, & Rotello, 2003), electron donating/withdrawing substituents at the 7- and 8-position alter the electronic properties of the flavin in the vicinity

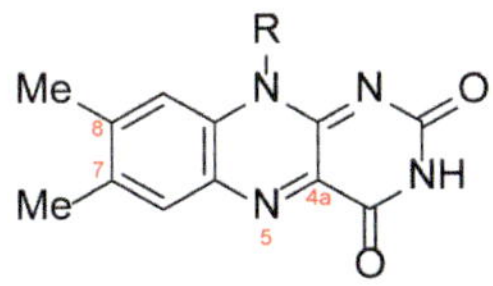

Scheme 1 Structure of the flavin coenzymes, highlighting the 7- and 8-positions that are normally exploited for linear free energy relationships, and the N5 and C4a positions that are typically involved in the chemical transformations mediate by the coenzyme. FMN = flavin monophosphate (riboflavin-5′-phosphate), FAD = flavin adenine dinucleotide.

of the N5 and C4a atoms of the coenzyme, which are typically involved directly in the chemistry steps of flavoenzyme-catalyzed reactions. Mechanistic studies employing a series of flavin analogues usually result in systematic perturbations of the spectroscopic and kinetic properties of flavoenzymes. These systematic perturbations can often be correlated with classical linear free energy relationships to elucidate details of the chemical mechanisms and transition states of flavoenzyme-catalyzed reactions.

This review will describe the systematic approach we and others have taken to prepare and utilize 7- and 8-substituted flavin analogues for the mechanistic analysis of flavoenzymes. We will first describe the procedure for the chemoenzymatic synthesis of the analogues starting from commercially available precursors. Next, we outline protocols for preparing apo-flavoenzymes, reconstituting apoprotein with flavin analogues, and performing quality control checks on the reconstituted enzymes. Finally, we discuss the theory behind linear free energy relationships (LFERs), the interpretation of the parameters derived from LFER studies, and how to apply these well-established relationships to illuminate details of flavoenzyme catalyzed reaction mechanisms. For alternative methods of flavin analogue preparation and the synthesis of other types of flavin analogues, the reader is directed to excellent earlier contributions by flavoenzyme pioneers (Edmondson & Ghisla, 1999; Ghisla & Massey, 1986; Hersh & Walsh, 1980; Murthy & Massey, 1997).

2. Chemoenzymatic synthesis of flavin analogues

2.1 Chemical synthesis of riboflavin analogues (overview)

The six riboflavin analogues depicted in Scheme 2 span a wide range of electron donating and electron withdrawing properties at the C7 or C8 positions and should provide a sufficiently broad range of properties for most

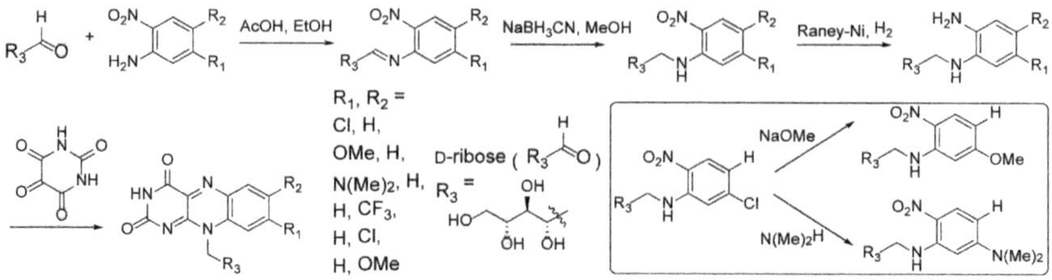

Scheme 2 Synthesis of riboflavin analogues (approach 1). Preparation of riboflavin analogues using D-ribose and substituted *o*-nitroanilines as starting materials.

applications. Indeed, we have found that this small series of analogues provided significant mechanistic insight into the type II isopentenyl diphosphate:dimethylallyl diphosphate isomerase (Thibodeaux, Chang, & Liu, 2010) and UDP-galactopyranose mutase (Sun, Ruszczycky, Chang, Thibodeaux, & Liu, 2012). If a more extensive set of 7- and 8-substituded analogues are desired, protocols for their preparation can be found elsewhere (Hasford & Rizzo, 1998; Legrand et al., 2003).

2.1.1 Chemical synthesis of riboflavin analogues (approach 1)

Following reported procedures (Hasford & Rizzo, 1998; Kasai, Fritz, & Matsui, 1987; Legrand et al., 2003), the riboflavin analogues depicted in Scheme 2 can be prepared by reacting commercially available *o*-nitroanilines with D-ribose to form the corresponding imines. The resulting crude imines are reduced to their corresponding amines under mild conditions using $NaBH_3CN$. Typically, 1.2–1.4 equivalents of *o*-nitroaniline is added to the ethanolic D-ribose solution (~0.1 M) at the room temperature and the reaction mixture is refluxed for ~48–72 h. Subsequently, the solvent is removed under vacuum. The crude mixture is then dissolved in methanol and 2.0 equivalents of $NaBH_3CN$ is added to the reaction mixture in small portions at room temperature. After stirring overnight, the reaction is quenched with acetone and subjected to silica gel column chromatography using dichloromethane containing 20% methanol as the eluent. Alternatively, recrystallization using ethanol can be applied to obtain the desired product. In the next step of the synthesis, the nitro group is converted to the corresponding analine using Raney-Ni as the catalyst under hydrogenation conditions. Briefly, the substrate is dissolved in methanol (~0.05–0.1 M) and the hydrogenation is carried out under an atmosphere of H_2 using commercially available Raney-Ni as the catalyst. After the reaction, the catalyst is removed by filtration and the filtrate is concentrated.

The crude diamine-containing product is then dissolved in glacial acetic acid (~0.2–0.3 M) and reacted with alloxan to give the riboflavin analogues. In this step, subsequent to addition of alloxan (1.1 equivalents) and boron oxide (2.0 equivalents) to the acidic solution, the reaction is brought to reflux for 3–4 h. After that, the reaction is brought back to room temperature and is allowed to stand at the same temperature overnight. During this time, the riboflavin crystals (which span a range of colors depending on the 7- and 8-substituents) will form and can be obtained by filtration. The crude riboflavin products are then subjected to recrystallization using ethanol and are ready for enzymatic conversion to the FAD and FMN levels with no further purification. Typical yields range from 15% to 25%. It is important to mention that, in our hands, direct reduction of the imine intermediate under the same conditions gives a mixture of products that cannot be easily separated by column chromatography or recrystallization. To introduce the methoxy and the *N,N*-dimethyl substituents, the halogenated intermediate (R_1 = Cl, R_2 = H) is subjected to nucleophilic aromatic substitution at the halogen *para* to the nitro group to yield the corresponding amine, which is then subjected to the same transformations discussed above to produce the final products (R_1 = OMe, R_2 = H or R_1 = N(Me)$_2$, R_2 = H).

2.1.2 Chemical synthesis of riboflavin analogues (approach 2)

In addition to the synthetic route described above, alternative approaches have been utilized to prepare riboflavin analogues bearing either regiospecifically labeled isotopes or various other functional groups starting from commercially available anilines (Mansurova, Koay, & Gartner, 2008; Mansurova, Simon, Salzmann, Marian, & Gartner, 2013; Neti & Poulter, 2016; Ortiz-Maldonado, Ballou, & Massey, 1999; Tishler, Pfister, Babson, Ladenburg, & Fleming, 1947). These approaches differ from Scheme 2 in that, subsequent to reductive amination of aniline and D-ribose, the resulting intermediate is converted to the diazo-containing product (Scheme 3). Tishler condensation with barbituric acid then yields the desired riboflavin analogues (Scheme 3, pathway *i*) (Neti & Poulter, 2016; Ortiz-Maldonado et al., 1999; Tishler et al., 1947). Alternatively, the ribitylated analines can be reacted with 6-chlorouracil, followed by acetylation of the hydroxy groups. The isoalloxazine is then constructed using the *N*-oxide as the key intermediate, which is reduced under sodium thiosulfate conditions (pathway *ii*) (Mansurova et al., 2008). In pathway *iii*, prior to reacting with 2,4,6-trichloropyrimidine, the OH groups of the ribitylated analines

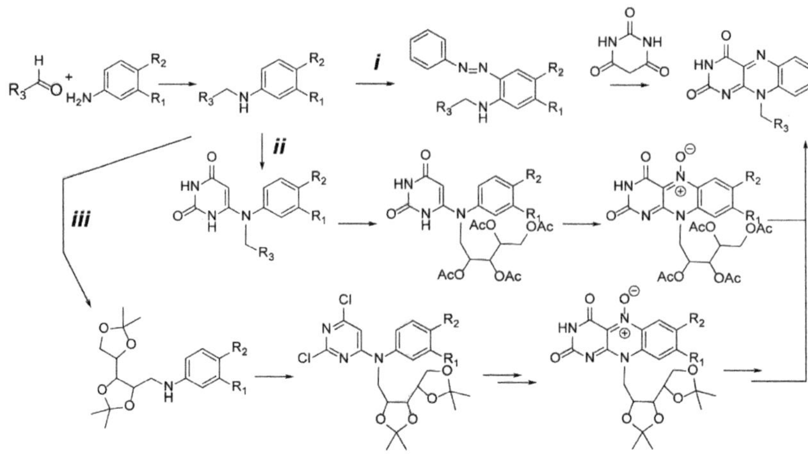

Scheme 3 Synthesis of riboflavin analogues (approach 2). Alternative approaches in the preparation of riboflavin analogues involving Tishler condensation (pathway *i*), reaction with 6-chlorouracil (pathway *ii*), or condensation with 2,4,6-trichloropyrimidine (pathway *iii*). In pathway *ii* and *iii*, the isoalloxazine was constructed using the *N*-oxide as the key intermediate, which is subsequently reduced using sodium thiosulfate.

are protected by the treatment of 2,2-dimethoxypropane. The reaction is then followed by substitution of chloride with benzyloxy groups and hydrogenation (Mansurova et al., 2013). Last, the isoalloxazine was constructed analogously as described in pathway *ii*.

2.2 Enzymatic conversion of riboflavin analogues to FAD and FMN (overview)

Following their chemical synthesis, the riboflavin analogues are enzymatically converted to their corresponding FAD forms by treatment with FAD synthetase from *Corynebacterium ammoniagenes* (Hersh & Walsh, 1980). The prokaryotic FAD synthetase is a bifunctional enzyme that catalyzes the ATP-dependent phosphorylation of riboflavin to form FMN, followed by the in situ adenylylation of the FMN intermediate using a second equivalent of ATP to form FAD (Scheme 4). The structure and mechanism of the *C. ammoniagenes* FAD synthetase have been well studied (Sebastian, Serrano, Velazquez-Campoy, & Medina, 2017), and owing to its high level expression as a His_6-tagged protein in *Escherichia coli*, this enzyme has been utilized by several labs in the preparation of FAD and FMN analogues. Moreover, it has recently been demonstrated that a construct comprising only the *C*-terminal riboflavin kinase domain of FAD synthetase is catalytically active (Herguedas et al., 2015), opening up the possibility of

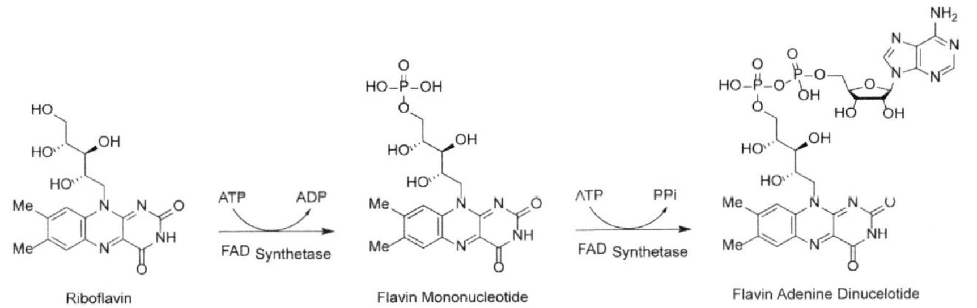

Scheme 4 Enzymatic conversion of riboflavin to FAD catalyzed by the bifunctional FAD synthetase enzyme from *Corynebacterium ammoniagenes*.

using this truncated FAD synthase variant to synthesize FMN analogues directly (i.e., without first going to the FAD level). It should also be mentioned that both riboflavin and FMN are known to inhibit FAD synthetase (Efimov, Kuusk, Zhang, & McIntire, 1998; Sebastian et al., 2017). Thus, during enzymatic preparation of FMN/FAD analogues, care should be taken to account for the inhibitory properties either by reducing the riboflavin analogue concentration as much as possible (the riboflavin K_d for FAD synthetase is ~80 nM) (Efimov et al., 1998), or by allowing sufficient time (usually 24–48 h) for the FMN/FAD analogue to accumulate in the reaction mixture. Despite this limitation, we have found that suitable quantities of the desired flavin analogue can usually be obtained with the procedure outlined below with little difficulty.

2.2.1 FAD synthetase reaction conditions

A typical 500 mL FAD synthetase reaction contains 50 μM riboflavin analogue (10–15 mg of riboflavin analogue), 1.5 mM ATP, 5 mM $MgCl_2$, and 1 μM FAD synthetase in 30 mM potassium phosphate buffer, pH 7.5. The riboflavin analogue should be added to the buffer first and this suspension should be warmed to approximately 70 °C to enhance the solubility of the riboflavin. After the majority of the riboflavin has been dissolved (which will be obvious as the color of the solution changes), lower the temperature of the reaction mixture to 37 °C before adding the ATP, $MgCl_2$, and FAD synthetase. Failure to do so may result in denaturation of the FAD synthetase. Incubate the reaction at 37 °C with shaking in the dark, or in an aluminum foil covered flask to prevent photodamage of the flavin. In general, always be sure to keep the flavin protected from light as much as possible. Depending on the exact concentrations of enzyme and flavin analogue, good conversion of the riboflavin to the FAD version usually requires about

24–48 h. If desired, the progress of the reaction can be monitored by HPLC coupled to UV-visible absorption spectroscopy. Typical HPLC conditions employed for the separation of flavin analogues are given in the following section.

2.2.2 Purification of FAD analogues

After the reaction has reached completion, filter the 500 mL reaction using a 10 kDa molecular weight cutoff membrane to remove the FAD synthetase, and apply the entire reaction mixture to a 2.5 × 25 cm DEAE sepharose anion exchange column (Cl^- form). Elute the column with three column volumes of H_2O, followed by a 1 L linear gradient of 0–0.2 M NaCl in 50 mM potassium phosphate buffer (pH 7.0). Depending on the extent of conversion, riboflavin, FMN, and FAD may all be present in the reaction mixture. The phosphate/diphosphate moieties of FMN/FAD will be ionized at pH 7.0 and will interact strongly with the anion exchange resin. In contrast, any residual riboflavin will be uncharged and should elute during the wash step or very early during the linear NaCl gradient. Combine the fractions containing FAD (based on their color), freeze and lyophilize to dryness. Following resuspension in a minimal volume of water, the FAD analogues can be purified and desalted by size exclusion chromatography on a 2.5 × 110 cm column manually packed with P2 Biogel (BioRad) using water as the eluent at a flow rate of 10 mL/h (delivered from a peristaltic pump). The experimenter should keep in mind that, for gel filtration chromatography, the sample volume loaded onto the column should ideally not exceed 1% of the total resin volume (this translates to approximately 5 mL for a 2.5 × 110 cm column). Collect the colored fraction as it elutes from the P2 column. If properly performed, the combination of DEAE and P2 chromatography will yield a desalted FAD preparation that is sufficiently pure to use for enzymatic assays. If desired, the FAD analogue can be further purified by reverse phase HPLC. For this approach, inject the concentrated, P2-purified sample into a C18 semiprep column (10 μm, 10 × 250 mm) and wash at a flow rate of 5 mL/min for 5 min with 95% solvent A (5 mM NH_4HCO_3, pH 7.0) and 5% solvent B (MeOH). Increase the concentration of MeOH to 100% over 20 min. This gradient typically provides good separation of FAD, FMN, and riboflavin, which have relative retention times of FAD < FMN < riboflavin on the C18 column. Caution should be employed because FAD is not strongly retained on C18 resin and will elute near the column dead volume.

2.2.3 Synthesis and purification of FMN analogues

If the FMN level of the flavin analogues is desired, dissolve the FAD analogues purified by P2 chromatography and/or HPLC in 150mL of 20mM potassium phosphate buffer (pH 7.6) along with 6mg of *Naja naja* snake venom phosphodiesterase (several products are available from Sigma). Incubate the reaction for 2h at room temperature with stirring. Remove the phosphodiesterase by filtering the reaction through a 10kDa MW cutoff and lyophilize the filtrate. Resuspend in 1–2mL of H_2O, and purify by P2 and/or by RP-HPLC as described above. Alternatively, as mentioned above, it should also be possible to produce the FMN analogues directly from the corresponding riboflavin using the isolated *C*-terminal riboflavin kinase domain of *C. ammoniagenes* FAD synthetase, though, to our knowledge, this has not yet been demonstrated for flavin analogues (Herguedas et al., 2015).

2.3 Preparation of apoflavoenzymes

Prior to studying the effects of flavin modification on the chemistry of flavoenzymes, the endogenous flavin must first be removed from the flavoprotein of interest. Below, we describe a simple method first developed by Vincent Massey and co-workers where the holoenzyme is dialyzed against a concentrated solution of halide ions at physiological pH (Massey & Curti, 1966; Nishina, Horiike, Shiga, Miyake, & Yamano, 1977). While this method has performed admirably for us, other flavoenzyme may require additional/more stringent conditions to remove the endogenous flavin coenzyme. For these more recalcitrant enzymes, the reader is directed to other literature that more thoroughly reviews the available methods for apoflavoprotein preparation (Hefti, Vervoort, & van Berkel, 2003; Husain & Massey, 1978). The overall goal is to remove the flavin without irreversible denaturation of the protein. For this reason, methods involving the use of a denaturing agent (such as urea or guanidinium chloride) should be avoided if possible, unless the flavoprotein of interest is known to have robust refolding kinetics. Following purification of the flavoprotein of interest, apoenzyme is prepared by dialysis of the as-purified enzyme against 6L of buffer containing 2M KBr, 10% glycerol, 50mM potassium phosphate (pH 7.5), followed by dialysis against 6L of storage buffer (50mM potassium phosphate, 10% glycerol, pH 7.5). The potassium phosphate can be replaced by other standard enzyme storage buffers. If necessary, small amounts of phosphodiesterase or phosphatase can be added to the dialysis buffer in order to degrade any FAD or FMN, respectively, that is, released from the enzyme (Muller & van Berkel, 1982). This will help to further push the flavin binding equilibrium toward the apoenzyme + free flavin state. If included, these hydrolytic enzymes need to be thoroughly removed by either centrifugal filtration or size exclusion chromatography before attempting to reconstitute the apoenzyme with flavin analogues. It is also advisable to monitor the secondary structure of the enzyme both before and after removal of the flavin using circular dichroism spectroscopy. If removal of the flavin leads to a drastic perturbation of the CD spectrum (or if protein precipitation is observed in the dialysis tubing), the apoenzyme should be reconstituted with the natural flavin and characterized using standard procedures before proceeding to reconstitution and studies with flavin analogues.

2.4 Reconstitution of apoenzyme with flavin analogues

After removal of the endogenous flavin, the apoflavoenzyme can be reconstituted individually with each flavin analogue. In many cases, reconstitution simply entails incubating the apoenzyme with the flavin for a long enough time period to ensure that the binding equilibrium has been established. For most enzymes, this is sufficiently fast and should be performed immediately before the reconstituted flavoenzyme is to be used for functional assays. In practice, one should dilute an aliquot derived from a concentrated sample of the flavin analogue (in water) into a buffered solution containing the enzyme. The FAD and FMN forms of flavins (including those shown in Scheme 2) are typically very soluble in water and stock solutions in low mM range are easily achievable. Though the exact numbers will vary depending on the enzyme and the flavin analogue, apoflavoenzymes typically bind to flavin analogues with K_d values in the ~10–100 μM range. It is also important to remember that the flavin binding affinity is usually significantly enhanced in the presence of all substrates under catalytically competent conditions (K_d values often in the 0.1–1 μM range or even lower). Thus, depending on the application, the enzyme and flavin analogue concentrations can be scaled accordingly. Enzyme kinetic assays will probably require less material, while measurements of the spectroscopic properties of the reconstituted enzyme will probably require more. To provide an approximate estimate for spectroscopic studies, the extinction coefficients across the UV-visible region of the spectrum for most of the flavin analogues utilized to date range from about 3000 to 15,000 M^{-1} cm^{-1} depending on the analogue and the oxidation state. Thus, concentrations in the 20–50 μM range are normally sufficient to obtain high-quality UV-visible absorption spectra. Prior to performing any detailed assays, the binding affinity of each flavin analogue to be used in the downstream analysis should be determined. Fortunately, the spectroscopic properties of flavins are often highly sensitive to the coenzyme environment, permitting flavin K_d values to be determined using either fluorescence or UV-visible absorption spectroscopy (Ghisla, 1980; Ghisla, Massey, Lhoste, & Mayhew, 1974; Munro & Noble, 1999). For these efforts, standard spectroscopic titration and equilibrium binding data fitting protocols should be followed. If kinetic assays are to be performed, the concentration of each flavin analogue should be varied while all other substrates are held at saturating concentrations, and the initial velocity of the reaction should be plotted vs. flavin concentration to ensure saturation.

3. Mechanistic analysis of flavoenzymes using linear free energy relationships

3.1 Overview

Once the desired set of flavin analogues is in hand and the enzyme of interest has been functionally reconstituted, linear free energy relationship (LFER) studies can be performed. Due to the dizzying array of chemical transformations catalyzed by flavoenzymes, it is impossible to outline a specific protocol for conducting LFER studies that will be generally applicable to all flavoenzymes. Instead, our discussion will focus on the origins of substituent effects and LFERs, and the physical meaning of the parameters extracted from the fit (which depends on the LFER being used). Our hope is that this discussion will assist readers in choosing the right LFER for their flavoenzyme and in interpreting the data they extract from their studies.

3.2 The origin of LFERs

The free energy change associated with a given reference reaction is shown in Eq. (1). The subscript "0" denotes a reference substituent. This equation is written for an equilibrium process, but the derivation for a kinetic process (assuming the validity of transition state theory) is identical, except that the reaction free energy term (ΔG) becomes an activation energy term ($\Delta G^{\ddagger}$) and the equilibrium constant (K) becomes a rate constant (k). When the same reaction is repeated with a reactant containing a different substituent (X), the free energy difference between the two reactions can be formulated as in Eq. (2), which illustrates that the free energy difference for the same reaction with two different substituents is directly proportional to the log ratio of the equilibrium (or rate) constants:

$$\Delta G_0 = -2.302RT\log(K_0) \tag{1}$$

$$\Delta\Delta G = \Delta G_0 - \Delta G_X = 2.302RT\log(K_X/K_0) \tag{2}$$

The effects of the *same substituents* (0 and X) on the free energy changes associated with a *different reaction of interest* can be written in an analogous manner (Eq. 3):

$$\Delta\Delta G' = \Delta G'_0 - \Delta G'_X = 2.302RT\log\left(K'_X/K'_0\right) \tag{3}$$

The relative effects of the same substituents on the two different reactions is then defined by the ratio of Eq. (3) to Eq. (2):

$$\Delta\Delta G'/\Delta\Delta G = \log\left(K'_X/K'_0\right)/\log(K_X/K_0) \tag{4}$$

It is important to note that the relationship in *Eq. (4) makes no assumptions on the nature of the reference reaction, the reaction of interest, or the structures of the reactant, products, or transition states involved*; it simply relates the relative effects of two specific substituents (0 and X) to the free energy changes associated with two specific reactions (the reference reaction and the reaction of interest). Rearrangement of Eq. (4) and definition of terms yields the more familiar form of LFERs (Eq. (5)):

$$\log K'_X = QC_X + \log K'_0 \tag{5}$$

Here, $C_X = \log(K_X/K_o)$ is termed the *substituent constant* for substituent X, where K_X and K_o are defined by a *reference reaction* (Eq. 1). As discussed below, the most important parameter for extracting mechanistic insight into the reaction of interest is the *sensitivity constant*, Q ($=\Delta\Delta G'/\Delta\Delta G$), which provides a quantitative relationship between the effects of the substituents on the reaction of interest relative to the reference reaction. *The interpretation of Q depends on the nature of the reference reaction and* how *the substituents used in the analysis affect that reference reaction.* The intercept term in Eq. (5) ($\log K'_0$) has no physical meaning and can be ignored. To apply an LFER in practice, equilibrium (or rate) constants are measured for a reaction of interest using a series of analogues with substituents "X." The logarithms of these values are then plotted vs. the substituent constants for that substituent (C_X). For enzymes, the rate constants to be correlated are typically steady state kinetic parameters (k_{cat} and k_{cat}/K_m) or observed rate constants from pre-steady state kinetic measurements. Equilibrium properties of interest could be, for example, the flavin redox potential (which is related to free energy via the Nernst equation) or a perturbation in a steady state spectroscopic signal. The data (plotted vs. C_X) are then fitted with a line to determine the slope (Q). The same data can often be correlated with different LFERs (using different substituent constants), to find the LFER that gives the best fit, or to elucidate additional mechanistic details of the reaction of interest (Ortiz-Maldonado et al., 1999; Thibodeaux et al., 2010). A more detailed explanation of substituent effects, LFERs, and their applications to understanding chemical mechanisms can be found in any physical organic chemistry textbook. We highly recommend the widely used *Modern Physical Organic Chemistry* by Anslyn and Dougherty (2006) which is very well written and contains many examples relevant to enzyme catalysis.

3.3 Hammett plots

3.3.1 Overview

To date, the most commonly employed LFERs in all of physical organic chemistry (and also in flavoenzyme chemistry) are Hammett analyses (Hammett, 1937), which broadly quantify electronic effects of substituents on reaction mechanisms. A variety of different reference reactions and substituent constants have been developed over the years to quantify different types of electronic effects (e.g., inductive effects, resonance effects, field effects, etc.) and extensive tabulations of Hammett substituent constants are available (Edmondson & Ghisla, 1999; Hansch, Leo, & Taft, 1991). The most common analysis employs the Hammett σ_x substituent constant, with its associated sensitivity constant, ρ (Eq. 6):

Hammett equation:

$$\log K'_X = \rho\sigma_x + \log K'_0 \quad (6)$$

The reference reaction used to define the Hammett σ_x values is the ionization of benzoic acid, where K_H and K_X are the ionization constants of benzoic acid and benzoic acid with substituent X, respectively (Scheme 5, Eq. 7):

$$\sigma_x = \log(K_X/K_H) \quad (7)$$

The negative charge that develops as benzoic acid is ionized cannot be delocalized by resonance into the benzene moiety; thus any electronic effects of altered benzene ring substituents (X) on the ionization constant of benzoic acid are attributable to inductive effect. Electron withdrawing

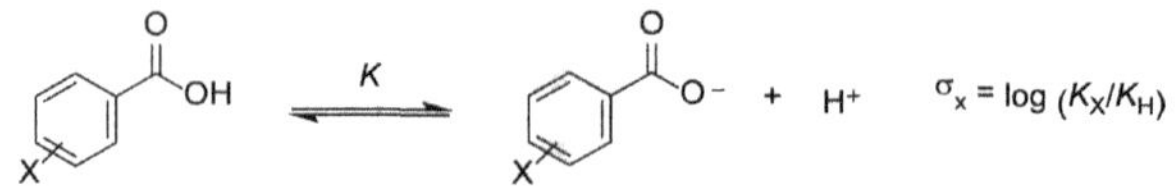

when X is an electron withdrawing group: $K_X > K_H$ and $\sigma_X > 0$
when X is an electron donating group: $K_X < K_H$ and $\sigma_X < 0$

Scheme 5 Benzoic acid ionization serves as the reference reaction for a Hammett analysis.

substituents that stabilize the negative charge on the benzoate anion lead to $K_X > K_H$, resulting in $\sigma_x > 0$. Conversely, electron donating substituents that destabilize the negative charge lead to $K_X < K_H$, resulting in $\sigma_x < 0$. A given substituent will be expected to effect the ionization constant differently depending on whether it is located in the *meta*- or *para*-position; thus, different substituent constants are needed for the *meta*- and *para*-substituents (σ_m and σ_p, respectively).

3.3.2 Interpretation of the Hammett ρ value

The Hammett equation (Eq. 6) relates the kinetic (or equilibrium) constants of the reaction of interest to the ionization of benzoic acid via the sensitivity constant, ρ. In short, *ρ quantifies the amount of charge that develops on the substituted molecule* in the product or transition state of the reaction of interest. When $\rho = 1$, the reaction of interest and the reference reaction accumulate similar amounts of negative charge (i.e., one formal negative charge). Equivalently, the reaction of interest could instead be losing one formal positive charge. In contrast, when $\rho = -1$, the reaction of interest is either gaining positive charge or losing negative charge. If the magnitude of $\rho < 1$, then either the formal charge is not fully formed in the transition state, or the charge is forming at a position in the substituted molecule that is more remote from the altered substituents than the carboxylate moiety of benzoic acid. In the latter case, the developing charge in the reaction of interest would not "feel" the inductive effect of the substituents as strongly as the developing charge in the reference reaction. In contrast, if the magnitude of $\rho > 1$, then either multiple formal charges of the same sign are forming in the reaction of interest, the charge is forming at a position in the substituted molecule that is closer to the altered substituents, or the surrounding medium somehow serves to magnify the inductive effects of the substituents relative to the reference reaction (which was carried out in water). Thus, the magnitude and sign of ρ determined from a Hammett plot provide information on the extent, location, and the type of charge accumulation in the products (if an equilibrium process is being studied) or in the transition state (if a kinetic process is being studied) of the reactant bearing the altered substituents.

3.3.3 Application of Hammett plots to study flavoenzymes

The Hammett σ_x substituent constant has been found to linearly correlate the kinetic, redox, and spectroscopic properties of a number of different flavoenzymes when the flavin is replaced with analogues substituted at the 7- and 8-position of the isoalloxazine (Abramovitz & Massey, 1976; Stewart & Massey, 1985; Sun et al., 2012; Thibodeaux et al., 2010; Yorita, Misaki, Palfey, & Massey, 2000; Yu et al., 2010). Without any further analysis, these empirical observations suggest that the 7- and 8-substituents affect flavoenzyme electronic properties largely via inductive effects, and that any charges that may accumulate over the reactive moieties of the flavin (N1, N5, and C4a) are—as expected—not strongly delocalized into the substituted benzene ring of the isoalloxazine. The 7- and 8-substituents can be considered to be *meta*- and *para*-, respectively, to the reactive N5 position of the flavin. This allows the corresponding σ_m and σ_p substituent constants (Hansch et al., 1991) to be used to correlate the flavoenzyme data using either Eq. (8) (the general case) or Eq. (9) if σ_m and σ_p contribute equally to ρ:

$$\log K_X = \sigma_m \rho_m + \sigma_p \rho_p + \log K_0 \quad (8)$$

$$\log K_X = (\sigma_m + \sigma_p)\rho + \log K_0 \quad (9)$$

In practice, LFER data should be fit with both equations and an analysis of variance (ANOVA) using a one-tailed *F*-test should be performed to determine if the inclusion of the extra parameter in Eq. (8) significantly improves the fit relative to Eq. (9). A nice discussion for how to perform this statistical test in the context of a flavoenzyme LFER is provided in the supporting information of our study on UDP-galactopyranose mutase (Sun et al., 2012). While most flavoenzyme LFERs performed to date can be adequately fit with Eq. (9), Edmonson and Ghisla have shown that the redox potentials of flavin analogues in solution are more accurately correlated with non-equivalent ρ_m and ρ_p (Eq. 8) (Edmondson & Ghisla, 1999).

3.3.4 Application of Hammett plots to study IDI-2

The type II isopentenyl diphosphate-dimethylallyl diphosphate isomerase (IDI-2) catalyzes a redox neutral 1,3-proton addition/elimination reaction

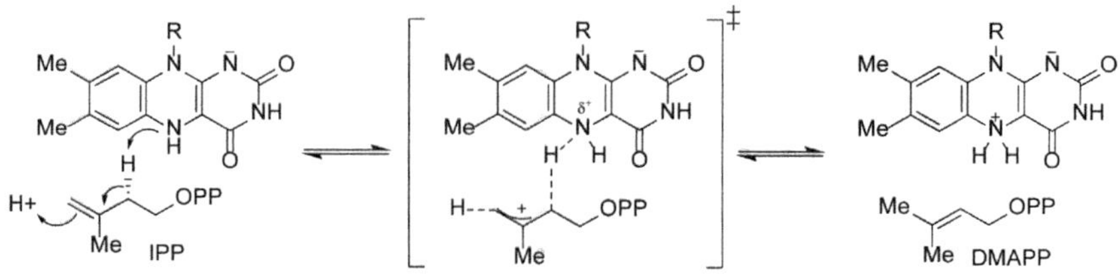

Scheme 6 Putative chemical mechanism for IDI-2, based on results obtained from a Hammett analysis of the steady state kinetic parameters. IPP = isopentenyl pyrophosphate, DMAPP = dimethylallyl pyrophosphate, PP = pyrophosphate. The large negative magnitude of ρ and relatively small substrate deuterium kinetic isotope effects suggest a late transition state, where the flavin N5 carries nearly a full positive charge.

(Scheme 6). Key mechanistic studies indicated that this enzyme requires reduced FMN for catalysis and that the substrate proton is migrating in the transition state that limits steady state turnover (Kaneda, Kuzuyama, Takagi, Hayakawa, & Seto, 2001; Thibodeaux, Mansoorabadi, Kittleman, Chang, & Liu, 2008). To investigate the potential role of the reduced flavin in the proton migration, a Hammett analysis was performed with 7- and 8-substituted FMN analogues (Thibodeaux et al., 2010). In this analysis, the steady state kinetic parameters (k_{cat} and k_{cat}/K_m) were measured with each flavin analogue and the data were correlated with the sum of $\sigma_m + \sigma_p$. Fitting of the LFER data to Eq. (9) yielded ρ values of -2.1 ± 0.3 for k_{cat} and -2.1 ± 0.5 for k_{cat}/K_m. These results provided several very crucial pieces of information regarding the chemical mechanism of IDI-2 (Thibodeaux et al., 2010). First, the results provided direct evidence for the involvement of the flavin in the rate limiting step(s), which had been shown previously to involve migration of the *pro R* proton of IPP (Thibodeaux et al., 2008). Second, the negative value of ρ suggested that the reduced flavin was either losing negative charge or gaining positive charge in the transition state. Finally, the large magnitude of ρ suggested that this charge accumulation/loss was occurring very close to the substituted benzyl moiety, such that the inductive effects of the substituents were being very strongly exerted. Solvent exclusion from the active site and/or slight changes in the conformation of the flavin upon substrate binding could also be helping to magnify the inductive effects of the substituents (Janczak & Poulter, 2016). Altogether, the LFER data pointed to a mechanism involving proton transfer from the substrate to the N5 atom of the flavin in the rate limiting transition state, and strongly supported a novel role for the flavin as a catalytic base. The large negative magnitude of ρ supports a late transition state for this proton transfer with substantial bonding of the labile proton to the flavin N5 atom (the putative site of charge accumulation). The late transition state is also consistent with the modest primary kinetic isotope effects measured for the proton transfer (Thibodeaux et al., 2008).

3.3.5 Application of Hammett plots to study UDP-galactopyranose mutase

UDP-galactopyranose mutase (UGM) catalyzes the reversible, redox neutral interconversion of UDP-galactopyranose and UDP-galactofuranose (Scheme 7). A variety of biochemical and mechanistic studies indicate that UGM cleaves the bond between the anomeric carbon and the phosphate oxygen of the substrate during turnover (Barlow, Girvin, & Blanchard, 1999) and employs a reduced FAD coenzyme to mediate the transformation via an iminium intermediate that is covalently linked to the N5 atom of the flavin (Mehra-Chaudhary, Dai, Sobrado, & Tanner, 2016; Soltero-Higgin, Carlson, Gruber, & Kiessling, 2004). To more closely investigate the role of the flavin in the chemical mechanism of UGM, the steady state turnover number (k_{cat}) was measured in the presence of several 7- and 8-substituted

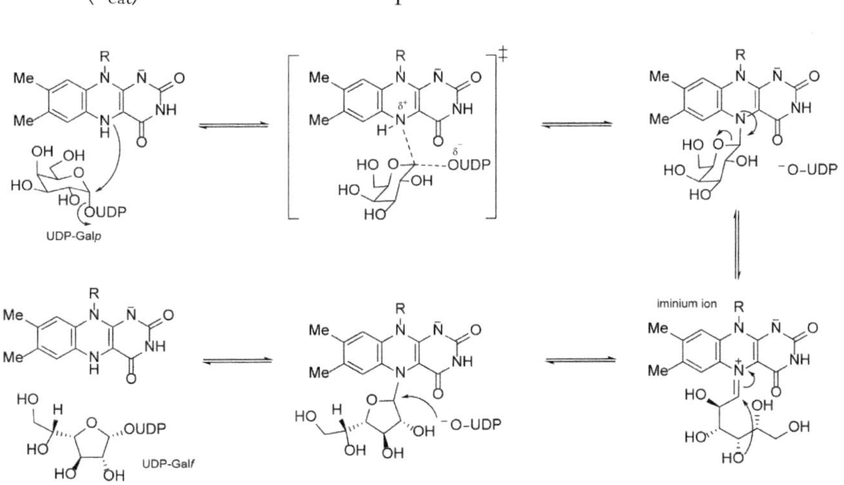

Scheme 7 Putative chemical mechanism for isomerization of UDP-Gal*p* and UDP-Gal*f* catalyzed by UDP-galactopyranose mutase (UGM). Positional isotope exchange experiments using reduced 5-deaza-FAD suggest that catalysis is initiated by direct nucleophilic addition of the flavin N5 onto the anomeric carbon of the substrate in an S_N2-type mechanism. This observation is supported by kinetic studies with 7- and 8-substituted FAD analogues, where Hammett plots of k_{cat} yielded a sensitivity constant (ρ) of −2.4, suggesting accumulation of positive charge on the flavin N5 in the rate determining step. The subsequent formation of the iminium ion may also contribute to k_{cat} under steady state turnover conditions and, hence, to the magnitude of the sensitivity constant, ρ.

FAD analogues (Sun et al., 2012). As with the IDI-2 enzyme discussed above, the rate constants were then correlated with the sum of the Hammett $\sigma_m + \sigma_p$ substituent constants, yielding a ρ value of -2.4 ± 0.4. The large negative ρ value supports a mechanism involving accumulation of positive charge on the flavin N5 atom. This result is consistent with a rate determining step involving an S_N2-type attack of the flavin N5 atom on the anomeric carbon of the substrate to trigger cleavage of the C1—O bond. Deprotonation of this initial flavin-substrate adduct and formation of the iminium ion trapped by Kiessling and co-workers also involve accumulation of positive charge on the flavin N5. Thus, iminium ion formation may also contribute to the large negative ρ value for k_{cat} if this step contributes to steady state rate limitation. The S_N2 mechanism for flavin N5-alkylation is also supported by the absence of C—O bond cleavage as measured by positional isotope exchange (PIX) experiments when reduced 5-deaza-FAD was used as the coenzyme (Sun et al., 2012), presumably because the nucleophile (i.e., the flavin N5 atom) is absent.

3.4 Brønsted plots

3.4.1 Overview

Brønsted plots are extremely useful LFERs that correlate various types of reactions to acidity and basicity. Most often, the reactions of interest studied with Brønsted plots involve general acid/base catalysis, nucleophilic addition, or the loss of leaving groups in the rate limiting step. The substituent constant used in Brønsted plots is the pK_a of the of the acid/base group that is involved in the reaction. Thus, for these LFERs, the reference reaction is the deprotonation of the acid/conjugate acid by water. The LFER for nucleophilicity (Eq. 10) has the sensitivity constant β_{Nuc} which has a positive slope. As the nucleophile becomes more basic (higher pK_a), it also becomes a better nucleophile:

$$\log k_x = \beta_{Nuc} pK_a + C \tag{10}$$

In contrast, the LFER for nucleofugality (Eq. 11) has the sensitivity constant β_{LG} and a negative slope, because as the leaving group gets more basic, the group is more difficult to eliminate and tends to stay bonded:

$$\log k_x = \beta_{LG} pK_a + C \tag{11}$$

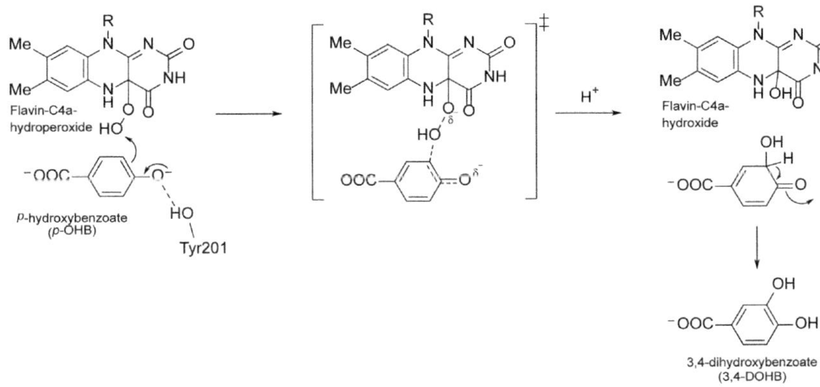

Scheme 8 Putative chemical mechanism for hydroxylation of *p*-OHB catalyzed by *para*-hydroxybenzoate hydroxylase (PHBH). Kinetic studies using a series of 8-substituted FAD analogues revealed a linear correlation between the rate of *p*-OHB hydroxylation and the pK_a of the flavin-C4-hydroxide leaving group. The value of the sensitivity constant determined from this analysis ($\beta_{LG} = -0.42$) is consistent with partial O—O bond cleavage in the transition state and an electrophilic aromatic substitution mechanism.

3.4.2 Application of the β_{LG} substituent constant to study p-hydroxybenzoate hydroxylase

p-Hydroxybenzoate hydroxylase (PHBH) is a flavin monooxygenase that catalyzes the hydroxylation of *p*-hydroxybenzoate (*p*-OHB) to form 3,4-dihydroxybenzoate (3,4-DOHB, Scheme 8). To clarify the mechanism of oxygen transfer from the flavin-C4a-hydroperoxy intermediate to the substrate, Ortiz-Maldonado et al. studied the effects of a series of 8-substituted FAD analogues on multiple steps in the PHBP reaction cycle (Ortiz-Maldonado et al., 1999). Using pK_a values calculated for the FAD-C4a-OH leaving group of each FAD analogue (Bruice, 1983), a β_{LG} value of -0.42 was obtained for the rate constant of the substrate hydroxylation step. This value indicates partial cleavage of the O—O bond of the FAD-C4a-OOH intermediate in the transition state, an observation that is consistent with the moderate electrophilicity of FAD-C4a-OOH and a chemical mechanism for *p*-OHB oxidation involving electrophilic aromatic substitution. The kinetic data for the hydroxylation step could also be correlated with a Hammett plot using a substituent constant (σ_I) that accounts for polarization of the O—O bond. The value of the sensitivity constant determined from this plot was also consistent with the electrophilic aromatic substitution mechanism shown in Scheme 8.

4. Summary and conclusion

The catalytic versatility of flavins has long been acknowledged and appreciated by mechanistic enzymologists. The facile substitution of the endogenous flavin with synthetic flavin analogues in many enzymes combined with linear free energy relationship analyses has provided a rare opportunity in enzyme catalysis to precisely manipulate the electronic properties of this functionally diverse coenzyme. Cumulatively, these studies have provided unprecedented insight into flavoenzyme transition state structures and reaction mechanisms. As new examples of unusual flavoenzyme catalyzed reactions continue to emerge (Adak & Begley, 2016, 2017; Piano, Palfey, & Mattevi, 2017; Teufel et al., 2013, 2015; White et al., 2015), the use of flavin analogues and LFERs is expected to continue to make a major contribution to our understanding of flavoenzyme chemistry.

References

Abramovitz, A. S., & Massey, V. (1976). Interaction of phenols with old yellow enzyme. Physical evidence for charge-transfer complexes. *The Journal of Biological Chemistry*, *251*(17), 5327–5336.

Adak, S., & Begley, T. P. (2016). Dibenzothiophene catabolism proceeds via a flavin-N5-oxide intermediate. *Journal of the American Chemical Society*, *138*(20), 6424–6426. https://doi.org/10.1021/jacs.6b00583.

Adak, S., & Begley, T. P. (2017). RutA-catalyzed oxidative cleavage of the uracil amide involves formation of a flavin-N5-oxide. *Biochemistry*, *56*(29), 3708–3709. https://doi.org/10.1021/acs.biochem.7b00493.

Anslyn, E. V., & Dougherty, D. A. (2006). *Modern physical organic chemistry*. Sausalito, CA: University Science Books.

Barlow, J. N., Girvin, M. E., & Blanchard, J. S. (1999). Positional isotope exchange catalyzed by UDP-galactopyranose mutase. *Journal of the American Chemical Society*, *121*, 6968–6969.

Bruice, T. C. (1983). Leaving group tendencies and the rates of mono-oxygen donation by hydrogen peroxide, organic hydroperoxides, and peroxycarboxylic acids. *Journal of the Chemical Society, Chemical Communications*, 14–15.

Edmondson, D., & Ghisla, S. (1999). Flavoenzyme structure and function. Approaches using flavin analogues. *Methods in Molecular Biology*, *131*, 157–179. https://doi.org/10.1385/1-59259-266-X:157.

Efimov, I., Kuusk, V., Zhang, X., & McIntire, W. S. (1998). Proposed steady-state kinetic mechanism for Corynebacterium ammoniagenes FAD synthetase produced by Escherichia coli. *Biochemistry*, *37*(27), 9716–9723. https://doi.org/10.1021/bi972817j.

Ghisla, S. (1980). Fluorescence and optical characteristics of reduced flavins and flavoproteins. *Methods in Enzymology*, *66*, 360–373.

Ghisla, S., & Massey, V. (1986). New flavins for old: Artificial flavins as active site probes of flavoproteins. *The Biochemical Journal*, *239*, 1–12.

Ghisla, S., Massey, V., Lhoste, J.-M., & Mayhew, S. G. (1974). Fluorescence and optical characteristics of reduced flavines and flavoproteins. *Biochemistry*, *13*(3), 589–597.

Hammett, L. P. (1937). The effect of structure upon the reactions of organic compounds. Benzene derivatives. *Journal of the American Chemical Society*, *59*, 96–103.

Hansch, C., Leo, A., & Taft, R. W. (1991). A survey of Hammett substituent constants and resonance and field parameters. *Chemical Reviews*, *91*, 165–195.

Hasford, J. J., & Rizzo, C. J. (1998). Linear free energy substituent effect in flavin redox chemistry. *Journal of the American Chemical Society*, *120*, 2251–2255.

Hefti, M. H., Vervoort, J., & van Berkel, W. J. (2003). Deflavination and reconstitution of flavoproteins. *European Journal of Biochemistry*, *270*(21), 4227–4242.

Herguedas, B., Lans, I., Sebastian, M., Hermoso, J. A., Martinez-Julvez, M., & Medina, M. (2015). Structural insights into the synthesis of FMN in prokaryotic organisms. *Acta Crystallographica. Section D, Biological Crystallography*, *71*, 2526–2542. https://doi.org/10.1107/S1399004715019641. Pt 12.

Hersh, L. B., & Walsh, C. (1980). Preparation, characterization, and coenzymic properties of 5-carba-5-deaza and 1-carba-1-deaza analogues of riboflavin, FMN, and FAD. *Methods in Enzymology*, *66*, 277–287.

Husain, M., & Massey, V. (1978). Reversible resolution of flavoproteins into apoproteins and fee flavins. *Methods in Enzymology*, *53*, 429–437.

Janczak, M. W., & Poulter, C. D. (2016). Kinetic and binding studies of Streptococcus pneumoniae type 2 isopentenyl diphosphate: Dimethylallyl diphosphate isomerase. *Biochemistry*, *55*(15), 2260–2268. https://doi.org/10.1021/acs.biochem.6b00087.

Kaneda, K., Kuzuyama, T., Takagi, M., Hayakawa, Y., & Seto, H. (2001). An unusual isopentenyl diphosphate isomerase found in the mevalonate pathway gene cluster from Streptomyces sp. strain CL190. *Proceedings of the National Academy of Sciences of the United States of America*, *98*(3), 932–937. https://doi.org/10.1073/pnas.020472198.

Kasai, S., Fritz, B. J., & Matsui, K. (1987). An improved synthesis of 8-amino-8-demethylriboflavin. *Bulletin of the Chemical Society of Japan*, *60*, 3041–3042.

Legrand, Y. M., Gray, M., Cooke, G., & Rotello, V. M. (2003). Model systems for flavoenzyme activity: Relationships between cofactor structure, binding and redox properties. *Journal of the American Chemical Society*, *125*(51), 15789–15795. https://doi.org/10.1021/ja036940b.

Mansoorabadi, S. O., Thibodeaux, C. J., & Liu, H. W. (2007). The diverse roles of flavin coenzymes-nature's most versatile thespians. *The Journal of Organic Chemistry*, *72*(17), 6329–6342. https://doi.org/10.1021/Jo0703092.

Mansurova, M., Koay, M. S., & Gartner, W. (2008). Synthesis and electrochemical properties of structurally modified flavin compounds. *European Journal of Organic Chemistry*, 5401–5406.

Mansurova, M., Simon, J., Salzmann, S., Marian, C. M., & Gartner, W. (2013). Spectroscopic and theoretical study on electronically modified chromophores in LOV domains: 8-Bromo- and 8-trifluoromethyl-substituted flavins. *Chembiochem*, *14*(5), 645–654. https://doi.org/10.1002/cbic.201200670.

Massey, V., & Curti, B. (1966). A new method of preparation of D-amino acid oxidase apoprotein and a conformational change after its combination with flavin adenine dinucleotide. *The Journal of Biological Chemistry*, *241*(14), 3417–3423.

Mehra-Chaudhary, R., Dai, Y., Sobrado, P., & Tanner, J. J. (2016). In crystallo capture of a covalent intermediate in the UDP-galactopyranose mutase reaction. *Biochemistry*, *55*(6), 833–836. https://doi.org/10.1021/acs.biochem.6b00035.

Muller, F., & van Berkel, W. J. (1982). A study on p-hydroxybenzoate hydroxylase from Pseudomonas fluorescens. A convenient method of preparation and some properties of the apoenzyme. *European Journal of Biochemistry*, *128*(1), 21–27.

Munro, A. W., & Noble, M. A. (1999). Fluorescence analysis of flavoproteins. *Methods in Molecular Biology*, *131*, 25–48. https://doi.org/10.1385/1-59259-266-X:25.

Murthy, Y. V., & Massey, V. (1997). Syntheses and applications of flavin analogs as active site probes for flavoproteins. *Methods in Enzymology*, *280*, 436–460.

Neti, S. S., & Poulter, C. D. (2016). Site-selective synthesis of (15)N- and (13)C-enriched flavin mononucleotide coenzyme isotopologues. *The Journal of Organic Chemistry*, *81*(12), 5087–5092. https://doi.org/10.1021/acs.joc.6b00640.

Nishina, Y., Horiike, K., Shiga, K., Miyake, Y., & Yamano, T. (1977). Effect of halide anions on the binding of FAD to D-amino acid oxidase and the tryptophanyl fluorescence of the apoenzyme. *Journal of Biochemistry, 81*(5), 1455–1463.

Ortiz-Maldonado, M., Ballou, D. P., & Massey, V. (1999). Use of free energy relationships to probe the individual steps of hydroxylation of p-hydroxybenzoate hydroxylase: Studies with a series of 8-substituted flavins. *Biochemistry, 38*(25), 8124–8137. https://doi.org/10.1021/bi990560e.

Piano, V., Palfey, B. A., & Mattevi, A. (2017). Flavins as covalent catalysts: New mechanisms emerge. *Trends in Biochemical Sciences, 42*(6), 457–469. https://doi.org/10.1016/j.tibs.2017.02.005.

Sebastian, M., Serrano, A., Velazquez-Campoy, A., & Medina, M. (2017). Kinetics and thermodynamics of the protein-ligand interactions in the riboflavin kinase activity of the FAD synthetase from Corynebacterium ammoniagenes. *Scientific Reports*, 7(1), 7281. https://doi.org/10.1038/s41598-017-07875-5.

Soltero-Higgin, M., Carlson, E. E., Gruber, T. D., & Kiessling, L. L. (2004). A unique catalytic mechanism for UDP-galactopyranose mutase. *Nature Structural & Molecular Biology, 11*, 539–543.

Stewart, R. C., & Massey, V. (1985). Potentiometric studies of native and flavin-substituted Old Yellow Enzyme. *The Journal of Biological Chemistry, 260*(25), 13639–13647.

Sun, H. G., Ruszczycky, M. W., Chang, W. C., Thibodeaux, C. J., & Liu, H. W. (2012). Nucleophilic participation of reduced flavin coenzyme in mechanism of UDP-galactopyranose mutase. *The Journal of Biological Chemistry, 287*(7), 4602–4608. https://doi.org/10.1074/jbc.M111.312538.

Teufel, R., Miyanaga, A., Michaudel, Q., Stull, F., Louie, G., Noel, J. P., et al. (2013). Flavin-mediated dual oxidation controls an enzymatic Favorskii-type rearrangement. *Nature, 503*(7477), 552–556. https://doi.org/10.1038/nature12643.

Teufel, R., Stull, F., Meehan, M. J., Michaudel, Q., Dorrestein, P. C., Palfey, B., et al. (2015). Biochemical establishment and characterization of EncM's flavin-N5-oxide cofactor. *Journal of the American Chemical Society, 137*(25), 8078–8085. https://doi.org/10.1021/jacs.5b03983.

Thibodeaux, C. J., Chang, W. C., & Liu, H. W. (2010). Linear free energy relationships demonstrate a catalytic role for the flavin mononucleotide coenzyme of the type II isopentenyl diphosphate:dimethylallyl diphosphate isomerase. *Journal of the American Chemical Society, 132*(29), 9994–9996. https://doi.org/10.1021/Ja104090m.

Thibodeaux, C. J., Mansoorabadi, S. O., Kittleman, W., Chang, W.-c., & Liu, H.-w. (2008). Evidence for the involvement of acid/base chemistry in the reaction catalyzed by the type II isopentenyl diphosphate/dimethylallyl diphosphate isomerase from *Staphylococcus aureus*. *Biochemistry, 47*, 2547–2558.

Tishler, M., Pfister, K., Babson, R. D., Ladenburg, K., & Fleming, A. J. (1947). The reaction between ortho-aminoazo compounds and barbituric acid—A new synthesis of riboflavin. *Journal of the American Chemical Society, 69*, 1487–1492.

White, M. D., Payne, K. A., Fisher, K., Marshall, S. A., Parker, D., Rattray, N. J., et al. (2015). UbiX is a flavin prenyltransferase required for bacterial ubiquinone biosynthesis. *Nature, 522*(7557), 502–506. https://doi.org/10.1038/nature14559.

Yorita, K., Misaki, H., Palfey, B. A., & Massey, V. (2000). On the interpretation of quantitative structure-function activity relationship data for lactate oxidase. *Proceedings of the National Academy of Sciences of the United States of America, 97*(6), 2480–2485. https://doi.org/10.1073/pnas.040559797.

Yu, Q., Schaub, P., Ghisla, S., Al-Babili, S., Krieger-Liszkay, A., & Beyer, P. (2010). The lycopene cyclase CrtY from *Pantoea ananatis* (formerly *Erwinia uredovora*) catalyzes an FADred-dependent non-redox reaction. *The Journal of Biological Chemistry, 285*, 12109–12120.

CHAPTER EIGHT

Vibrational spectroscopy of flavoproteins

James N. Iuliano[a], **Jarrod B. French**[a,b], **Peter J. Tonge**[a,c,d,*]
[a]Department of Chemistry, Stony Brook University, Stony Brook, NY, United States
[b]Department of Biochemistry and Cell Biology, Stony Brook University, Stony Brook, NY, United States
[c]Department of Radiology, Stony Brook University, Stony Brook, NY, United States
[d]Center for Advanced Study of Drug Action, Stony Brook University, Stony Brook, NY, United States
*Corresponding author: e-mail address: peter.tonge@stonybrook.edu

Contents

Abstract

The flavin cofactor performs many functions in the cell based on the ability of the isoalloxazine ring to undergo one- or two-electron reduction and form covalent adducts with reactants such as amino acids. In addition, the strong visible absorption of the cofactor is also the basis for flavin-dependent photoreceptors. Vibrational spectroscopy is uniquely suited to studying the mechanism of flavoproteins since the frequency of the vibrational modes is very sensitive to the electronic structure and environment of the isoalloxazine ring. This chapter describes the mechanistic information that can be gained using vibrational spectroscopy as well experimental challenges and approaches that are used to obtain and interpret the complex data contained in a vibrational spectrum.

Methods in Enzymology, Volume 620
ISSN 0076-6879
https://doi.org/10.1016/bs.mie.2019.03.011

1. Introduction

1.1 Vibrational spectroscopy: Basic theory and instrumentation

There are many texts that describe the fundamental principles of vibrational spectroscopy (Colthup, Daly, & Wiberley, 1975). Here we give only an outline of the basic concepts in order to provide a foundation for the application of vibrational spectroscopy to the study of flavoproteins. In vibrational spectroscopy, incident electromagnetic radiation exchanges energy with a molecule that results in a change in vibrational energy. Therefore, vibrational spectroscopy is a method for detecting and studying the vibrational frequencies in a molecule. Vibrations arise through motions of the atoms in the molecule and can be broadly categorized into stretching, bending, rocking, twisting, and wagging motions. These motions have different frequencies in the vibrational spectrum and their exact position is very sensitive to both the structure and environment of the molecule. Based on classical theory, in which nuclei are represented as points with mass connected to each other by springs, a molecule has 3N-6 (3N-5 for a linear molecule) normal modes of vibration. In addition, due to anharmonicity, additional bands such as overtones and combinations can also be present, increasing the complexity of the vibrational spectrum. Thus, even a small molecule has many vibrational bands, and the challenge with vibrational spectroscopy is to obtain the vibrational spectrum of only that part of the molecule that is of interest—for example, the flavin chromophore in a flavoprotein.

Vibrational spectra are commonly obtained using two primary methods: infrared spectroscopy (IR) and Raman spectroscopy. IR spectroscopy involves the absorption of infrared light by a molecule and vibrational modes will appear in the IR spectrum provided that there is a change in dipole moment during the vibration. In contrast, Raman spectroscopy utilizes a monochromatic light source such as a laser, which can be of any wavelength but is usually a visible frequency (~400–750 nm). In Raman spectroscopy the incident light is scattered by the sample leading to an exchange of energy if there is a change in polarizability during the bond vibration. The vibrational spectrum is then given by the difference in frequency between the incident light and Raman scattered light. If the excitation frequency is coincident with an electron transition, the vibrational bands associated with the electron transition can be greatly enhanced in intensity (10^4–10^6)

in a process known as resonance Raman spectroscopy. Because IR and Raman spectroscopy operate via different selection rules, bands that are observed using one technique may be weak or absent in the spectrum obtained by the other technique and vice versa. Thus, the two methods are complimentary. Table 1 summarizes some of the advantages and limitations of each technique. Finally, vibrational spectroscopy is a very fast technique so that the vibrational spectrum is a snapshot of all species that are present in solution. Consequently, there is no conformational averaging as can occur in NMR spectroscopy. In addition, the timescale of vibrational spectroscopy also means that very fast processes can be monitored such as activation of a photoreceptor on the picosecond timescale (Wilbrandt, Pagsberg, Hansen, & Weisberg, 1975).

Infrared spectra are normally acquired using a Fourier transform infrared (FTIR) spectrometer (Fig. 1), in which a polychromatic light source is passed through an interferometer (commonly a Michelson interferometer) which separates wavelengths of light in time. The beam is passed through the sample and transmitted light is analyzed using a detector. The signal from the detector is Fourier-transformed to give an absorption spectrum.

Table 1 Techniques for measuring vibrational spectra.

	Advantages	**Disadvantages**
Infrared	• Information from both chromophore and protein can be obtained • Quantitative: Band intensity is proportional to concentration • Accurate subtractions using fixed pathlength cells	• Strong water absorption • Additional steps are needed to reveal bands from the region of the molecule that are of interest such as isotope editing or light minus dark subtraction • High concentration, ~1 mM required in most cases
Raman	• Observation of chromophore modes without interference from protein modes • The Raman spectrum of water is relatively weak	• Band intensity is dependent on polarizability and cannot be used to determine sample concentration
Resonance Raman	• Strong enhancement of chromophore modes enabling lower concentrations to be used	• Sample fluorescence can obscure the Raman spectrum • Light absorption can lead to the photochemical generation of new species and/or sample damage

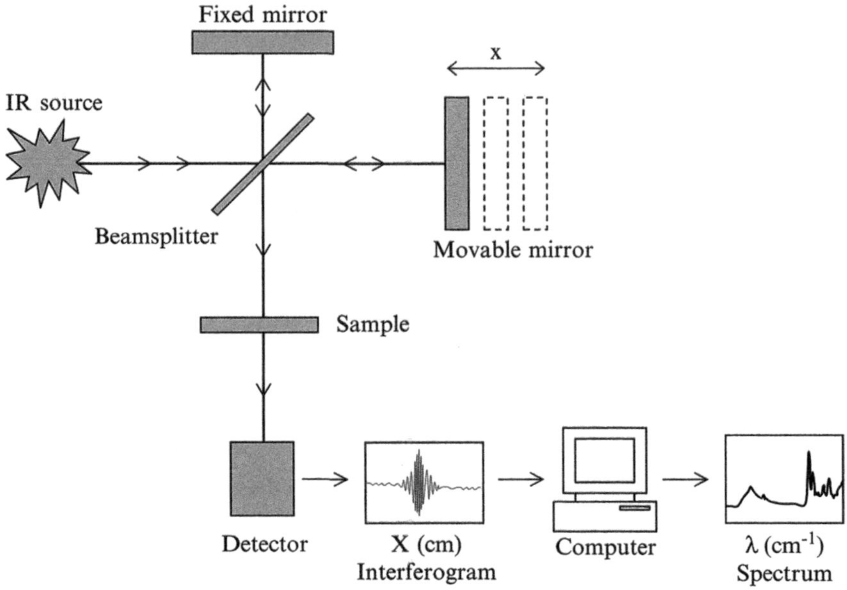

Fig. 1 Schematic Layout of a FTIR Spectrometer. FTIR spectrometers use an interferometer which creates an interferogram from an IR source. After passing through the sample the interferogram is Fourier transformed to give the IR spectrum (Ojeda & Dittrich, 2012).

The sensitivity of commercial FTIR spectrometers can be improved by incorporating a liquid nitrogen cooled HgCdTe (MCT) detector. An advantage of FTIR spectroscopy is that the method is quantitative, so that concentrations can be determined provided that the extinction coefficient and pathlength of the sample is known.

Raman spectroscopy is a scattering technique in which light from a monochromatic source such as a visible laser is focused into the sample (Fig. 2). Scattered light is then collected at 90° to the incident light and focused into a spectrograph in which a grating is used to disperse light onto a CCD detector. A notch filter with near zero transmission at the laser wavelength is used in the collection arm of the optical path to remove scattered light with the same frequency as the excitation source (Rayleigh light). Raman spectroscopy often uses red excitation sources (e.g., 785 nm) to minimize sample fluorescence, an approach that has been aided by the development of high quantum efficiency detectors such as a back-thinned deep-depletion CCD for NIR Raman spectroscopy (Rudik, Bell, Tonge, & Thorpe, 2000).

Each method has its strengths and limitations (Table 1). FTIR spectroscopy is normally performed with high sample concentrations of 1–2 mM. In addition, water has a strong IR spectrum and samples are often prepared

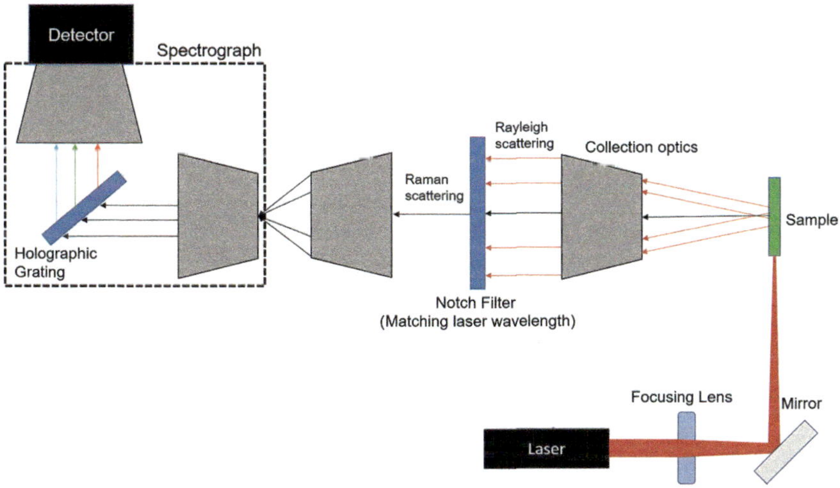

Fig. 2 Schematic Layout of a Raman spectrometer. Scattered light is collected at 90° to the excitation axis and focused onto the entrance slit of a spectrometer equipped with a holographic transmission grating. The grating disperses the spectrum across the exit slit where it is analyzed using a CCD camera. The use of a single grating spectrometer is made possible by a notch filter which efficiently rejects the Rayleigh scattered light.

in D_2O to shift the H—O—H bending vibration from 1650 to 1200 cm^{-1} (D-O-D) thus giving a "window" in the spectrum where important bands from the solute (flavoprotein) can be observed. The vibrational spectrum between 1700 and 1300 cm^{-1} is dominated by vibrational modes associated with the amide backbone of the protein, the amide I, II and III bands, which can be used to estimate the secondary structure content of the protein. However, additional steps are needed to visualize bands from the region of interest such as isotope editing (Deng, Vedad, Desamero, & Callender, 2017; Dong et al., 2001) or, in the case of light-activated proteins, using light minus dark difference spectroscopy. In this regard the use of a fixed pathlength cell enables accurate difference spectra to be obtained.

In contrast to IR spectroscopy, the relationship between Raman band intensity and sample concentration is more complex since the intensity of Raman scattering is dependent on polarizability. Thus, the peak intensity cannot be directly related to concentration. However, the Raman spectrum of water is relatively weak, so that spectra in both H_2O and D_2O can be readily obtained. In addition, Raman spectroscopy has a major advantage if the portion of the structure that is of interest is a chromophore such as the flavin. While resonance Raman spectroscopy can be deployed in this situation, using an excitation frequency that is in resonance with the electronic transition of the chromophore, as noted above this can lead to unwanted fluorescence and photochemistry (Tonge, Pusztai, White, Wharton, & Carey, 1991). Instead, red excitation can be used (e.g., 785 nm). Although the excitation frequency lies outside the electronic transition, the conjugated π electron system of the chromophore is much more polarizable than the protein and hence vibrational bands associated with the chromophore are much more intense than those associated with protein modes (Carey & Tonge, 1995).

1.2 The vibrational spectrum of the isoalloxazine ring

The chemistry of FMN and FAD is associated with the isoalloxazine ring which can exist in several different ionization/redox states (Fig. 3). Since vibrational band frequencies are very sensitive to structure and environment, vibrational spectroscopy is a powerful method for analyzing the interactions within the flavoprotein binding site that modulate the chemistry of the isoalloxazine ring. The flavin cofactor has several distinct vibrational modes that can be detected in both IR and Raman spectra, and the FTIR spectrum of FMN dissolved in D_2O is shown in Fig. 4. The C2=O (blue) and C4=O (red) carbonyl groups (Fig. 3) have relatively intense IR absorbances at ~1660–1680 and ~1690–1710 cm^{-1}, respectively. The carbonyl frequencies are highly sensitive to hydrogen bonding from solvent or protein amino acids, which is especially useful in determining structural rearrangements that are

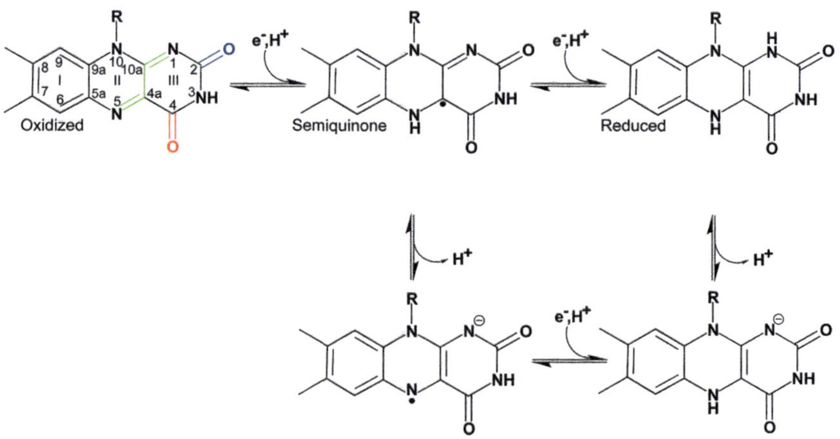

Fig. 3 Ionization and redox states of the isoalloxazine ring. The numbering of the isoalloxazine ring is shown for the oxidized form. *Figure adapted from Conrad, K. S., Manahan, C. C., & Crane, B. R. (2014). Photochemistry of flavoprotein light sensors.* Nature Chemical Biology, *10(10), 801–809.*

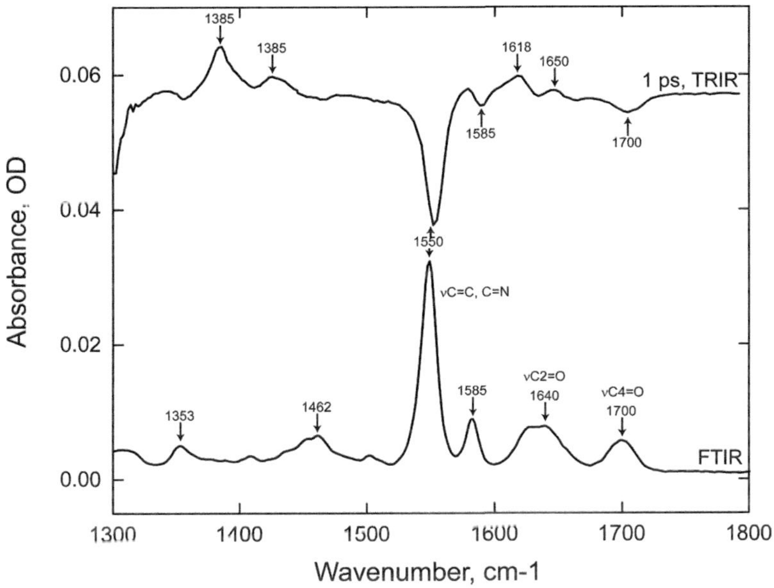

Fig. 4 IR spectra of FMN. The FTIR spectrum of FMN in D_2O (2 mM) after subtraction of the solvent spectrum is shown at the bottom. The light minus dark time-resolved IR (TRIR) spectrum of FMN in D_2O obtained at a delay time of 1 ps after excitation at 450 nm is shown in the upper trace. Negative bands (bleaches) are vibrational modes depopulated following absorption of light and are thus associated with the ground state of the isoalloxazine chromophore. Positive bands are associated with the FMN excited state or species formed by photoexcitation.

associated with biological activity. Another useful vibrational mode arises from motions of N5=C4a—C10a=N1 (green) and is found at $\sim$1550 cm^{-1} in both IR and Raman spectra. This delocalized mode is sensitive to several structural modifications of the isoalloxazine ring including protonation of N5 or N1 upon reduction to the semiquinone and/or covalent adduct formation at C4a during catalysis (Piano, Palfey, & Mattevi, 2017). Normal modes have been assigned for the oxidized, semiquinone and reduced forms of the isoalloxazine ring and a selection of these assignments are given in Table 2.

2. Applications of vibrational spectroscopy to flavoproteins

This section will outline the various methods that have been used to obtain vibrational spectra of flavoproteins. This includes Raman and resonance Raman spectroscopy as well as FTIR spectroscopy and is

Table 2 Selected normal mode assignments of the isoalloxazine ring.

Oxidized (Bowman & Spiro, 1981)		Semiquinone (Sugiyama, Nisimoto, Mason, & Loehr, 1985)		Reduced (Zheng, Carey, & Palfey, 2004)	
1713	C4=O	1532	C4a-N5, C10a-N1	1714	C2=O, N1—H, N3-H
1645	C2=O, C10a=N1	1388	C2-N1	1638	C4=O, N1—H, N5-H
1615	C4=O, C2=O, C10a=N	1268	Ring II and III	1615	C4a=C10a
1511	C4a—N5, N10—C10a, C10a—N1, C4a—C10a	1227	Ring II and III	1560	Ring I, N5-H
1451	C4a—N5, C10a—N1			1539	N1—H, C4a=C10a
1394	N1—C2				
1379	N10—C10a				
1269	N3—H				

followed by a brief introduction to more recent approaches such as time-resolved infrared spectroscopy (TRIR), two-dimensional (2D) IR spectroscopy, and femtosecond stimulated Raman spectroscopy (FSRS). Some examples of flavoproteins that have been studied using vibrational spectroscopy are given in Table 3.

2.1 Raman and resonance Raman spectroscopy of flavoproteins

Many early vibrational spectroscopy experiments on biological systems used resonance Raman (RR) spectroscopy to take advantage of the dramatic increase in Raman band intensity when the laser excitation frequency was in resonance with a chromophore in the sample. Although this was an ideal approach to selectively observe vibrational modes associated with the chromophore, RR spectroscopy was often hampered by strong fluorescence from the chromophore which obscured the Raman spectrum. The flavin chromophore was no exception and initial RR spectra of flavoproteins acquired by excitation into the 450 nm electronic transition were

Table 3 Flavoproteins that have been studied using vibrational spectroscopy.

Protein	**Key results**
D-Amino Acid Oxidase (DAAO)	• Assignment of reduced flavin spectrum, role of C4=O in catalysis (Nishina, Sato, Miura, Matsui, & Shiga, 1998)
Medium-chain Acyl-CoA Dehydrogenase (MCAD)	• Effect of ligand binding and redox state on C2=O and C4=O vibrations (Nishina et al., 2007)
Glucose Oxidase (GOX)	• TRIR of FAD radicals, assignment of excited state spectra (Lukacs et al., 2012)
Photolyase	• Assignment of the resonance Raman spectra of flavin semiquinone (Schelvis et al., 2003)
Light, oxygen, voltage (LOV)	• Observation of Cys Thiol-FMN C4a adduct formation from a reactive triplet state (Swartz, Wenzel, Corchnoy, Briggs, & Bogomolni, 2002)
Blue-light using flavin (BLUF)	• Hydrogen bond network rearrangement causes perturbations in carbonyl modes upon blue light activation (Masuda, Hasegawa, Ishii, & Ono, 2004)

severely impacted by the intrinsic fluorescence of the isoalloxazine ring (Benecky et al., 1979). Several approaches were taken in an attempt to minimize sample fluorescence including the use of UV excitation (Copeland & Spiro, 1986), and coherent anti-Stokes Raman spectroscopy (CARS) (Dutta & Nestor, 1977). CARS avoids fluorescence since anti-Stokes Raman scattering is at higher frequency than the excitation frequency whereas fluorescence emission occurs at lower frequency than the absorbed light with the caveat that anti-Stokes scattering is much lower in intensity than Stokes scattered light. Dutta et al. used CARS to obtain the RR spectrum of FAD in solution and bound to glucose oxidase (GOX) and were able to assign several vibrational modes from the isoalloxazine ring (Dutta & Nestor, 1977). The authors were also able to determine the frequency shifts associated with binding of free FAD to GOX and to estimate the hydrogen deuterium exchange (HDX) rate in deuterated buffer.

Nishina and coworkers subsequently used RR spectroscopy to measure the vibrational spectra of riboflavin and several derivatives bound to riboflavin binding protein (RBP) isolated from egg white (Nishina et al., 1978). In this approach, fluorescence was avoided due to the quenching effect of RBP on riboflavin fluorescence. These studies were followed by more sophisticated assignment of spectra using isotope labeling, and it was determined that ring III of the isoalloxazine ring is involved in interactions with RBP (Kitagawa et al., 1979; Nishina, Shiga, et al., 1980). Investigation of the charge-transfer state of old yellow enzyme using RR spectroscopy yielded crucial information about flavoenzyme catalysis and it was suggested, based on resonance enhancement of modes assigned to N5 and C4a, that nucleophilic attack would most likely occur at these two atoms (Nishina, Kitagawa, Shiga, Watari, & Yamano, 1980).

Kim and Carey extended the work on riboflavin and RBP using red excitation at 647.1 nm to obtain the off resonance Raman spectrum and an instrument based on a holographic notch filter, transmission grating spectrometer and a liquid nitrogen cooled red optimized CCD camera (Kim, Owen, & Carey, 1993). Even though the excitation frequency was not in resonance with the chromophore, the Raman spectrum of riboflavin bound to RBP was still dominated by vibrational bands arising from the isoalloxazine ring due to pre-resonance enhancement arising from the polarizability of the chromophore. In addition, off-resonance excitation minimized fluorescence from the sample. Raman spectra of riboflavin in solution and bound to RBP enabled resolution of a carbonyl feature assigned to C4=O in the protein that was previously unresolvable due to low sensitivity (Fig. 5) (Kim & Carey, 1993).

A similar approach using 752 nm excitation was used to study solvent access to the bound flavin in *p*-hydroxybenzoate hydroxylase (PHBH) and, in combination with DFT calculations, led to the assignment of frequency shifts resulting from solvent exposure and an estimate of the relative populations of the "in" and "out" conformations of the flavin (Zheng, Dong, Palfey, & Carey, 1999). These studies ultimately revealed that the flavin carbonyl vibrations, specifically that arising from the C4=O group at 1712 cm^{-1}, are highly sensitive and undergo shifts of >10 cm^{-1} upon binding to the protein.

More recently, Schelvis and coworkers reported RR spectra of the photolyase enzyme from *Escherichia coli*, which binds the neutral semiquinone form of FAD, and GOX, in which FAD can be converted from the oxidized form to $FADH^{\bullet}$ and $FAD^{\bullet-}$ (Schelvis, Pun, Goyal, & Sokolova, 2006; Schelvis et al., 2003). Both studies yielded information on the hydrogen

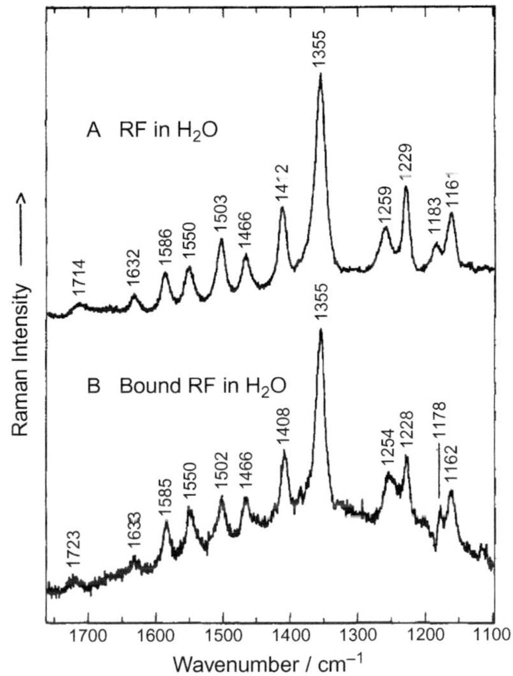

Fig. 5 Raman spectra of riboflavin in solution and bound to RBP. Advances in optics and detector technology enabled the observation of the Raman spectrum of bound riboflavin at 200 μM concentration. *Figure taken from Kim, M., & Carey, P. R. (1993). Observation of a carbonyl feature for riboflavin bound to riboflavin-binding protein in the red-excited raman spectrum.* Journal of the American Chemical Society, 115*(15), 7015–7016.*

bonding environment of the flavin in the protein. The semiquinone form of FAD does not have strong fluorescence enabling subtraction of the weak fluorescence background.

2.2 FTIR spectroscopy of flavoenzymes and light-activated proteins

In contrast to Raman and RR spectroscopy, the IR spectrum of a chromophore bound to a protein cannot be selectively enhanced by resonance or pre-resonance effects. Instead the IR spectrum of the chromophore comprises only a small fraction of the overall IR spectrum of the protein spectrum. Thus, IR spectroscopy of proteins primarily depends on difference methods in which the IR spectrum of the protein is subtracted from a spectrum of the protein after the structure of the chromophore has been perturbed or after a ligand has bound to the protein. This approach is greatly facilitated by Fourier transform IR spectrometers which can acquire and average IR spectra very rapidly with high sensitivity. In flavoproteins, IR difference spectroscopy has enabled the observation of vibrational modes that have been perturbed due to either redox changes or photoconversion. This approach was first applied to cytochrome *c* oxidase where the IR spectrum of the protein was subtracted from a second spectrum acquired after irradiation with a tungsten lamp to generate a light adapted state (Alben, Moh, Fiamingo, & Altschuld, 1981). The difference spectrum (ΔA) is calculated using the following equation:

$$\Delta A = -\log\left(\frac{T_{light}}{T_{dark}}\right) = A_{light} - A_{dark}$$

where T_{light} is the transmitted IR during or post irradiation and T_{dark} is the transmitted IR pre-irradiation. This is the classic light minus dark (L-D) FTIR spectrum.

FTIR difference spectroscopy has been applied to both light activated flavoproteins and flavoenzymes. Swartz and coworkers used L-D FTIR spectroscopy to study light state formation in the Light Oxygen Voltage (LOV) domain of oat phototropin, and determined that Cys-adduct formation from a reactive triplet state perturbs the N5=C4a vibrational mode (Salomon, Christie, Knieb, Lempert, & Briggs, 2000; Swartz et al., 2002). Additionally, L-D FTIR spectroscopy was also applied to the Blue Light Utilizing FAD (BLUF) photoreceptors, in which photoexcitation leads to a rearrangement in a hydrogen bonding network that surrounds the FAD (Hasegawa, Masuda, & Ono, 2005). From these studies it was determined that the change in hydrogen bonding of the FAD C4=O correlates with the observed red-shift in the visible absorbance spectrum of the FAD transition at 450 nm.

Difference FTIR is not restricted to light-activated proteins and has been applied to studying changes in the redox state of the flavin. A spectroelectrochemical cell, which consisted of a gold electrode attached to CaF_2 windows, was developed for the study of redox sensitive proteins in solution (Moss, Nabedryk, Breton, & Mäntele, 1990). This cell was used to measure the reduced form of FAD in solution and FAD bound to pyruvate oxidase (POX), glucose oxidase (GOX), and D-amino acid oxidase (DAAO) where it was found that POX-bound FAD adopts a bent conformation over the entire isoalloxazine ring in the reduced state (Fig. 6) (Wille et al., 2003).

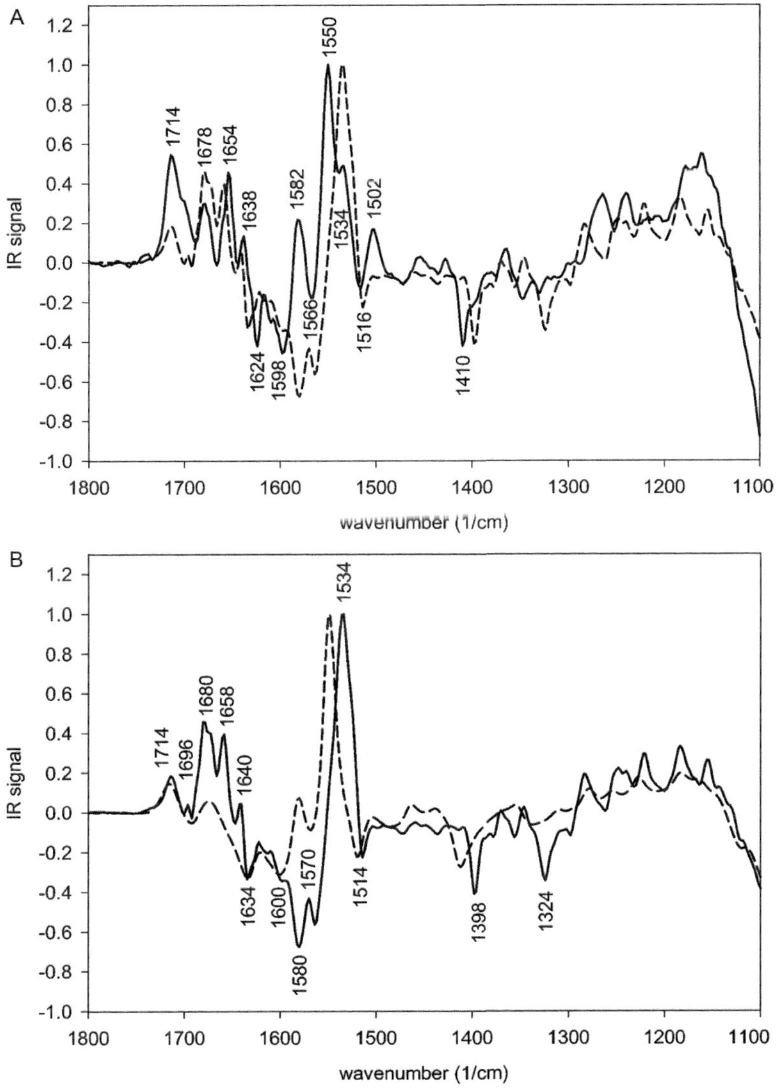

Fig. 6 Redox triggered difference FTIR spectra of (A) GOX and (B) POX. The spectrum of FAD in solution is overlaid on each protein spectrum in dashed lines. The frequency of the strong transient at $\sim$1550 cm^{-1} is sensitive to differences in solvent and protein interactions. *Figure taken from Moss, D., Nabedryk, E., Breton, J., & Mäntele, W. (1990). Redox-linked conformational changes in proteins detected by a combination of infrared spectroscopy and protein electrochemistry. Evaluation of the technique with cytochrome c.* European Journal of Biochemistry, 187*(3), 565–572.*

2.3 Time-resolved infrared (TRIR) spectroscopy of flavoprotein photoreceptors

TRIR combines IR difference spectroscopy with the high time-resolution achievable in vibrational spectroscopy. Most techniques employ a pump-probe approach in which a visible or infrared pump is used to initiate photochemistry or a temperature change in the sample. TRIR spectroscopy was used to study oxidized FAD and the lumiflavin anion in solution by Kondo et al. (2006)), and then extended to the BLUF protein AppA in which the effect of the protein matrix on the FAD vibrational modes was determined and markers for light-state formation were assigned (Stelling, Ronayne, Nappa, Tonge, & Meech, 2007). TRIR also enables the kinetics of excited state decay and ground state recovery as well as the formation and decay of intermediates to be observed, and TRIR datasets are commonly analyzed using global fitting software to extract spectra associated with each component of the fit (Snellenburg, Laptenok, Seger, Mullen, & van Stokkum, 2012). This approach allowed both excited state and ground state vibrational modes to be assigned in combination with DFT calculations. A complete assignment of the TRIR spectrum of AppA was achieved using isotope labeling and reconstituting FAD isotopologues into the protein (Haigney et al., 2011).

TRIR can also be used to characterize the reduced and semi-reduced states and intermediates of flavin in protein and in solution. Reduced FAD was studied by adding sodium dithionite or sodium oxalate to the buffered solution and irradiating the FAD sample for 30 min under anaerobic conditions prior to measurement (Zhao et al., 2011). This work was extended to GOX, in which photoexcitation results in the formation of the anionic radical $FAD^{\bullet -}$ and the corresponding Trp^{+} which decay through internal conversion on the picosecond timescale (Lukacs et al., 2012). In the BLUF protein PixD it was shown that TRIR can separate proton coupled electron transfer (PCET) into distinct steps in which the initial formation of $FAD^{\bullet -}$ is then followed by a proton transfer to yield $FADH^{\bullet}$ (Gil, Laptenok, Iuliano, et al., 2017) (Fig. 7). In the case of the LOV domain protein family, it has been shown that the excited state spectra are sensitive to small changes in hydrogen bonding around the chromophore despite the highly conserved binding pocket (Iuliano et al., 2018).

2.4 Two-dimensional infrared spectroscopy

Multi-dimensional spectroscopy has revolutionized the field of NMR, in which 2D techniques generate high resolution structural information

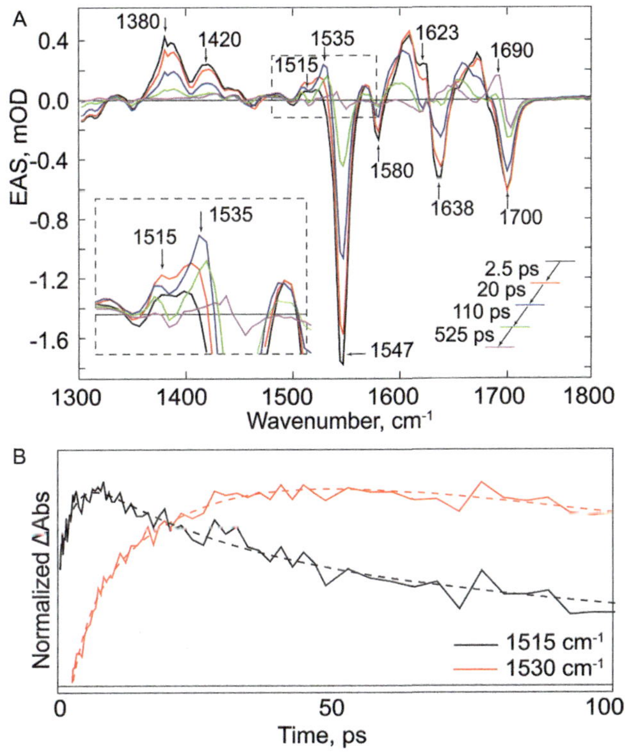

Fig. 7 TRIR of the BLUF protein PixD. (A) EADS derived from global fitting of TRIR data reveals intermediates assigned to $FAD^{\bullet -}$ (1515 cm^{-1}) and $FADH^{\bullet}$ (1535 cm^{-1}). (B) The change in intensity as a function of time is plotted for the two species showing that $FADH^{\bullet}$ (1530 cm^{-1}) forms as $FAD^{\bullet -}$ (1515 cm^{-1}) decays. *Figure adapted from Gil, A. A., Laptenok, S. P., French, J. B., Iuliano, J. N., Lukacs, A., Hall, C. R., et al. (2017). Femtosecond to millisecond dynamics of light induced allostery in the avena sativa LOV domain.* The Journal of Physical Chemistry B, 121*(5), 1010–1019; Gil, A. A., Laptenok, S. P., Iuliano, J. N., Lukacs, A., Verma, A., Hall, C. R., et al. (2017). Photoactivation of the BLUF protein PixD probed by the site-specific incorporation of fluorotyrosine residues.* Journal of the American Chemical Society, 139*(41), 14638–14648.*

in complex protein systems. In two-dimensional infrared spectroscopy (2DIR), coupling between vibrational modes can be observed on any timescale with high time-resolution. This provides information on bond environment, distance, and bond angle through electrostatic and hydrogen bonding interactions which manifest as off diagonal cross peaks in the 2D spectrum (Hamm & Zanni, 2011). 2DIR has already been applied to several enzyme-substrate systems and advances in laser technology and IR detectors have advanced to a point where lower, more physiological-relevant protein concentrations can be studied (Bagchi, Boxer, & Fayer, 2012; Pagano, Guo, Kohen, & Cheatum, 2016). Although 2DIR has so far not been used to study flavoproteins, El Khoury and coworkers have applied 2DIR to the isolated flavin in solution. In this study a spectroelectrochemical cell was used to measure the 2DIR spectra of oxidized and reduced FAD in solution where it was found that reduced FAD exhibits stronger carbonyl coupling to the isoalloxazine ring modes (El Khoury, Van Wilderen, & Bredenbeck, 2015). These developments in 2DIR and its application to FAD in solution hold promise for the extension of this method to the flavin chromophore in more complex systems such as flavoproteins.

2.5 Femtosecond stimulated Raman spectroscopy

Femtosecond stimulated Raman spectroscopy (FSRS) utilizes the high time and spectral resolution of vibrational spectroscopy to analyze excited state chromophore modes. This is achieved by using several laser pulses: the first is an actinic or visible pulse usually $<100\,\text{fs}$, the second is a Raman pulse overlapping with an electronic transition in the chromophore (the excited state absorbance of the flavin is $\sim 700\,\text{nm}$), and a white-light probe pulse to measure the Raman scattering on top of the transient absorption generated by the actinic pulse (Kukura, McCamant, & Mathies, 2007). FSRS was first applied to flavins by Weigel and coworkers where they assigned excited state spectra using DFT calculations and observed vibrational modes sensitive to deuteration (Weigel et al., 2011). Shortly after, Hall and coworkers used FSRS to study the flavin bound to the BLUF domain of AppA in both the dark and photoexcited light states, establishing that FSRS can be used to study ultrafast chromophore dynamics in proteins (Hall et al., 2017).

3. Methods to interpret vibrational spectra

This section summarizes experimental and computational methods that aid in the interpretation of vibrational spectroscopy data. Isotope labeling is the gold standard for the assignment of vibrational bands and can be applied to either the flavin itself or to the protein. In addition, computational methods such as density functional theory (DFT) are used for the isolated chromophore while QM/MM methods are used for the protein-bound flavin.

3.1 Selective isotope labeling

The synthetic method developed by Tishler has been exploited to develop selectively labeled flavins that can be incorporated into proteins, replacing the natural flavin chromophore (Bacher, 1986; Tishler et al., 1947). Recently, Illarionov and coworkers developed a biosynthetic method for converting purine bases to FMN (Illarionov et al., 2015). This approach was exploited in the assignment of the TRIR spectrum of AppA. For example, reconstitution of FAD labeled with carbon-13 at either the C2=O ([2-$^{13}C_1$]-FAD) or C4=O/C10a ([4,10a-$^{13}C_2$]-FAD) into the BLUF domain of AppA allowed unambiguous assignment of a bleach at 1700 cm^{-1} to C4=O and a bleach at 1663 cm^{-1} to C2=O (Haigney et al., 2011). This approach was also used in the study of the BLUF protein BlsA (Brust et al., 2013) and the LOV protein AsLOV2 (Fig. 8) (Gil, Laptenok, French, et al., 2017). In addition to AsLOV2 reconstituted with site-specifically labeled FAD isotopologues, AsLOV2 was also expressed in ^{13}C-glucose enriched media in a bacterial strain lacking the ability to synthesize flavin. The isotope-labeled apo protein was subsequently reconstituted with unlabeled FAD, enabling assignment of modes arising from the protein (Gil, Laptenok, French, et al., 2017).

3.2 Computational methods

In addition to isotope labeling, vibrational spectra calculations play a key role in vibrational band assignments. Density functional theory (DFT) is a common quantum mechanical method implemented in software packages such as Gaussian that can be used to estimate the vibrational spectra of small organic molecules such as the isoalloxazine ring. DFT calculates the electronic structure which is then used to calculate other properties, such as the vibrational frequencies of the molecule. DFT calculations can use several different basis sets, which are groups of Gaussian functions that describe the molecular orbitals of a molecule, to run the calculation and, in addition to ground state spectra, excited state spectra can be also be calculated using time-domain (TD)-DFT (Klaumünzer, Kröner, & Saalfrank, 2010). Choice of the basis set depends on computing power and accuracy required. For example, split-valence basis sets are constructed in increasing complexity at the expense of computing power and time. Zhao and coworkers used DFT to calculate the vibrational spectra of FAD in the gas phase and simulated a solution phase measurement by adding water molecules around the flavin thereby enabling assignment of the spectra of the reduced forms of FAD (Fig. 9) (Zhao et al., 2011). The calculations were performed using the B3LYP/6-31G basis set to calculate shifts due to H/D exchange.

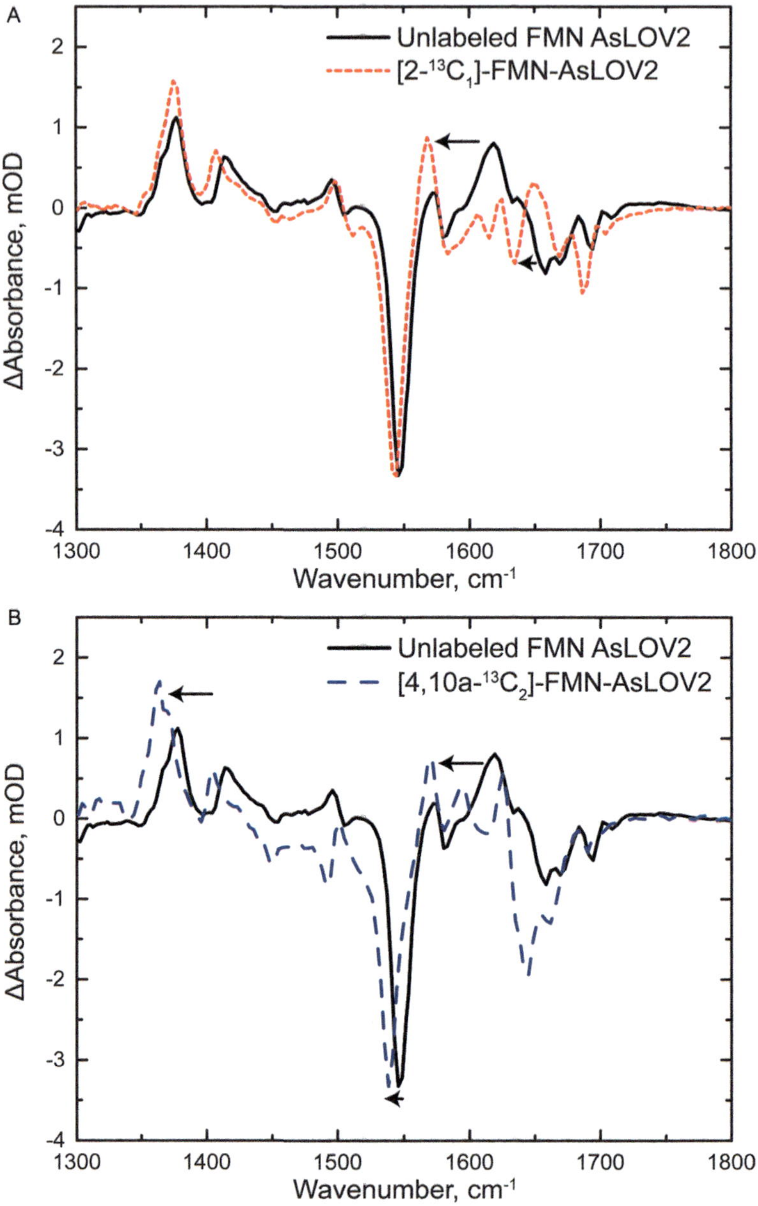

Fig. 8 Effect of isotope labeling on the TRIR spectra of AsLOV2. (A) [2-$^{13}C_1$]-FMN and (B) [4,10a-$^{13}C_2$]-FMN were reconstituted into AsLOV2. The 1 ps TRIR spectrum reveals that bands associated with the C2═O and C4a—C10a modes are red-shifted in the labeled samples. *Figure adapted from Gil, A. A., Laptenok, S. P., French, J. B., Iuliano, J. N., Lukacs, A., Hall, C. R., et al. (2017). Femtosecond to millisecond dynamics of light induced allostery in the avena sativa LOV domain.* The Journal of Physical Chemistry B, 121*(5), 1010–1019; Gil, A. A., Laptenok, S. P., Iuliano, J. N., Lukacs, A., Verma, A., Hall, C. R., et al. (2017). Photoactivation of the BLUF protein PixD probed by the site-specific incorporation of fluorotyrosine residues.* Journal of the American Chemical Society, 139*(41), 14638–14648.*

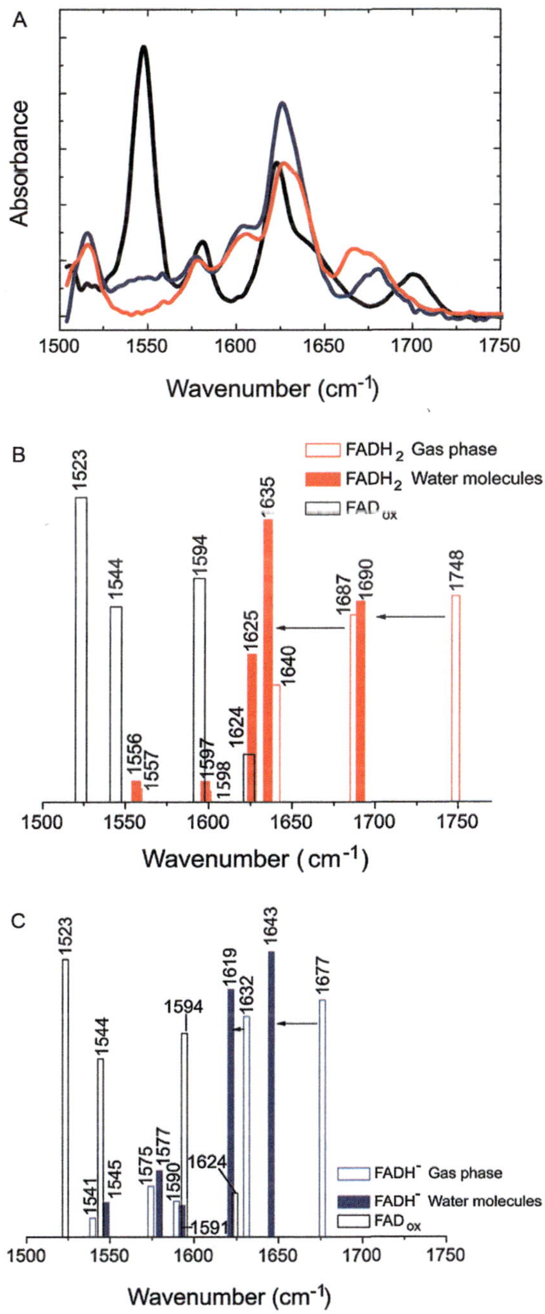

Fig. 9 FTIR and DFT spectra of FAD in solution. (A) FTIR spectra of FAD_{ox} (*black*), $FADH_2$ (*red*), and $FADH^-$ (*blue*) in phosphate buffer. DFT calculations in the gas phase and in water are shown for (B) $FADH_2$ and (C) $FADH^-$. *Figure adapted from Zhao, R.-K., Lukacs, A., Haigney, A., Brust, R., Greetham, G. M., Towrie, M., et al. (2011). Ultrafast transient mid IR to visible spectroscopy of fully reduced flavins.* Physical Chemistry Chemical Physics, *13(39), 17642.*

QM/MM is especially useful in predicting the magnitude of frequency shifts induced by isotope labeling of the flavin chromophore as well as protein modes hydrogen bonded to the isoalloxazine ring. This approach was exploited by Domratcheva and coworkers in the assignment of vibrational modes associated with a conserved glutamine that are coupled to the flavin in BLUF photoreceptors (Domratcheva, Hartmann, Schlichting, & Kottke, 2016). Overall, the combination of biochemical approaches that enable site specific labeling and computational methods to predict frequencies affords much higher information content from vibrational spectroscopy studies.

4. General experimental considerations

4.1 Concentration determination

While methods such as the BCA assay can be used to determine protein concentration when the extinction coefficient is not known, in most cases the concentration of the flavoprotein can be determined with the visible absorption spectrum of the flavin chromophore and the associated extinction coefficient (Table 4) (Whitby, 1953). In this case it is critical to ensure that there is no unbound FAD or FMN, which can be removed either using a size exclusion or desalting column, or by ultrafiltration using a spin column with a semipermeable membrane which retains the protein but not the free chromophore. In the latter approach, the absorbance of the flow-through can be checked to ensure that no unbound cofactor is present. These methods are generally successful since FAD or FMN has high (nM) affinity for the protein and with mM protein concentrations the equilibrium lies strongly in favor of the complex.

4.2 Protection from ambient light

For flavin photoreceptors it is important to protect the sample from light before and during any measurements. Purification can be conducted in the dark to ensure that the protein is in the dark state. In addition, sufficient

Table 4 Extinction coefficients (Whitby, 1953)

	450 nm	375 nm	260 nm
Riboflavin	12,200 M^{-1} cm^{-1}	10,600 M^{-1} cm^{-1}	27,700 M^{-1} cm^{-1}
FMN	12,200 M^{-1} cm^{-1}	10,400 M^{-1} cm^{-1}	27,100 M^{-1} cm^{-1}
FAD	11,300 M^{-1} cm^{-1}	9300 M^{-1} cm^{-1}	37,000 M^{-1} cm^{-1}

time must be allowed for the photoexcited protein to return to the dark state if multiple measurements are being conducted. The rate of light to dark state recovery can vary widely from minutes to hours. In general, all flavoproteins should be kept in the dark even if they do not have obvious photosensory properties.

4.3 Buffer for protein stabilization

Careful choice of buffer is necessary for vibrational spectroscopy experiments. The buffer salts and solvents used should not have absorbances overlapping with modes from the isoalloxazine ring or protein. Commonly used buffers include Tris, sodium phosphate, or potassium phosphate supplemented with either sodium chloride or potassium chloride. An additional consideration for FTIR spectroscopy is the solvent used for the buffer. As noted above, H_2O strongly absorbs IR light, especially in the 1600–1700 cm^{-1} region which overlaps with important vibrational bands from the chromophore such as the carbonyl modes. Therefore, H_2O is normally replaced by D_2O. Solvent exchange can be accomplished by lyophilizing the protein and redissolving in D_2O, or using a desalting column equilibrated with D_2O buffer or by successive rounds of ultrafiltration in which the sample is repetitively diluted with D_2O buffer and then reconcentrated. Successful exchange can be determined by obtaining an IR spectrum of the ultrafiltration column flow through. While some enzymes require glycerol in buffer for stabilization, high concentrations of glycerol will interfere with acquisition of Raman spectra (Schelvis et al., 2003).

4.4 Sample size

FTIR spectroscopy of proteins in solution is most commonly performed using a demountable transmission cell in which the sample is placed between two BaF_2 or CaF_2 windows with a spacer to set the pathlength. Choice of pathlength depends on solvent: a very small pathlength (6–12 μm) should be used for strongly absorbing solvents such as H_2O while a 50 μm spacer is commonly used for measurements in D_2O. Sample volume is normally 30–50 μL when a 50 μm spacer is used. Solution cells come in two main types in which the cell is assembled around the sample or using a refillable cell that does not have to be disassembled between samples. The latter is recommended if the spectra of two different protein samples are being subtracted since this ensures that the pathlength for the two measurements are identical. However, larger volumes are needed to fill refillable cells and care must be taken when cleaning and drying these cells so that the protein solution flows evenly into the cell without generating bubbles which can cause serious problems in IR data collection. Flow cells are used for photochemically sensitive samples such as photoreceptors, and in this case sample volume is 1–2 mL so that the protein solution can be recirculated through the cell. Such experiments require large amounts of protein since sample concentration is 1–2 mM. In addition, the cell is rastered in the beam to minimize unwanted photochemistry. In addition to solution cells, attenuated total reflectance (ATR) uses a crystal surface as the sample cell such that only molecules that are in contact with the crystal contribute to the vibrational spectrum: the advantage of this approach is that it is amenable to membrane proteins which can be deposited as films, while the smaller pathlength reduces the absorbance of the solvent such as H_2O.

Raman spectroscopy is more flexible in terms of sampling modality. Samples can be analyzed in capillaries or quartz cuvettes while it is also possible to obtain Raman spectra directly from single crystals mounted on a goniometer (Kalp, Totir, Buynak, & Carey, 2009). Raman crystallography yields valuable information on the ligand in the protein crystal and can also be used to assess photodamage during X-ray data collection and to analyze structural differences between solid and solution phase samples (Røhr, Hersleth, & Kristoffer Andersson, 2010). Raman spectra from small samples and cells can also be acquired using a Raman microscope (Carey, 2006), an approach that can also be used for imaging. The amount of sample necessary for Raman spectroscopy is dependent on the size of the cuvette (50–500 μL) and generally sample concentrations range from 0.2 to 1 mM.

References

Alben, J. O., Moh, P. P., Fiamingo, F. G., & Altschuld, R. A. (1981). Cytochrome oxidase (a3) heme and copper observed by low-temperature Fourier transform infrared spectroscopy of the CO complex. *Proceedings of the National Academy of Sciences of the United States of America*, *78*(1), 234–237.

Bacher, A. (1986). Heavy riboflavin synthase from *Bacillus Subtilis*,. In F. Chytil & D. B. McCormick (Eds.), *Methods in enzymology* (pp. 192–199). Elsevier.

Bagchi, S., Boxer, S. G., & Fayer, M. D. (2012). Ribonuclease S dynamics measured using a nitrile label with 2D IR vibrational echo spectroscopy. *The Journal of Physical Chemistry B*, *116*(13), 4034–4042.

Benecky, M., Li, T. Y., Schmidt, J., Frerman, F., Watters, K. L., & McFarl, J. (1979). Resonance Raman study of flavins and the flavoprotein fatty acyl coenzyme A dehydrogenase. *Biochemistry*, *18*(16), 3471–3476.

Bowman, W. D., & Spiro, T. G. (1981). Normal mode analysis of lumiflavin and interpretation of resonance Raman spectra of flavoproteins. *Biochemistry*, *20*(11), 3313–3318.

Brust, R., Haigney, A., Lukacs, A., Gil, A., Hossain, S., Addison, K., et al. (2013). Ultrafast structural dynamics of BlsA, a photoreceptor from the pathogenic bacterium *Acinetobacter baumannii*. *Journal of Physical Chemistry Letters*, *5*(1), 220–224.

Carey, P. R. (2006). Raman crystallography and other biochemical applications of Raman spectroscopy. *Annual Review of Physical Chemistry*, *57*(1), 527–554.

Carey, P. R., & Tonge, P. J. (1995). Unlocking the secrets of enzyme power using Raman spectroscopy. *Accounts of Chemical Research*, *28*(1), 8–13.

Colthup, N. B., Daly, L. H., & Wiberley, S. E. (1975). *Introduction to infrared and Raman spectroscopy* (2nd ed.). New York: Academic Press, Inc.

Copeland, R. A., & Spiro, T. G. (1986). Ultraviolet resonance Raman spectroscopy of flavin mononucleotide and flavin adenine dinucleotide. *The Journal of Physical Chemistry*, *90*(25), 6648–6654.

Deng, H., Vedad, J., Desamero, R. Z. B., & Callender, R. (2017). Difference FTIR studies of substrate distribution in triosephosphate isomerase. *The Journal of Physical Chemistry B*, *121*(43), 10036–10045.

Domratcheva, T., Hartmann, E., Schlichting, I., & Kottke, T. (2016). Evidence for tautomerisation of glutamine in BLUF blue light receptors by vibrational spectroscopy and computational chemistry. *Scientific Reports*, *6*(1), 22669.

Dong, J., Wan, Z.-L., Chu, Y.-C., Nakagawa, S. N., Katsoyannis, P. G., Weiss, M. A., et al. (2001). Isotope-edited Raman spectroscopy of proteins: A general strategy to probe individual peptide bonds with application to insulin. *Journal of the American Chemical Society*, *123*(32), 7919–7920.

Dutta, P. K., & Nestor, J. R. (1977). Resonance coherent anti-Stokes Raman scattering spectra of fluorescent biological chromophores: Vibrational evidence for hydrogen bonding of flavin to glucose oxidase and for rapid solvent exchange. *Proceedings of the National Academy of Sciences of the United States of America*, *74*(10), 4146–4149.

El Khoury, Y., Van Wilderen, L. J. G. W., & Bredenbeck, J. (2015). Ultrafast 2D-IR spectroelectrochemistry of flavin mononucleotide. *The Journal of Chemical Physics*, *142*(21), 212416.

Gil, A. A., Laptenok, S. P., French, J. B., Iuliano, J. N., Lukacs, A., Hall, C. R., et al. (2017). Femtosecond to millisecond dynamics of light induced allostery in the *Avena sativa* LOV domain. *The Journal of Physical Chemistry B*, *121*(5), 1010–1019.

Gil, A. A., Laptenok, S. P., Iuliano, J. N., Lukacs, A., Verma, A., Hall, C. R., et al. (2017). Photoactivation of the BLUF protein PixD probed by the site-specific incorporation of fluorotyrosine residues. *Journal of the American Chemical Society*, *139*(41), 14638–14648.

Haigney, A., Lukacs, A., Zhao, R. K., Stelling, A. L., Brust, R., Kim, R. R., et al. (2011). Ultrafast infrared spectroscopy of an isotope-labeled photoactivatable flavoprotein. *Biochemistry*, *50*(8), 1321–1328.

Hall, C. R., Heisler, I. A., Jones, G. A., Frost, J. E., Gil, A. A., Tonge, P. J., et al. (2017). Femtosecond stimulated Raman study of the photoactive flavoprotein AppABLUF. *Chemical Physics Letters*, *683*, 365–369.

Hamm, P., & Zanni, M. T. (2011). *Concepts and methods of 2D infrared spectroscopy*. Cambridge University Press.

Hasegawa, K., Masuda, S., & Ono, T. A. (2005). Spectroscopic analysis of the dark relaxation process of a photocycle in a sensor of blue light using FAD (BLUF) protein Slr1694 of the cyanobacterium *Synechocystis* sp. PCC6803. *Plant & Cell Physiology*, *46*(1), 136–146.

Illarionov, B., Zhu, F., Eisenreich, W., Bacher, A., Weber, S., & Fischer, M. (2015). Preparation of flavocoenzyme isotopologues by biotransformation of purines. *The Journal of Organic Chemistry*, *80*(5), 2539–2544.

Iuliano, J. N., Gil, A. A., Laptenok, S. P., Hall, C. R., Tolentino Collado, J., Lukacs, A., et al. (2018). Variation in LOV photoreceptor activation dynamics probed by time-resolved infrared spectroscopy. *Biochemistry*, *57*(5), 620–630.

Kalp, M., Totir, M. A., Buynak, J. D., & Carey, P. R. (2009). Different intermediate populations formed by tazobactam, sulbactam, and clavulanate reacting with SHV-1 β-lactamases: Raman crystallographic evidence. *Journal of the American Chemical Society*, *131*(6), 2338–2347.

Kim, M., & Carey, P. R. (1993). Observation of a carbonyl feature for riboflavin bound to riboflavin-binding protein in the red-excited Raman spectrum. *Journal of the American Chemical Society*, *115*(15), 7015–7016.

Kim, M., Owen, H., & Carey, P. R. (1993). High-performance Raman spectroscopic system based on a single spectrograph, CCD, notch filters, and a Kr+ laser ranging from the near-IR to near-UV regions. *Applied Spectroscopy*, *47*(11), 1780–1783.

Kitagawa, T., Kyogoku, Y., Nishina, Y., Yamano, T., Ohishi, N., Takai-Suzuki, A., et al. (1979). Resonance Raman spectra of carbon-13- and nitrogen-15-labeled riboflavin bound to egg-white flavoprotein. *Biochemistry*, *18*(9), 1804–1808.

Klaumünzer, B., Kröner, D., & Saalfrank, P. (2010). (TD-)DFT calculation of vibrational and vibronic spectra of riboflavin in solution. *The Journal of Physical Chemistry B*, *114*(33), 10826–10834.

Kondo, M., Nappa, J. J., Ronayne, K. L., Stelling, A. L., Tonge, P. J., & Meech, S. R. (2006). Ultrafast vibrational spectroscopy of the flavin chromophore. *The Journal of Physical Chemistry B*, *110*(41), 20107–20110.

Kukura, P., McCamant, D. W., & Mathies, R. A. (2007). Femtosecond stimulated Raman spectroscopy. *Annual Review of Physical Chemistry*, *58*(1), 461–488.

Lukacs, A., Zhao, R. K., Haigney, A., Brust, R., Greetham, G. M., Towrie, M., et al. (2012). Excited state structure and dynamics of the neutral and anionic flavin radical revealed by ultrafast transient mid-IR to visible spectroscopy. *The Journal of Physical Chemistry B*, *116*(20), 5810–5818.

Masuda, S., Hasegawa, K., Ishii, A., & Ono, T. A. (2004). Light-induced structural changes in a putative blue-light receptor with a novel FAD binding fold sensor of blue-light using FAD (BLUF); Slr1694 of *Synechocystis* sp. PCC6803. *Biochemistry*, *43*(18), 5304–5313.

Moss, D., Nabedryk, E., Breton, J., & Mäntele, W. (1990). Redox-linked conformational changes in proteins detected by a combination of infrared spectroscopy and protein electrochemistry. Evaluation of the technique with cytochrome c. *European Journal of Biochemistry*, *187*(3), 565–572.

Nishina, Y., Kitagawa, T., Shiga, K., Horiike, K., Matsumura, Y., Watari, H., et al. (1978). Resonance Raman spectra of riboflavin and its derivatives in the bound state with egg riboflavin binding proteins. *Journal of Biochemistry*, *84*(4), 925–932.

Nishina, Y., Kitagawa, T., Shiga, K., Watari, H., & Yamano, T. (1980). Resonance Raman study of flavoenzyme-inhibitor charge-transfer interactions. *Journal of Biochemistry*, *87*(3), 831–839.

Nishina, Y., Sato, K., Miura, R., Matsui, K., & Shiga, K. (1998). Resonance Raman study on reduced flavin in purple intermediate of flavoenzyme: Use of [4-carbonyl-18O]-enriched flavin. *Journal of Biochemistry*, *124*(1), 200–208.

Nishina, Y., Sato, K., Setoyama, C., Tamaoki, H., Miura, R., & Shiga, K. (2007). Intramolecular and intermolecular perturbation on electronic state of FAD free in solution and bound to flavoproteins: FTIR spectroscopic study by using the C=O stretching vibrations as probes. *Journal of Biochemistry*, *142*(2), 265–272.

Nishina, Y., Shiga, K., Horiike, K., Tojo, H., Kasai, S., Yanase, K., et al. (1980). Vibrational modes of flavin bound to riboflavin binding protein from egg white. *Journal of Biochemistry*, *88*(2), 403–409.

Ojeda, J. J., & Dittrich, M. (2012). Fourier transform infrared spectroscopy for molecular analysis of microbial cells. In A. Navid (Ed.), *Methods in molecular biology. Microbial systems biology* (pp. 187–211). Totowa, NJ: Humana Press.

Pagano, P., Guo, Q., Kohen, A., & Cheatum, C. M. (2016). Oscillatory enzyme dynamics revealed by two-dimensional infrared spectroscopy. *Journal of Physical Chemistry Letters*, 7(13), 2507–2511.

Piano, V., Palfey, B. A., & Mattevi, A. (2017). Flavins as covalent catalysts: New mechanisms emerge. *Trends in Biochemical Sciences*, *42*(6), 457–469.

Røhr, Å. K., Hersleth, H. P., & Kristoffer Andersson, K. (2010). Tracking flavin conformations in protein crystal structures with Raman spectroscopy and QM/MM calculations. *Angewandte Chemie (International Ed. in English)*, *49*(13), 2324–2327.

Rudik, I., Bell, A. F., Tonge, P. J., & Thorpe, C. (2000). 4-Hydroxycinnamoyl-CoA: An ionisable probe of the active site of medium chain acyl-CoA dehydrogenase. *Biochemistry*, *39*, 92–101.

Salomon, M., Christie, J. M., Knieb, E., Lempert, U., & Briggs, W. R. (2000). Photochemical and mutational analysis of the FMN-binding domains of the plant blue light receptor, phototropin. *Biochemistry*, *39*(31), 9401–9410.

Schelvis, J. P. M., Pun, D., Goyal, N., & Sokolova, O. (2006). Resonance Raman spectra of the neutral and anionic radical semiquinones of flavin adenine dinucleotide in glucose oxidase revisited. *Journal of Raman Specroscopy*, *37*(8), 822–829.

Schelvis, J. P. M., Ramsey, M., Sokolova, O., Tavares, C., Cecala, C., Connell, K., et al. (2003). Resonance Raman and UV-Vis spectroscopic characterization of FADH(center dot) in the complex of photolyase with UV-damaged DNA. *The Journal of Physical Chemistry B*, *107*(44), 12352–12362.

Snellenburg, J. J., Laptenok, S. P., Seger, R., Mullen, K. M., & van Stokkum, I. H. M. (2012). Glotaran: A Java-based graphical user interface for the R Package TIMP. *Journal of Statistical Software*, *49*(3), 1–22.

Stelling, A. L., Ronayne, K. L., Nappa, J., Tonge, P. J., & Meech, S. R. (2007). Ultrafast structural dynamics in BLUF domains: Transient infrared spectroscopy of AppA and its mutants. *Journal of the American Chemical Society*, *129*(50), 15556–15564.

Sugiyama, T., Nisimoto, Y., Mason, H. S., & Loehr, T. M. (1985). Flavins of NADPH-cytochrome P-450 reductase: Evidence for structural alteration of flavins in their one-electron-reduced semiquinone states from resonance Raman spectroscopy. *Biochemistry*, *24*(12), 3012–3019.

Swartz, T. E., Wenzel, P. J., Corchnoy, S. B., Briggs, W. R., & Bogomolni, R. A. (2002). Vibration spectroscopy reveals light-induced chromophore and protein structural changes in the LOV2 domain of the plant blue-light receptor phototropin 1. *Biochemistry*, *41*(23), 7183–7189.

Tishler, M., Pfister, K., Babson, R. D., Ladenburg, K., Fleming, A. J., & Ladenburg, K. (1947). The reaction between o-aminoazo compounds and barbituric acid. A new synthesis of riboflavin. *Journal of the American Chemical Society*, *69*(6), 1487–1492.

Tonge, P. J., Pusztai, M., White, A. J., Wharton, C. W., & Carey, P. R. (1991). Resonance Raman and Fourier transform infrared spectroscopic studies of the acyl carbonyl group in [3-(5-methyl-2-thienyl)acryloyl]chymotrypsin: Evidence for artifacts in the spectra obtained by both techniques. *Biochemistry*, *30*(19), 4790–4795.

Weigel, A., Dobryakov, A., Klaumünzer, B., Sajadi, M., Saalfrank, P., Ernsting, N. P., et al. (2011). Femtosecond stimulated Raman spectroscopy of flavin after optical excitation. *The Journal of Physical Chemistry B*, *115*(13), 3656–3680.

Whitby, L. G. (1953). A new method for preparing flavin-adenine dinucleotide. *The Biochemical Journal*, *54*(3), 437–442.

Wilbrandt, R., Pagsberg, P., Hansen, K. B., & Weisberg, C. V. (1975). Fast resonance Raman spectroscopy of a free radical. *Chemical Physics Letters*, *36*(1), 76–78.

Wille, G., Ritter, M., Friedemann, R., Mäntele, W., Hübner, G., Ma, W., et al. (2003). Redox-triggered FTIR difference spectra of FAD in aqueous solution and bound to flavoproteins. *Biochemistry*, *42*(50), 14814–14821.

Zhao, R.-K., Lukacs, A., Haigney, A., Brust, R., Greetham, G. M., Towrie, M., et al. (2011). Ultrafast transient mid IR to visible spectroscopy of fully reduced flavins. *Physical Chemistry Chemical Physics*, *13*(39), 17642.

Zheng, Y., Carey, P. R., & Palfey, B. A. (2004). Raman spectrum of fully reduced flavin. *Journal of Raman Specroscopy*, *35*(7), 521–524.

Zheng, Y., Dong, J., Palfey, B. A., & Carey, P. R. (1999). Using raman spectroscopy to monitor the solvent-exposed and "buried" forms of flavin in p-hydroxybenzoate hydroxylase. *Biochemistry*, *38*(51), 16727–16732.

CHAPTER NINE

Measuring electronic structure properties of flavins and flavoproteins by electronic Stark spectroscopy

Robert J. Stanley*, Cornelius J. van Galen
Department of Chemistry, Temple University, Philadelphia, PA, United States
*Corresponding author: e-mail address: rstanley@temple.edu

Contents

Abstract

The optical spectrum of a flavoprotein is one of its signature properties. No two flavoprotein absorption spectra are exactly alike as each encodes the details of the interaction of the flavin cofactor electronic structure with the specific protein binding pocket. Electronic Stark spectroscopy has the potential to elucidate these interactions with high sensitivity, at low cost, and requiring minimal technical sophistication. In this chapter we will outline the theoretical basis for Stark spectroscopy and describe the construction of

Methods in Enzymology, Volume 620
ISSN 0076-6879
https://doi.org/10.1016/bs.mie.2019.03.012

the Stark spectrometer. Step-by-step instructions are given for acquiring and interpreting Stark spectra to retrieve difference moments of the flavin ground versus excited state charge distributions.

Abbreviations

2-MTHF 2-methyl-tetrahydrofuran
AC alternating current
BHT butylated hydroxytoluene
DFT density functional theory
DNA deoxyribonucleic acid
FAD flavin adenine dinucleotide
Fl_{OX} oxidized flavin
FMN flavin mononucleotide
FWHM full width at half maximum height
HOMO highest occupied molecular orbital
HV high voltage
HVA high voltage amplifier
ITO indium tin oxide
LIA lock-in amplifier
LN2 liquid nitrogen
LT low temperature
LTA low temperature absorption
LUMO lowest unoccupied molecular orbital
NADPH reduced nicotinamide adenine dinucleotide phosphate
OYE old yellow enzyme
PL photolyase
S/N signal to noise
TD-DFT time-dependent density functional theory
TPARF tetraphenylacetyl-riboflavin
UV ultraviolet

1. Introduction

1.1 Motivation

Flavoproteins perform a vast array of biochemical reactions in and out of the cell (Lienhart, Gudipati, & Macheroux, 2013; Macheroux, Kappes, & Ealick, 2011). Of paramount significance is the fact that flavins can transfer either one or two electrons, making them more versatile than nicotinamide cofactors (e.g. NADPH). Flavin reduction potentials are modulated by proteins to accommodate a wide variety of chemistries (Stankovich, 1991). It is widely agreed, at least for noncovalently bound flavins, that the flavin binding site in the protein is how this modulation is accomplished. Hence, the biochemical characterization of a flavoenzyme is incomplete without reporting the absorption spectrum of the flavoprotein, preferably in as many oxidation states as are accessible. Many studies have focused on how protein contacts, conserved water, etc., affect the flavin reduction potential and spectrum in a number of individual flavoenzymes. As of yet, there does not appear to be a way of predicting these properties by sequence or structure.

For the past 20 years, we have been focused on the light-driven function of photoflavoproteins (Gindt et al., 2016; Kodali, Siddiqui, & Stanley, 2009; Lee, Kodali, Stanley, & Matsika, 2016; MacFarlane IV & Stanley, 2001, 2003; Narayanan, Singh, Kodali, Moravcevic, & Stanley, 2017; Pauszek, Kodali, Siddiqui, & Stanley, 2011, 2016; Stanley & MacFarlane IV, 2000), and flavin dyads (Pauszek et al., 2013; Yu et al., 2012). In either case, it is not the ground state electronic structure that is of interest but, rather, the *excited* electronic states. Measuring this excited state charge distribution affords a rationality for how light-activated flavins interact with substrates or electron donors/acceptors.

Stark spectroscopy affords a determination of difference dipoles $\Delta\vec{\mu}_{eg} = \vec{\mu}_e - \vec{\mu}_g$, and difference polarizabilities $\Delta\vec{\alpha}_{eg} = \vec{\alpha}_e - \vec{\alpha}_g$ that characterize the (g)round- and (e)xcited-state permanent moments of the flavin charge distribution, $\rho(\vec{r})$. Moving toward richer fields, we hypothesize that the ground state UV/vis absorption spectrum of all flavoproteins has imprinted on it the effect of the flavin binding-cavity *electrostatic field*, $\vec{E}_{cav}$, and that Stark spectroscopy can be used to compute $\vec{F}_{cav}$. This will require Stark and absorption spectra taken in several oxidation/protonation states. We have already accomplished this for *E. coli* DNA photolyase (Kodali et al., 2009).

In this chapter, we provide protocols to build a Stark spectrometer as well as how to interpret the data obtained from it. To this end we start with some basic electrostatics.

1.2 Solute charge distributions

To approach the problem, consider a dipolar ellipsoidal molecule as solute in a simple dipolar solvent. We want to explore the effect of solvation on the energy of the solute through optical spectroscopy, so we assume that the molecule absorbs visible light. We also assume that the dipolar ellipsoidal molecule is a "point-dipole." That is, we view (or measure) the electrostatic properties of the molecule at sufficient distance such that the details of molecular structure (bond length, angle, etc.) are negligible.

The electrostatic effects of the electronic structure (or charge distribution, $\rho(\vec{r})$) of the molecule are transmitted at some distance $\vec{r}$ from the molecule by the electric field generated by its charge distribution, $\vec{E}_{solute}$. $\vec{E}_{solute}$ produced by $\rho(\vec{r})$ is the divergence of the electrostatic gradient, φ, as

$$\vec{E}_{solute} = -\vec{\nabla} \cdot \varphi_{solute} \tag{1}$$

where the electrostatic gradient arises from the charge distribution of the molecule,

$$\varphi_{solute} = \frac{1}{4\pi\varepsilon_0} \int_V \frac{\rho\left(\vec{r}'\right)}{\left|\vec{r} - \vec{r}'\right|} dv' \tag{2}$$

Here φ can be expanded in terms of the monopole, dipole, and quadrupole moments of $\rho(\vec{r})$, giving

$$\varphi_{solute}\left(\vec{r}\right) = \frac{1}{4\pi\varepsilon_0}\left[\frac{Q}{\vec{r}} + \frac{\vec{\mu}\cdot\vec{r}}{\left|\vec{r}\right|^3} + \sum_{i=1}^{3}\sum_{j=1}^{3}\frac{1}{2}\frac{\left|\vec{r}_i\right|\left|\vec{r}_j\right|}{\left|\vec{r}\right|^5}\alpha_{ij}\right] \tag{3}$$

where the first term is the gradient arising from a monopole or ionic species ($Q = 0$ for a neutral molecule), the second term is the gradient due to the dipole moment of the charge distribution $\vec{\mu}$, and the third term is the gradient contribution from the quadrupole moment or polarizability (Reitz, Milford, & Christy, 1979). The individual elements of the polarizability tensor $\vec{\vec{\alpha}}$ represent the degree to which the molecule can be polarized by an external electric field. $\vec{\vec{\alpha}}$ is a 3×3 tensor,

$$\vec{\vec{\alpha}} = \begin{pmatrix} \alpha_{xx} & \alpha_{xy} & \alpha_{xz} \\ \alpha_{yx} & \alpha_{yy} & \alpha_{yz} \\ \alpha_{zx} & \alpha_{zy} & \alpha_{zz} \end{pmatrix} \tag{4}$$

representing the polarizability along duples of the Cartesian coordinates ($ij = xx$, xy, xz, etc.). All these moments are defined relative to the center of mass of the molecule.

The dipolar electric field at some point in space, *r*, generated by the molecule is given by (Reitz et al., 1979)

$$\vec{E}_{dipole} = \frac{1}{4\pi\varepsilon_0}\left[\frac{3\,\vec{r}\cdot\vec{\mu}}{\left|\vec{r}\right|^5}\vec{r} - \frac{\vec{\mu}}{\left|\vec{r}\right|^3}\right] \tag{5}$$

It is instructive to consider the magnitude of these effects. The computed ground state dipole moment of flavin is about $\vec{\mu}_g = 7$ Debye (7 D) and points roughly along the long axis of the molecule (Hall, Orchard, & Tripathy, 1987) (two opposite charges separated by 0.21 Å gives 1 D or 3.36×10^{-30} C·m).

All the field contributions to $\vec{E}_{solute} = \vec{E}_{ion} + \vec{E}_{dipole} + \vec{E}_{polarizability}$ are "imprinted" on the solvent. This signature of the charge distribution can be observed in the absorption spectrum as shown next.

1.3 Solvent effects on the absorption spectrum: Solvatochromism

In a solvent, the volume and electric field of the solute will push and pull solvent molecules generating a roughly ellipsoidal cavity (Liptay, 1969). The electric field of the molecule(s) create(s) an oriented solvent shell(s). We assume that the total interaction of the solute with the solvent can be represented by the sum of pairwise interactions (ignoring hydrogen bonds for the moment). If there are no ions present, then the strongest interaction will be two interacting dipoles, *A* and *B*, the energy of which is given by (Cantor & Schimmel, 1980):

$$E_{dipole} = \frac{1}{4\pi\varepsilon_0}\left(\frac{\vec{\mu}_A\cdot\vec{\mu}_B}{\left|\vec{r}_{AB}\right|^3} - \frac{3\left(\vec{\mu}_A\cdot\vec{r}_{AB}\right)\left(\vec{\mu}_B\cdot\vec{r}_{AB}\right)}{\left|\vec{r}_{AB}\right|^5}\right) \tag{6}$$

where, again, distance and orientation play a major role. The solvent dipoles will align around the dipolar solute to minimize the overall energy. The more polar the solvent, the greater this solvent electric field. If the solute were removed instantaneously, then this anisotropic solvent electric field represents the reaction field. This is the basis for the phenomenon known as solvatochromism, where the solvent reaction electric field, $\vec{F}_R$, interacts with the dipolar solute leading to a change in solute energy levels relative to the vacuum. Bear in the mind that above the freezing point of the solvent, there is no one orientation, but a fluctuating solvent cavity that gives rise to an average electric field.

The reaction field due to the oriented solvent is responsible for solvatochromic band shifts, ΔE_{shift}, where $\Delta E_{Shift} = -\vec{\mu}_e \cdot \vec{F}_R - \left(-\vec{\mu}_g \cdot \vec{F}_R\right) = -\Delta\vec{\mu}_{eg} \cdot \vec{F}_R$, where $\Delta\vec{\mu}_{eg} = \vec{\mu}_e - \vec{\mu}_g$. An underlying assumption of this model is that the reaction field $\vec{F}_R$ is unchanged during the photon absorption event. $\Delta\vec{\mu}_{eg}$ is a very important quantity, as it tells us about the *change* in dipole moments between ground and excited states. In photobiology, it is the excited state electronic structure of the chromophore that dictates reactivity. $\Delta\vec{\mu}_{eg}$ supplies not only the magnitude of charge displacement, but also the *direction* as well. Examples of this will be shown below. Even so, as ground state flavinologists, why should we care about $\Delta\vec{\mu}_{eg}$ since most flavoproteins are not photoactivated.

To answer this let's look again at solvatochromism, an effect known to most biochemists. The color-shifting of a biological chromophore is often noted in the literature. Taking oxidized flavin (Fl_{OX}) as the most frequent example, when the Fl_{OX} spectrum shifts to the red in a more polar solvent we ascribe this to a larger solvent reaction field producing a lower flavin energy. The magnitude of this shift depends mostly on $\left|\Delta\vec{\mu}_{eg}\right|$. In contrast, a less polar solvent produces a blue shift. Why not stop here. Isn't this all the information we can glean from optical absorption spectroscopy?

The cubic term of this expansion (Eq. 3) gives the polarizability of the molecule. As chemists we think of polarizability as the ease with which charge can be polarized or redistributed by an external electric field. This field could be the reaction field, an externally applied electric field, or it could be due to charges in the solvent (an ion, a charged amino acid residue, etc.) that are close enough to the solute to induce a dipole through $\vec{\mu}_{ind} = \vec{\vec{\alpha}} \cdot \vec{F}_R$, where the energy of interaction is $E_{ind} = \vec{F}_R \cdot \vec{\vec{\alpha}} \cdot \vec{F}_R$. The vectoral nature of this interaction is always stabilizing ($E_{ind} < 0$). When the chromophore absorbs light energy, we can expect that the polarizability of the excited state is larger than the ground state, $\vec{\vec{\alpha}}_e > \vec{\vec{\alpha}}_g$. Intuitively this makes sense since the electronic structure of the ground state is the best compromise between electrons' attraction to the nuclei and their mutual repulsion (the highest occupied molecular orbital or HOMO). Any increase in electronic energy will place electrons in more dispersed orbitals (LUMO,

LUMO + 1, etc.). Being further from the nuclei they will more easily influenced by external electric fields. Therefore, the difference polarizability, $\Delta\vec{\alpha}_{eg} = \vec{\alpha}_e - \vec{\alpha}_g$ will always lead to a lower transition energy of the molecule (a red shift).

1.4 Application of an external electric field: Electrochromism

So far, we have seen how the solvent reacts to a dipolar solute, and that energy shifts that can be expected depending on the polarity of the solvent. These *solvatochromic* effects have been well studied and there is a firm theoretical basis for their interpretation. However, there is no way to measure the *direction* of the solvent reaction field, only its magnitude. Further, other interactions, most notably hydrogen bonding between solvent and solute, can produce specific spectral perturbations that are difficult to separate from reaction field effects.

If an *external* electric field, $\vec{F}_{ext}$, is applied then all of the above arguments are in force, with the significant advantage that the direction of $\vec{F}_{ext}$ can be changed by the experimenter to separate out effects due to the reaction field and specific interactions. This effect is termed *electrodichroism*, in that the transmittance of the sample as a function of wavelength will change depending on the magnitude and direction of the applied external field. Most importantly, $\vec{F}_{ext}$ will provide information about the magnitude and direction of the change in charge distribution in the solute.

Fig. 1 illustrates the basic principle for a solute with ground and excited state dipole moments $\vec{\mu}_g$ and $\vec{\mu}_e$. With no external field ($\vec{F}_{ext} = 0$) absorption of a photon of wavelength λ_{ex} will lead to an absorption band. The band width will depend on a number of factors, primarily solvent-induced conformational inhomogeneous broadening and temperature.

Application of the electric field will cause the dipolar ground and excited states to change in energy. Because $\Delta E = -\vec{\mu} \cdot \vec{F}_{ext}$ molecules whose $\vec{\mu}$ is aligned parallel with the field will shift to lower energy while those antiparallel to the applied field will shift up in energy. For an unoriented ensemble this will produce a broadening of the absorption band, the width of which depends on the magnitude of the solute dipoles. A similar argument can be made for molecules where $\vec{\mu}_g$ and $\vec{\mu}_e = 0$ but $|\vec{\alpha}_g|$ or $|\vec{\alpha}_e| > 0$ then molecules will also energy shift in the electric field. Since this is an induced effect the

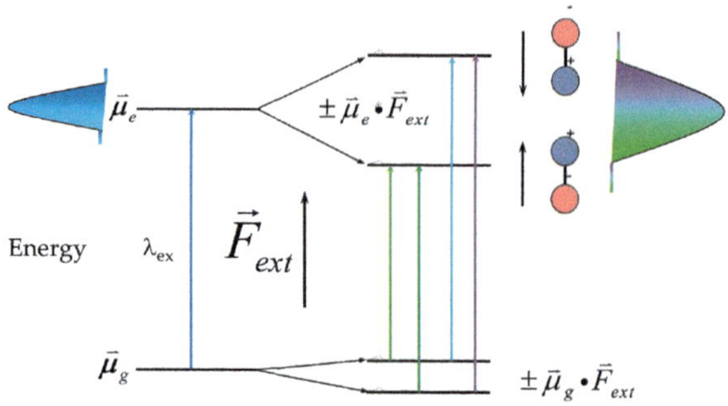

$$\Delta E = h\Delta\nu \approx -\Delta\vec{\mu}_{ge}\bullet\vec{F}_{ext} - \frac{1}{2}\vec{F}_{ext}\bullet\Delta\overline{\overline{\alpha}}_{ge}\bullet\vec{F}_{ext}$$

Fig. 1 A simplified Jablonski diagram showing the effect of an external electric field on an optical transition. Both ground and excited states change energy in response to the field. Here only the extrema of the effect are drawn.

shift will always be to lower energy. Finally, if the molecules are free to rotate during the application of $\vec{F}_{ext}$ and if $\left|\vec{\mu}_g\right| > 0$ then the molecules will pole in the field to achieve the lowest energy. Since the permanent and transition dipoles are fixed relative to each other in the molecular frame, this poling could result in an apparent increase or decrease in extinction coefficient.

The effect of the applied field on the absorption spectrum for these three cases is shown in Fig. 2.

It is critical to remember that these permanent moments of the charge distribution, whether in the ground or excited state, are locked (referenced) to the *transition dipole moment* of the molecule, $\vec{m}_{eg}$ connecting these states. Quantum mechanically, $\vec{m}_{eg} = \int \vec{\Psi}_e^*(-e\vec{r})\vec{\Psi}_g ds$, and $\left|\vec{m}_{eg}\right|^2$ is proportional to the familiar molar decadic extinction coefficient, ε. As the definition suggests, $\vec{m}_{eg}$ is a transient dipole moment much like a dipolar antenna found in common electronic devices. The probability, P, of absorption of radiation will depend on the relative orientations of the polarization vector of the light, $\vec{p}_0$, and $\vec{m}_{eg}$ such that $P \propto \left(\vec{p}_0 \cdot \vec{m}_{eg}\right)^2 = \left(\left|\vec{p}_0\right|\left|\vec{m}_{eg}\right|\cos\theta\right)^2$ where θ is the angle between these two vectors. Electromagnetic radiation polarized in the direction of $\vec{m}_{eg}$ will cause the ground state charge distribution to oscillate at the Bohr frequency, $\nu_{eg} = \Delta E_{eg}/h = (E_e - E_g)/h$, while radiation polarized perpendicularly to $\vec{m}_{eg}$ will not be coupled (detected) into the molecular

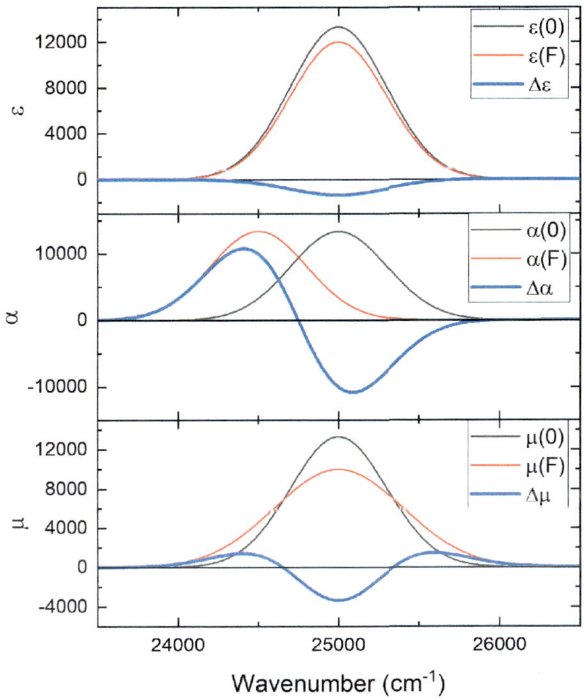

Fig. 2 Simulations of the Stark signal assuming a Gaussian band shape. Panel (A) shows the effect of an applied field on an ensemble of molecules that are unoriented and mobile. A decrease in the extinction coefficient is observed. Panel (B) shows the field effect on the polarizability, $|\Delta\alpha| \neq 0$, is obtained. Panel (C) shows the effect if $|\Delta\mu| \neq 0$.

charge cloud. This process is termed *photoselection* and is particularly relevant to non-oriented ensembles of molecules interacting with polarized light, as described below. In essence, the $\cos^2\theta$ term confers a p-orbital-like spatial probability distribution for the absorption of a polarized photon, selecting out those molecules most closely oriented along $\vec{p}_0$.

This photoselected population will have a defined orientation to the externally applied electric field, $\vec{F}_{ext}$. The angle between these two directions is $\chi = \angle\vec{p}_0, \vec{F}_{ext}$ and it affords the transition from the laboratory frame of reference into the molecular frame through $\vec{p}_0 \cdot \vec{m}_{eg}$(photoselection) and $\zeta_A = \angle\vec{m}_{eg}, \Delta\vec{\mu}_{eg}$. The final reference to the difference dipole is set by the experimentalist as shown in Fig. 3.

Once this connection from the laboratory frame to the molecular frame is made, $\vec{m}_{eg}$ becomes our molecular reference vector. Fig. 4 shows how the

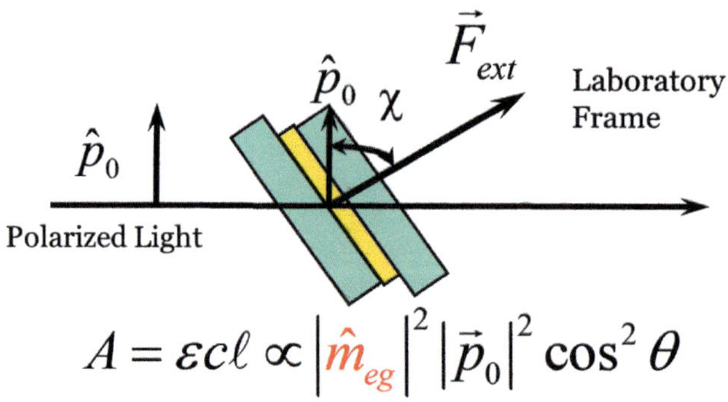

Fig. 3 The orientation of the polarized probe light to the sample cuvette as seen from above with the probe light polarized in the plane of the page. χ is the angle between the polarization vector of the light $\vec{p}_0$ and the applied electric field $\vec{F}_{ext}$. At normal incidence, $\chi = 90^\circ$, $\vec{p}_0$ can have almost any orientation with respect to $\vec{F}_{ext}$. Note too that the path length changes as the sample is rotated from normal incidence. At 45° the path length increases by a factor of $\sqrt{2}$.

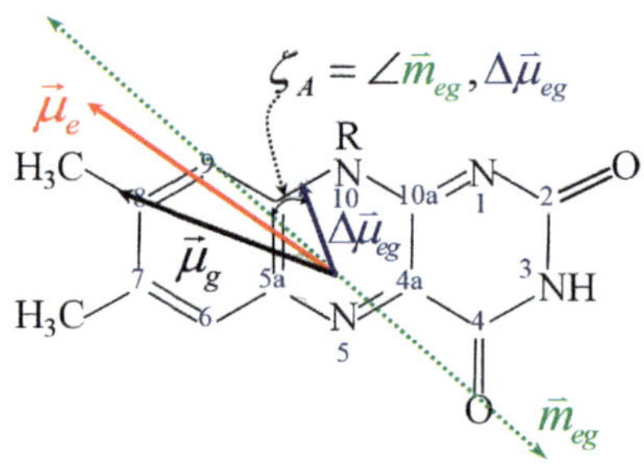

Fig. 4 Permanent dipoles and the transition dipole in the molecular frame of reference, where χ_A is the angle between $\vec{m}_{eg}$ and $\Delta\vec{\mu}_{eg}$.

transition dipole moment relates to the permanent and difference dipole moments for a flavin in a simple solvent. The extraction of these vectors is the outlined in the next section.

1.5 Liptay analysis of Stark spectra

The Liptay formalism (Liptay, 1974) treats the electric field as a perturbation to the electronic states of the molecule through dipolar and induced dipolar couplings, as described above. If it is assumed that the electric field-induced perturbation does not affect the wave function of the molecule appreciably then the line shape of the spectrum, $S(\bar{\nu})$ (e.g. a Gaussian function as used in Fig. 2), may be expanded in a Taylor series in terms of derivatives of the zero-field line shape,

$$\Delta S\left(\overline{\nu} : \overrightarrow{F}_{ext}\right) = S\left(\overline{\nu} : \overrightarrow{F}\right) - S(\overline{\nu} : 0) = S(\overline{\nu}) - \frac{dS}{d\overline{\nu}}\Delta\overline{\nu} + \frac{1}{2}\frac{d^2S}{d\overline{\nu}^2}(\Delta\overline{\nu})^2 + \cdots \quad (7)$$

Note that the transition energy, $\overline{\nu} = 1/\lambda$, is defined as wavenumbers in reciprocal centimeters, or cm^{-1}. This equation is generally truncated after the third term, resulting in the sum of the zeroth, first and second derivatives of the absorption spectrum. These terms are weighted by coefficients A_χ, B_χ, and C_χ, which are the connection between the observed line shape changes and the desired quantum-mechanical properties of the solute. A_χ, B_χ, and C_χ reflect the degree of electric field poling, polarizability response, and dipolar response to $\overrightarrow{F}_{ext}$. Since the amplitude of the line shape is proportional to the molar decadic extinction coefficient as a function of energy, $\varepsilon/\overline{\nu}$, the electric field-induced change in these moments of the charge distribution can be written as:

$$\left(\frac{\Delta\varepsilon}{\overline{\nu}}\right) = \left(f\overrightarrow{F}_{ext}\right)^2 \left\{ A_\chi\left(\frac{\varepsilon(\overline{\nu})}{\overline{\nu}}\right) + \frac{B_\chi}{15ch}\frac{\partial\left(\frac{\varepsilon(\overline{\nu})}{\overline{\nu}}\right)}{\partial\overline{\nu}} + \frac{C_\chi}{30c^2h^2}\frac{\partial^2\left(\frac{\varepsilon(\overline{\nu})}{\overline{\nu}}\right)}{\partial\overline{\nu}^2} \right\} \quad (8)$$

where c is the speed of light, h is Planck's constant, f is the local field correction factor (see below), and $\varepsilon/\overline{\nu}$ is the energy-weighted extinction (absorption spectrum) in wavenumbers.

The first term A_χ leads to a change in the intensity of the band. For immobilized molecules, as is typical in a Stark experiment, A_χ is negligible since the molecules are not able to orient with respect to the applied field. A_χ contains terms (not shown) including the transition moment polarizability and hyperpolarizability where, as the name implies, $\overrightarrow{m}$ has significant polarizability as well as dipolar character. Cases exist in which A_χ cannot be ignored, such as the Stark analysis of conjugated polyenes (Bublitz, Ortiz, Marder, & Boxer, 1997) or, charge-transfer flavin conjugates (Pauszek, Kodali, Caldwell, et al., 2013). As an example however, for a molecule with a typical transition polarizability of 100 Å^3 in a typical applied field of 3×10^5 V/cm, the induced transition dipole $\overrightarrow{m}_{eg}^{ind} = A_{eg} \cdot \overrightarrow{F}_{ext} \sim 0.03D$.

The B_χ term is the degree of shift in the absorption band due to the *induced* dipole moment by $\overrightarrow{F}_{ext}$, resulting in a first derivative line shape. This is due to the difference polarizability, $\Delta\overrightarrow{\alpha}$, of the molecule. The B_χ coefficient is, approximately:

$$B_\chi \approx \frac{5}{2} Tr\Delta \vec{\alpha} + \left(3\cos^2\chi - 1\right)\left(\frac{3}{2}\hat{m}\cdot\Delta\vec{\alpha}\cdot\hat{m} - \frac{1}{2}Tr\Delta\vec{\alpha}\right) \tag{9}$$

where $Tr\Delta\vec{\alpha}$ is the average difference polarizability (the sum of the diagonal elements of $\Delta\vec{\alpha} = \vec{\alpha}_e - \vec{\alpha}_g$, see Eq. 4) and $\hat{m}\cdot\Delta\vec{\alpha}\cdot\hat{m}$ is the component of $\Delta\vec{\alpha}$ along the direction of the transition dipole moment $\hat{m}$, here given as a unit vector. B_χ also contains terms related to the transition polarizability. For the same reasons listed above these terms are generally neglected.

The third term C_χ is a second derivative line shape caused by a broadening of bands due to positive and negative energy shifts of the isotropically oriented molecules with respect to the field. C_χ is related to the difference dipole moment $\Delta\vec{\mu}_{eg}$ and, in most cases, dominates the Stark spectrum. C_χ has the form:

$$C_\chi = \left|\Delta\vec{\mu}\right|^2\left\{5 + \left(3\cos^2\chi - 1\right)\left(3\cos^2\zeta_A - 1\right)\right\} \tag{10}$$

where ζ_A is the angle between the transition dipole moment and the difference dipole moment as described above.

To bring all of this into focus for the flavinologist, consider the flavin *N*(3)-methyl,*N*(10)-isobutylisoalloxazine (a.k.a. *N*(3)-flavin, a generous gift of Prof. Vincent Rotello, Univ. of Mass., Amherst). This flavin is very useful because it is soluble over a wide range of polarities. The low temperature absorption spectra of *N*(3)-flavin in ethanol, *n*-butanol, 2-MTHF, and toluene are shown in Fig. 5.

As a guide, the two dotted lines in Fig. 5 approximately mark the low energy edge $S_0 \rightarrow S_1$ and $S_0 \rightarrow S_2$ transitions of the oxidized flavin in ethanol, respectively. The effects of the solvent reaction field on the electronic energies of the molecule are evident though not easily interpreted. The general rule is that as the polarity of the solvent decreases the transition energy should increase ($\vec{F}_R$ is decreasing with polarity). Simple solvents such as ethanol and butanol engage in hydrogen bonding (H-bonding) and are not really that simple. Even 2-MTHF has H-bonding capacity. As a solvatochromic reference for nonpolar solutes toluene is useful. Its interactions with most molecules are minimal, and it forms an optically clear glass reliably.

A comparison of numerical derivatives of the absorption spectrum with a Stark spectrum will serve to illustrate the information content inherent in

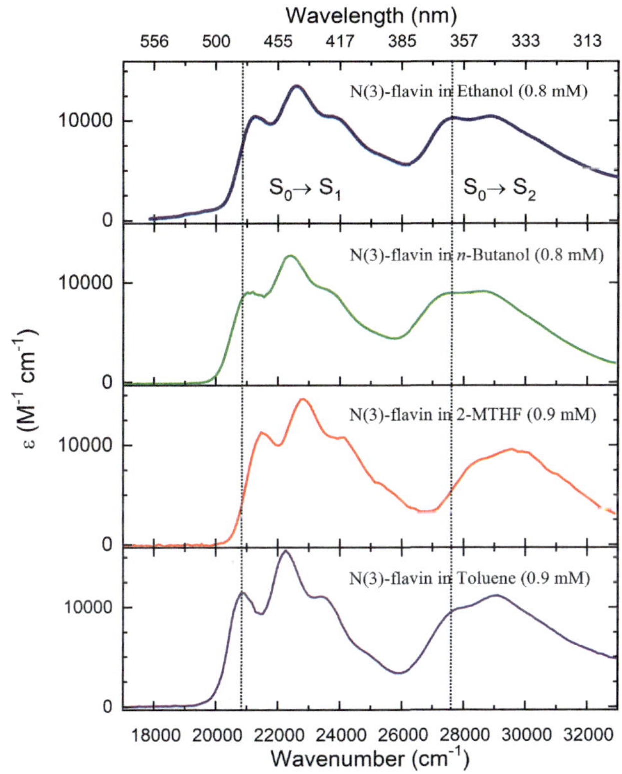

Fig. 5 Low temperature (77 K) absorption spectra of *N*(3)-methyl,*N*(10)-isobutylisoalloxazine in solvents of decreasing polarity.

Stark spectroscopy. Consider Fig. 6 which replots the 77 K absorption spectrum of *N*(3)-methyl-TPARF in a toluene glass, this time plotted by wavelength rather than energy. In panels (B) and (C), the first and second numerical derivatives of A are plotted. In panel (D), the Stark spectra are plotted for two different values of χ. All the plots are *energy-weighted* (the extinction is divided by the wavenumber, $\bar{\nu}$, as required by the Liptay analysis).

The second derivative of the absorption spectrum matches the Stark spectrum in shape indicating that charge redistribution is mostly dipolar (C_χ, Eq. 10) in nature. The lack of agreement in the amplitudes for the two different transitions highlights the ability of Stark spectroscopy to separate transitions with different charge redistributions. Here $|\Delta\vec{\mu}_{20}| > |\Delta\vec{\mu}_{10}|$ since the Stark spectrum for the S_{20} transition is much larger in amplitude than the corresponding range in panel (C), while the S_{10} transition is of

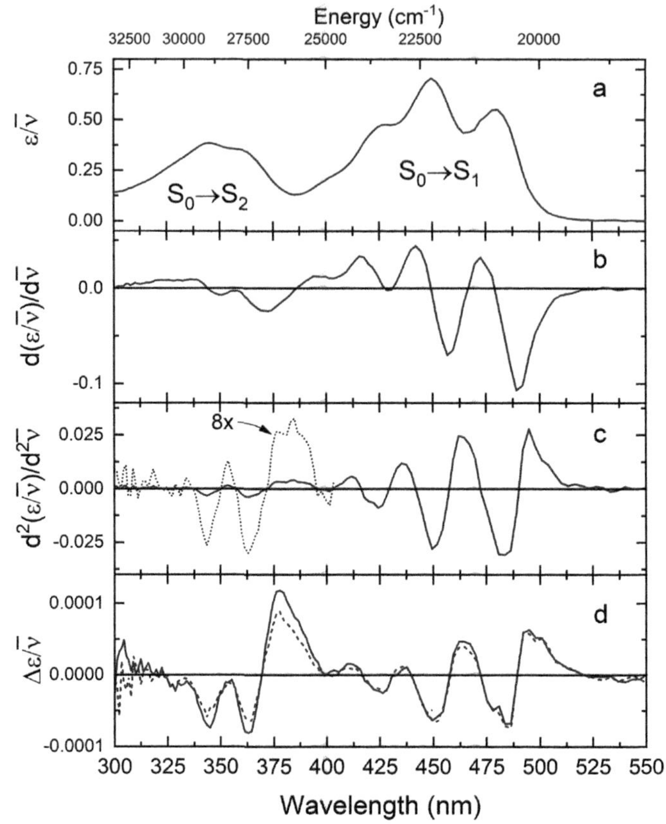

Fig. 6 LT energy-weighted absorption spectrum (A), with first (B), and second (C) derivatives for 3.3 mM *N*(3)-TPARF in toluene. Panel (D) shows the $\chi = 55°$ (–) and 90° (- - -) Stark spectra.

the small relative magnitude. This part of the second derivative spectrum was enlarged eightfold, i.e. $C_\chi^{20} \sim 8C_\chi^{10}$ so that $|\Delta\vec{\mu}_{20}| \sim \sqrt{8}|\Delta\vec{\mu}_{10}|$.

We exploited this sensitivity to different overlapping transitions to demonstrate unequivocally that the one-transition assumption that the near-U*V*/visible absorption band of reduced anionic flavin was actually two overlapping transitions (Pauszek et al., 2016). This result has made conclusions drawn on the one-band assumption in need of revision (Kao et al., 2008; Zhao et al., 2011; Zhong, 2015), including our 2003 ultrafast study on the mechanism of DNA repair by photolyase (MacFarlane IV & Stanley, 2003).

It is also worth noting that vibronic structure in the absorption spectrum is amplified by the derivative nature of Stark spectroscopy. In fact, we have been able to resolve two overlapping progressions in the S_{10} transition as "dimples" at c. 500 and 478 nm superimposed on the more obvious progression.

Nowhere is this more evident than in the Stark spectra of flavoproteins. Since flavin binding sites are generally nonpolar and conformational freedom is substituted by distinct amino acid contacts, a remarkable unique report of this environment is *imprinted* on the Stark spectrum. This can be seen by comparing the spectra of old yellow enzyme (Hopkins & Stanley, 2003) with that of DNA photolyase (Kodali et al., 2009).

In Fig. 7(A) and (C), the 77 K ground state absorption spectra are plotted vs wavenumber (energy). The oxidized proteins show characteristic transitions from $S_0 \rightarrow S_1$ and $S_0 \rightarrow S_2$ (~20,000 and 21,200 cm^{-1} for OYE and PL, respectively). Overlapping vibronic progressions as also evident, splitting the major features into doublets. These features are more evident in the Stark spectra (normalized to field strength of 1 MV/cm) as shown in Fig. 7(B) and (D). The apparent second derivative nature of the signal suggests that the $\Delta\vec{\mu}_{f0}$ dominates charge redistribution and that $\Delta\vec{\mu}_{20} > \Delta\vec{\mu}_{10}$. The amplitude of the $S_0 \rightarrow S_2$ is significantly larger than that obtained for the $S_0 \rightarrow S_1$.

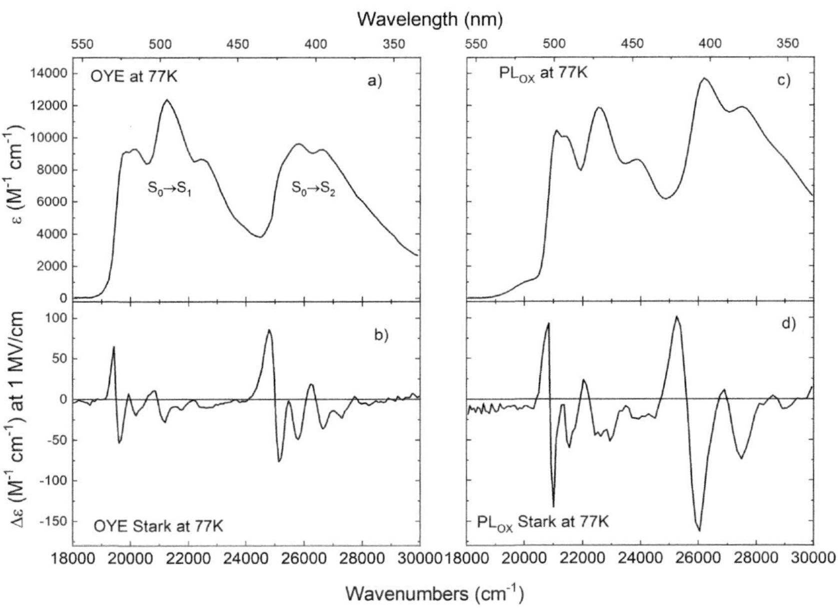

Fig. 7 77 K absorption (A/C) and Stark spectra (B/D) of OYE/DNA photolyase. The Stark spectra were taken with $\chi = 53 \pm 2°$ and normalized to 10^6 V/cm for comparison.

What is also intriguing is the difference in band splitting for $S_0 \rightarrow S_2$ transition. In OYE a series of doublets are seen around 25,000–28,000 cm^{-1}. In DNA photolyase these bands appear to merge together. By comparison, the band splitting pattern in the $S_0 \rightarrow S_1$ transition are nearly identical. We believe that these unique Stark spectra reflect differences in contacts between the protein and its flavin, as well as the protein reaction field interacting with the dipolar flavin. A long-term goal of our work is to use flavoprotein Stark spectroscopy to "read out" this information and correlate it with structure and function.

2. Electronic Stark spectroscopy: Materials and methods

2.1 Solvent selection

2.1.1 Overview

The first step in the performance of an electronic Stark spectroscopy experiment is the selection of an adequate solvent in which to dissolve the chromophore to be studied. In addition to fully solvating the desired chromophore, the selected solvent must form an optically transparent glass at the cryogenic sample temperature of 77 K characteristic of spectral acquisition. For free organic chromophores such as riboflavin, FMN, or FAD, a number of nonpolar, aprotic organic solvents are available which adequately solvate chromophore molecules, form an optically transparent glass at 77 K, and are of very low dielectric constant. Furthermore, such solvents are limited in their ability to form hydrogen bonds with solute chromophores, reducing the multiplicity of environmental electrostatic or dispersive perturbations contributing to inhomogeneous broadening of chromophore absorption bands. Organic solvents used in our group, with advantages and disadvantages, are as follows.

2.1.2 2-Methylpentane

This alkane is used primarily to dissolve chromophores possessing extended aliphatic moieties making dissolution in more polar organic solvents difficult. It is totally unable to hydrogen bond with solute chromophores, possesses a low dielectric constant, and forms an acceptably transparent glass at 77 K. The primary disadvantage of this solvent is its high vapor pressure. Bubbles form in the sample almost immediately following loading, necessitating rapid immersion in the LN2 dewar cavity immediately following sample loading. If bubbles do form in the cuvette, the absence of dielectric media in the space occupied by the bubbles will allow shorting of the system and prevent Stark signal acquisition.

2.1.3 Toluene

Toluene is more viscous than 2-methylpantane, allowing more careful, if not leisurely, sample preparation, is unable to form hydrogen bonds, and possesses a low dielectric constant. Significant disadvantages, however, include absorption below 300 nm, and the fact that toluene must be thoroughly dried by distillation onto hot molecular sieves in order to permit formation of a transparent glass. Aside from the drying requirement, toluene is an excellent solvent for chromophores with negligible absorption below 300 nm.

2.1.4 2-MTHF

2-MTHF is capable of hydrogen bonding, possesses a vapor pressure approaching that of 2-methylpentane, and is of a higher dielectric constant than either 2-methylpentane or toluene, but reliably forms nearly perfectly transparent optical glasses at 77 K even at large path lengths, up to 1 mm for a low temperature absorption spectroscopy experiment. An important caveat when working with 2-MTHF is that it is often shipped containing the peroxide-formation inhibitor butylated hydroxytoluene (BHT). BHT absorbs at 280 nm, and must be removed prior to use with chromophores absorbing into the UV.

2.1.5 Ethanol, 1-butanol

These two solvents are of high dielectric constant, but may be used when solubility considerations necessitate a solvent that is either of greater polarity or able to form hydrogen bonds. The greater extent of 1-butanol's alkane component contributes to a lower vapor pressure and lower dielectric constant relative to ethanol, but ethanol has been observed to permit formation of optically transparent glasses of greater path length than 1-butanol. If a path length of 25 μm is to be used, then 1-butanol is preferable, but for path lengths beyond 25 μm ethanol must be used.

2.1.6 Solvents for biological macromolecules

Enzymes and DNA require the use of aqueous solvents in order to limit denaturation of the macromolecular native structure. The most significant disadvantage of aqueous solvents is their high dielectric, necessitating the use of small path length cuvettes to recover field magnitude. Aqueous buffers typical to biochemistry are incapable of forming optically transparent glasses at 77 K. One option for adapting biochemical aqueous buffers to use in Stark spectroscopy is to combine the buffer solution with 8 M lithium chloride. 8 M LiCl solutions hold decent fields, form high quality glasses, and are easy to work with. The high ionic strength of the solvent, however, perturbs the electrostatic interactions that stabilize the structures of biological macromolecules, potentially denaturing the native structures of the species desired to be studied.

Perturbation of native structural states may be avoided by using a composite solvent consisting of a 35:65 volume-to-volume mixture of aqueous buffer and glycerol. The glass formed using this solvent is of inferior quality to that formed by 8 M LiCl, and glycerol may disrupt pi–pi interactions, but the reduced relative dielectric constant of the aqueous glycerol solution allows higher field magnitudes to be generated within the cuvette, contributing to sharper Stark signals.

2.2 Cuvette construction

2.2.1 Overview

The Stark cuvette is an optically transparent parallel plate capacitor, with the space between the two plates filled with the solvated chromophore to be studied. The plates are approximately 2.5 cm square pieces of glass or quartz slides of thickness 0.5 to 0.7 mm surfaced on one side by a coating that is transparent and conductive. The plates are separated from one another by strips of an insulating material, typically kapton. The distance between the plates is set by the thickness of the insulating material, for example, 25 to 50 μm. Greater path lengths hinder formation of optical glasses of sufficient transparency. By staggering the plates, it is possible to connect each plate to a separate electrical lead, permitting required charging of the plates during Stark spectrum acquisition.

Construction of high quality Stark cuvettes requires patience, dexterity, and practice. Until familiarity with cuvette construction is gained, it is advisable to practice making a few cuvettes with just solvent in order to identify difficult steps and arrive at a personalized method for cuvette construction, especially if the chromophore to be studied is expensive or difficult to obtain. Cuvette construction is accomplished in the following manner.

2.2.2 Cuvette materials

The first step in cuvette construction is to obtain the necessary materials. The two plates of the cuvette capacitor are slides of approximate dimensions 2.5 cm × 2.5 cm × 0.7 mm. The height and length of the slides to be used may vary depending on the inner diameter of the cryostat to be used in a particular instrumental set-up. The width of the slides, on the order of

0.5–0.7 mm, is not customizable in the lab. Thinner slides transmit light better but are more fragile, making use of thicker slides the safer and less frustrating choice. The slides are typically purchased in larger sections that may cut to the desired size and shape in the lab using a diamond scribe. The material of the slide to be used in a particular Stark experiment depends on the range of wavelengths required to be transmitted through the cuvette and sample. If the studied chromophore absorbs solely in the visible range (i.e. above 320 nm), then float glass slides are the most economical choice. If transitions between 300 and 320 nm will be interrogated spectroscopically, then boro-aluminosilicate glasses will suffice.

For chromophores absorbing in the UV range to a lower limit of 200 nm, more expensive quartz slides must be used. In order to permit charging of the capacitor plates, the slides must be coated on the inside surface of the cuvette with a conducting and optically transparent material. For chromophores absorbing above 260 nm, a 15–30 nm coating of indium tin oxide (ITO) is the conductive coating of choice. Although a thicker coating of ITO would improve conductivity and lower resistance, thus maximizing the frequency response of the electric field applied to the sample, diminished transmission at shorter wavelengths of the accessible range reduces sensitivity below 300 nm. For chromophores absorbing in the UV range down to and beyond 260 nm, inconel-coated fused silica neutral density filters transmit light with constant attenuation of approximately 50% to below 260 nm.

Non-adhesive kapton tape of thickness 25 μm or 50 μm must be obtained to form the insulating spacers separating the two slides of the cuvette. The spacers should be cut to a width that is approximately 20% that of the slides, and a length that is perhaps 4 mm longer than the slides.

2.2.3 Cuvette assembly

The next step in cuvette construction is to clean the slides and insulating spacers and verify the side possessing the conductive surface. Wearing clean, powderless examination gloves, hold each slide or spacer in one corner and, with a folded optical cleaning wipe saturated with acetone and held using a plastic hemostat or tweezers, wipe each side of the slide or spacer lightly in a single motion. Change the corner by which the slide or spacer is held and repeat. After cleaning, place each cuvette component onto a clean optical wipe and verify the conductive side using a multimeter set to measure resistance. The surface of the slide coated with the conductive surface will

Fig. 8 (Left) The bottom slide with the Kapton spacers. (Right) The top slide.

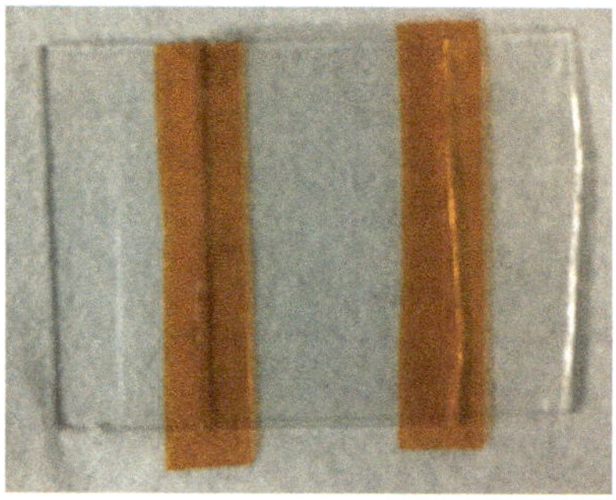

Fig. 9 The assembled cuvette.

register a non-zero, non-infinite resistivity, while the non-conducing side will register an infinite resistivity that may be reported on the device by a zero (Figs. 8 and 9).

After verifying the sides that are conductive, place one slide on the optical wipe with the conductive surface up. Position one spacer on top of the slide such that the center of the spacer is positioned immediately upon the right edge of the slide. Position the second spacer on top of the slide such that the center of the spacer is approximately 30% of the length of the slide from the left edge of the slide.

Place the second slide onto the spacers with the conductive surface facing down. The left edge of the top slide should be positioned immediately above the center of the left spacer, and the right edge of the top slide should project over the right edge of the bottom slide by approximately 20% of the length of the top slide. The initially assembled cuvette may then be picked up carefully using a hemostat or tweezers and placed onto the immobilized cold finger platform (Fig. 10). If the components shift while moving them, their relative positions may be adjusted before attaching the cuvette to the cold

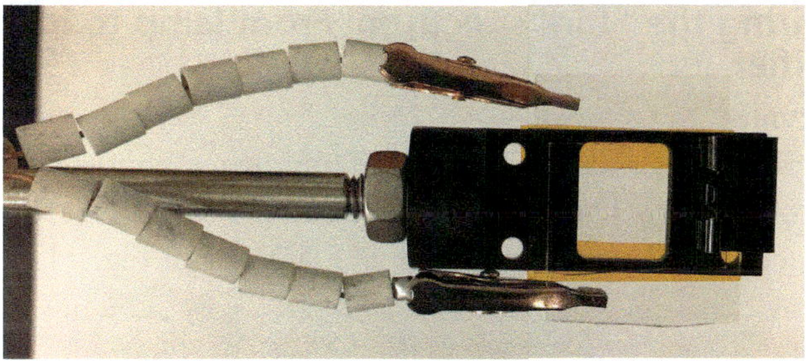

Fig. 10 The assembled cuvette attached to the cold finger platform and leads.

finger platform using a small binder clip. The arms of the binder clip may then be removed, and the extended surfaces of the top and bottom slide connected to the cold finger's electrical leads.

At this point a micrometer can be used to measure the path length of the cuvette if the slide thicknesses are known. A multimeter should then be used to determine if the cuvette has capacitance. For ITO-coated boro-aluminosilicate slides with a plate separation of 50 μm, the measured capacitance is typically on the order of approximately 300 pF. If the cuvette possesses capacitance, it may be filled using a syringe by capillary action (fill volume approximately 50 μL) and immersed in the liquid nitrogen bath of the cryostat.

2.2.4 Samples in organic solvents

Depending on the volatility of the organic solvent used, filling should be done as near to the cryostat as possible. If bubbles form in the cuvette between filling and immersion, the capacitor will likely fail upon application of the electric field and have to be reassembled.

2.2.5 Samples in aqueous solvents

Aqueous samples are sufficiently viscous to preclude loading the cuvette via capillary action. Instead, the sample must be applied as a droplet using a syringe to the bottom plate before the top plate of the cuvette is positioned. After the sample is applied, cuvette construction proceeds in the same manner described above, with the exception that care must be taken when attaching the cuvette to the cold finger platform so that the space occupied by the sample is not perturbed to the point that the sample "spills," and that the outer surfaces of the cuvette may need to be wiped prior to immersion in the case that the sample has spilled to a small extent.

2.3 Acquiring the Stark spectrum: From lamp to lock-in amplifier

2.3.1 Overview

The Stark spectrometer currently in use in our group consists of a number of discrete electronic devices and optical elements able to be optimized for study of specific chromophores. They are laid out in Fig. 11.

2.3.2 Light sources

The first component of the spectrometer is a 300 W xenon arc lamp (Hamamatsu L-2479), power supply/housing (Oriel 66485/69911, see Fig. 11A). A xenon arc lamp is used because of its stable intensity throughout the wavelength range from 200 to 800 nm, covering the UV/vis absorption of flavins. We have also used Hg lamps, but their spectral output is more "jagged" than Xe and the intrinsic stability is lower. New Xe sources have been explored as well (e.g. EQ series from Energetiq Inc.) and are recommended for those setting up a spectrometer *de novo*.

In general, the higher the power output of the lamp, the larger the arc size, thus limiting the coupling efficiency into most monochromators.

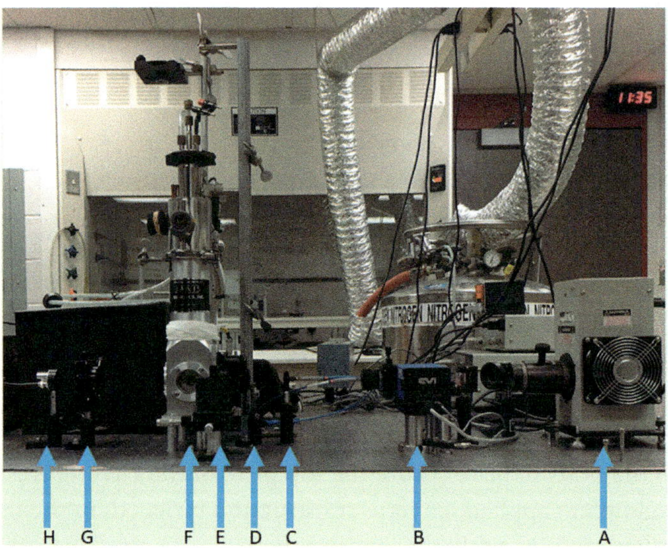

Fig. 11 An overview of the Stark spectrometer in use in our group, including the (A) xenon arc lamp and light intensity controller, (B) 1/8-m monochromator, (C) fiber optic outlet, (D) Glan–Taylor polarizer, (E) plano-convex focusing lens, (F) cryostat, (G) double convex focusing lens, and (H) photodiode. The photodiode Faraday cage and current amplifier are not shown in this image.

Short-arc lamps are preferred for their low noise and smaller image formation on the slit of the monochromator. However, lamps with output power below 150 W will probably not produce sufficient light on the sample to obtain high S/N spectra. A light intensity controller (Oriel 68945) is used to stabilize beam intensity on the scale of minutes to hours in order to limit lamp drift.

2.3.3 Monochromator

The second component of the spectrometer is a 1/8-m monochromator (CM110, Spectral Products Inc.). Almost any computer-controllable monochromator will work, as the absorption bands of flavin at 77 K are typically tens of nanometers broad (cf. Figs. 5 and 6). For example, bands in the Stark spectra of OYE (Fig. 7B) have a FWHM ~7 nm. We typically use 0.6–1.2 mm slits (bandpass of 2–4 nm, depending on the grating) to maximize light intensity while sufficiently resolving the line shape of the transitions.

Most monochromators scan in equal wavelength increments (e.g. $\Delta\lambda = 2$ nm/step), but the scanned transitions are linear in energy, not wavelength. Therefore, we have chosen to scan in equal wavenumber increments (e.g. form 300 to 600 nm with $\Delta\bar{\nu} = 111$ cm^{-1}). In this case, the step size in wavelength at 300 nm will be $\Delta\lambda(300\text{ nm}) = 1$ nm while $\Delta\lambda(600\text{ nm}) = 4$ nm. A careful simulation of this approach suggests that any artifacts are below the level of detection under typical conditions (data not shown).

2.3.4 Optics

The third, fourth, and fifth components of the spectrometer are a fiber optic cable (400 μm, UV/SR-VIS High OH Content Ocean Optics OOS-006872-09, Fig. 12C) connected to the exit port of the monochromator, a collimating telescope, a 1 cm quartz Glan–Taylor polarizer (Fig. 12D), and a simple fused silica plano-convex focusing lens (Fig. 11E). The fiber optic cable connects to the exit port of the monochromator by means of a Spectral Products F/# Matching Fiber Optic Adapter (CVI AF-CML) and depolarizes the light exiting the monochromator. A collimating telescope at the exit of the fiber (Fig. 12C) must be fused silica as well and produces a ~5 mm uniformly illuminated beam. This combination removes much of the spatial "structure" inherent in the lamp-monochromator. The light passes through the Glan–Taylor polarizer set to the desired polarization. The polarized light is then focused by the plano-convex lens (F/# $<$ 2) onto the sample.

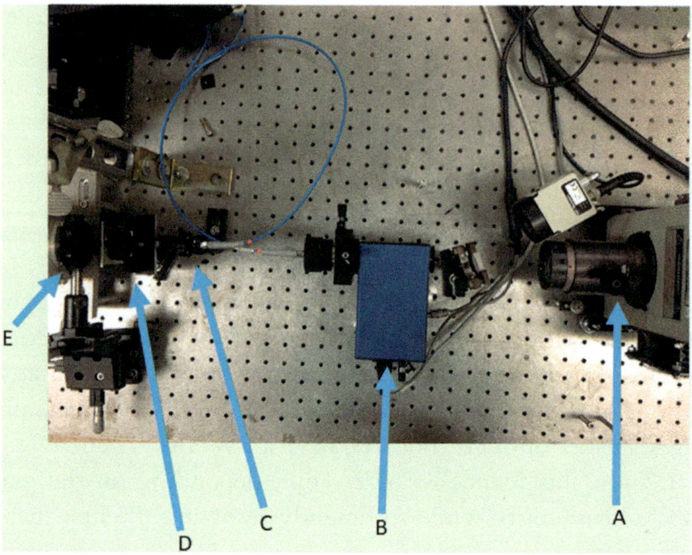

Fig. 12 A close-up of the first half of the Stark spectrometer set-up. The camera next to the lamp and mirror between the lamp and monochromator are components of the light intensity controller system. The plano-convex lens (E) is positioned so closely to the cryostat in order to focus the beam exiting the polarizer to a single point on the sample held within the cryostat.

It is desirable to have as small a spot size on the frozen glass as possible. Often frozen glasses fracture and a smaller probe beam diameter will allow good spectra to be obtained without the scattering produced by these fractures. Using an optical fiber with a diameter of 400 μm, about the width of the exit slit, and a low F/# focusing lens (Fig. 12E) will produce this condition.

2.3.5 Dewar or cryostat

The sixth component of the spectrometer is the sample assembly within the cryostat. We have used several dewars over the past 20 years, including an inexpensive H.S. Martin quad-window immersion dewar that can provide an optically quiet environment so long as the space above the LN2 is purged with helium. Our current two chamber cryostat is based on the design of Andrews and Boxer (2000) and is manufactured by JANIS Research. The unit has four fused silica windows on each chamber (Fig. 13F). The outer chamber contains a sacrificial LN2 reservoir used to reduce thermal currents in the inner immersion chamber of the dewar, thus minimizing optical noise generated by LN2 bubbling or swirling across the optical path. The standard operating procedure for filling the cryostat is as follows.

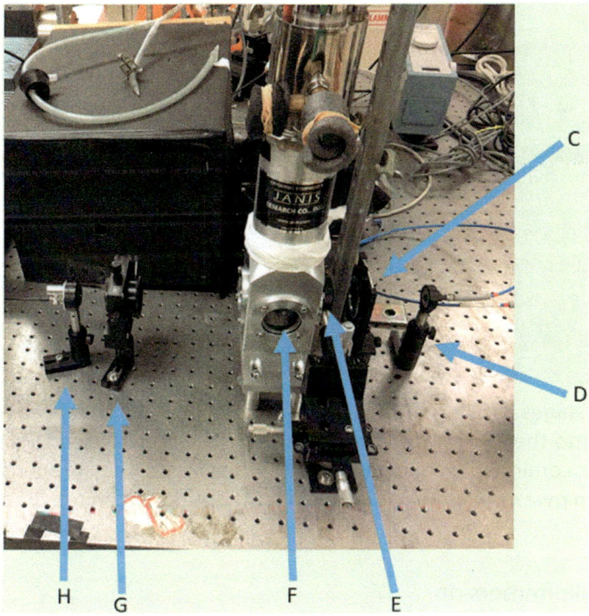

Fig. 13 A close-up of the second half of the Stark spectrometer set-up.

2.3.5.1 Filling the cryostat

(a) Purge the inner and outer dewars with anhydrous nitrogen gas for at least 40 min prior to filling with LN2 (Fig. 14A). We use a length of tygon tubing to carry the purge gas to the bottom of the inner dewar to ensure effective purging.

(b) While purging the inner dewar the outer dewar is filled with LN2 via gravity from an external transfer dewar. The filling apparatus (Fig. 14B) consists of a plastic funnel connected to a thin metal tube extending to the bottom of the outer reservoir. A conical piece of filter paper is placed in the funnel to prevent ice from entering the outer dewar. As the outer dewar fills, the cryostat as a whole cools slowly, and vaporized LN2 escapes from the outer dewar exit port. When the outer dewar is filled with LN2, excess LN2 will spurt from this port. After filling is completed the outer dewar's filling hole should be stoppered.

(c) The inner dewar is filled using a custom piece of glassware utilizing a glass frit filter to draw ice-free LN2 into the inner dewar under vacuum (Fig. 14C). After the inner dewar is filled, it is essential to stopper the top of the inner dewar with a two-holed stopper and continue to purge the inner dewar to prevent snow formation in the LN2.

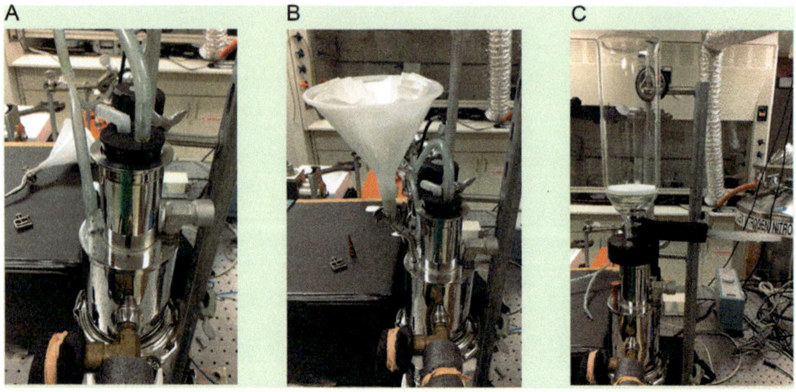

Fig. 14 Three stages of filling the cryostat with LN2. (A) The tubing carrying anhydrous nitrogen gas into the inner and outer dewars. (B) The funnel used for filling the outer dewar and the continued purging of the inner dewar. (C) The stoppered outer dewar and the custom pyrex LN2 transfer funnel with frit positioned on top of the inner dewar.

2.3.5.2 Sample immersion

The cuvette is mounted on the cold finger/flange (Fig. 15) that consists of a long steel rod with an in-house made platform attached to the end (Fig. 15D). The cuvette is attached to the nylon platform using binder clips, as discussed above.

The cold finger can be rotated to select the angle χ using a rotation stage (top of support rod in Fig. 11, Newport Inc.). Copper wire leads insulated using glass and ceramic tubes (Fig. 15B–E, Kimball Physics Inc.) are attached to the rod and connect to the cuvette via small flat-tipped alligator clips. A small amount of indium foil is applied between the clips and cuvette to maximize contact (not shown). The copper leads are attached to the rod in such a manner as to limit relative movement of the alligator clips and cuvette, limiting the probability of disconnect upon rotation to a new angle (Fig. 15C). *Warning: A high voltage hazard is present depending on the connectors used between the cold finger flange and the HVA (cf.* Fig. 15A and E*). Double check that the HVA is off before connecting or disconnecting the cold finger.* Following immersion of the cuvette in the cryostat LN2 reservoir, the sample glass should be allowed to anneal for ~45 min to attain good glass quality and stability.

2.3.6 Photodiode and current preamplifier

There are many photodiodes available. Our current favorite is the Hamamatsu S1226-44BQ, operating in photocurrent (low noise) mode. A Si photodiode

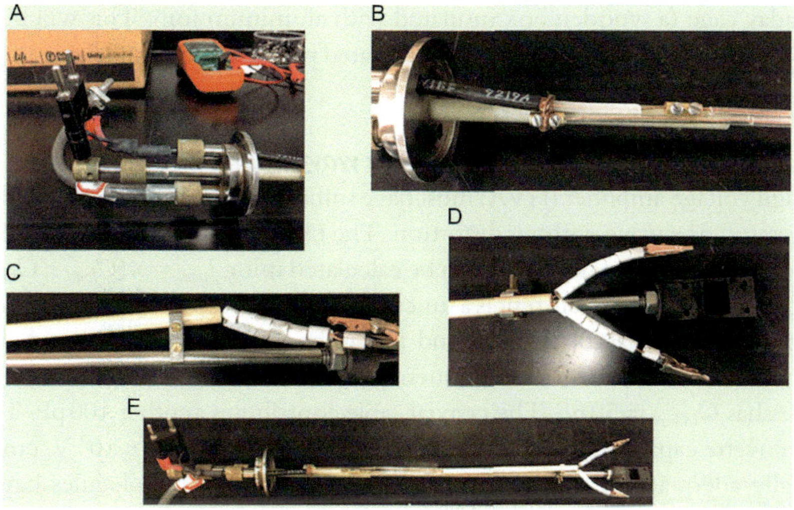

Fig. 15 Different views of the assembled cold finger. Panel (A) leads entering the external banana plugs are derived from a piece of partially stripped coaxial cable, with the cable core used to carry the output from the high voltage amplifier to one plate of the cuvette capacitor. The separated cable braid is used to ground the cuvette. (B) The partially stripped coaxial cable on the inner side of the cold finger flange. The copper connects shown in (B) above are used for convenience. Copper wires are jacketed completely in glass tubes. The ceramic pieces in (C) and (D) permit isolating rotational motion of the central rod from the transverse relative motion of the alligator clips, reducing the likelihood of disconnecting the cuvette. The black nylon platform is visible in (D) and was custom-machined.

is used for measurements from 190 to 1100 nm while a SiC photodiode (SGlux) is used for measurements from 220 to 360 nm as it conveniently rejects visible light without a loss of sensitivity. A low F/# fused silica double convex lens (Fig. 11G) focuses light transmitted by the sample onto a photodiode (Fig. 11H).

The photocurrent is amplified 10^5–10^7-fold using a low noise current amplifier (Stanford Research Systems SR570 or Keithley 427) in direct current or low pass mode. Some photodiodes however, can be purchased with an integrated current preamplifier (e.g. OSI UDT-455UV). Care must be taken in setting the gain of the amplifier so that signal components above $2f$ are amplified without distortion. A good way to check for this is to verify that the LIA output voltage of the Stark signal exactly scales with the amplifier gain.

Since the HVA produces significant through-space background noise at the applied frequency we have enclosed the photodiode in a makeshift

Faraday cage (a wooden box insulated with aluminum foil). This will block AC pickup from the HV leads. An integrated photodiode/amplifier is also an excellent way to avoid pickup.

2.3.7 *High voltage amplifier and sine wave sources*

A high voltage amplifier (HVA) must have sufficient current and bandwidth to generate a sine wave without distortion. The HVA sees the cuvette as a capacitive load, and the peak current can be calculated using $I_{peak} = \pi(C_{load} + C_{HVA}) fV_{p-p}$, where the peak current is a function of the load capacitance C_{load}, C_{HVA} is the HVA internal capacitance, and $V_{p\text{-}p}$ is the peak-to-peak voltage desired from the HVA at the desired frequency *f*. For example, the TREK 609E-6 HVA has $C_{HVA} = 50\,\text{pF}$. The coaxial cable contributes another 100 pF. Typical cuvette capacitance is $C_{load} = 400\,\text{pF}$. Field strengths of 4×10^5 V/cm are usually adequate to obtain good S/N Stark spectra for molecules having $|\Delta\mu| > 1$ D. For a 25 μm path length cuvette $V_{p\text{-}p} = 2000\,\text{V}$. Therefore $I_{peak} = \pi(400pF + 50pF)(10^{-12}F/1pF)(1.5x10^3 s^{-1})(2000\,\text{V}) = 4.2\,\text{mA}$. The Trek HVA can deliver up to 10 mA and has a large signal bandwidth of >6 kHz with only 1% distortion. For the kinds of cuvettes and solvents we use it is important to have $f \geq 1.5$ kHz. The higher the frequency the lower the noise, $N(N \propto 1/f)$.

A sine wave generator is necessary as the input to the HVA. The fidelity and stability of the generator are very important, since variations in the generator will generate noise as $\left|\vec{F}_{ext}\right|^2$. We have found that the SR830 LIA provides a pure sine wave that can be adjusted either manually or under computer control.

2.4 Acquisition of the Stark spectra

The acquisition of Stark data is facilitated using an in-house computer program written in Labview (LV 2015, National Instruments). The LV program interfaces with the LIA amplifier, as well as a 16-bit data acquisition board (NI PCI-6036E). This program will be described later but first the physical components of the data acquisition are outlined.

2.4.1 *Lock-in amplifier*

The LIA (SR830) is used as the central clock and detects and digitizes both the signal and quadrature of the signal (*X* and *Y*, respectively). With proper phasing only the signal is important, but recording/displaying both *X* and *Y* channels will expedite setting the LIA phase. The internal sine wave generator is manually set to control both *f* and the amplitude of the sine wave (in volts rms, V_{RMS}, such that the zero-peak voltage $V_{0-\text{Peak}} = \sqrt{2}V_{rms}$). This variable sine wave is amplified by a TREK 609E-6 high voltage amplifier and connected as described using the cold finger's copper leads to the two plates of the cuvette capacitor. Light transmitted by the sample is modulated at *f*, however, the resultant linear Stark effect will cancel due to symmetry, and therefore the signal is detected at 2*f*, allowing the LIA to isolate the Stark signal with a sensitivity of 10^{-6}.

Signal to noise is maximized during Stark spectrum acquisition by maximizing the frequency and voltage of the applied electric field. We operate our Stark experiments at a frequency of 3.5 kHz derived from the LIA sine wave generator. Because the modulation of the sample transmittance due to the electrochromism of the chromophore is small, even at high chromophore concentrations, a single cuvette affords both the electrochromic signal, $\Delta A\left(\vec{F}_{ext}\right) = \Delta\varepsilon\left(\vec{F}_{ext}\right)c\ell$ where $\Delta\varepsilon\left(\vec{F}_{ext}\right) = \frac{2\sqrt{2}\,I\left(\vec{F}_{ext}\right)}{2.303\,I\left(\vec{F}_{ext}=0\right)c\ell}$ and $I_{REF}(\lambda)$, *as long as* $\Delta I \ll I_{REF}$. $I_{REF}(\lambda)$ is obtained using a NI DAQ analog–digital converter (ADC) with 16-bit digitization. Since LIA signals rarely rise above 10 μV for a 4–5 V amplified photocurrent, this requirement is met. The differential extinction can be computed using Eq. (11):

$$\Delta\varepsilon\left(\vec{F}_{ext}\right) = 2\sqrt{2}I\left(\vec{F}_{ext}\right) = \frac{2\sqrt{2}I\left(\vec{F}_{ext}\right)}{2.303I\left(\vec{F}_{ext}=0\right)c\ell} \tag{11}$$

2.4.2 Phasing the LIA signal

Phasing of the LIA is challenging because the signal can be positive or negative depending on whether the transition is mostly dominated by changes in μ or α (see Fig. 2). The SR830 LIA affords measurement of both X and Y components of the signal. It is strongly recommended that a preliminary scan be performed to set the proper phase angle.

(a) Using the room or low temperature absorption spectrum as a guide, set the monochromator to the maximum absorbance, λ_{max}, of the red-most absorption band. Doing so will minimize signal cancelation artifacts due to overlapping bands.

(b) If the LIA has an auto-phase function then set the LIA time constant to 3–5 × the time constant to be used for data acquisition. Auto-phase after the signal has settled. This will give the maximum positive X output. If no auto-phase is available, then find the phase angle that gives the maximum positive signal.

(c) Run a scan and verify that the quadrature is indeed randomly wandering about $Y=0$. All the signal should now be in the X component.

(d) Once this condition has been met examine the spectrum carefully. If $\Delta\mu$ dominates the Stark spectrum then the maximum in the signal will occur at the $\lambda_{\max}$ (see Fig. 2C). However, the signal should be negative so change the LIA phase by $\pm 180°$ so that a maximum negative signal is achieved. If instead the maximum signal has red-shifted compared to $\lambda_{\max}$ then $\Delta\alpha$ dominates the signal which should be positive (see Fig. 2B) and the LIA phase is correctly set.

2.4.3 Optimizing the signal

Because the magnitude of the Stark signal is proportional to the square of the magnitude of the applied field, the magnitude of the applied field should be maximized to improve signal to noise of acquired Stark spectra. After immersion and annealing, the HV is slowly applied to the cuvette. Fields of up to $F_{ext} \geq 0.5$ MV/cm can be achieved in the best samples. The applied V_{RMS} required from the lock-in sine generator for this field and a 50 μm cuvette is $V_{RMS} = V/\left(\sqrt{2}g\right) = (5 \times 10^5\text{V/cm})(5 \times 10^{-3}\text{cm})/\left(\sqrt{2} \times 10^3\right) = 1.768$ V, where g is the gain of the HVA (1000 × in our case).

2.5 Acquiring the low temperature absorption spectrum

2.5.1 Overview

As discussed above, interpreting Stark spectra via the Liptay formalism requires an absorption spectrum measured under the same conditions as the Stark spectrum. The absorption spectrum is calculated as:

$$A(\lambda) = \log \frac{\langle I_0(\lambda) \rangle}{\langle I(\lambda) \rangle} \tag{12}$$

where $\langle I_0(\lambda) \rangle$ and $\langle I(\lambda) \rangle$ are the averaged reference and sample transmittances. Because of the use of separate sample and reference cuvettes, the acquisition of LTA spectra is the most technically difficult component of the Stark experiment. As such, signal to noise of the low temperature absorption spectrum must be maximized by limiting noise introduced by differential scattering of light by the sample and reference cuvettes through especial effort to obtain high quality glasses that are free of bubbles and cracks, and to limit the accumulation of snow within the inner dewar. To reduce the likelihood of artifacts entering into the data the Stark spectrometer is modified to acquire the low temperature absorption (LTA) spectrum in the following manner.

2.5.2 The absorption cuvette

The sample solution should be at the same concentration used for Stark spectrum acquisition. The cuvette is modified by eliminating the extending portions of the two plates used during Stark spectrum acquisition to connect the copper leads. While this modification serves no technical purpose, it makes the cuvette more compact and thus less likely to be snagged and disrupted during lowering into the cryostat. Short-path fused silica cuvettes can also be used. It is advantageous to extend the optical path length when working with weakly absorbing solutes. For example, a 50 μm cuvette filled with 1 mM flavin will only have an absorbance of 0.055 OD. Measuring the LTA in path lengths of 0.1–0.5 mm is a workaround for this. Following immersion of the cuvette in the cryostat LN2 reservoir, the reference and sample glasses should be allowed to anneal for 45 min prior to interrogation to maximize glass quality.

2.5.3 Configuring the absorption spectrometer

The Stark spectrometer is modified by removing the Glan–Taylor polarizer to increase the incident light intensity. The cuvette is positioned normal to incident light. A kHz optical chopper is added at the output of the fiber optic. The blade spacing for high frequency choppers is small, ~5 mm, which puts constraints on the width of the probe beam. This highlights an advantage of using the fiber optic described above. The chopper is run at its maximum frequency given the constraints imposed by the beam size. The LIA Amplifier is operated in "R" mode ($R = \sqrt{(X^2 + Y^2)}$). The amplified signal from the spectrometer photodiode is detected at the chopper frequency ω (not 2ω). The downstream components are otherwise identical to the Stark spectrometer set-up.

2.5.4 Data acquisition parameters

We typically acquire and average three separate sample and reference spectra at the maximum achievable LIA sensitivity. The choice of time constant always depends on field frequency, but also depends on the frequency of the noise spectrum. For the absorption spectrum it is often advantageous to use a shorter time constant (e.g. 100 ms) but increase the number of scans so the same data acquisition time is obtained as with a smaller number of scans using a 1 s time constant. The reason for this is that there are fluctuations in the dewar on a longer time period than 1 s, and increasing the number of scans while reducing the LIA time constant will reduce this "slow" noise through averaging.

3. Calculations

3.1 Overview

Absorption spectra must be converted to extinction, $\varepsilon(\lambda) = A(\lambda)/(c_{corr} \times l)$ (see Eq. 11). Here c_{corr} is the solute concentration corrected for solvent contraction. Most organic solvents contract 15–25% during immersion, as determined by measuring the height of solvent in a standard 1 mm quartz cuvette before and after freezing.

The calculation of $\Delta\varepsilon$ as a function of applied field and χ has been reviewed elsewhere, as have the fitting procedures (Pauszek et al., 2016; Pauszek, Kodali, & Stanley, 2013; Pauszek & Stanley, 2014; Stanley, 2001).

3.2 Least squares fitting of the Stark and absorption spectra

Having acquired the low temperature absorption and Stark spectra, electro-optical charge redistribution parameters are determined by simultaneously fitting all measured spectra using a custom MATLAB executable that is freely available upon request. The general strategy employed by our method proceeds as follows:

The low temperature absorption spectrum is modeled as a superposition of Gaussian functions that lacks physical meaning but is easily differentiated analytically. First and second analytical derivatives of the low temperature absorption spectrum are computed and their Gaussian parameters (three per band) are optimized to fit the LTA.

The collected Stark and LTA spectra are then fitted simultaneously. After determining sets of A_χ, B_χ, and C_χ for Stark spectra measured at $\geq 2\ \chi$-values (and perhaps multiple field strengths as well) it is possible to find unique values to the electro-optical charge redistribution parameters described in Eqs. (8)–(10). The solute does not experience the full strength of the applied external field due to the solvent cavity shape and dielectric properties. This attenuation if termed the *local field correction factor, f*. Several methods have been employed to estimate this factor and have been referenced previously. When the shape of the solvent cavity is difficult to estimate then it is still useful to report the Stark parameters in terms of *f*. Ratios of difference dipoles can then lead to use comparisons.

3.3 The ζ_A degeneracy problem

Because the Liptay formalism permits determination of only the relative orientation of the difference permanent dipole and transition dipole moments,

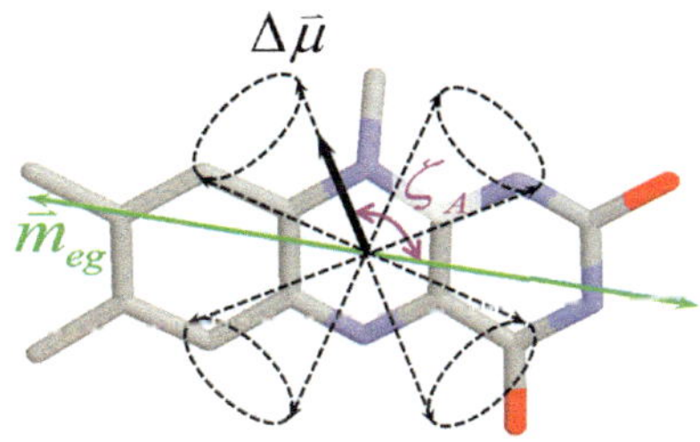

Fig. 16 The ζ_A degeneracy problem. The multiple cones represent mathematically correct solutions for the orientation of $\Delta\mu$ relative to $\vec{m}_{eg}(\zeta_A = \angle\vec{m}_{eg}, \Delta\vec{\mu}_{eg})$.

and not the absolute orientations of the two dipole moments within the structural context of the chromophore, theoretical calculations are needed to determine the absolute orientations of the difference permanent dipole and transition dipole moments. This conundrum is illustrated in Fig. 16. Quantum calculations and chemical intuition must be employed to find the most likely ζ_A.

DFT and TD-DFT calculations at the TD/B3LYP/6-311 + G(d,p) level of theory are typically sufficient to determine the absolute orientations of the fitted ζ_A and $\Delta\vec{\mu}_{eg}$ within the molecular frame of reference. For flavins though, higher levels of theory are sometimes useful. These approaches and results have been described in detail in previous papers (Kodali, Kistler, Matsika, & Stanley, 2008; Kodali, Kistler, Narayanan, Matsika, & Stanley, 2010; Kodali, Narayanan, & Stanley, 2012; Kodali et al., 2009; Pauszek III, Kodali, & Stanley, 2014; Pauszek, Kodali, Caldwell, et al., 2013; Pauszek et al., 2016; Rohwer et al., 2018). We believe that we are on the threshold of elucidating how the unique electrostatic cavity fields in proteins affect flavin energetics and catalysis.

4. Summary

Stark spectroscopy can be implemented in almost any laboratory at relatively low cost. It is unique in the information it can retrieve about cofactors in proteins and the impact of the environment surrounding them (Boxer, 2009; Bublitz & Boxer, 1997; Fried, Bagchi, & Boxer, 2014; Fried & Boxer, 2015; Wu & Boxer, 2016).

Acknowledgments

We wish to thank Professor Ronald Koder (CUNY, City College, NY) for the generous gift of *N*(3)-TPARF. This work was supported, in part, by a NASA Exobiology grant 80NSSC17K0033.

References

Andrews, S. S., & Boxer, S. G. (2000). A liquid nitrogen immersion cryostat for optical measurements. *Review of Scientific Instruments*, *71*(9), 3567–3569.

Boxer, S. G. (2009). Stark realities. *Journal of Physical Chemistry B*, *113*(10), 2972–2983.

Bublitz, G. U., & Boxer, S. G. (1997). Stark spectroscopy: Applications in chemistry, biology, and materials science. *Annual Review of Physical Chemistry*, *48*, 213–242.

Bublitz, G. U., Ortiz, R., Marder, S. R., & Boxer, S. G. (1997). Stark spectroscopy of donor/acceptor substituted polyenes. *Journal of the American Chemical Society*, *119*(14), 3365–3376.

Cantor, C. R., & Schimmel, P. R. (1980). Biophysical chemistry. *Vol. 2*. New York: W.H. Freeman and Company.

Fried, S. D., Bagchi, S., & Boxer, S. G. (2014). Extreme electric fields power catalysis in the active site of ketosteroid isomerase. *Science*, *346*(6216), 1510–1514. https://doi.org/10.1126/science.1259802.

Fried, S. D., & Boxer, S. G. (2015). Measuring electric fields and noncovalent interactions using the vibrational stark effect. *Accounts of Chemical Research*, *48*(4), 998–1006. https://doi.org/10.1021/ar500464j.

Gindt, Y. M., Edani, B. H., Olejnikova, A., Roberts, A. N., Munshi, S., & Stanley, R. J. (2016). The missing electrostatic interactions between DNA substrate and Sulfolobus solfataricus DNA photolyase: What is the role of charged amino acids in thermophilic DNA binding proteins? *Journal of Physical Chemistry B*, *120*(39), 10234–10242. https://doi.org/10.1021/acs.jpcb.6b07201.

Hall, L. N., Orchard, B. J., & Tripathy, S. K. (1987). The structure and properties of flavins: Molecular orbital study based on totally optimized geometries. I. Molecular geometry investigations. *International Journal of Quantum Chemistry*, *31*, 195–216.

Hopkins, N., & Stanley, R. J. (2003). Measurement of the electronic properties of the flavoprotein old yellow enzyme (OYE) and the OYE:P-cl phenol charge-transfer complex using stark spectroscopy. *Biochemistry*, *42*(4), 991–999.

Kao, Y.-T., Saxena, C., He, T.-F., Guo, L., Wang, L., Sancar, A., et al. (2008). Ultrafast dynamics of flavins in five redox states. *Journal of the American Chemical Society*, *130*(39), 13132–13139.

Kodali, G., Kistler, K. A., Matsika, S., & Stanley, R. J. (2008). 2-Aminopurine excited state electronic structure measured by stark spectroscopy. *Journal of Physical Chemistry B*, *112*(6), 1789–1795.

Kodali, G., Kistler, K. A., Narayanan, M., Matsika, S., & Stanley, R. J. (2010). Change in electronic structure upon optical excitation of 8-vinyladenosine: An experimental and theoretical study. *Journal of Physical Chemistry A*, *114*(1), 256–267.

Kodali, G., Narayanan, M., & Stanley, R. J. (2012). Excited-state electronic properties of 6-methylisoxanthopterin (6-MI): An experimental and theoretical study. *Journal of Physical Chemistry B*, *116*(9), 2981–2989. https://doi.org/10.1021/jp2110083.

Kodali, G., Siddiqui, S. U., & Stanley, R. J. (2009). Charge redistribution in oxidized and semiquinone E. coli DNA photolyase upon photoexcitation: Stark spectroscopy reveals a rationale for the position of Trp382. *Journal of the American Chemical Society*, *131*(13), 4795–4807.

Lee, W., Kodali, G., Stanley, R. J., & Matsika, S. (2016). Coexistence of different electron transfer mechanisms in the DNA repair process by photolyase. *Chemistry—A European Journal*, *22*(32), 11371–11381. https://doi.org/10.1002/chem.201600656.

Lienhart, W. D., Gudipati, V., & Macheroux, P. (2013). The human flavoproteome. *Archives of Biochemistry and Biophysics*, *535*(2), 150–162. https://doi.org/10.1016/j.abb.2013.02.015.

Liptay, W. (1969). Electrochromism and solvatochromism. *Angewandte Chemie International Edition*, *8*(3), 177–188.

Liptay, W. (1974). Dipole moments and polarizabilities of molecules in excited electronic states. In E. C. Lim (Ed.), *Vol. 1. Excited states* (pp. 129–229). New York: Academic Press.

MacFarlane, A. W., IV, & Stanley, R. J. (2001). Evidence of powerful substrate electric fields in DNA photolyase: Implications for thymidine dimer repair. *Biochemistry*, *40*(50), 15203–15214.

MacFarlane, A. W., IV, & Stanley, R. J. (2003). Cis-Syn thymidine dimer repair by DNA photolyase in real time. *Biochemistry*, *42*(28), 8558–8568.

Macheroux, P., Kappes, B., & Ealick, S. E. (2011). Flavogenomics—A genomic and structural view of flavin-dependent proteins. *FEBS Journal*, *278*(15), 2625–2634. https://doi.org/10.1111/j.1742-4658.2011.08202.x.

Narayanan, M., Singh, V. R., Kodali, G., Moravcevic, K., & Stanley, R. J. (2017). An ethenoadenine FAD analog accelerates UV dimer repair by DNA photolyase. *Photochemistry and Photobiology*, *93*(1), 343–354. https://doi.org/10.1111/php.12684.

Pauszek, R. F., III, Kodali, G., & Stanley, R. J. (2014). Excited state electronic structures of 5,10-methenyltetrahydrofolate and 5,10-methylenetetrahydrofolate determined by stark spectroscopy. *Journal of Physical Chemistry A*, *118*(37), 8320–8328. https://doi.org/10.1021/jp501143u.

Pauszek, R. F., Kodali, G., Caldwell, S. T., Fitzpatrick, B., Zainalabdeen, N. Y., Cooke, G., et al. (2013). Excited state charge redistribution and dynamics in the donor-pi-acceptor flavin derivative ABFL. *Journal of Physical Chemistry B*, *117*(49), 15684–15694. https://doi.org/10.1021/jp406420h.

Pauszek, R., Kodali, G., Siddiqui, M. S., & Stanley, R. J. (2011). Excited state charge redistribution in reduced flavins obtained by stark spectroscopy: One lump or two? In *Paper presented at the proceedings of the 17th international symposium on flavins and flavoproteins, Berkely, CA USA*.

Pauszek, R. F., Kodali, G., Siddiqui, M. S. U., & Stanley, R. J. (2016). Overlapping electronic states with nearly parallel transition dipole moments in reduced anionic flavin can distort photobiological dynamics. *Journal of the American Chemical Society*, *138*(45), 14880–14889. https://doi.org/10.1021/jacs.6b06449.

Pauszek, R. F., Kodali, G., & Stanley, R. J. (2013). Stark spectroscopy of flavins. In E. S. & S. W. (Eds.), *Methods in molecular biology*. New York: Humana Press.

Pauszek, R. F., & Stanley, R. J. (2014). A "how-to" guide to the stark spectroscopy of flavins and flavoproteins. In S. Weber & E. Schleicher (Eds.), *Vol. 1146. Flavins and flavoproteins: Methods and protocols* (pp. 443–466).

Reitz, J. R., Milford, F. J., & Christy, R. W. (1979). *Foundations of electromagnetic theory*. Addison Wesley Publishing Company.

Rohwer, E. J., Akbarimoosavi, M., Meckel, S. E., Liu, X. S., Geng, Y., Daku, L. M. L., et al. (2018). Dipole moment and polarizability of tunable intramolecular charge transfer states in heterocyclic pi-conjugated molecular dyads determined by computational and stark spectroscopic study. *Journal of Physical Chemistry C*, *122*(17), 9346–9355. https://doi.org/10.1021/acs.jpcc.8b02268.

Stankovich, M. T. (1991). Redox properties of flavins and flavoproteins. In F. Müller (Ed.), *Vol. I. Chemistry and biochemistry of flavoenzymes*. Boca Raton: CRC Press.

Stanley, R. J. (2001). Advances in flavin and flavoprotein optical spectroscopy. *Antioxidants & Redox Signaling*, *3*(5), 847–866.

Stanley, R. J., & MacFarlane, A. W., IV. (2000). Ultrafast excited state dynamics of oxidized flavins: Direct observations of quenching by purines. *Journal of Physical Chemistry A*, *104*(30), 6899–6906.

Wu, Y. F., & Boxer, S. G. (2016). A critical test of the electrostatic contribution to catalysis with noncanonical amino acids in ketosteroid isomerase. *Journal of the American Chemical Society*, *138*(36), 11890–11895. https://doi.org/10.1021/jacs.6b06843.

Yu, X., Eymur, S., Singh, V., Yang, B. Q., Tonga, M., Bheemaraju, A., et al. (2012). Flavin as a photo-active acceptor for efficient energy and charge transfer in a model donor–acceptor system. *Physical Chemistry Chemical Physics*, *14*(19), 6749–6754. https://doi.org/10.1039/c2cp40073a.

Zhao, R. K., Lukacs, A., Haigney, A., Brust, R., Greetham, G. M., Towrie, M., et al. (2011). Ultrafast transient mid IR to visible spectroscopy of fully reduced flavins. *Physical Chemistry Chemical Physics*, *13*(39), 17642–17648. https://doi.org/10.1039/c1cp22097g.

Zhong, D. (2015). Electron transfer mechanisms of DNA repair by photolyase. In M. A. Johnson & T. J. Martinez (Eds.), *Vol. 66. Annual review of physical chemistry* (pp. 691–715).

CHAPTER TEN

EPR spectroscopy on flavin radicals in flavoproteins

Daniel Nohr, Stefan Weber, Erik Schleicher*
Institut für Physikalische Chemie, Albert-Ludwigs-Universität Freiburg, Freiburg, Germany
*Corresponding author: e-mail address: erik.schleicher@physchem.uni-freiburg.de

Contents

Abstract

Flavin semiquinone redox states are important intermediates in a broad variety of reactions catalyzed by flavoproteins. As paramagnetic states they can be favorably probed by EPR spectroscopy in all its flavors. This review summarizes recent results in the characterization of flavin radicals. On the one hand, flavin radical states, e.g., trapped as reaction intermediates, can be characterized using modern pulsed EPR methods to unravel their electronic structure and to gain information about the surrounding environment and its changes on protein action. On the other hand, short-lived intermediate flavin radical states generated, e.g., photochemically, can be followed by time-resolved EPR, which allows a direct tracking of flavin-dependent reactions with a temporal resolution reaching nanoseconds.

1. Introduction

Flavins (FL) catalyze many different reactions of physiological importance (Fraaije & Mattevi, 2000; Macheroux, Kappes, & Ealick, 2011; Massey, 2000). Riboflavin, flavin mononucleotide (FMN), and flavin

Methods in Enzymology, Volume 620
ISSN 0076-6879
https://doi.org/10.1016/bs.mie.2019.03.013

adenine dinucleotide (FAD) have the 7,8-dimethyl isoalloxazine ring system in common, but differ in the side chain attached to N(10). With their three redox states with various degrees of protonation, fully oxidized, one-electron reduced ($FL^{\bullet-}$ and $FLH^{\bullet}$, Fig. 1) and fully reduced (FLH^{-} and FLH_2), flavins can be involved in one-electron and two-electron transfer reactions (Massey, 2000). For example, flavin semiquinones were one of the first enzymatic radicals identified in electron transfer reactions of flavoproteins and reactions of dihydroflavins with molecular oxygen (Massey, 1994). Flavin semiquinones function in terminal electron transport complexes by facilitating electron transfer between obligatory two-electron-reducing agents (e.g., NADH), and one-electron acceptors (e.g., iron sulfide centers) (Frey, 2001). Moreover, flavin semiquinones play important roles as intermediates in a number of light-activated processes ranging from blue-light photoreception (Chaves et al., 2011; Losi & Gärtner, 2012; Nohr, Rodriguez, Weber, & Schleicher, 2015; Paulus, Bajzath, Weber, & Schleicher, 2013) and magnetoreception (Hore & Mouritsen, 2016) via DNA photorepair (Zhong, 2015) to recently discovered decarboxylation reactions (Sorigué et al., 2017).

For decades, protein-bound flavin radicals have been investigated using EPR spectroscopy and special emphasis was given to unravel the electronic structure of their isoalloxazine moiety. This is because knowledge of the spin density distribution is considered essential to fully understand the chemical reactivity of the isoalloxazine ring and its catalytic versatility in different protein surroundings. Although flavin radicals can in principle exist without a protein environment, it turns out that FL/FMN/FAD radicals are barely stabilized in water as rapid disproportionation reactions between individual flavin radicals inhibit their accumulation (Ehrenberg, Müller, & Hemmerich, 1967). A number of chemically modified flavin derivatives in organic solvents (see, e.g., Ehrenberg et al., 1967; Kurreck et al., 1984; Tan &

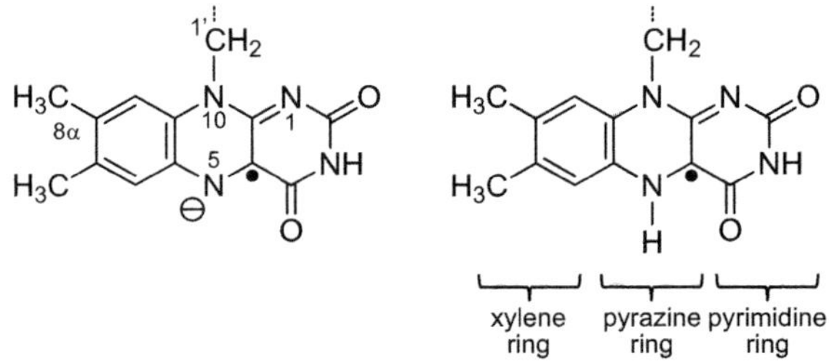

Fig. 1 Chemical structures, names of the individual heteroaromatic rings, and IUPAC numbering scheme of anionic (left) and neutral (right) flavin radicals.

Webster, 2012; Walker & Ehrenberg, 1969; Walker, Ehrenberg, & Lhoste, 1970; Weilbacher et al., 1988) have been characterized some time ago using EPR. Most recently, flavin radicals could be stabilized and first characterized in an aqueous agarose matrix, but the vast majority of data were obtained from protein-bound flavins (Rostas et al., 2018), and thus, this review is focused on protein-bound flavin radicals. It is divided into the following sections. First, **g**-tensor measurements of flavoproteins and the modulation of the principal values of **g** by the protein surroundings are discussed. Then, recent examples of application of pulsed EPR spectroscopy, in particular of pulsed electron-nuclear double resonance (ENDOR) spectroscopy, will be reviewed, and finally, time-resolved EPR spectroscopy, which is most favorably used to study photo-generated triplet states and radical pairs, will be introduced and explained.

2. The electronic structure of flavin radicals

Early EPR studies on flavoproteins were mainly focused on the identification of paramagnetic flavin redox states by using conventional continuous-wave (cw) EPR instrumentation at X-band frequencies (~9–10 GHz). Single inhomogeneously broadened EPR signals centered at $g_{iso} = 2.0034$ (see, e.g., Edmondson, 1985) were obtained, independent of whether FMN or FAD radicals were investigated. In addition, it was often the aim to unravel the protonation state of the respective flavin radical (Fig. 1). Anionic flavin radicals that are deprotonated at N(5) show peak-to-peak linewidths of up to 1.5 mT, whereas neutral flavin radicals exhibit larger spectral widths of around 1.8–2.0 mT, because of the additional large and anisotropic hyperfine coupling from the H(5) proton (Weber, Kay, Bacher, Richter, & Bittl, 2005). However, because hydrogen bonding of surrounding amino acids to either N(5) in $FL^{\bullet -}$ or from H(5) in $FLH^{\bullet}$ contributes divergently in different protein environments to the EPR resonance width, an unambiguous assignment of the protonation state is often not possible based on the EPR linewidth alone. Moreover, other spectroscopic methods such as optical spectroscopy are also able to identify the protonation state of flavin radicals (Massey, 2000) and thus, can be used in addition to EPR.

In line with recent progress on EPR instrumentation concomitant with enhanced spectral resolution, recent studies have been targeted on precisely measuring the **g**-tensor of flavin radicals to correlate it to the chemical structure of flavin semiquinones (Kay et al., 2005; Okafuji, Schnegg, Schleicher,

Möbius, & Weber, 2008; Schnegg, Kay, et al., 2006; Schnegg, Okafuji, et al., 2006). Because **g** principal values of organic radicals only deviate slightly from the free-electron value, $g_e = 2.00232$, rather large magnetic fields and correspondingly high microwave frequencies are required to resolve the small **g** anisotropies of flavin radicals that are mostly obscured by hyperfine couplings at standard X-band frequencies (Savitsky & Möbius, 2009). In general, the **g**-tensor reflects the overall electronic structure of the paramagnetic isoalloxazine ring, and thus, should be able to differentiate between different flavin radical species (e.g., non-covalently versus covalently bound, and neutral versus anionic radicals).

In Table 1, **g**-tensors obtained from high-magnetic-field/high-microwave-frequency EPR spectra of a number of flavin radicals are depicted. In detail, the neutral FAD radical of *Escherichia coli* CPD photolyase (*Ec*PL) (Fuchs et al., 2002) is compared with anionic and neutral flavin radicals of *Aspergillus niger* glucose oxidase (GO) (Okafuji et al., 2008) and NADH-ubiquinone oxidoreductase (NQR) (Barquera et al., 2003). In addition, a radical of an isoalloxazine-4(a)-methionine adduct within a photoreceptor protein environment is listed (Bittl et al., 2003). When comparing the **g**-tensors of anionic and neutral flavin radicals it is evident that the g_{iso}, g_x and g_z principal values are very similar, and only small differences of the g_y values can be detected: The latter are slightly larger in anionic radicals, and consequently, the shapes of such **g**-tensors are more rhombic than those of neutral radicals. But despite these minor differences, the g_{iso} values of non-covalently bound flavins seem mostly unaffected toward changes in the isoalloxazine's local surroundings (Schnegg, Okafuji, et al., 2006) (Table 1), which is in clear contrast to other organic radicals such as tyrosines,

Table 1 EPR parameters of different protein-bound flavin radicals.

	Anionic flavin radicals		Neutral flavin radicals			Modified flavins
	GO (Okafuji et al., 2008)	**NQR (Barquera et al., 2003)**	**GO (Okafuji et al., 2008)**	**NQR (Barquera et al., 2003)**	***Ec*PL (Fuchs et al., 2002)**	**FL-N(5)-methionine-adduct radical (Bittl, Kay, Weber, & Hegemann, 2003)**
g_{iso}	**2.00345(3)**	**2.00355(2)**	**2.0035(1)**	**2.00337(2)**	**2.00336(6)**	**2.00397(6)**
g_x	2.00429(3)	2.00436(2)	2.0043(1)	2.00425(2)	2.00431(5)	2.00554(5)
g_y	2.00389(3)	2.00402(2)	2.0036(1)	2.00360(2)	2.00360(5)	2.00391(5)
g_z	2.00216(3)	2.00228(2)	2.0021(1)	2.00227(2)	2.00217(7)	2.00247(7)

tryptophans or nitroxides; the latter are frequently used in spin-label EPR studies (Klare & Steinhoff, 2009; Lubitz, Lendzian, & Bittl, 2002; Stoll, 2011). These organic radicals typically exhibit broader distributions of **g**-principal values, depending on their immediate surrounding (Klare, 2013). The chemically altered flavin radical investigated here shows an almost fully rhombic **g**-tensor (Table 1), and thus, can be easily distinguished from unmodified flavin radicals. In sum, the **g** tensor of unmodified flavin radicals are quite "robust" and show only subtle differences in different flavoproteins; however, chemical modifications are clearly reflected by an altered **g** matrix.

In addition, high-field EPR spectra often show hyperfine couplings of two of the nitrogen atoms, namely, N(5) and N(10). It has to be mentioned that only one component, $A_{||}$, of the assumed strongly axial tensor is larger than the EPR linewidth and can be resolved in conventional EPR experiments, whereas the other principal value, $A_{\perp}$, is expected to be very small. Values of around [~50 ~0] MHz and [~30 ~0] MHz are obtained for N(5) and N(10), respectively (Okafuji et al., 2008), independent of the investigated protein.

3. Protein-cofactor interactions in flavin radicals

3.1 ENDOR spectroscopy

Pulsed (Davies-type) ENDOR spectroscopy (for an in-depth review of pulse sequence and theory, see Cox, Nalepa, Pandelia, Lubitz, & Savitsky, 2015; Harmer, 2016) reveals hyperfine couplings of a molecule by inducing NMR transitions within a paramagnetic species and surrounding hyperfine-coupled nuclei and detecting them via EPR. For this purpose, a radio-frequency (rf) source and amplifier are used. In comparison to continuous-wave EPR, a static magnetic field is applied; mostly, the ENDOR spectrum is recorded at the magnetic field position where the signal is at maximum, or, if different orientations of the hyperfine tensors should be probed, at the respective **g**-principal values. The Davies-ENDOR pulse sequence starts with a 180° (i.e., π) microwave pulse to invert the magnetization of the electronic spin system, generating an "inverted" spin system, in which NMR transitions can be induced by application of a 180° (i.e, π) rf pulse of varying frequency following the first microwave pulse. In case of resonance, the rf pulse will again invert the magnetization, thus reducing the net magnetization. The pulse sequence is completed by a standard Hahn echo (π/2–π) microwave pulse sequence. The final spectrum shows the inverted echo intensity as a function of the radio frequency. This allows a direct readout of the type of coupled nucleus by its nuclear Larmor frequency. A possible limitation of this technique is the relaxation time of the electronic spin system, which has to be long enough to apply the rf pulse that is relatively long as compared to the microwave pulses; however, this is only a minor problem when working with organic radicals in general and with flavin radicals in particular.

In general, hyperfine couplings of a particular nucleus can be directly read out from ENDOR spectra as pairs of resonance lines that are, according to the condition $\nu_{\mathrm{ENDOR}} = |\nu_{\mathrm{N}} \pm A/2|$, either equally spaced about the magnetic-field dependent nuclear Larmor frequency ν_{N} and separated by the hyperfine coupling constant A (for the case $\nu_{\mathrm{N}} > |A/2|$), or centered around $A/2$ and separated by $2\nu_{\mathrm{N}}$ (for $\nu_{\mathrm{N}} < |A/2|$). In principal, ENDOR spectra can be recorded from any nuclei as long it has a nuclear spin. Practically, most studies use proton ENDOR as it provides most information without the necessity for labeling with, e.g., ^{13}C or ^{15}N. Importantly, hyperfine couplings of nuclei are directly correlated to its electron-spin density and as a result, are sensitive probes of the local electronic structure of a paramagnetic molecule. Moreover, hyperfine couplings are quite sensitive to changes in their environment; therefore, altered hyperfine couplings can be used to gain structural information on the close surroundings of a paramagnetic molecule.

Two exemplary proton ENDOR spectra, one originating from the FADH$^{\bullet}$ of *Xenopus laevis* (6-4) photolyase, and the other from the FAD$^{\bullet-}$ of *Drosophila melanogaster* cryptochrome, are shown in Fig. 2. The spectrum of a neutral flavin radical can be divided into five parts (Fig. 2, upper panel): (i) the strong central matrix-ENDOR signal reaching from around 13 to 16 MHz, originating from weakly coupled protons such as backbone or water protons close to the flavin, or those directly attached to the isoalloxazine ring, namely, H(3), H(7α) and H(9). (ii) Hyperfine couplings at around 12 and 17 MHz are assigned to the proton H(6). (iii) The most intensive features in a proton ENDOR spectrum arise from the protons of the methyl group at C(8), which can be detected at 10–12 and 17–19 MHz. (iv) In most published spectra, at least one of the two β-protons attached to C(1′) of the ribityl side chain is observed as small shoulders at 9–10 and 19–20 MHz (Schleicher et al., 2010). Please note that the hyperfine couplings of the two β-protons strongly depend on the dihedral angle between the H(1′) protons and the isoalloxazine plane, and thus, their values can differ significantly (Heller & McConnell, 1960). (v) In protonated flavin radicals, a broad rhombic feature can be detected reaching from 20 up to 34 MHz, which is assigned to H(5).

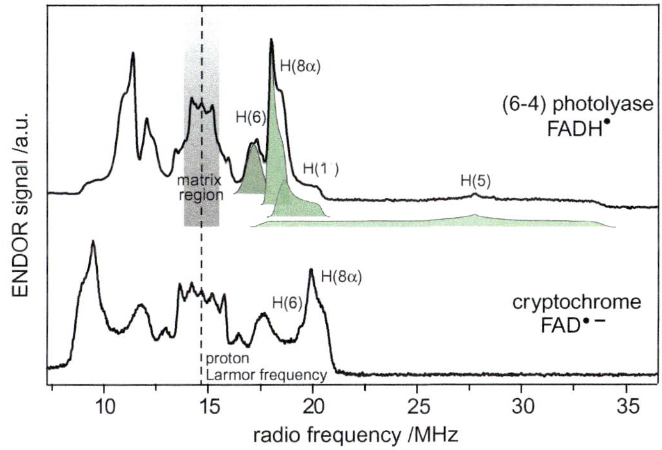

Fig. 2 Davies ENDOR spectra of FADH$^{\cdot}$ in *Xenopus laevis* (6-4) photolyase (upper panel) and FAD$^{\cdot-}$ in *Drosophila melanogaster* cryptochromes (lower panel). Selected hyperfine tensors are highlighted in green. *Adapted from Schleicher, E., Wenzel, R., Ahmad, M., Batschauer, A., Essen, L.-O., Hitomi, K., et al. (2010). The electronic state of flavoproteins: investigations with proton electron–nuclear double resonance,* Applied Magnetic Resonance, 37, *339–352.*

Despite the protonation state of the flavin radical that can be directly read out from the respective ENDOR spectrum, significantly different proton hyperfine couplings (in particular those from H(8α) and H(6)) can be observed upon transforming the anionic FAD$^{\cdot-}$ into the neutral FADH$^{\cdot}$ radical. This is because protonation of N(5) results in a redistribution of the unpaired electron-spin density from the less polar xylene ring of the isoalloxazine moiety toward the pyrazine and pyrimidine rings. Unfortunately, however, the resonances of a number of proton hyperfine couplings such as those from H(1′), H(6) and H(8α) overlap in anionic flavin radicals, which makes an unambiguous assignment rather difficult. One possibility to assist the assignment is to use calculated hyperfine couplings from DFT calculations (García et al., 2002), another is to reduce the number of overlapping hyperfine couplings in anionic flavin radicals by selective deuterium labeling; however, no experimental study has been published so far.

As mentioned above, ENDOR spectroscopy is able to unravel subtle changes in the direct vicinity of the cofactor. For instance, one study focused on the location and the protonation state of two conserved histidine residues close to the FAD cofactor in (6-4) photolyase from *Xenopus laevis* (Schleicher et al., 2007). Both amino acids were considered essential for the strongly pH-dependent enzymatic DNA repair reaction (Hitomi et al., 2001);

however, no structural information was available at that time. Two mutations, His354Ala and His358Ala (*Xenopus laevis* numbering scheme), were examined using ENDOR spectroscopy at two pH values, and compared to the wild-type enzyme. The spectra of both the WT and the mutant lacking His354 showed distinct differences between pH 6 and pH 9.5, mostly due to altered hyperfine couplings of H(8α) and H(1′). ENDOR spectra of His358Ala, on the other hand, showed no pH dependence. Therefore, His358, but not His354, changes its protonation state upon decreasing the pH. Moreover, different pH-dependent changes in the H(1′) hyperfine coupling demonstrated that His354 has to be closer to the ribityl side chain than His358 (Schleicher et al., 2007); this finding was later confirmed with the availability of the protein's crystal structure (Hitomi et al., 2009; Maul et al., 2008). The proposed reaction mechanism requires an initially protonated histidine, which only His354 is able to fulfill.

In the second example, subtle protein-cofactor interactions were probed in the so-called light-oxygen-voltage (LOV) domains. LOV domains are small protein modules containing an FMN cofactor. They are widespread in nature and serve as versatile blue-light photoreceptor modules (Christie, 2007; Losi & Gärtner, 2012; Möglich, Yang, Ayers, & Moffat, 2010; Zoltowski & Gardner, 2011). After absorption of a blue-light photon, the FMN cofactor undergoes efficient intersystem crossing within nanoseconds to yield the triplet state of FMN (Kennis et al., 2003; Swartz et al., 2001). Subsequently, a covalent thioether bond between atom C(4a) of the isoalloxazine ring and a conserved nearby cysteine residue is formed (a so-called cysteinyl-4a-adduct) (Salomon, Christie, Knieb, Lempert, & Briggs, 2000; Salomon et al., 2001), presumably via an intermediate radical pair (Schleicher et al., 2004). The photoreaction is fully reversible, and the signaling state thermally reverts to the dark state. Despite closely similar amino-acid sequence and almost identical tertiary structure, individual LOV domains differ markedly in the kinetics of their photocycle. Dark-state recovery times between a few seconds up to hours have been reported (e.g., Zoltowski, Vaccaro, & Crane, 2009), which could be modulated drastically via point mutations (e.g., Jones, Feeney, Kelly, & Christie, 2007; Raffelberg, Mansurova, Gärtner, & Losi, 2011).

Assuming that subtle changes in the cofactor environment, which alter the stability of the carbon–sulfur bond in the cysteinyl-4a-adduct, modulate the reaction speed of dark-state recovery, a study has been designed where parts of the micro-environment in the close vicinity of the FMN cofactor were altered and, via temperature-dependent ENDOR spectroscopy, the influence on the protein's reactivity was elucidated (Brosi et al., 2010). For this purpose, a LOV2 domain from *Avena sativa* with a cysteine-to-alanine mutation (*As*LOV2 Cys450Ala) was used. This mutation prevents the formation of the C4a adduct and instead leads to a meta-stable FMNH$^{\bullet}$, which has been identified to serve as a reaction intermediate analog (Kay et al., 2003). In detail, ENDOR spectroscopy detected a unique spectral behavior of this mutation as the 8α-methyl-group rotational motion was slowed down upon cooling already starting at the rather elevated temperature of 110 K (Brosi et al., 2010). To identify the amino acids that are responsible for this behavior, various protein variants with point mutations in the direct vicinity of the 8α methyl-group were examined. Alterations of three amino acids clearly led to changed temperature behavior and the hyperfine couplings of the three arrested 8α protons shifted depending on the individual mutant. Additionally, Asn425 was identified as a key amino acid for the stability of the signaling state of LOV domains (Brosi et al., 2010). Kinetic measurements confirmed that the *As*LOV2 Asn425Cys sample has a sevenfold shorter adduct-state life time as compared to the wild-type protein, most likely due to a slight reorientation of the FMN cofactor with respect to the conserved cysteine, which in the end destabilizes the intrinsically weak C—S bond of the cysteinyl-4a-adduct. This type of study was recently repeated with an engineered LOV photoreceptor YF1 using optical and EPR spectroscopy (Diensthuber et al., 2014).

In sum, proton ENDOR spectroscopy allows an easy discrimination between the two protonation states, and allows probing the molecular wave function of the paramagnetic flavin via the determination of electron-spin densities that are directly related to the respective hyperfine couplings. Moreover, the example studies highlighted here demonstrate that the surroundings of the flavin cofactor can indeed be probed via ENDOR spectroscopy, and in combination with side-directed mutagenesis, both structural and mechanistic conclusions can be drawn. In this context it has to be pointed out that the hyperfine tensor is in principle sensitive enough to detect even the slightest changes of the cofactor environment, which makes this method highly applicable for enzymatic investigations containing flavin radical intermediates.

3.2 Examples of advanced pulsed EPR spectroscopy

Pulsed EPR methods such as electron spin-echo envelope modulation (ESEEM) and hyperfine sublevel correlation (HYSCORE) are excellent tools for detecting and assigning large and anisotropic hyperfine couplings (see Prisner, Rohrer, & MacMillan, 2001; Stoll, 2017; van Doorslaer, 2017 for details on theory and application of both methods). However, only a few examples on flavin radicals have been published so far (Martínez, Alonso, Gómez-Moreno, & Medina, 1997; Martínez, Alonso, & Medina, 2012; Martínez et al., 2016; Medina et al., 1999; Medina, Vrielink, & Cammack, 1997). Here, in particular hyperfine tensors of the neutral flavin radical in *Anabaena* flavodoxin have been investigated and the hyperfine tensor of N(5) could be unraveled. However, a clear-cut assignment of the other nitrogens was not possible based on the HYSCORE analysis alone. The reason for this is that two of the nitrogens in the isoalloxazine ring (Fig. 1), N(5) and N(10), harbor large electron spin densities, whereas the other two, N(1) and N(3), do not. Correspondingly, experimental ESEEM and HYSCORE data are strongly impaired from overlapping resonances, in particular from those containing additional quadrupole interactions (the ^{14}N nucleus has spin quantum number $I=1$) and thus, inhibit an unambiguous assignment. One possibility for future studies on flavin radicals is to selectively label one or two of the nitrogens with ^{15}N ($I=½$). Although the Larmor precessional frequencies of ^{14}N and ^{15}N nuclei are rather similar, the missing quadrupole interaction in ^{15}N is expected to reduce the complexity of overlapping resonances significantly, thus enabling a disentanglement of individual hyperfine couplings.

4. Long-range interactions in flavoproteins

Pulsed electron–electron double resonance (PELDOR, also frequently called DEER) has emerged as an immensely popular tool for the measurement of inter-spin distances, e.g., in proteins and in DNA (Berliner, Eaton, & Eaton, 2000; Jeschke, 2012; Jeschke, Pannier, Godt, & Spiess, 2000). When applied to conformational analyses of proteins, PELDOR usually involves attaching spin labels at specific positions (preferably at cysteine residues) using site-directed spin labeling as most proteins do not contain intrinsic radicals (Berliner, 2010; Klare, 2013). The disadvantage of this procedure is, besides the necessity to produce various point mutants for selective spin labeling, that the structure or the function of the protein could be affected by the spin label, and that the rather flexible spin probe often results in broad and therefore difficult to analyze distance distributions (Polyhach, Bordignon, & Jeschke, 2011). In proteins that natively contain a flavin or another redox-active chromophore, one option to reduce this problem is to generate a radical state. It was shown, for example, that the FAD bound to the *Paracoccus denitrificans* ETF protein can be used in such a way, if reduced to its anionic radical form (Swanson et al., 2009). In this study, the authors were able to detect two separate conformations of the protein by analyzing the distance distributions between the flavin radical and one exogeneously introduced nitroxide spin label. An earlier study focused on the determination of the flavin-to-flavin distance in augmenter of liver regeneration dimers (Kay, Elsässer, Bittl, Farrell, & Thorpe, 2006). Here, the authors could determine a 2.6-nm distance between two neutral flavin radicals. The rather localized spin density distribution in flavin radicals around C(4a) and N(5) harbors a high potential for this PELDOR-based approach to enable the study of dimerization/multimerization states and conformational changes in flavoproteins (Nohr et al., 2015).

5. Time-resolved methods for investigating radical pairs

The previous paragraphs emphasized that one main interest of EPR characterizations of (meta)stable flavin radicals is to elucidate their electronic structures and their close surroundings. However, when trying to understand the photochemistry of flavoproteins, characterization of short-lived excited-state flavin radical species is also of great importance. After light excitation and formation of primary excited flavin states, which are typically formed within picoseconds or nanoseconds, secondary reactions can proceed. For example, excited singlet-state flavins can form longer-lived triplet states or initiate electron-transfer reactions (either as donor or as acceptor molecule), which in turn determine the fate of subsequent reactions and finally, the activity of the respective flavoprotein (Losi, Gardner, & Möglich, 2018; Nohr et al., 2015).

Several light-active flavoproteins have been identified during the last centuries. The largest families are the blue-light dependent photoreceptors containing three different classes of flavoproteins, namely, the LOV domains, the BLUF domains, and the cryptochromes (Crys) (Losi & Gärtner, 2012; Zoltowski & Gardner, 2011). Closely related to Crys are photolyases (PLs), which are DNA repair enzymes (Chaves et al., 2011; Zhong, 2015). Although all classes harbor the same isoalloxazine chromophore, they perform different reactions upon light excitation. PLs engage in two light-dependent reactions: If the FAD cofactor is in its fully reduced state, an electron can be transferred to a DNA lesion resulting in the repair of the damaged DNA (Sancar, 1992). On the other hand, if FAD is not fully reduced, electron transfer can occur via a chain of conserved tryptophans (the so-called tryptophan triad) in PLs and Crys (Aubert, Vos, Mathis, Eker, & Brettel, 2000; Giovani, Byrdin, Ahmad, & Brettel, 2003; Li, Heelis, & Sancar, 1991). The resulting intermediate [FAD···Trp] radical pair can either recombine, or, if the tryptophanyl radical is reduced by an external electron donor, the FAD remains in a one-electron reduced state. The latter reaction is proposed to form the signaling state in Crys (Banerjee et al., 2007; Bouly et al., 2007; Hoang et al., 2008). Intermediate radical pairs are found in LOV and BLUF domains, too. In LOV domains, a radical pair consisting of a conserved cysteine and the FMN in its excited triplet state is formed, which reacts further to a covalent adduct (see also Section 3.1). In BLUF domains, the intermediate radical pair consists of a conserved tyrosine and the FAD in its excited singlet state, which results in a rearrangement of the amino acids close to the FAD (Gauden et al., 2007, 2006).

Two prominent EPR methods, which have been proven to be supremely helpful while trying to understand excited flavin radical reactions, are transient EPR (trEPR) (Kim & Weissman, 1976; Weber, 2017) and the pulsed out-of-phase ESEEM (oopESEEM) (Salikhov, Kandrashkin, & Salikhov, 1992; Tang, Thurnauer, & Norris, 1994) method. TrEPR is favorable to study short-lived paramagnetic intermediates, such as triplet states (Kowalczyk, Schleicher, Bittl, & Weber, 2004; Schleicher et al., 2004; Weber et al., 2011) and singlet-born radical pairs (Bittl & Weber, 2005; Bittl & Zech, 2001; Fursman & Hore, 1999; Nohr et al., 2017; Santabarbara et al., 2005; Tang et al., 1994), by measuring directly the electron-spin polarization as a function of time at a fixed value of the external magnetic field. On the other hand, oopESEEM is used best for the characterization of spin-correlated radical pairs by measuring the modulation of the echo intensity as a function of the pulse separation time in the out-of-phase channel of the spectrometer's signal detection. The frequency modulation is directly correlated to the exchange and dipolar interactions, and is largely unaffected by Zeeman or hyperfine interactions. In the following, the reader will be given a basic understanding of these methods while simultaneously presenting recent results from the field of light-active flavoprotein investigations.

5.1 Transient EPR characterization of flavins

In general, from a trEPR spectrum of an electron-spin-polarized system, information on the excited state, such as the Zeeman interaction, hyperfine couplings, spin-spin-interactions (if the system has more than one paramagnetic center), and the zero-field splitting (in case of a high-spin system with $S > ½$), is obtained. A typical trEPR spectrum comprises of absorptive and emissive features, which, at least for singlet-born radical pairs, in sum equal to zero when integrated over the entire magnetic field range (Weber et al., 2010).

The triplet state of a flavin can form after light excitation of the fully oxidized ground state via intersystem crossing and shows a high oxidation potential (Grodowski, Veyret, & Weiss, 1977). The spectral width of the triplet signal reflects the strong mutual interaction of the unpaired electron spins in the triplet configuration. Because they are both localized on the same molecule, the spin–spin interactions are strong, and hence, trEPR spectra of flavin triplet states are rather broad. A trEPR spectrum of a triplet state is mainly characterized by its **g** tensor, the zero-field splitting parameters D and E of the dipole–dipole interaction tensor **D**, and the zero-field populations of the three triplet sublevels. FMN and FAD in frozen aqueous solutions show broader EPR spectra for lower pH than for higher pH (Kowalczyk et al., 2004), which hints toward the influence of the protonation state to the triplet-state parameters. Moreover, altered triplet spectra in a number of BLUF domains demonstrated that the triplet parameters are again sensitive to the protein environment of the FAD cofactor (Weber et al., 2011).

Typically, trEPR spectra of radical pairs are not easy to interpret by mere inspection as the number of EPR parameters that influence their spectral shapes renders analyses rather intricate. To give the reader an example of how quickly such spectra turn complex, even when only two spins are involved, and to demonstrate the influence of subtle parameter changes on the observed spectrum, spectral simulations were performed based on the theory of correlated coupled radical pairs (Closs, Forbes, & Norris, 1987; Hore, Hunter, McKie, & Hoff, 1987) (Fig. 3, the respective simulation parameters are listed in Table 2). The first three simulated spectra show two uncoupled radical species that overlap due to identical or very similar **g** tensors. Simulation 5 demonstrates how easy the spectra become complex and difficult to understand by mere inspection. Please note that spectral simulations of experimental radical pairs require more than 10 parameters (**g**-tensors, dipolar and exchange spin-spin couplings, angles between the two radicals, line-width, and a number of hyperfine couplings) (Nohr et al., 2016).

Most recently published studies on flavin-dependent, light-induced radical pairs used proteins from the Cry and PL family. Due to the rather short

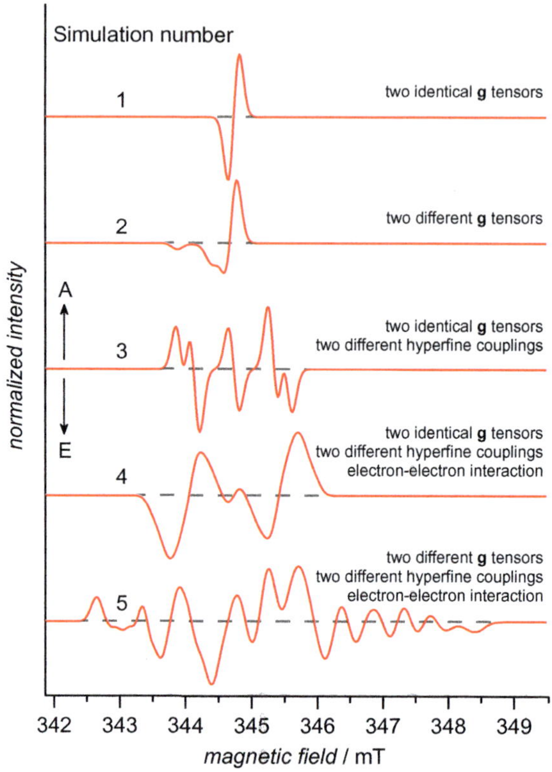

Fig. 3 Tr-EPR theory. While trEPR spectra of uncoupled biradicals (simulation 1–3) and simple radical-pair systems (simulation 4 and 5) can be described, or at least anticipated, this figure shall give the reader an example of how quickly such spectra become rather complex, even with only a few spins involved. It demonstrates that minor changes in the parameters can have a large influence on the observed spectrum. Simulation parameters are listed in Table 2.

distance of around 20 Å between the radical-pair partners, being flavin as the electron acceptor and a tryptophan or tyrosine residue as the electron donor, most of these studies were only able to make qualitative assumptions about the individual species. The reason is situated in the rather large values of the dipolar coupling (D) and the exchange coupling (J) of the spin-spin-interaction, which overlay all other interactions in the spectrum. This is not necessarily bad, as outcomes about alternative electron-transfer routes in mutants of DASH Crys from *Xenopus laevis* (Biskup et al., 2013, 2009; Weber et al., 2010) and *Synechocystis* species (Biskup et al., 2011) demonstrated. These studies identified a shift of the trEPR spectrum on the magnetic-field axis caused by the simple difference in the **g**-tensor of the

Table 2 Simulation parameters used for the simulations depicted in Fig. 3 (top to bottom).

	Simulation 1	**Simulation 2**	**Simulation 3**	**Simulation 4**	**Simulation 5**
$g_{1,x}$	2.000	2.002	2.000	2.000	2.003
$g_{1,y}$	2.000	2.000	2.000	2.000	2.008
$g_{1,z}$	2.000	2.000	2.000	2.000	2.000
$A_{1,x}$/mT	–	–	0.800	0.800	0.400
$A_{1,y}$/mT	–	–	0.800	0.800	0.700
$A_{1,z}$/mT	–	–	0.800	0.800	1.000
Line broadening/mT	0.200	0.200	0.200	0.200	0.200
$g_{2,x}$	2.000	2.000	2.000	2.000	1.980
$g_{2,y}$	2.000	2.000	2.000	2.000	2.000
$g_{2,z}$	2.000	2.005	2.000	2.000	1.995
$A_{2,x}$/mT	–	–	1.200	1.200	0.300
$A_{2,y}$/mT	–	–	1.200	1.200	0.500
$A_{2,z}$/mT	–	–	1.200	1.200	0.700
Line broadening/mT	0.200	0.200	0.200	0.200	0.300
D/mT	–	–	–	−0.500	−0.250
J/mT	–	–	–	0.150	–

A_1 is always a nitrogen spin coupling with radical 1 (**g**-tensor $\mathbf{g_1}$), A_2 is always a hydrogen spin coupling with radical 2 (**g**-tensor $\mathbf{g_2}$). All simulations were performed using a microwave frequency of 9.65 GHz.

electron-donor molecule as the acceptor was in all cases FAD. Thereby, discrimination between tryptophan as final donor in the wild-type protein and tyrosine in a mutant, which lacked one of the tryptophan triad residues, became possible. The knowledge about which tryptophan has been replaced by redox inactive phenylalanine helped immensely to identify the correct tyrosine residue.

Based on a sequence alignment it was recently proposed that animal cryptochromes and (6-4) photolyases have an extended electron-transfer pathway, which spans over four instead of three amino-acid residues (Müller, Yamamoto, Martin, Iwai, & Brettel, 2015). By using trEPR this suggestion has been verified spectroscopically for the animal cryptochrome from *Drosophila melanogaster* (*Dm*Cry) and the (6-4) photolyase from *Chlamydomonas reinhardtii* (*Cr*Cry) (Franz et al., 2018; Nohr et al., 2016). Compared to the known spectra from cryptochromes of the DASH-type, the spectra from *Dm*Cry and *Cr*Cry are much more complex and exhibit vivid hyperfine structures. This observation pointed toward a larger distance between the radical-pair partners, because lower D and J values allow other EPR parameters to actually show up. More importantly, however, the resolution of this distinct pattern of absorptive and emissive signals allowed an in-depth and quantitative spectral simulation of the full coupling tensor including $\mathbf{g}$ of both radicals, $\mathbf{A}$ of the strongly coupled nuclei, and D and J for the spin-spin-interaction. Supported by DFT calculations, this offered new insights into the coupling nuclei and to the extent they interact with the respective electron spin. Because of the involvement of magnetic nuclei in the mixing of singlet and triplet radical-pair configurations, this information is very important to tackle the question, if a magnetic molecule can fulfill the requirements of a light-induced molecular compass (Hore & Mouritsen, 2016). Additionally, the determined $\mathbf{g}$-tensors allowed to identify a tyrosine rather than a tryptophan residue to be the final electron donor in the electron-transfer chain in *Cr*Cry (Nohr et al., 2016). In contrast to the wild-type samples, mutant proteins, in which the proclaimed final electron donor was exchanged with redox-inactive phenylalanine residues, again showed spectra similar to those of the cryptochrome-DASH subclade with a shorter electron-transfer pathway (Biskup et al., 2009). Additionally, the $\mathbf{g}$-tensor shifts on the magnetic-field axis suggested tryptophan as final electron donor in both cases, supporting the assumption that even with the native final donor being inactivated the remaining triad is still active in these proteins (Fig. 4).

5.2 OopESEEM investigation of flavin-containing radical pairs

From a technical point of view, an oopESEEM experiment is a simple two-pulse ESEEM sequence, in which the pulse separation time τ is gradually prolonged. For a stable radical, the different τ values lead to a different intensity of the induced echo in the in-phase channel mostly due to hyperfine interactions, while the out-of-phase channel shows no signal at all. This is different for spin-correlated radical pairs, which also exhibit an echo modulation in the out-of-phase channel (Tang et al., 1994). If the phase was correctly adjusted and the radical started in a singlet-precursor state, the modulation period of this out-of-phase signal is solely dependent on the spin-spin-interaction values D and J (Salikhov et al., 1992), and therefore, allows a direct access to these values without the interference of other EPR parameters, see formula below (Santabarbara et al., 2005).

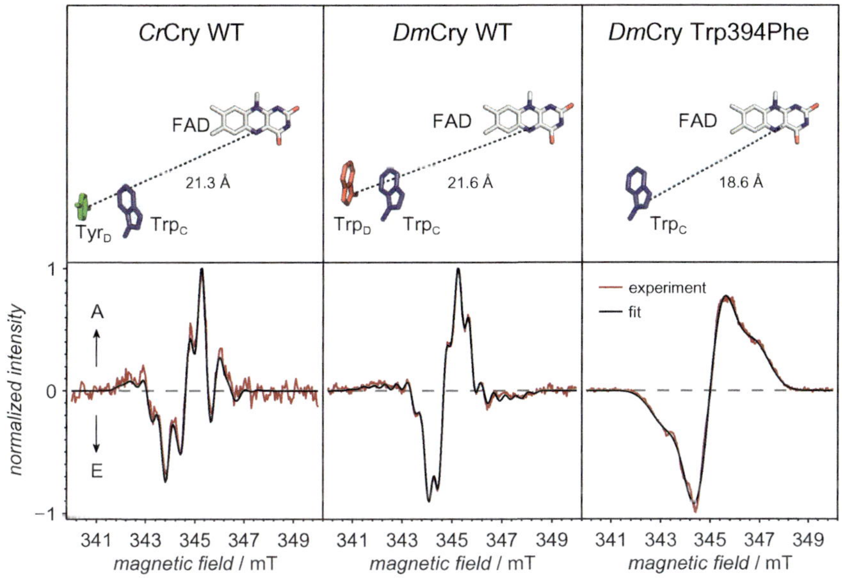

Fig. 4 Top: Distance between and composition of radical-pair partners in different samples from the Cry/PI family. Bottom: Respective normalized X-band trEPR spectra (red) at 277 K including spectral simulation (black). For experimental and simulation details, see Nohr et al. (2016). Distances between radical-pair partners were measured between points of highest electron-spin density in the radical state (C(4a)–N(5) in isoalloxazine, C(3) in tryptophan and C(1) in tyrosine; Bleifuss et al., 2001) of the following crystal structures: PDB-IDs: *Cr*Cry 6FN0 and *Dm*Cry 4JZY.

The additional parameters are H, which accounts for the amplitude of the echo by including $\mathbf{g}$ and $\mathbf{A}$. τ is the pulse separation time, chosen by the operator; τ_p, also called the phase-memory time, on the other hand is necessary to describe the weakening of the echo signal for longer values of τ. FrC and FrS are Fresnel functions.

$$S(\tau) = 2H\sqrt{\frac{\pi^3}{D\tau}}\exp\left(-\frac{\tau}{\tau_p}\right)\left[\sin\left(\frac{2(D+3J)\tau}{3}\right)\mathrm{FrC}\left(\sqrt{\frac{4D\tau}{\pi}}\right) - \cos\left(\frac{2(D+3J)\tau}{3}\right)\mathrm{FrS}\left(\sqrt{\frac{4D\tau}{\pi}}\right)\right]$$

In the case of investigating cryptochromes and photolyases, oopESEEM experiments are ideal to measure the distance between flavin and its radical-pair partner by determining D and J (Nohr et al., 2017). By determining this

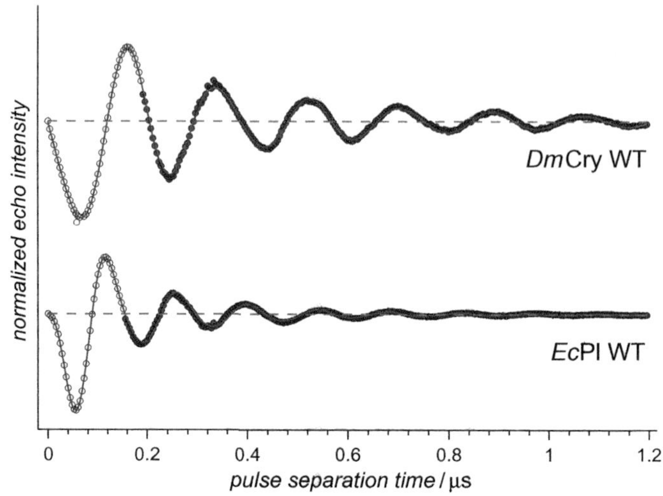

Fig. 5 Q-band oopESEEM spectra of *Dm*Cry WT (confirmed tetrad, longer distance between radical pair partners) and *Ec*Pl WT (confirmed triad, shorter distance between radical-pair partners) measured at 80 K. For experimental and simulation details, see Nohr et al. (2017). *Adapted from Nohr, D., Paulus, B., Rodriguez, R., Okafuji, A., Bittl, R., Schleicher, E. et al. (2017). Determination of radical–radical distances in light-active proteins and their implication for biological magnetoreception.* Angewandte Chemie, International Edition, 56, *8550–8554.*

distance and having structural information at hand, the identification of the terminal electron donor in the photoreduction of FAD is rather straightforward. This has been achieved with the above-mentioned *Dm*Cry as an example system for a functioning tetrad, and with *Ec*PL as an example system for a functioning triad (Fig. 5). In case of the cryptochrome, a distance of (22.4 ± 0.5) Å was found, matching the 21.6 Å between C(4a)–N(5) of flavin and C(3) of the fourth tryptophan, both points with rather high spin density, as foretold by the crystal structure. Even better is the match between the measured distance of (19.5 ± 0.5) Å by oopESEEM and the predicted distance of 20.1 Å in the crystal structure for the third tryptophan in the photolyase (Nohr et al., 2017).

6. Outlook

In recent years, a wealth of information on flavin radicals in flavoproteins has been obtained by EPR methods. In general, this technique has proven to be well suited for studying flavin radicals as reaction intermediates, because the method observes only the cofactor and its direct vicinity. Therefore, it is highly selective to changes of the active site of a flavoprotein, and these changes can be probed, e.g., via ENDOR spectroscopy (see Section 3.1). Because of the intrinsically non-destructive method, EPR spectroscopy can be applied not only in aqueous solution, but also in complex samples such as intact cells; a number of examples have been published (Banerjee et al., 2007; Bouly et al., 2007; Engelhard et al., 2014; Hoang et al., 2008).

Comparisons of EPR spectra from various flavoprotein radicals revealed that the **g**-tensor, the global probe of the flavin radical, is quite robust and is not altered particularly in different protein surroundings. Larger deviations of the principal values of the **g**-tensor have so far only been detected for chemically altered flavin radicals (see Section 2). It will be interesting to see if the EPR parameters of flavoproteins with newly discovered, naturally occurring, modified flavins, such as prenylated flavins (Marshall, Payne, & Leys, 2017), 8-formyl FAD (Robbins, Souffrant, Hamelberg, Gadda, & Bommarius, 2017) or flavin-N(5) oxides (Teufel et al., 2015), which can be stabilize radical states, are significantly different compared to those of unmodified flavin radicals.

Large proton and nitrogen hyperfine couplings, such as the ones from N(5) and H(8α), have been extracted for a number of flavin radicals, and their differences have been discussed. However, experimental information on hyperfine couplings of, e.g., the carbon backbone is still lacking. In order to correlate changes in the electronic structure of different flavoprotein radicals with their reactivity, it is essential to have full experimental access to the electron spin density. This can be achieved by using stable isotopologs, such as ^{2}H, ^{13}C or ^{15}N. In detail, various selectively labeled flavins need to be synthesized, incorporated into proteins, and the produced flavin radicals need to be analyzed spectroscopically. Although this is a challenging task, the influence of individual protein-cofactor interactions could then be directly correlated with altered electron-spin densities of the 7,8-dimethyl isoalloxazine moiety, which in turn would shed light on the catalytic variability in flavoproteins.

But not only the direct vicinity of a radical may be accessed by EPR, larger distance interactions between several electron spins can be detected as well. This provides the means to investigate conformational changes and coupled radical pairs. Finally, trEPR has emerged as a tool to analyze time-dependent dynamics in electron-transfer processes. It will be interesting to see if trEPR spectroscopy can also help to unravel the reaction mechanism of newly discovered light-dependent flavoproteins (Sorigué et al., 2017).

References

Aubert, C., Vos, M. H., Mathis, P., Eker, A. P. M., & Brettel, K. (2000). Intraprotein radical transfer during photoactivation of DNA photolyase. *Nature (London), 405*, 586–590.

Banerjee, R., Schleicher, E., Meier, S., Muñoz Viana, R., Pokorny, R., Ahmad, M., et al. (2007). The signaling state of *Arabidopsis* cryptochrome 2 contains flavin semiquinone. *The Journal of Biological Chemistry, 282*, 14916–14922.

Barquera, B., Morgan, J. E., Lukoyanov, D., Scholes, C. P., Gennis, R. B., & Nilges, M. J. (2003). X- and W-band EPR and Q-band ENDOR studies of the flavin radical in the Na^+-translocating NADH:quinone oxidoreductase from *Vibrio cholerae*. *Journal of the American Chemical Society, 125*, 265–275.

Berliner, L. J. (2010). From spin-labeled proteins to in vivo EPR applications. *European Biophysics Journal, 39*, 579–588.

Berliner, L. J., Eaton, G. R., & Eaton, S. S. (2000). *Distance measurements in biological systems by EPR*. New York: Kluwer Academic/Plenum Publishers.

Biskup, T., Hitomi, K., Getzoff, E. D., Krapf, S., Koslowski, T., Schleicher, E., et al. (2011). Unexpected electron transfer in cryptochrome identified by time-resolved EPR spectroscopy. *Angewandte Chemie, International Edition, 50*, 12647–12651.

Biskup, T., Paulus, B., Okafuji, A., Hitomi, K., Getzoff, E. D., Weber, S., et al. (2013). Variable electron transfer pathways in an amphibian cryptochrome. Tryptophan versus tyrosine-based radical pairs. *The Journal of Biological Chemistry, 288*, 9249–9260.

Biskup, T., Schleicher, E., Okafuji, A., Link, G., Hitomi, K., Getzoff, E. D., et al. (2009). Direct observation of a photoinduced radical pair in a cryptochrome blue-light photoreceptor. *Angewandte Chemie, International Edition, 48*, 404–407.

Bittl, R., Kay, C. W. M., Weber, S., & Hegemann, P. (2003). Characterization of a flavin radical product in a C57M mutant of a LOV1 domain by electron paramagnetic resonance. *Biochemistry, 42*, 8506–8512.

Bittl, R., & Weber, S. (2005). Transient radical pairs studied by time-resolved EPR. *Biochimica et Biophysica Acta, 1707*, 117–126.

Bittl, R., & Zech, S. G. (2001). Pulsed EPR spectroscopy on short-lived intermediates in photosystem I. *Biochimica et Biophysica Acta, 1507*, 194–211.

Bleifuss, G., Kolberg, M., Pötsch, S., Hofbauer, W., Bittl, R., Lubitz, W., et al. (2001). Tryptophan and tyrosine radicals in ribonucleotide reductase: A comparative high-field EPR study at 94 GHz. *Biochemistry, 40*, 15362–15368.

Bouly, J.-P., Schleicher, E., Dionisio-Sese, M., Vandenbussche, F., Van der Straeten, D., Bakrim, N., et al. (2007). Cryptochrome blue-light photoreceptors are activated through interconversion of flavin redox states. *The Journal of Biological Chemistry, 282*, 9383–9391.

Brosi, R., Illarionov, B., Mathes, T., Fischer, M., Joshi, M., Bacher, A., et al. (2010). Hindered rotation of a cofactor methyl group as a probe for protein–cofactor interaction. *Journal of the American Chemical Society, 132*, 8935–8944.

Chaves, I., Pokorny, R., Byrdin, M., Hoang, N., Ritz, T., Brettel, K., et al. (2011). The cryptochromes: Blue light photoreceptors in plants and animals. *Annual Review of Plant Biology, 62*, 335–364.

Christie, J. M. (2007). Phototropin blue-light receptors. *Annual Review of Plant Biology, 58*, 21–45.

Closs, G. L., Forbes, M. D. E., & Norris, J. R. (1987). Spin-polarized electron paramagnetic resonance spectra of radical pairs in micelles. Observation of electron spin-spin interactions. *The Journal of Physical Chemistry, 91*, 3592–3599.

Cox, N., Nalepa, A., Pandelia, M.-E., Lubitz, W., & Savitsky, A. (2015). Pulse double-resonance EPR techniques for the study of metallobiomolecules. *Methods in Enzymology, 563*, 211–249.

Diensthuber, R. P., Engelhard, C., Lemke, N., Gleichmann, T., Ohlendorf, R., Bittl, R., et al. (2014). Biophysical, mutational, and functional investigation of the chromophore-binding pocket of light-oxygen-voltage photoreceptors. *ACS Synthetic Biology*, *3*, 811–819.

Edmondson, D. E. (1985). Electron-spin-resonance studies on flavoenzymes. *Biochemical Society Transactions*, *13*, 593–600.

Ehrenberg, A., Müller, F., & Hemmerich, P. (1967). Basicity, visible spectra, and electron spin resonance of flavosemiquinone anions. *European Journal of Biochemistry*, *2*, 286–293.

Engelhard, C., Wang, X., Robles, D., Moldt, J., Essen, L.-O., Batschauer, A., et al. (2014). Cellular metabolites enhance the light sensitivity of *Arabidopsis* cryptochrome through alternate electron transfer pathways. *Plant Cell*, *26*, 4519–4531.

Fraaije, M. W., & Mattevi, A. (2000). Flavoenzymes: Diverse catalysts with recurrent features. *Trends in Biochemical Sciences*, *25*, 126–132.

Franz, S., Ignatz, E., Wenzel, S., Zielosko, H., Putu, E. P. G. N., Maestre-Reyna, M., et al. (2018). Structure of the bifunctional cryptochrome aCRY from *Chlamydomonas reinhardtii*. *Nucleic Acids Research*, *46*, 8010–8022.

Frey, P. A. (2001). Radical mechanisms of enzymatic catalysis. *Annual Review of Biochemistry*, *70*, 121–148.

Fuchs, M., Schleicher, E., Schnegg, A., Kay, C. W. M., Törring, J. T., Bittl, R., et al. (2002). The g-tensor of the neutral flavin radical cofactor of DNA photolyase revealed by 360-GHz electron paramagnetic resonance spectroscopy. *The Journal of Physical Chemistry B*, *106*, 8885–8890.

Fursman, C. E., & Hore, P. J. (1999). Distance determination in spin-correlated radical pairs in photosynthetic reaction centres by electron spin echo envelope modulation. *Chemical Physics Letters*, *303*, 593–600.

García, J. I., Medina, M., Sancho, J., Alonso, P. J., Gómez-Moreno, C., Mayoral, J. A., et al. (2002). Theoretical analysis of the electron spin density distribution of the flavin semiquinone isoalloxazine ring within model protein environments. *The Journal of Physical Chemistry. A*, *106*, 4729–4735.

Gauden, M., Grinstead, J. S., Laan, W., van Stokkum, I. H. M., Avila-Perez, M., Toh, K. C., et al. (2007). On the role of aromatic side chains in the photoactivation of BLUF domains. *Biochemistry*, *46*, 7405–7415.

Gauden, M., van Stokkum, I. H. M., Key, J. M., Lührs, D. C., van Grondelle, R., Hegemann, P., et al. (2006). Hydrogen-bond switching through a radical pair mechanism in a flavin-binding photoreceptor. *Proceedings of the National Academy of Sciences of the United States of America*, *103*, 10895–10900.

Giovani, B., Byrdin, M., Ahmad, M., & Brettel, K. (2003). Light-induced electron transfer in a cryptochrome blue-light photoreceptor. *Nature Structural Biology*, *10*, 489–490.

Grodowski, M. S., Veyret, B., & Weiss, K. (1977). Photochemistry of flavins. II. Photophysical properties of alloxazines and isoalloxazines. *Photochemistry and Photobiology*, *26*, 341–352.

Harmer, J. R. (2016). Hyperfine spectroscopy—ENDOR. *eMagRes*, *5*, 1493–1514.

Heller, C., & McConnell, H. M. (1960). Radiation damage in organic crystals. II. Electron spin resonance of $(CO_2H)CH_2CH(CO_2H)$ in β-succinic acid. *The Journal of Chemical Physics*, *32*, 1535–1539.

Hitomi, K., DiTacchio, L., Arvai, A. S., Yamamoto, J., Kim, S.-T., Todo, T., et al. (2009). Functional motifs in the (6-4) photolyase crystal structure make a comparative framework for DNA repair photolyases and clock cryptochromes. *Proceedings of the National Academy of Sciences of the United States of America*, *106*, 6962–6967.

Hitomi, K., Nakamura, H., Kim, S.-T., Mizukoshi, T., Ishikawa, T., Iwai, S., et al. (2001). Role of two histidines in the (6–4) photolyase reaction. *The Journal of Biological Chemistry*, *276*, 10103–10109.

Hoang, N., Schleicher, E., Kacprzak, S., Bouly, J.-P., Picot, M., Wu, W., et al. (2008). Human and Drosophila cryptochromes are light activated by flavin photoreduction in living cells. *PLoS Biology*, *6*, e160.1559–e160.1569.

Hore, P. J., Hunter, D. A., McKie, C. D., & Hoff, A. J. (1987). Electron paramagnetic resonance of spin-correlated radical pairs in photosynthetic reactions. *Chemical Physics Letters*, *137*, 495–500.

Hore, P. J., & Mouritsen, H. (2016). The radical-pair mechanism of magnetoreception. *Annual Review of Biophysics*, *45*, 299–344.

Jeschke, G. (2012). DEER distance measurements on proteins. *Annual Review of Physical Chemistry*, *63*, 419–446.

Jeschke, G., Pannier, M., Godt, A., & Spiess, H. W. (2000). Dipolar spectroscopy and spin alignment in electron paramagnetic resonance. *Chemical Physics Letters*, *331*, 243–252.

Jones, M. A., Feeney, K. A., Kelly, S. M., & Christie, J. M. (2007). Mutational analysis of phototropin 1 provides insights into the mechanism underlying LOV2 signal transmission. *The Journal of Biological Chemistry*, *282*, 6405–6414.

Kay, C. W. M., Elsässer, C., Bittl, R., Farrell, S. R., & Thorpe, C. (2006). Determination of the distance between the two neutral flavin radicals in augmenter of liver regeneration by pulsed ELDOR. *Journal of the American Chemical Society*, *128*, 76–77.

Kay, C. W. M., Schleicher, E., Hitomi, K., Todo, T., Bittl, R., & Weber, S. (2005). Determination of the g-matrix orientation in flavin radicals by high-field/high-frequency electron-nuclear double resonance. *Magnetic Resonance in Chemistry*, *43*, S96–S102.

Kay, C. W. M., Schleicher, E., Kuppig, A., Hofner, H., Rüdiger, W., Schleicher, M., et al. (2003). Blue light perception in plants. Detection and characterization of a light-induced neutral flavin radical in a C450A mutant of phototropin. *The Journal of Biological Chemistry*, *278*, 10973–10982.

Kennis, J. T. M., Crosson, S., Gauden, M., van Stokkum, I. H. M., Moffat, K., & van Grondelle, R. (2003). Primary reactions of the LOV2 domain of phototropin, a plant blue-light photoreceptor. *Biochemistry*, *42*, 3385–3392.

Kim, S. S., & Weissman, S. I. (1976). Detection of transient electron paramagnetic resonance. *Journal of Magnetic Resonance*, *24*, 167–169.

Klare, J. P. (2013). Site-directed spin labeling EPR spectroscopy in protein research. *Biological Chemistry*, *394*, 1281–1300.

Klare, J. P., & Steinhoff, H.-J. (2009). Spin labeling EPR. *Photosynthesis Research*, *102*, 377–390.

Kowalczyk, R. M., Schleicher, E., Bittl, R., & Weber, S. (2004). The photo-induced triplet of flavins and its protonation states. *Journal of the American Chemical Society*, *126*, 11393–11399.

Kurreck, H., Bock, M., Bretz, N., Elsner, M., Kraus, H., Lubitz, W., et al. (1984). Fluid solution and solid-state electron nuclear double resonance studies of flavin model compounds and flavoenzymes. *Journal of the American Chemical Society*, *106*, 737–746.

Li, Y. F., Heelis, P. F., & Sancar, A. (1991). Active site of DNA photolyase: Tryptophan-306 is the intrinsic hydrogen atom donor essential for flavin radical photoreduction and DNA repair in vitro. *Biochemistry*, *30*, 6322–6329.

Losi, A., Gardner, K. H., & Möglich, A. (2018). Blue-light receptors for optogenetics. *Chemical Reviews*, *118*, 10659–10709.

Losi, A., & Gärtner, W. (2012). The evolution of flavin-binding photoreceptors: An ancient chromophore serving trendy blue-light sensors. *Annual Review of Plant Biology*, *63*, 49–72.

Lubitz, W., Lendzian, F., & Bittl, R. (2002). Radicals, radical pairs and triplet states in photosynthesis. *Accounts of Chemical Research*, *35*, 313–320.

Macheroux, P., Kappes, B., & Ealick, S. E. (2011). Flavogenomics—A genomic and structural view of flavin-dependent proteins. *The FEBS Journal*, *278*, 2625–2634.

Marshall, S. A., Payne, K. A. P., & Leys, D. (2017). The UbiX-UbiD system: The biosynthesis and use of prenylated flavin (prFMN). *Archives of Biochemistry and Biophysics*, *632*, 209–221.

Martínez, J. I., Alonso, P. J., Gómez-Moreno, C., & Medina, M. (1997). One- and two-dimensional ESEEM spectroscopy of flavoproteins. *Biochemistry*, *36*, 15526–15537.

Martínez, J. I., Alonso, P. J., & Medina, M. (2012). The electronic structure of the neutral isoalloxazine semiquinone within *Anabaena* flavodoxin: New insights from HYSCORE experiments. *Journal of Magnetic Resonance*, *218*, 153–162.

Martínez, J. I., Frago, S., Lans, I., Alonso, P. J., García-Rubio, I., & Medina, M. (2016). Spin densities in flavin analogs within a flavoprotein. *Biophysical Journal*, *110*, 561–571.

Massey, V. (1994). Activation of molecular oxygen by flavins and flavoproteins. *The Journal of Biological Chemistry*, *269*, 22459–22462.

Massey, V. (2000). The chemical and biological versatility of riboflavin. *Biochemical Society Transactions*, *28*, 283–296.

Maul, M. J., Barends, T. R. M., Glas, A. F., Cryle, M. J., Domratcheva, T., Schneider, S., et al. (2008). Crystal structure and mechanism of a DNA (6–4) photolyase. *Angewandte Chemie, International Edition*, *47*, 10076–10080.

Medina, M., Lostao, A., Sancho, J., Gómez-Moreno, C., Cammack, R., Alonso, P. J., et al. (1999). Electron-nuclear double resonance and hyperfine sublevel correlation spectroscopic studies of flavodoxin mutants from *Anabaena* sp. PCC 7119. *Biophysical Journal*, *77*, 1712–1720.

Medina, M., Vrielink, A., & Cammack, R. (1997). Electron spin echo envelope modulation studies of the semiquinone anion radical of cholesterol oxidase from *Brevibacterium sterolicum*. *FEBS Letters*, *400*, 247–251.

Möglich, A., Yang, X., Ayers, R. A., & Moffat, K. (2010). Structure and function of plant photoreceptors. *Annual Review of Plant Biology*, *61*, 21–47.

Müller, P., Yamamoto, J., Martin, R., Iwai, S., & Brettel, K. (2015). Discovery and functional analysis of a 4th electron-transferring tryptophan conserved exclusively in animal cryptochromes and (6-4) photolyases. *Chemical Communications*, *51*, 15502–15505.

Nohr, D., Franz, S., Rodriguez, R., Paulus, B., Essen, L.-O., Weber, S., et al. (2016). Extended electron-transfer pathways in animal cryptochromes mediated by a tetrad of aromatic amino acids. *Biophysical Journal*, *111*, 301–311.

Nohr, D., Paulus, B., Rodriguez, R., Okafuji, A., Bittl, R., Schleicher, E., et al. (2017). Determination of radical–radical distances in light-active proteins and their implication for biological magnetoreception. *Angewandte Chemie, International Edition*, *56*, 8550–8554.

Nohr, D., Rodriguez, R., Weber, S., & Schleicher, E. (2015). How can EPR spectroscopy help to unravel molecular mechanisms of flavin-dependent photoreceptors? *Frontiers in Molecular Biosciences*, *2*[Art.-No. 49].

Okafuji, A., Schnegg, A., Schleicher, E., Möbius, K., & Weber, S. (2008). G-tensors of the flavin adenine dinucleotide radicals in glucose oxidase: A comparative multifrequency electron paramagnetic resonance and electron–nuclear double resonance study. *The Journal of Physical Chemistry B*, *112*, 3568–3574.

Paulus, B., Bajzath, C., Weber, S., & Schleicher, E. (2013). Flavoproteins and blue light reception in plants. In R. Hille, S. Miller, & B. Palfey (Eds.), *Complex flavoproteins, dehydrogenases and physical methods*: *Vol. 2. Handbook of flavoproteins* (pp. 361–392). Berlin; Boston: Walter de Gruyter GmbH.

Polyhach, Y., Bordignon, E., & Jeschke, G. (2011). Rotamer libraries of spin labelled cysteines for protein studies. *Physical Chemistry Chemical Physics*, *13*, 2356–2366.

Prisner, T., Rohrer, M., & MacMillan, F. (2001). Pulsed EPR spectroscopy: Biological applications. *Annual Review of Physical Chemistry*, *52*, 279–313.

Raffelberg, S., Mansurova, M., Gärtner, W., & Losi, A. (2011). Modulation of the photocycle of a LOV domain photoreceptor by the hydrogen-bonding network. *Journal of the American Chemical Society*, *133*, 5346–5356.

Robbins, J. M., Souffrant, M. G., Hamelberg, D., Gadda, G., & Bommarius, A. S. (2017). Enzyme-mediated conversion of flavin adenine dinucleotide (FAD) to 8-formyl FAD in formate oxidase results in a modified cofactor with enhanced catalytic properties. *Biochemistry*, *56*, 3800–3807.

Rostas, A., Einholz, C., Illarionov, B., Heidinger, L., Al Said, T., Bauss, A., et al. (2018). Long-lived hydrated FMN radicals: EPR characterization and implications for catalytic variability in flavoproteins. *Journal of the American Chemical Society*, *140*, 16521–16527.

Salikhov, K. M., Kandrashkin, Y. E., & Salikhov, A. K. (1992). Peculiarities of free induction and primary spin echo signals for spin-correlated radical pairs. *Applied Magnetic Resonance*, *3*, 199–216.

Salomon, M., Christie, J. M., Knieb, E., Lempert, U., & Briggs, W. R. (2000). Photochemical and mutational analysis of the FMN-binding domain of the plant blue light receptor, phototropin. *Biochemistry*, *39*, 9401–9410.

Salomon, M., Eisenreich, W., Dürr, H., Schleicher, E., Knieb, E., Massey, V., et al. (2001). An optomechanical transducer in the blue light receptor phototropin from *Avena sativa*. *Proceedings of the National Academy of Sciences of the United States of America*, *98*, 12357–12361.

Sancar, A. (1992). Photolyase: DNA repair by photoinduced electron transfer. In P. E. - Mariano (Ed.), *Advances in electron transfer chemistry* (pp. 215–272). London: JAI Press.

Santabarbara, S., Kuprov, I., Fairclough, W. V., Purton, S., Hore, P. J., Heathcote, P., et al. (2005). Bidirectional electron transfer in photosystem I: Determination of two distances between $P_{700}{}^{+}$ and $A_1{}^{-}$ in spin-correlated radical pairs. *Biochemistry*, *44*, 2119–2128.

Savitsky, A., & Möbius, K. (2009). High-field EPR. *Photosynthesis Research*, *102*, 311–333.

Schleicher, E., Hitomi, K., Kay, C. W. M., Getzoff, E. D., Todo, T., & Weber, S. (2007). Electron nuclear double resonance differentiates complementary roles for active site histidines in (6-4) photolyase. *The Journal of Biological Chemistry*, *282*, 4738–4747.

Schleicher, E., Kowalczyk, R. M., Kay, C. W. M., Hegemann, P., Bacher, A., Fischer, M., et al. (2004). On the reaction mechanism of adduct formation in LOV domains of the plant blue-light receptor phototropin. *Journal of the American Chemical Society*, *126*, 11067–11076.

Schleicher, E., Wenzel, R., Ahmad, M., Batschauer, A., Essen, L.-O., Hitomi, K., et al. (2010). The electronic state of flavoproteins: Investigations with proton electron–nuclear double resonance. *Applied Magnetic Resonance*, *37*, 339–352.

Schnegg, A., Kay, C. W. M., Schleicher, E., Hitomi, K., Todo, T., Möbius, K., et al. (2006). The g-tensor of the flavin cofactor in (6–4) photolyase: A 360 GHz/12.8 T electron paramagnetic resonance study. *Molecular Physics*, *104*, 1627–1633.

Schnegg, A., Okafuji, A., Bacher, A., Bittl, R., Fischer, M., Fuchs, M. R., et al. (2006). Towards an identification of chemically different flavin radicals by means of their *g*-tensor. *Applied Magnetic Resonance*, *30*, 345–358.

Sorigué, D., Légeret, B., Cuiné, S., Blangy, S., Moulin, S., Billon, E., et al. (2017). An algal photoenzyme converts fatty acids to hydrocarbons. *Science*, *357*, 903–907.

Stoll, S. (2011). High-field EPR of bioorganic radicals. *Electronparamagnetic resonance*: *Vol 22* (pp 107–154): The Royal Society of Chemistry

Stoll, S. (2017). Pulse EPR. *eMagRes*, *6*, 23–38.

Swanson, M. A., Kathirvelu, V., Majtan, T., Frerman, F. E., Eaton, G. R., & Eaton, S. S. (2009). DEER distance measurement between a spin label and a native FAD semiquinone in electron transfer Flavoprotein. *Journal of the American Chemical Society*, *131*, 15978–15979.

Swartz, T. E., Corchnoy, S. B., Christie, J. M., Lewis, J. W., Szundi, I., Briggs, W. R., et al. (2001). The photocycle of a flavin-binding domain of the blue light photoreceptor phototropin. *The Journal of Biological Chemistry*, *276*, 36493–36500.

Tan, S. L. J., & Webster, R. D. (2012). Electrochemically induced chemically reversible proton-coupled electron transfer reactions of riboflavin (vitamin B_2). *Journal of the American Chemical Society*, *134*, 5954–5964.

Tang, J., Thurnauer, M. C., & Norris, J. R. (1994). Electron spin echo envelope modulation due to exchange and dipolar interactions in a spin-correlated radical pair. *Chemical Physics Letters*, *219*, 283–290.

Teufel, R., Stull, F., Meehan, M. J., Michaudel, Q., Dorrestein, P. C., Palfey, B., et al. (2015). Biochemical establishment and characterization of EncM's Flavin-N5-oxide cofactor. *Journal of the American Chemical Society*, *137*, 8078–8085.

van Doorslaer, S. (2017). Hyperfine spectroscopy: ESEEM. *eMagRes*, *6*, 51–70.

Walker, W. H., & Ehrenberg, A. (1969). Optical and ESR spectra of flavin neutral and cation radicals in unpolar solvents. *FEBS Letters*, *3*, 315–318.

Walker, W. H., Ehrenberg, A., & Lhoste, J. M. (1970). ^{13}C studies on flavin free radicals. *Biochimica et Biophysica Acta*, *215*, 166–175.

Weber, S. (2017). Transient EPR. *eMagRes*, *6*, 255–270.

Weber, S., Biskup, T., Okafuji, A., Marino, A. R., Berthold, T., Link, G., et al. (2010). Origin of light-induced spin-correlated radical pairs in cryptochrome. *The Journal of Physical Chemistry B*, *114*, 14745–14754.

Weber, S., Kay, C. W. M., Bacher, A., Richter, G., & Bittl, R. (2005). Probing the N(5) H bond of the isoalloxazine moiety of flavin radicals by X- and W-band pulsed electron–nuclear double resonance. *ChemPhysChem*, *6*, 292–299.

Weber, S., Schroeder, C., Kacprzak, S., Mathes, T., Kowalczyk, R. M., Essen, L.-O., et al. (2011). Light-generated paramagnetic intermediates in BLUF domains. *Photochemistry and Photobiology*, *87*, 574–583.

Weilbacher, E., Helle, N., Elsner, M., Kurreck, H., Müller, F., & Allendoerfer, R. D. (1988). ^{1}H, ^{2}H, ^{19}F, ^{14}N ENDOR and TRIPLE resonance investigations of substituted flavin radicals in their different protonation states. *Magnetic Resonance in Chemistry*, *26*, 64–72.

Zhong, D. (2015). Electron transfer mechanisms of DNA repair by photolyase. *Annual Review of Physical Chemistry*, *66*, 691–715.

Zoltowski, B. D., & Gardner, K. H. (2011). Tripping the light fantastic: Blue-light photoreceptors as examples of environmentally modulated protein–protein interactions. *Biochemistry*, *50*, 4–16.

Zoltowski, B. D., Vaccaro, B., & Crane, B. R. (2009). Mechanism-based tuning of a LOV domain photoreceptor. *Nature Chemical Biology*, *5*, 827–834.

CHAPTER ELEVEN

Applications of molecular modeling to flavoproteins: Insights and challenges

Emil Sjulstok[a], Ilia A. Solov'yov[a], Peter L. Freddolino[b,*]
[a]Department of Physics, Chemistry and Pharmacy, University of Southern Denmark, Odense, Denmark
[b]Department of Biological Chemistry and Department of Computational Medicine & Bioinformatics, University of Michigan, Ann Arbor, MI, United States
*Corresponding author: e-mail address: petefred@umich.edu

Contents

Abstract

The general field of molecular simulation provides a wide spectrum of methods for studying the structure and function of biomolecules. Depending on the scale and question of interest, appropriate approaches may range from *ab initio* quantum mechanical calculations (when detailed aspects of and changes in electronic structure must be considered) to Brownian dynamics and coarse-grained molecular dynamics (to track large scale conformational motions, diffusion, and inter-molecular interactions). The entire range of molecular simulation methods has been fruitfully applied to a range of flavoenzymes, allowing researchers to address everything from the specific intermediates involved in the photoreactions of flavin chromophore-containing light sensors, to

Methods in Enzymology, Volume 620
ISSN 0076-6879
https://doi.org/10.1016/bs.mie.2019.03.014

the very long timescale motions induced by covalent modifications to bound flavin. The unique challenge posed by flavoproteins to all types of molecular simulation arises from the chemistry of the flavin isoalloxazine moiety, which presents an unusually large delocalized electron system which must be carefully treated in order to represent its contributions to the overall behavior of the system. Here we outline the particular considerations required for appropriate treatment of flavoproteins in simulations ranging from electronic structure calculations to long-timescale modeling of flavoprotein conformational transitions.

1. Introduction

Dramatic increases in the amount of available computing power for research purposes over the last 50 years have driven the development of a broad range of computational tools to supplement experimental approaches in biochemistry and molecular biology. Molecular modeling offers a range of approaches to the study of protein structure, function, and dynamics, offering increasingly accurate methods to obtain high-resolution insight into biomolecular function (recently dubbed the "computational microscope," Lee, Hsin, Sotomayor, Comellas, & Schulten, 2009). As the function and dynamics of flavoenzymes typically revolve around the reactivity and interactions of their flavin cofactor, molecular modeling calculations have proven particularly valuable in revealing the precise mechanisms that govern the interactions of the flavin with its protein surroundings and any substrates that are present.

Flavoenzymes do, however, present several unique challenges at every commonly-used level of theory. The isoalloxazine ring of flavins (Fig. 1A) presents a large, delocalized electron system which cannot be easily decomposed—thus, all electronic structure calculations must consider the entire isoalloxazine moiety, and parameterization into non-quantum mechanical models must fit a high-dimensional set of parameters to a highly constrained system. In addition, flavoenzymes make use of flavins with a massive range of covalent modifications (e.g., Fig. 1B; Fagan & Palfey, 2010) and electronic structures (redox states and spin configurations; Heelis, 1982), typically requiring higher-level quantum mechanical (QM) calculations and re-parameterization of non-QM models for each molecular state.

In this review we present key considerations that are crucial for molecular modeling of flavoproteins, covering QM calculations for the study

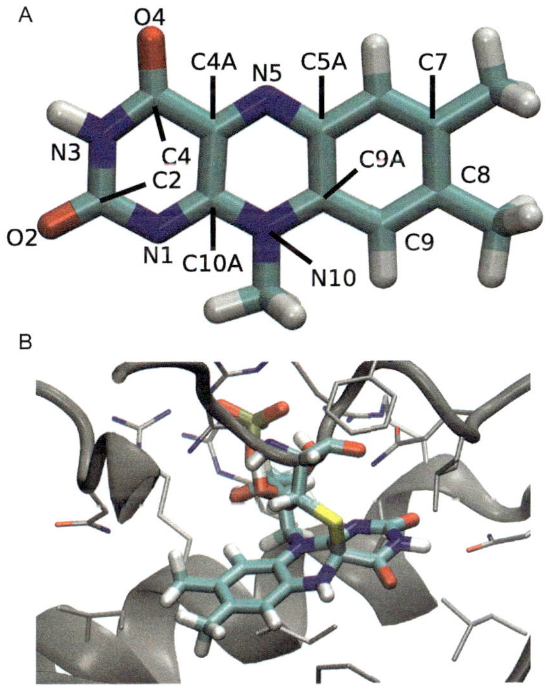

Fig. 1 Overview of flavin structure in the context of flavoenzymes. (A) QM-optimized structure of oxidized lumiflavin (isoalloxazine with a methyl group at the N10 position), a commonly used model compound in molecular modeling calculations of flavoproteins. Atom naming as shown in this figure is used throughout the text. (B) Covalent photoadduct of flavin mononucleotide with the LOV2 domain of *A. sativa* Phot1 (licorice representation with cyan carbon) in the context of the binding site. Formation of a covalent bond between a cysteine residue of the protein and the C4A atom of the flavin results in an altered conformation of the chromophore, breaking the planarity of the ring system, and protonation of the N5 atom.

of spectral properties and chemical reactions (Section 2); coupling of the electronic structure and reactions of the flavin with the overall mechanics of the flavoprotein using hybrid quantum mechanics/molecular mechanics (QM/MM) calculations (Section 3); study of longer timescale conformational motions using molecular dynamics (MD) calculations (Section 4); and extension of molecular simulations to otherwise intractable time and length scales using coarse-graining and related approaches (Section 5). At each level of resolution, we also provide representative examples of prior successful applications that demonstrate both the power of molecular modeling approaches to contribute to studies of flavoenzymes, and highlight key considerations which must be addressed in such applications. We put a particular emphasis on the study of flavin-based photoreceptor proteins, largely due to the volume of work that has been done in this particular area, but similar methods should be applicable to other classes of flavoenzymes.

2. Quantum mechanical methods

The absorption and emission properties of flavins can nowadays be studied using a wide range of computational methods. Density functional theory (DFT) (Hohenberg & Kohn, 1964; Kohn & Sham, 1965) is often employed for the ground-state calculations because of its computational efficiency and reasonable accuracy. Alternatively, Møller–Plesset perturbation theory (MP2) (Pople, Binkley, & Seeger, 1976) or the coupled-cluster (CC) method (Christiansen, Koch, & Jørgensen, 1995) can be utilized. The energies of the excited-states of the flavin and its derivatives are often computed with the time-dependent DFT (TD-DFT) method (Runge & Gross, 1984), which in its formulation is, in principle, similar to the ground state DFT analog.

Flavin excited states could also be studied using the configuration interaction (CI) theory. In its simplest formulation CI theory offers CI with single excitations (CIS), which typically overestimates the excitation energies of the chromophore. This problem is, however, improved in the so-called spin-orbit scaling SOS-CIS method (Foresman, Head-Gordon, Pople, & Frisch, 1992; Rhee & Head-Gordon, 2007), which yields a reasonable agreement between computed and experimental spectra (Khrenova, Nemukhin, Grigorenko, Krylov, & Domratcheva, 2010). Alternative CI-based analogs include the multi-reference CI method which uses the DFT Kohn-Sham molecular orbitals (DFT-MRCI) (Grimme & Waletzke, 1999) and the symmetry-adapted cluster configuration interaction (SAC-CI) method (Hasegawa, Bureekaew, & Nakatsuji, 2007; Nakatsuji, 1999).

Another class of pure quantum mechanical methods that was applied for studying flavin derivatives is based on the complete active space self-consistent field (CASSCF) method (Roos, 1972). This method combines the configuration interaction and self-consistent field theories and has been widely utilized in photochemical studies of organic molecules. The multi-reference CASSCF wavefunction permits accounting for the critically important static electron correlations, but overestimates the excitation energies because the so-called dynamic electron correlations are largely neglected. Therefore, the multi-reference second-order perturbation theory (PT2) methods have been developed, and among others include CASPT2 (Roos, Andersson, Fülscher, et al., 1988), XMCQDPT2 (Granovsky, 2011) and NEVPT2 (Angeli, Cimiraglia, Evangelisti, Leininger, & Malrieu, 2001) frameworks.

When studying a flavin chromophore quantum mechanically, the DFT method is a commonly preferred choice for obtaining the ground-state equilibrium geometries, which should then be used for calculations of the chromophore absorption properties with a more accurate excited-state method. The computational demand for establishing the optimized geometry of the flavin excited state is significantly higher as it typically has to be obtained using either the TD-DFT or the CASSCF approach; multi-reference PT2 methods are still hardly applicable to that problem due to the high computational cost. Combined approaches for optimizing geometry in one quantum mechanical method, while computing the properties with a different approach have shown to be efficient and were utilized by several groups during the last years. Specifically to establish the properties of the excited flavin chromophore, Climent, Gonzalez-Luque, Merchan, and Serrano-Andres (2006), Domratcheva, Fedorov, and Schlichting (2006), Udvarhelyi and Domratcheva (2011), and Solov'yov, Domratcheva, Moughal Shahi, and Schulten (2012) relied on PT2-CASSCF protocols, while Salzmann, Silva, Thiel, and Marian (2009); Salzmann, Tatchen, and Marian (2008), Sadeghian and Schütz (2007) and Sadeghian, Bocola, and Schütz (2008, 2010) used MRCI-TD-DFT and CC2-TD-DFT methods, respectively.

It is also important to consider the basis set used for the wavefunction expansion when studying flavins with pure quantum mechanical methods. Medium-sized basis sets, such as, e.g., 6-31G*, DZVP, TZVP, or cc-pVDZ, are capable of reproducing experimental observations for flavins rather satisfactorily (Domratcheva et al., 2006; Solov'yov et al., 2012; Udvarhelyi & Domratcheva, 2011), but may need to be extended once the highly correlated methods, mentioned above, are used.

Flavoproteins are often activated by light, and the corresponding dark and light states are typically defined based on the observed spectroscopic and sensory properties. The terminology usually suggests that flavoproteins in the dark state absorb light at a certain characteristic wavelengths, whereas the light state initiates the signaling cascade. For some flavoproteins the crystal structures have been determined for both conformations, and, therefore, these structures can be used directly for computational modeling of the spectroscopic properties. In many cases, the protein dynamics in solution are experimentally observed with tools that in-turn are linking the spectral shifts to the structural changes of the protein. To compute the corresponding spectrum of the protein in the aqueous environment, molecular dynamics (MD) studies are usually combined with quantum-chemical calculations. Computational spectroscopy of flavoproteins thus usually includes the following steps: (i) different conformations of the investigated flavoprotein are defined through MD snapshots and (ii) quantum-chemical models for excitation energy computations are selected and the spectra are computed. At this stage, the computed spectra may be averaged over many snapshots according to the MD statistics to indirectly account for dynamical effects in the protein and model spectral-line broadening. Molecular cluster models comprising flavin in its immediate molecular surrounding are often found useful (Domratcheva et al., 2006; Solov'yov et al., 2012; Solov'yov, Domratcheva, & Schulten, 2014; Udvarhelyi & Domratcheva, 2011) to determine flavin absorption properties using a pure quantum mechanical approach. The cluster models are first geometry optimized with the coordinates of the terminal atoms held fixed in space in order to mimic the mechanical embedding of the cluster in the protein. The IR and UV-vis spectra can then be computed for the cluster model using, for example, TD-DFT methods. TD-DFT satisfactorily reproduce the most prominent spectral features of the oxidized flavin, such as, for example the double-bond stretching frequencies in the ground state and the UV-vis absorption spectrum maxima (Sadeghian et al., 2008, 2010), but it should be noted that TD-DFT methods severely underestimate the excitation energies of the states that correspond to the electron transfer processes, that are a rather common feature in many flavoproteins (Sadeghian et al., 2010, 2008). This TD-DFT artifact must, therefore, be carefully checked in every calculation, and the corresponding correction has to be applied (Nielsen, Nørby, Kongsted, & Solov'yov, 2018).

2.1 Application to *Arabidopsis thaliana* cryptochrome-1a

To highlight the use of some of the methods mentioned above, here we present an example of a pure quantum mechanical study of cryptochrome-1a from the plant *Arabidopsis thaliana*. The cryptochrome active site model shown in Fig. 2 was constructed from the *Arabidopsis thaliana* cryptochrome crystal structure (PDB code 1U3C) (Brautigam et al., 2004). The FAD chromophore modeled by riboflavin and several amino-acid side chains were included into the computational models. The C atoms in the amino acid side chains were terminated by hydrogens and were fixed to the same

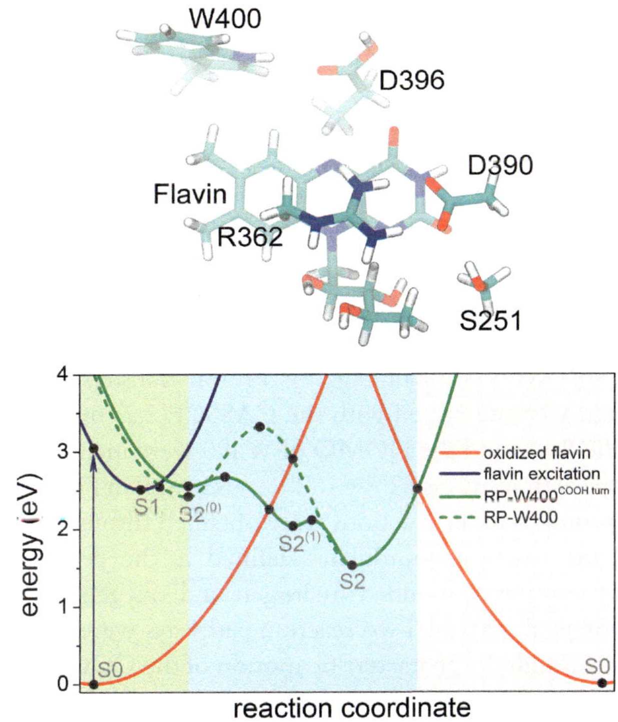

Fig. 2 *Arabidopsis thaliana* cryptochrome-1a active site model. Formation of the FADH$^{\bullet}$ +W400$^{\bullet}$ radical pair is modeled by riboflavin, and the side chains of S251, R362, D390, D396, and W400 residues terminated by hydrogens. Calculated potential energy profiles of the key electronic states describing cryptochrome photoactivation are shown in the lower part of the figure. The energy of oxidized flavin is shown in red, of excited flavin in blue, and of the radical pair state RP-W400 in green. Filled circles represent computed energies while lines show schematic potential energy surfaces. The colored background distinguishes electron transfer step (light green), proton transfer steps (light blue) and coupled electron-proton transfer step (pink).

position as in the cryptochrome crystal structure. Riboflavin was terminated by a methyl group with the C-atom fixed as in the crystal structure. Geometry optimization of the active site models in the electronic ground and excited states was performed with the state-averaged (equal weights) CASSCF method employing a protocol developed earlier (Solov'yov et al., 2012; Udvarhelyi & Domratcheva, 2011). At the located optimized geometries the excitation spectra were computed using the perturbation theory-based XMCQDPT2 method (Granovsky, 2011). The 6-31G* basis set was used in all calculations to expand the electronic wavefunctions.

In the calculations the CASSCF wavefunctions were selected according to the principal-orbital complete active space approach (Domratcheva, 2011; Domratcheva et al., 2006; Udvarhelyi & Domratcheva, 2011), where the single-electron excitations are described by including two molecular orbitals in the CASSCF active space.

Formation of the $FADH^{\bullet -} + W400^{\bullet +}$ radical pair was studied in a cryptochrome active site model involving riboflavin, S251, R362, D390, D396, and W400 (see Fig. 2). Photo-induced electron transfer was described using the CASSCF(4,3) method, where the HOMO and the LUMO of the flavin and the HOMO of W400 were included in the CASSCF active space (transitions $S0 \rightarrow S1 \rightarrow S2^{(0)}$ in Fig. 2). Proton-transfers and radical pair recombination were described with the CASSCF(2,2) method, where the LUMO of the flavin and the HOMO of W400 were included in the CASSCF active space (transitions $S2^{(0)} \rightarrow S2^{(1)} \rightarrow S2 \rightarrow S0$ in Fig. 2). The energy barriers for proton transfer reactions were obtained through relaxed energy scans along the reaction coordinates defined as the decreasing distance between non-covalently bonded hydrogen and oxygen/nitrogen atoms; see (Solov'yov et al., 2012). Two reaction pathways were considered here, which differ through the characteristic motion of the COOH– group of the D396 residue (see green solid and dashed lines in Fig. 2). The rotation of this group could severely impact on the energetics of cryptochrome activation through lowering the protonation barriers.

3. QM/MM approximation

Long-range electrostatic effects in flavoproteins can be accounted for using the hybrid QM/MM approach. QM/MM coupling schemes account for the electrostatic interactions between the quantum-mechanically modeled chromophore and the classically modeled protein. In addition, the classical non-bonding van der Waals interactions between the QM and MM atoms are computed with standard force field parameters. Since the interactions between the flavin chromophore and neighboring amino acids play a decisive role in photochemistry, including the respective amino acid side chains in the quantum-mechanical part of the QM/MM model is essential.

3.1 Hybrid QM/MM MD simulations

A direct coupling of protein dynamics with QM calculations can be achieved by a simultaneous evolution of the molecular and electronic degrees of freedom in the system. When studying flavoenzymes dynamically, it is thus important to account for the entire structure of the protein matrix; to achieve at least nanosecond time resolution in the dynamical simulations and treat hundreds of atoms quantum chemically, a number of approximations have to be made. Hierarchically, the first approximation that comes to mind is the hybrid QM/MM approach. In general, the biggest bottle neck when performing QM/MM calculations is the quantum mechanical part. Below we outline how the nuclear and electronic degrees of freedom can be efficiently coupled to permit investigations of electron transfer reactions in flavoproteins. When performing QM/MM computations the system has first to be split in a QM region, where the electron transfer occurs, and the MM region, which can be described with a classical molecular mechanics forcefield. To expedite the QM calculations, fragment molecular orbital (FMO) approach can be applied to the QM region, which allows for linear scaling between the computational time and the system size. FMO involves splitting the quantum subsystem into individual fragments, and performing quantum chemical calculations on each of these fragments. Such an approach allows for very efficient calculation of molecular properties, and can be used to treat very large systems (Lüdemann, Solov'yov, Kubar, & Elstner, 2015; Sjulstok, Lüdemann, Kubar, Elstner, & Solov'yov, 2018). As only the frontier orbitals are then considered quantum mechanically, the remaining electronic structure is described by molecular mechanics. These frontier orbitals typically represent the highest occupied molecular orbitals (HOMO's) and allow to model efficiently the excess electron or the electron hole in the system, e.g., the charge transfer reaction in flavoproteins. The interaction between the excess electron, electron hole and neutral fragments is typically modeled via point charge electrostatics, while the same assumption applies to the environment and the MM region, which interact with the QM region via point charges. The approximations outlined above require careful benchmarking, which has been performed earlier for several flavoproteins (see, e.g., Lüdemann et al., 2015; Woiczikowski, Steinbrecher, Kubar, & Elstner, 2011).

The general idea of dynamic QM/MM simulations is to propagate the electronic wavefunction, jointly with the nuclear degrees of freedom. This method, as described above, is commonly referred to as the DFT tight binding method (Elstner & Seifert, 2014; Seifert, 2007), or DFTB for short. Here the electronic structure is computed using DFTB, while the nuclear degrees of freedom are propagated using molecular dynamics; note that this approach is specifically useful for modeling charge transfer reactions, that are rather common in flavoenzymes (see e.g., Lüdemann et al., 2015; Sjulstok et al., 2018). To employ DFTB, it is useful to start with an expression for the total energy of the system. In the case of charge migration inside a flavoprotein, e.g., cryptochrome, the electron hole, not the excess electron, should be tracked, so the energy expression has to be written for the N − 1 electron system, with the density ρ. The formalism has been derived previously (Kubar & Elstner, 2010; Lüdemann, Woiczikowski, Kubar, Elstner, & Steinbrecher, 2013), by applying a Taylor like expansion of the charge neutral electron system, with the density ρ_0, up to the second order:

$$E^{N-1}[\rho] = E^N[\rho_0] - \langle \Psi | \hat{H}_0 | \Psi \rangle + E^{2\text{nd}} \tag{1}$$

here E^N is the total energy of the charge neutral system, $\langle \Psi | \hat{H}_0 | \Psi \rangle$ is the energy of the one electron in the HOMO orbital Ψ, which is subtracted from the total energy. $E^{2\text{nd}}$ is the second order term for the energy, which we will return to later.

The first term $E^N[\rho_0]$ in Eq. (1) can be identified as the classical energy of the system, which is handled through classical molecular mechanics approximation, while the two remaining terms account for the QM part. The wavefunction Ψ, can be expanded as a set of orbitals localized to the fragments of the QM region, and the expansion reads as

$$\Psi = \sum_n a_n \varphi_n, \tag{2}$$

where φ_n represents the HOMO orbitals of the fragments and a_n are the expansion coefficients. The site energies of the QM subsystem fragments can then be computed as $\epsilon_m = \left\langle \phi_m \middle| \widehat{H_0} \middle| \phi_m \right\rangle$, which correspond to the ionization potentials of the fragments, while the electronic couplings between the fragments are given by $T_{mn} = \left\langle \phi_m \middle| \widehat{H_0} \middle| \phi_n \right\rangle$.

The second order energy term in Eq. (1) can be approximated using the monopole approximation, as done earlier (Kubar & Elstner, 2010; Lüdemann et al., 2013), yielding the final expression for the energy of the system:

$$E^{N-1}[\rho] \approx E^{\text{MM}} - \sum_{mn} a_m^{\star} a_n H_{mn} - \frac{1}{2} \sum_{mn} |a_m|^2 |a_n|^2 \Gamma_{mn}. \tag{3}$$

where $H_{mm} = \epsilon_m$, are the diagonal elements of the Hamiltonian, being the site energies, and the off-diagonal elements are the electronic couplings $H_{mn} = T_{mn}$. Γ_{mm} is the Hubbard parameter of the fragment, the derivative of the site energy with respect to the charge, while the off-diagonal elements are the inverse distance between the atoms of the fragments.

The electronic degrees of freedom are propagated using the time-dependent Schrödinger equation.

$$i\hbar\frac{\partial \Psi}{\partial t} = \widehat{H}\Psi, \tag{4}$$

while the nuclear degrees of freedom are evolving classically following the derivatives of the energy of an atom α with respect to its coordinate.

$$M_\alpha \ddot{\mathbf{R}} = -\frac{\partial E^{N-1}[\rho]}{\partial \boldsymbol{R}_\alpha} \tag{5}$$

In this way, the electronic and nuclear degrees of freedom are computed simultaneously in an efficient way, allowing for dynamical QM/MM-based simulations on the scale of nanoseconds.

3.1.1 Applications of hybrid QM/MM to cryptochromes

To showcase the use of the DFTB QM/MM method, we highlight an example where it was used in *Arabidopsis thaliana* cryptochrome-1a to study the electron transfer reactions occurring upon photo activation, which involved the flavin moiety of cryptochrome, as well as the triad of tryptophan amino acids that seem to be conserved in cryptochromes from different species (Lüdemann et al., 2015). The ability to include the entire biomolecular structure and the surrounding water, in the calculations of the electron transfer reactions showed the importance of water interacting with the third tryptophan. When excluding water from the calculation, the energetics of the electron transfer change drastically, and is no longer favorable. When including water, the electron transfer rates calculated are in agreement with experimental data.

Alternative electron transfer reactions have also been studied in *Xenopus laevis* cryptochrome DASH (Sjulstok et al., 2018). In this protein the tryptophan triad has been mutated experimentally (Biskup et al., 2011, 2013) and radical pair formation was still observed, involving either tryptophan or tyrosine amino acid residues; see Fig. 3. By tracking the electron directly by the means of the computational microscope, it is possible to illuminate the exact amino acids that are involved in the electron transfer, and further understand the possible pathways. The right panel in Fig. 3 illustrates the occupancy of the three tryptophan residues in *Xenopus laevis* cryptochrome DASH over time and reveals that the terminal tryptophan W_C acquires a radical character through donating an electron to the flavin moiety in sub-nanoscale times (Sjulstok et al., 2018). Experimental techniques cannot

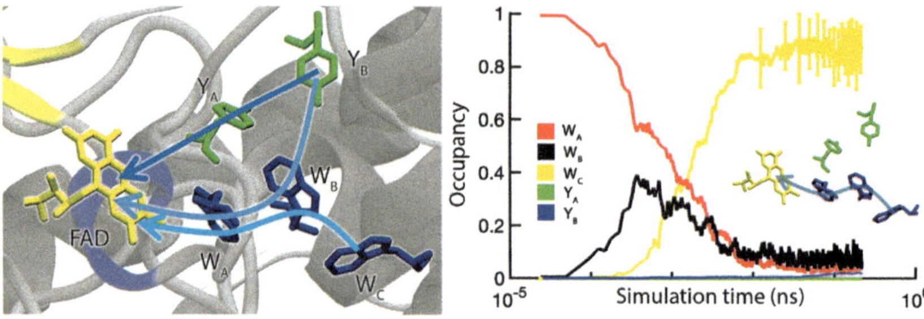

Fig. 3 Electron transfer pathways in *Xenopus laevis* cryptochrome DASH. The pathways that electron transfer can utilize in this protein have been studied using QM/MM MD simulations, where it was found that alternative electron transfer through closely located tyrosines are possible. In the left side of the figure, the arrows indicate some of the electron transfers found, while the right side shows an example of how the electron site occupancy changes over time in the wildtype protein. The electron-hole is initially on the first tryptophan, W_A, indicating electron transfer to FAD has occurred, and the electron is then transferred to the last tryptophan, W_C, through an intermediary state at W_B. Note that the two tyrosine amino acids are not involved in the wild-type electron transfer.

provide such a detailed analysis and evidence of the specific amino acids involved in cryptochrome activation, and instead deliver more indirect evidence through the analysis of transient absorption spectroscopy data analysis (Biskup et al., 2011; Biskup et al., 2013; Solov'yov & Schulten, 2012).

3.2 Polarizable embedding

A different method to include the molecular environment, at a higher level of theory compared to hybrid QM/MM MD method is the polarizable embedding approach. Polarizable embedding (PE) is a QM/MM method where response properties of a molecular system can be calculated using embedding potentials (Olsen, Aidas, & Kongsted, 2010; Olsen & Kongsted, 2011; Sneskov, Schwabe, Kongsted, & Christiansen, 2011), which are derived from quantum chemical calculations.

The PE method relies on splitting the molecular system of interest into two regions, one that is treated quantum mechanically, the QM region, and another that is represented through an embedding potential, here denoted as the environment. This method is rather general and can be applied to many different systems, including flavoproteins.

The environment is typically represented through an embedding potential, which is made up of atom-centered multipoles, representing the electrostatic potential of the protein environment. The multipoles are calculated using molecular fractionation with conjugated caps (MFCC), where the environment is split into capped fragments and conjugate caps, a procedure which has been standardized for proteins (Olsen, 2012; Olsen & Kongsted, 2011; Olsen, List, Kristensen, & Kongsted, 2015). The Hamiltonian for a system of N fragments can be written as:

$$\hat{H}_{tot} = \hat{H} + \hat{V} = \sum_{a=1}^{N} \hat{H}^{a} + \sum_{a>b}^{N} \hat{V}^{ab} \tag{6}$$

where $\hat{H}$ is the Hamiltonian of a single fragment, while $\hat{V}$ is the interaction between fragments, described by the interaction operator $\hat{V}^{ab}$ (see also Fig. 4). The main contributions to the interaction energy can be found by using time-independent perturbation theory, where the interaction operator is used as the perturbation. The energy is thus expanded as

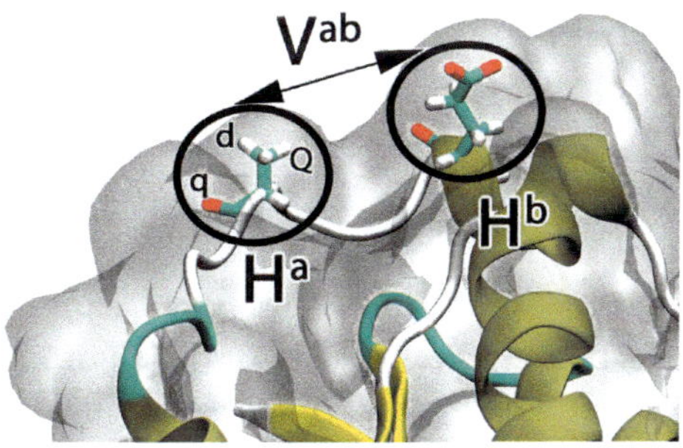

Fig. 4 Example of the elements that comprise the Hamiltonian in the PE framework. H^a and H^b are the Hamiltonians of the single fragments, a and b. V^{ab} is the interaction operator between the fragments, which interact through electrostatics described by a multipole expansion with charges q, dipoles d, and quadrupoles Q.

$$E = E^0 + E^1 + E^2 + \cdots, \tag{7}$$

where E^0 is the unperturbed energy, which is simply the sum of the fragment energies,

$$E^0 = \sum_{a=1}^{N} E^a. \tag{8}$$

Here the unperturbed energy consists of the one- and two-electron integrals and the nuclear repulsion within fragment a only.

The first order correction term in Eq. (7) turns out to be the electrostatic interactions between the fragments,

$$E^1 = \sum_{a>b}^{N} \left\langle ab \middle| \hat{V}^{ab} \middle| ab \right\rangle = \sum_{a>b}^{N} E_{\text{elec.}}^{ab} \tag{9}$$

where $|a\rangle$ and $|b\rangle$ are the wavefunctions for fragments a and b respectively. This is the classical electrostatic interaction energy, which can be calculated through multipole moment expansion.

The third energy term in Eq. (7) is the first-order response caused by the perturbation, which reads as

$$E^2 = \sum_{a=1}^{N} E_{\text{induced}}^{a} + \sum_{a>b}^{N} E_{\text{disp}}^{ab} \tag{10}$$

and represents the induction energy, the polarization of fragment a, and the dispersion related energies. In the PE model the dispersion energies are not taken into account, so the second order correction is essentially containing only the linear polarization term.

The final expression for the energy thus becomes

$$E = \sum_{a=1}^{N} E^a + \sum_{a>b}^{N} E_{\text{elec}}^{ab} + \sum_{a=1}^{N} E_{\text{induced}}^{a}. \tag{11}$$

In practice, first the embedding potential, consisting of multipole moments and polarizabilities, is calculated for the environment. The embedding potential is then included in the calculation of the desired properties of the QM region, through the interactions with the fragments in the environment (Reese, List, Kongsted, & Solov'yov, 2016; Schwabe, Olsen, Sneskov, Kongsted, & Christiansen, 2011; Sjulstok, Olsen, & Solov'yov, 2015). An embedding potential typically consists of multipoles expanded up the order of quadrupoles, that is, point charges, dipoles, quadrupoles and polarizabilities which has been shown to accurately describe the environment (Olsen et al., 2015). The formalism works for both wavefunction based methods and density based methods such as DFT.

3.2.1 *Application of polarizable embedding to* Arabidopsis thaliana *cryptochrome-1a*

The polarizable embedding method was used to find the energetics of the electron transfer reaction in *Arabidopsis thaliana* cryptochrome-1a with the protein environment, water and ions represented through the full electrostatic potential (Sjulstok et al., 2015). To obtain the energetics of the ground

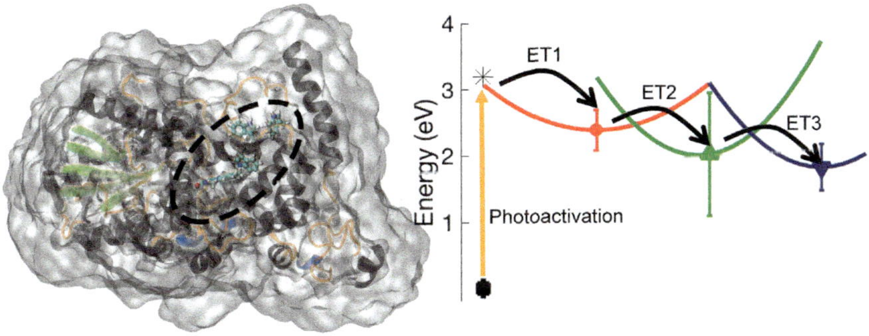

Fig. 5 Energies for the electron transfer states in *Arabidopsis thaliana* cryptochrome-1a. Using the polarizable embedding methodology, the energetics of the electron transfer states in the protein were found, while the entire protein and surrounding water influences the electron transfer reaction. The active site of the protein participating in the electron transfer reactions ET1–ET3 is highlighted in the left panel and includes the flavin moiety jointly with the three tryptophan amino acid residues that drive the activation of the protein upon flavin photoactivation.

state and the three electron transfer states, multiple protein configurations were taken into account, since the thermal fluctuations of the protein induce changes in the electronic energy, as seen from the error bars in Fig. 5.

By tuning of the accuracy of the embedding potential it was possible to conclude that the polarization term in Eq. (11) is crucial for an accurate description of the energetics and thus the electron transfer in *Arabidopsis thaliana* cryptochrome-1a, and specifically for the characterization of their excited state nature (Sjulstok et al., 2015).

4. Atomistic molecular dynamics simulations

Molecular mechanics calculations provide a very commonly used level of representation for biomolecular systems, treating atoms as point masses/charges which interact via simplified potentials known as force fields. Molecular mechanics force fields typically include classical electrostatics, van der Waals and repulsive interactions (often lumped together into the Lennard-Jones potential), and some set of harmonic and periodic terms that describe two-body, three-body, and four-body interactions between chemically bonded atoms. A few commonly used force fields include CHARMM (Best et al., 2012), AMBER (Ponder & Case, 2003), and OPLS (Robertson, Tirado-Rives, & Jorgensen, 2015); a vast literature exists, and should be consulted, on the philosophies, versions, strengths, and weaknesses of these and others (e.g., Cerutti, Freddolino, Duke Jr, & Case, 2010; Lindorff-Larsen et al., 2012; Robustelli, Piana, & Shaw, 2018).

Molecular mechanics force fields themselves provide only a function for calculating potential energy based on atomic coordinates; they may be applied in simple geometry optimizations (potential energy minimization), specialized conformational searches such as molecular docking calculations, conformational sampling and thermodynamic calculations via Monte Carlo simulations, or conformational sampling and time-resolved dynamics propagation using molecular dynamics simulations. We focus our interest on the use of classical, atomistic molecular dynamics calculations, which enable efficient conformational sampling, qualitative evaluation of the effects of changes to proteins themselves or their ligands and substrates, identification of key conformational changes induced by changes to the environment of a biomolecule, or calculations of statistical mechanical properties such as conformational free energy surfaces or binding constants. MD simulations may be applied productively to flavoenzymes as with any other protein, so long as appropriate considerations are taken to account for the unusual character of the flavin cofactor.

4.1 Parameterization of flavin compounds using classical force fields

The central concern in any atomistic MD simulation of flavoproteins must of course be the flavin itself. Whereas most components of a flavoprotein (amino acids, solvent components, and even adenosine derivatives) exist in most commonly applied MD force fields (e.g., recent AMBER and CHARMM releases), the isoalloxazine moiety so central to flavin chemistry is typically not present, and thus any practitioner wishing to simulate flavoproteins must either find reliable parameters in the literature compatible with the force field that they wish to apply to the rest of their system, or perform the parameterization themselves (note that several such parameterization exist in the literature for various forcefields (e.g., Freddolino, Gardner, & Schulten, 2013; Freddolino, Dittrich, & Schulten, 2006; Solov'yov et al., 2012, 2014 for CHARMM, Lüdemann et al., 2015; Sjulstok et al., 2018 for AMBER).

Two additional aspects of the isoalloxazine moiety increase the difficulty associated with parameterizing these compounds. First, the three-ring delocalized electron system present in unmodified isoalloxazines cannot be easily decomposes into smaller groups (as would typically be done in many parameterization procedures), and requires optimization of charge distributions and bonded term parameters to satisfy many constraints across the length of the ring system. Second, flavoproteins make use of a huge range of flavin variants in which either covalent bonds to the isoalloxazine are formed (e.g., Fig. 1B), or the electronic structure of the isoalloxazine has changed (see above, Section 2.1); any such changes will necessitate substantial reparameterization effort across the entire isoalloxazine, as any changes in electronic structure will propagate across the ring system. Thus, even the parameterization of several flavin-containing molecules outlined above by no means exhausts the range of chemical structures and electronic states that occur in flavoproteins.

As an example illustrating both the challenges associated with an initial parameterization of an isoalloxazine-containing moiety for all-atom MD simulations, and the re-parameterization efforts needed to account for covalent modifications, we take as a case study the FMN chromophore contained by LOV domains. As noted above, LOV domains typically contain a fully oxidized FMN molecule in the dark state, but upon illumination, form a covalent bond between the C4A atom of the isoalloxazine and a nearby cysteine residue, which subsequently decays thermally over time scales of minutes to hours (Kennis et al., 2003; Kottke, Heberle, Hehn, Dick, & Hegemann, 2003; Losi, Polverini, Quest, & Gärtner, 2002; Swartz et al., 2001). Several generations of parameters for FMN and the FMN-LOV domain photoadduct have been presented in the literature (Freddolino et al., 2013, 2006), and at times even generalized force field parameters have been applied to the chromophore (Bocola, Schwaneberg, Jaeger, & Krauss, 2015), which simplifies the parameterization process but likely at the cost of accuracy. Here, we demonstrate the parameterization using the CHARMM general force field (CGenFF; Vanommeslaeghe et al., 2010) and the VMD Force Field Toolkit plugin (FFTK; Mayne, Saam, Schulten, Tajkhorshid, & Gumbart, 2013), which together provide a streamlined interface for obtaining and refining parameters for any molecule of interest. While the electronic structure of isoalloxazine does not permit decomposition of the three-ring system into smaller components for parameterization, we can at least separate the flavin mononucleotide chromophore into a lumiflavin moiety (that is, isoalloxazine with a single methyl group at the N10 position; Fig. 1A) and the ribityl-phosphate moiety, the latter of which can be handled by transfer of parameters from other nucleotides present in the CHARMM force field.

The essential logic of parameterization of lumiflavin using CGenFF and FFTK is as follows:

1. Obtain initial parameter guesses using the CGenFF program.
2. Fix parameters that can be reasonably transferred, and identify parameters requiring refinement.

3. Obtain an appropriate quantum mechanically (QM) optimized geometry and corresponding Hessian matrix for the molecule of interest.
4. Calculate optimal interactions (QM distance and energy) between polar groups on the isoalloxazine and a water molecule.
5. Optimize the charge distribution to reproduce the molecular dipole moment and water interaction energies and distances.
6. Optimize bond and angle parameters to reproduce the vibrational spectrum provided by the QM Hessian matrix.
7. Perform QM geometry scans of rotatable bonds requiring parameterization to obtain a target potential energy surface for rotations.
8. Optimize dihedral and improper dihedral terms to reproduce the QM potential energy surface.
9. *Re*-optimize the geometry using the newly obtained MD parameters.
10. Repeat steps 4–8 using an MD-optimized structure (but keeping with the QM Hessian matrix) until parameters converge.

We comment below on key considerations for each of these steps, and their practical application to lumiflavin.

4.1.1 Obtain initial parameter guesses using the CGenFF program

A mol2 file containing model coordinates, bond orders, and atom types for the lumiflavin molecule can be obtained using any molecular editing program (e.g., ChemDraw, or the VMD Molefacture plugin) and then uploaded to the CGenFF server to obtain initial parameter guesses (Vanommeslaeghe, Raman, & MacKerell Jr, 2012; Vanommeslaeghe and MacKerell Jr., 2012).

Upon doing so, the user will be presented with a complete topology file and set of parameters for the lumiflavin molecule; the first few lines of atom-level parameters in the obtained output are:

```
RESI LMN   0.000
GROUP          ! CHARGE CH_PENALTY
ATOM N1  NG2D1  -0.758 ! 174.706
ATOM C2  CG2O6   0.876 ! 112.516
ATOM O2  OG2D1  -0.390 ! 8.989
ATOM N3  NG2S1  -0.232 ! 31.688
```

Note that the last, commented field indicates the "penalty" score assigned by the CGenFF server, and ought to be interpreted in line with the admonitions of the CGenFF program itself: "Penalties lower than 10 indicate the analogy is fair; penalties between 10 and 50 mean some basic validation is recommended; penalties higher than 50 indicate poor analogy and mandate extensive validation/optimization." Considered in this light, it is clear that the estimated charges for the lumiflavin group are not trustworthy, and will require optimization based on quantum mechanical calculations.

Similar CGenFF output will likewise be obtained for bonded parameters; a few example lines from the lumiflavin output are:

```
BONDS
CG2O6  NG2D1 500.00  1.3100 ! genera , from CG2N1 NG2D1, penalty= 15
CG2R64 NG2D1 305.00  1.4140 ! genera , from CG2R64 NG2S1, penalty= 160
ANGLES
NG2D1 CG2O6 OG2D1  60.00 125.70 ! genera , from NG2S1 CG2O6 OG2D1,
penalty= 16
CG2O6 NG2D1 CG2R64 50.00 108.00 ! genera , from CG2N1 NG2D1 CG331,
penalty= 82.9
```

Here the actual forcefield parameters occur before the ! mark, and comments after that mark includes the atom types that were used in suggesting the parameterization by analogy. As with the charges, the presence of several high-penalty entries suggests the need for manual optimization of many of the parameters.

4.1.2 Fix parameters that can be reasonably transferred, and identify parameters requiring refinement

After having obtained initial parameter guesses from CGenFF, it is necessary to critically inspect the initial parameters to determine which can be used as is, and which require further optimization. Even penalties in the intermediate 10–50 range should not be above suspicion; for example, the CG2O6-NG2D1 bonded term suggested above, which would be applied to the N1-C2 bond of the lumiflavin, is in fact derived by analogy to a carbon-nitrogen bond in methyl-guanidinium (as can be determined by inspection of the CG2N1-NG2D1 interactions present in model compounds in the CGenFF topology files). As the analogy is in our view strained, due to the absence of a ring system in the model compound as compared with the case in lumiflavin, we would consider this and similar cases as suspect and in need of additional optimization.

At the end of the parameter inspection process described above, one will be left with three classes of parameters:

- Those usable directly from the CGenFF parameter files based on atom type assignments. Such parameters will not even appear in the CGenFF output because they are directly present in the default CGenFF parameter files; for example, in lumiflavin, the C2 and O2 atoms are assigned types CG2O6 and OG2D1, respectively; because CGenFF already contains bonded parameters between those atom types, no further work is typically needed.

- Those parameterized by analogy and found to be reasonable. Any bonded parameters with low penalty scores, or those with intermediate penalty scores which are deemed upon inspection to be reasonable, should fall into this category.
- Parameters that were assigned by analogy and are deemed to require optimization, based on a combination of the penalty scores assigned and manual inspection of the source of the parameters.

Typically the three categories of parameters noted above should be split into three separate files, to simplify the process of optimizing the parameters in the third category while not perturbing the others.

4.1.3 Obtain an appropriate QM-optimized geometry and corresponding Hessian matrix for the molecule of interest

The process of performing an appropriate quantum mechanical geometry optimization, followed by calculation of the Hessian matrix for bonded parameter optimization, can be handled automatically (e.g., in the "Opt. Geometry" and "Calc. Bonded" tabs, respectively, of the FFTK graphical user interface (GUI). Care should be taken by the user to confirm that the calculations are performed at an appropriate level of theory for the force field of interest (for CGenFF, geometry optimization and Hessian matrix calculation should be at MP2/6-31G* for most molecules of interest; Vanommeslaeghe et al., 2010). In addition, the user must manually inspect the run logs to confirm that the geometry optimization converges to a local minimum and not a saddle point, and that the Hessian matrix calculation finishes successfully. In our experience, the use of the Opt = Tight parameter in Gaussian may be necessary for lumiflavin and related compounds to ensure convergence to a local minimum.

4.1.4 Calculate optimal interactions (QM distance and energy) between polar groups on the isoalloxazine and a water molecule

Unlike many other molecular mechanics force fields, the CHARMM force field requires partial charges to be calculated to reproduce quantum mechanical interaction energies and distances for interactions of polar groups on the molecule of interest with water—in particular, the partial charges (and technically the Lennard-Jones parameters as well, although these rarely need to be altered) are tuned such that interactions of the parameterized molecule with TIP3P water (Jorgensen, Chandrasekhar, Madura, Impey, & Klein, 1983) reproduce the quantum mechanical calculations. The process of setting up the necessary calculations using the Gaussian QM package are handled by the "Water Int." tab of FFTK, which automates the otherwise-tedious process of setting up appropriate water interaction locations (other, similar tools for running such calculations are also available, such as GAAMP (Huang & Roux, 2013), which uses a somewhat different objective function for the charge optimization). At the stage of water interaction calculations, two principle considerations are required: first, which atoms should have interactions with water calculated? And second, what orientations of water should be used?

In answer to the first question, typically a complete parameterization would mandate interaction of water with all accessible hydrogen atoms or hydrogen bond donors in the molecule of interest. In parameterization of lumiflavin and closely related molecules, however, the situation is simplified somewhat by the fact that the CHARMM force field specifies standardized charges for several atom types; for example, methyl groups in many contexts have charges of +0.27 and −0.09 assigned to the carbon and hydrogens, respectively; similarly, C–H pairs in aromatic rings without major substitutions are assigned charges of +0.115/−0.115. After application of these standard charges, only roughly half of the lumiflavin molecule even has charges that can in principle be modified, yielding five interaction sites worth considering (Fig. 6A).

To answer the second question, that is, what orientations of water should be considered, several approaches are possible. A minimalistic strategy would involve selection of a single appropriate water interaction geometry for each hydrogen, excluding those with fixed charges specified by the force field (e.g., the positions in Fig. 6B)–in principle such interactions could be considered for all accessible atoms, including for example the nonpolar hydrogens of the ring system, but here we follow the advice of Vanommeslaeghe et al. (2010) in terms of pruning the considered set of sites. Due to the fairly close spacing of polar sites on the isoalloxazine moiety, in practice, several non-equivalent water interaction geometries are possible for each polar site (see, e.g., Fig. 6C, where the QM-derived target interaction energies and distances for the two shown conformations of water interacting with N5 differ by more than 6 kcal/mol and more than 0.4 Å, respectively). While some case for consistency could be made by either simply choosing one of the possible geometries at random for each site, or performing the calculations for all possible geometries and including them all in the partial charge fitting, we suggest a third path for many flavin

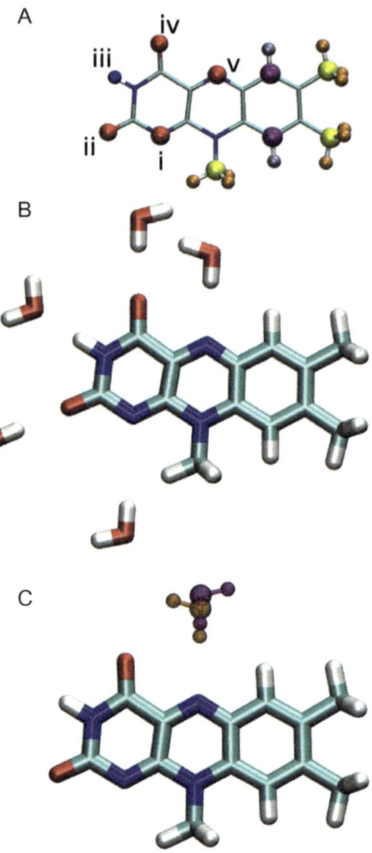

Fig. 6 Selection of (and constraints on) sites for lumiflavin parameterization. Purple/ice blue and yellow/orange spheres label atoms with a class of fixed charges specified by the CHARMM forcefield. The five sites labeled with roman numerals, in constrast, are reasonable sites for water interaction calculation (labeled with Roman numerals i–v); hydrogen bond donors are labeled with blue spheres, and acceptors with red spheres. (B) Typical water interaction conformations for all appropriate polar sites on lumiflavin. (C) Comparison of possible water orientations (orange vs purple) for interaction with the N5 atom of lumiflavin.

derivatives: consider multiple water orientations for each site, but constrain them to the subset that represents a relevant environment for the biological question of interest. In the case of the water conformations shown in Fig. 6C, the orange water has a vastly stronger interaction energy due to simultaneous interactions with N5 and O4, which pushes the fit toward a much more negative partial charge on N5 than would be suggested based on the purple conformation (it may not be possible to identify a charge distribution that simultaneously reproduces the QM energies/distances for all orientation geometries). While both orientations can be used simultaneously, leading to fitting of an intermediate set of parameters that averages between the two extremes another alternative is to prune the water interaction orientations based on the intended usage—for example, for simulations of a protein where the active site of a flavoenzyme is not expected to be hydrated, and hydrogen bond donors interacting with the N5 site could not simultaneously interact with O4 in the way that a water molecule would, it would be appropriate to fit only based on the purple conformation (plus additional geometries designed to interact optimally with O4 and the other polar sites shown in Fig. 6B).

An alternative that sidesteps the complications of choosing water geometries listed above, the point charge distributions can instead be assigned using electrostatic potential fitting methods (e.g., Bayly, Cieplak, Cornell, & Kollman, 1993). Such methods have been applied in the past to flavins (Freddolino et al., 2006; Ganguly, Thiel, & Crane, 2017; Luo, Andricioaei, Xie, & Karplus, 2006) and a variety of other difficult-to-parameterize molecules, although one should note that the resulting charges do not strictly fit in with the CHARMM parameterization procedure.

4.1.5 Optimize the charge distribution to reproduce the molecular dipole moment and water interaction energies and distances

Once quantum mechanical interaction distances and energies have been calculated in the prior step, the free partial charges over the isoalloxazine group must be optimized to reproduce those values, and the dipole moment of the molecule, as closely as possible (n.b. some scaling is typically applied to the QM results; see (Vanommeslaeghe et al., 2010) for details). The optimization routine provided in the VMD FFTK plugin typically suffices for this purpose, but we note that due to the number of free parameters (partial charges) present in flavin model compounds, in our experience it is preferable to perform multiple simulated annealing runs in parallel with slow annealing schedules, in order to find the most optimal charge distribution to reproduce the quantum mechanical values.

4.1.6 Optimize bond and angle parameters to reproduce the vibrational spectrum provided by the QM Hessian matrix

Optimization of the bond/angle force constants and equilibrium values to reproduce the quantum mechanical Hessian matrix is typically used as a straightforward approach to obtaining bonded parameters, aiming to minimize the deviation from QM values in both the equilibrium geometry and energy upon small deformations. Isoalloxazine and its derivatives pose the special challenge that to avoid strained rings, the CHARMM force field specifies that the sum of all equilibrium angle values around a planar center in a six membered ring should be 360, and likewise, the sum of interior equilibrium angles for all members of a planar six-membered ring should be 720 (plain isoalloxazine contains 17 such constraints, although in any particular case some will likely be resolved by parameters transferred by analogy). The bond/angle optimizer provided with FFTK does not, at present, support these constraints. As an alternative, we have successfully used the COBYLA optimizer (Powell, 1994) to perform a constrained optimization of the bond/angle parameters while satisfying all fixed angle sums arising from the rings. A final optimization with the constraints relaxed or removed may be justified to improve agreement with the QM data, as in our experience with flavin parameterizations it is not always possible to satisfy the standard CGenFF targets for deviations in equilibrium bonds and angles (0.025 Å distance deviation, three angle deviation) and the ring angle sum constraints simultaneously.

4.1.7 Perform QM geometry scans of rotatable bonds requiring parameterization to obtain a target potential energy surface for rotations

The next step in a typical parameterization is to perform a quantum mechanical scan of possible values of all dihedrals in the system to obtain target potential energy surfaces for rotatable bonds. This procedure is appropriate for lumiflavin as well, except that whereas for a typical molecule the torsions would be calculated for a full range of possible values, the highly constrained triple ring system of the isoalloxazine moiety does not permit any substantial torsional twists within the ring itself. Thus, we recommend a reduced scanning window (±30 degrees at 5 degree increments) for torsions with central atoms in the isoalloxazine ring system. A harmonic-type potential energy surface should be readily observable from such calculations in most cases.

4.1.8 Optimize dihedral and improper dihedral terms to reproduce the QM potential energy surface

Once QM potential energy surfaces for the torsions have been obtained, FFTK (or similar software) can be used to optimize the associated torsion terms to reproduce those surfaces. Unlike many of the steps above, we typically encounter no unusual difficulties in parameterization of flavin derivatives at this step.

Based both on the chemistry of the groups involved (compared with similar cases in the distributed CGenFF parameters), and QM calculations, we find that it is often justified to apply improper dihedral terms centered on the isoalloxazine C2, C4, and N10 atoms. While not directly incorporated into the FFTK interface, such calculations may be applied straightforwardly simply by performing a small QM potential energy scan varying the value of the improper dihedral of interest, followed by directly fitting the resulting potential energy surface with the improper dihedral term of interest while holding all other parameters constant.

4.1.9 Re-optimize the geometry using the newly obtained MD parameters, and iteratively re-optimize to convergence

After all of the above steps are completed, a draft set of parameters for the molecule of interest will have been obtained. However, CGenFF usage guidelines suggest that the appropriate next step is to re-optimize the original geometry using the newly obtained molecular mechanics force field parameters and charges, and then perform repeated iterations of all steps from water interaction calculation to geometry re-optimization until the parameters appear reasonably converged.

The parameterization process described above must be repeated for each molecule of interest, owing to the relative lack of modularity of the flavin moiety, which prevents the typical approach of transferring newly obtained parameters for a compound of interest to provide most of the data required for a related compound. Once parameterization is completed for an appropriate flavin cofactor (and any other unusual ligands or substrates), molecular dynamics simulations of flavoproteins can be pursued using standard methods. The interpretation of any molecular mechanics-based simulations on flavoproteins must be interpreted with some humility, because the complexities of the electronic structure of these cofactors likely cannot be fully reproduced by classical force fields. In the future, more complex models (e.g., polarizable force fields using Drude oscillators; Lamoureux & Roux, 2003) will likely provide a more nuanced description of the flavin cofactor. As discussed in the examples below, however, even simple fixed-charged force fields, when carefully parameterized and applied, have permitted considerable insight into flavoprotein behavior.

4.2 Activation and photocycle of phototropin LOV domains

The LOV (Light-, Oxygen-, and Voltage-sensitive) domain is a highly versatile, flavin-containing photosensory module found in proteins arising from all domains of life (Herrou & Crosson, 2011). While their exact behavior differs depending on their structural contexts, in all cases LOV domains transform light-induced covalent bond formation between the flavin chromophore and a conserved cysteine residue, into domain-scale structural changes that activate downstream functional activities. The robustness and versatility of LOV domains is also illustrated by their inclusion in engineered constructs for use in optogenetics and synthetic biology (e.g., Möglich & Moffat, 2010). LOV domains, and particularly the LOV2 domains of plant phototropins, have been extensively studied using molecular modeling calculation. Formation of the typical flavin-cysteine adduct, the A'α and Jα helices dissociate from the LOV2 domain core, yielding an active state of the overall phototropin that releases repression of an attached kinase domain (Fig. 7A; kinase not shown) (Harper, Neil, & Gardner, 2003; Zayner, Antoniou, French, Hause, & Sosnick, 2013; Zayner, Antoniou, & Sosnick, 2012), providing an emblematic example case of how LOV domains transduce a single chemical bond formation at their active site into large-scale conformational changes stretching across protein domains.

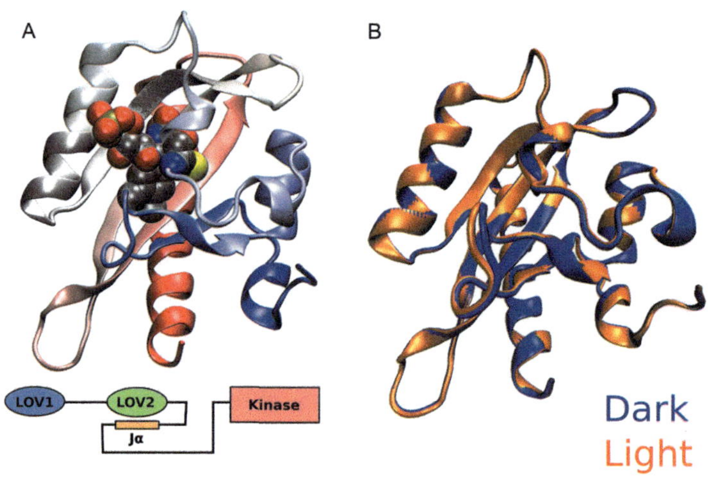

Fig. 7 Light-activated conformational changes of the *A. sativa* Phot1 LOV2 domain. (A) Overview of the lit state of the LOV2 domain, shown covalently bound to a flavin mononucleotide chromophore, above a schematic of the LOV2 domain's place in the overall Phot1 structure. Coloring runs blue to red from N terminus to C terminus. Photoactivation is known to cause dissociation of the N-terminal (A'α) and C-terminal (Jα) helices from the LOV2 domain, and subsequently allows kinase activation, as described in the test. (B) Superposition of crystal structures of the dark and lit states of *A. sativa* Phot1 LOV2 (PDB codes 2V1A and 2V1B (Halavaty & Moffat, 2007), showing a 0.2 Angstrom C_α root mean squared deviation between the crystal structures.

Molecular dynamics calculations on LOV2 domains have demonstrated the chromophore-proximal conformations induced by the photoreaction, allowed rationalization of the effects of mutations to the chromophore binding site on the photocycle, and demonstrated the path through which conformational change propagates from the chromophore to the key A′α and Jα helices (Freddolino et al., 2013; Freddolino et al., 2006; Peter, Dick, & Baeurle, 2010; Song et al., 2011; Zayner & Sosnick, 2014; Zayner et al., 2012, 2013;), as well as revealing similar conformational changes in related proteins such as the *N. crassa* light sensor VVD (Ganguly et al., 2017). Molecular modeling calculations have proven especially important to advancing our understanding of the behavior of LOV domains because crystallographic structures tend to show few changes between the dark and lit states (see, e.g., Fig. 7B), in contrast with NMR data and other experimental assays which demonstrate that in solution large-scale structural changes occur after the photoreaction.

4.3 Interaction of Photolyase With DNA

Another illustration of the use of MD techniques in studying flavoproteins is exemplified here for the so-called DNA photolyase (Heelis, Kim, Okamura, & Sancar, 1993; Kao et al., 2008; Kavakli & Sancar, 2004; Sancar, 2003). DNA photolyase is an enzyme that assists UV-damage repair in DNA molecules. For this process the DNA photolyase binds to the damaged DNA site and initiates a series of cyclic electron transfers to and from DNA that restores the correct structure of DNA. MD simulations were used to investigate the difference in DNA binding for photolyase and cryptochrome (Jepsen & Solov'yov, 2017). These proteins are highly homologous but serve very different functions, and only photolyase binds to UV-damaged DNA. By comparing the electrostatic interaction energies of cryptochrome and DNA with that of photolyase and DNA it is found that subtle changes in the binding site of the proteins, lead to a clear preference for DNA binding. Fig. 8 illuminates how the interaction energy between the DNA and photolyase/cryptochrome is distributed. It is evident that a difference of over 2400 kcal/mol can be seen by comparing the two proteins.

5. Coarse-grained and continuum methods

While atomistic representations give a common and immensely useful resolution for a wide range of molecular modeling applications, at present timescales beyond tens of microseconds are inaccessible without heroic

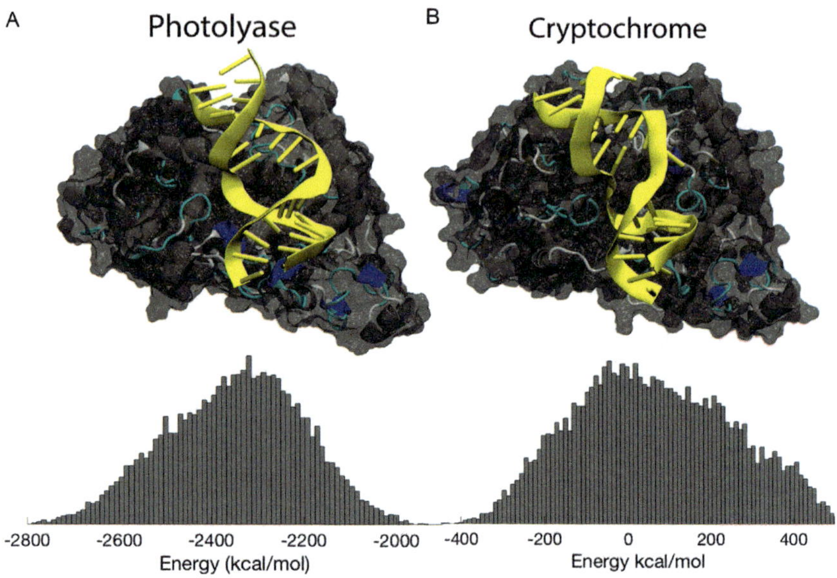

Fig. 8 Interaction energies of DNA with photolyase and cryptochrome. Through MD simulations of cryptochrome and photolyase the non-bonded interaction energy with DNA can be found and plotted in the form of distributions in the lower part of the figure. This shows the DNA prefers binding to photolyase compared to cryptochrome, despite their highly similar structure.

levels of usage of specialized supercomputers such as, e.g., the Anton2 system (Shaw et al., 2014). A variety of simplified approaches have thus been developed to represent biomolecular systems, for example by lumping many atoms into individual particles (often called "beads") (Marrink, Risselada, Yefimov, Tieleman, & De Vries, 2007), treating the biomolecule as a sort of elastic network or mesh (Arkhipov, Freddolino, & Schulten, 2006), or using a simplified potential to calculate low-frequency modes of motion of the biomolecule and thus determine likely directions of conformational change (Bahar, Lezon, Bakan, & Shrivastava, 2009). While such approaches could in principle be applied to flavoproteins, in general the importance of the fine-grained details of the flavin cofactor, and its interaction with its protein surroundings, reduces the applicability of such approaches. Nevertheless, careful applications of clever methods of reduced system modeling, mainly using MD simulations as a starting point to project onto likely longer- and larger-timescale motions, have been fruitfully applied to a variety of flavoproteins; we give here a few representative examples.

5.1 Topological network models

To elucidate the early post-photoreaction structural changes in flavin-binding proteins one can evaluate the data from classical MD simulations in terms of topological models derived from graph representations of the protein. In addition to the now popular contact maps, a graph representation that is simultaneously informed by the correlation of displacements in addition to the usual spatial adjacency. Such representation can reveal efficient reorganization networks, which could mediate the signaling in a more directed, purposeful way than networks predicted on the basis of contact maps (Kattnig, Nielsen, & Solovýov, 2018). Such an approach builds upon two pillars: the use of covariance matrices to analyze the correlated dynamics of protein residues from molecular dynamics, which has for example been described by Hünenberger, Mark, and Vangunsteren (1995) as well as Karplus and Ichiye (1996) and Ichiye and Karplus (1991), and network representations of the protein structure (Paola, Ruvo, Paci, Santoni, & Giuliani, 2013) and dynamics (Kasahara, Fukuda, & Nakamura, 2014; Yoon, Lee, Park, & Wu, 2018). In terms of the graph theoretical analysis, the approach described here is similar to that of Kasahara et al. (2014).

In the topological network models of cryptochromes and photolyase, the different interaction sites of the flavin adenine dinucleotide cofactor are typically represented by several residues corresponding to the flavin, the ribityl chain, the diphosphate, the ribose and the adenine, respectively (Kattnig et al., 2018). The protein 3D structural information is mapped onto a graph in which nodes represent residues/sites and edges indicate the existence of neighborhood relationships. Two fragments are considered to be adjacent, i.e., neighbors, if the average distance of their centers of mass is smaller than 8 Å. Subsequently, topological relationships between residues are explored by graph-theoretical tools such as the geodesic distance (the number of edges in a shortest path connecting two nodes), vertex degree (number of edges incident to a vertex, i.e., number of adjacent residues), the local clustering coefficient (a measure of the link-density in the node's neighborhood) and the betweenness centralities (a measure of the importance of a node defined as the number of shortest paths that pass through a vertex relative to the total number of shortest paths) (Kattnig et al., 2018). Following (Gaci, 2010), nodes with a degree exceeding 3/2 times the average degree can be classified as hubs. For a comprehensive introduction to network representations of protein structures, the inclined reader is referred to (Paola et al., 2013).

5.2 Application to European robin cryptochrome 4

Contact networks provide an alternative representation of the structural features of proteins. For the sake of illustration, the graph representations of a European robin cryptochrome 4 were generated from the distance matrices of the dark and activated states (Kattnig et al., 2018). Cryptochrome 4 is different from the related protein for plant and insect as it has been recently found to process important properties needed to endow migratory birds with a chemical compass (Günther et al., 2018; Hore & Mouritsen, 2016). A graphical representation of the contact network of both states in European robin cryptochrome is shown in Fig. 9. Characteristic nodes have been highlighted by colors; mobile sites have been drawn in yellow. Some of the mobile sites are obviously associated with peripheral loops or extensions of the protein. The CTT part of European robin cryptochrome 4 is

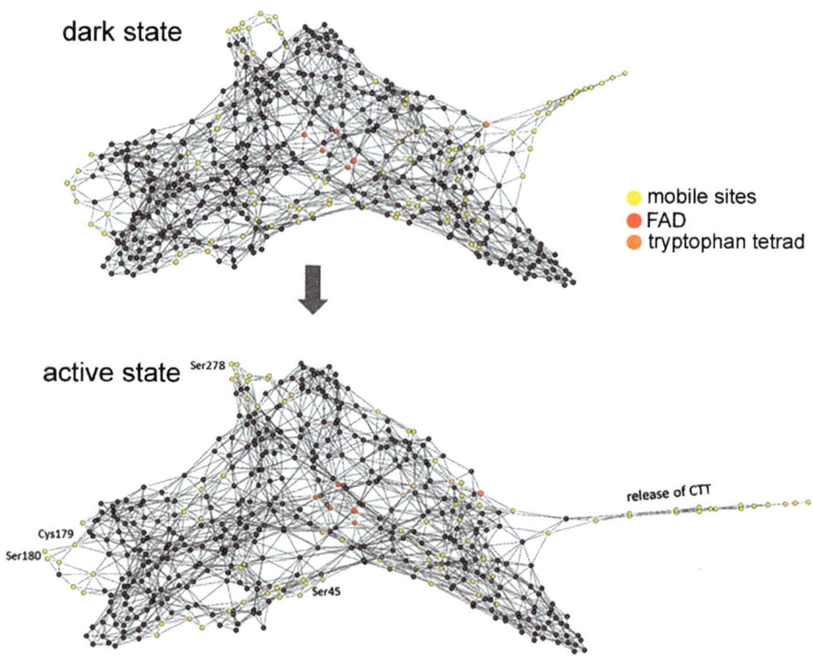

Fig. 9 Illustration of topological network analysis in avian cryptochrome. Cryptochrome 4 from European robin contact networks of the dark state (top) and the activated state (bottom). Mobile sites have been highlighted in yellow. The FAD cofactor (five effective fragments) and the tryptophan tetrad are shown as red and orange circles, respectively. *Figure adopted from Kattnig, D. R., Nielsen, C., & Solovýov, I. A., 2018. Molecular dynamics simulations disclose early stages of the photoactivation of cryptochrome 4.* New Journal of Physics, *20, 083018.*

particularly prominent, as is the loop in between the helices H6 and H7 at the N-terminal side of the protein and the loop connecting the helices H9 and H10. The prominently reorganizing site surrounding residue Ser45 is located at the bottom of the graph, although not at its topological surface. Many mobile residues appear to be scattered over the network without following an obvious pattern. Comparing the two plots in Fig. 9 reveals the structural changes associated with the photo-activation. The release of the CTT is striking as are the reorganizations at residues 180 and 278, which appear to assume a more "exposed" position. The mobile residues around Ser45 are moved to the bottom surface of the graph. These changes are in accordance with the picture of an opening of the protein structure, which has also been postulated to accompany the activation of other cryptochromes (Kattnig et al., 2018).

5.3 Example application of kinetic Monte Carlo to phototropins

Another example that we wish to bring here is related to phototropins. One of the central questions in the field of phototropin dynamics is that of the long timescale behavior of interactions between the two LOV domains and the kinase domain (as described in Section 4.2)—in particular, what conformational motions govern LOV2 Jα dissociation in the lit state, and how is that transduced to changes in kinase activity? While conventional atomistic MD simulations have provided insight into early motions induced by the photoreaction (see above), the likely conformational transitions giving rise to destabilization of the interface have eluded direct simulation, due to the ms or longer timescale of helix dissociation and kinase activation.

The importance of fine atomistic details of the chromophore binding site precludes the use of most bead-based coarse-graining methods for LOV domain dynamics, and likewise, the need for a large conformational transition prohibits the use of most elastic network models. As an alternative, Peter, Dick, Stambolic, and Baeurle (2014) applied an innovative approach in which short periods of MD simulations were interspersed with extended Monte Carlo moves allowing large conformational motions to occur with probabilities informed by the fine grained dynamics. By combining detailed sampling of local conformational spaces with pseudo-Lévy flights between local minima, the authors were able to effectively perform a simulation coarse-grained in time instead of via particle count reduction, thus observing conformational transitions corresponding to light activation of *C. reinhardtii*

Phototropin 1 and its component pieces that would ordinarily be inaccessible to atomistic MD calculations. The kinetic Monte Carlo approach described here is conceptually similar to the Markov state decompositions often applied to protein folding calculations (Pande, Beauchamp, & Bowman, 2010), and similar approaches likewise ought to be applicable to flavoproteins. We also note that enhanced sampling techniques such as parallel tempering (replica exchange) simulations (Sugita & Okamoto, 1999) and the use of accelerated molecular dynamics (Hamelberg, Mongan, & McCammon, 2004) are also in principle applicable to flavoproteins, but we consider these methods to fall under the broader domain of atomistic molecular mechanics simulations, and the considerations in Section 4.2 are fully applicable.

6. Concluding remarks

Molecular modeling provides an immensely powerful set of tools for translating essential physical principles—drawn primarily from quantum mechanics and statistical mechanics—to obtain high resolution insight on structure-function relationships in biomolecules. Studies of flavoproteins using computational methods ranging from *ab initio* quantum mechanics to coarse-grained models of long-timescale dynamics have guided both the interpretation and design of experiments on a broad range of systems, contributing fundamentally to our understanding of how those systems carry out their biological functions. As we have reviewed here, nearly all commonly applied methods in molecular modeling may be fruitfully used in the study of flavoproteins, so long as proper attention is paid to the unique chemistry and interactions of the flavin-containing cofactor itself. We expect that as both computing hardware and the theoretical underpinnings of the approaches described here continue to advance, the contributions of molecular modeling calculations to the overall studies of flavoenzymes and other flavoproteins will likewise continue to grow.

Acknowledgments

The authors are supported by the United States' National Institutes of Health (Grant R35GM128637, to PLF), the Lundbeck Foundation, and the Danish Council for Independent Research (IAS and ES). The authors are also grateful to DeiC National HPC Center (SDU) and the Extreme Science and Engineering Discovery Environment (XSEDE), which is supported by National Science Foundation grant number ACI-1548562 through allocation MCB140220, for providing computational resources necessary for some of the calculations.

References

Angeli, C., Cimiraglia, R., Evangelisti, S., Leininger, T., & Malrieu, J. (2001). Introduction of n-electron valence states for multireference perturbation theory. *The Journal of Chemical Physics, 114*, 10252–10264.

Arkhipov, A., Freddolino, P. L., & Schulten, K. (2006). Stability and dynamics of virus capsids described by coarse-grained modeling. *Structure, 14*(12), 1767–1777.

Bahar, I., Lezon, T. R., Bakan, A., & Shrivastava, I. H. (2009). Normal mode analysis of biomolecular structures: Functional mechanisms of membrane proteins. *Chemical Reviews, 110*(3), 1463–1497.

Bayly, C. I., Cieplak, P., Cornell, W., & Kollman, P. A. (1993). A well-behaved electrostatic potential based method using charge restraints for deriving atomic charges: The RESP model. *The Journal of Physical Chemistry, 97*(40), 10269–10280.

Best, R. B., Zhu, X., Shim, J., Lopes, P. E. M., Mittal, J., Feig, M., et al. (2012). Optimization of the additive CHARMM all-atom protein force field targeting improved sampling of the backbone phi, psi and side-chain chi1 and chi2 dihedral angles. *Journal of Chemical Theory and Computation, 8*, 3257–3273.

Biskup, T., Hitomi, K., Getzoff, E. D., Krapf, S., Koslowski, T., Schleicher, E., et al. (2011). Unexpected electron transfer in cryptochrome identified by time-resolved EPR spectroscopy. *Angewandte Chemie (International Ed. in English), 50*, 12647–12651.

Biskup, T., Paulus, B., Okafuji, A., Hitomi, K., Getzoff, E. D., Weber, S., et al. (2013). Variable electron transfer pathways in an amphibian cryptochrome: Tryptophan versus tyrosine-based radical pairs. *The Journal of Biological Chemistry, 288*, 9249–9260.

Bocola, M., Schwaneberg, U., Jaeger, K.-E., & Krauss, U. (2015). Light-induced structural changes in a short light, oxygen, voltage (LOV) protein revealed by molecular dynamics simulations—Implications for the understanding of LOV photoactivation. *Frontiers in Molecular Biosciences, 2*, 55.

Brautigam, C. A., Smith, B. S., Ma, Z., Palnitkar, M., Tomchick, D. R., Machius, M., et al. (2004). Structure of the photolyase-like domain of cryptochrome 1 from *Arabidopsis thaliana*. *Proceedings of the National Academy of Sciences of the United States of America, 101*, 12142–12147.

Cerutti, D. S., Freddolino, P. L., Duke, R. E., Jr., & Case, D. A. (2010). Simulations of a protein crystal with a high resolution X-ray structure: Evaluation of force fields and water models. *The Journal of Physical Chemistry. B, 114*(40), 12811–12824.

Christiansen, O., Koch, H., & Jørgensen, P. (1995). The second-order approximate coupled cluster singles and doubles model CC2. *Chemical Physics Letters, 243*, 409–418.

Climent, T., Gonzalez-Luque, R., Merchan, M., & Serrano-Andres, L. (2006). Theoretical insight into the spectroscopy and photochemistry of isoalloxazine, the flavin core ring. *The Journal of Physical Chemistry. A, 110*, 13584–13590.

Domratcheva, T. (2011). Neutral histidine and photoinduced electron transfer in DNA photolyases. *Journal of the American Chemical Society, 133*, 18172–18182.

Domratcheva, T., Fedorov, R., & Schlichting, I. (2006). Analysis of the primary photocycle reactions occurring in the light, oxygen, and voltage blue-light receptor by multiconfigurational quantum-chemical methods. *Journal of Chemical Theory and Computation, 2*, 1565–1574.

Elstner, M., & Seifert, G. (2014). Density functional tight binding. *Philosophical Transactions of the Royal Society of London A: Mathematical, Physical and Engineering Sciences, 372*(2011), 20120483.

Fagan, R. L., & Palfey, B. A. (2010). 7.03—Flavin-dependent enzymes. In H.-W. B. Liu & L. Mander (Eds.), *Comprehensive natural products II* (pp. 37–113). Oxford: Elsevier.

Foresman, J., Head-Gordon, M., Pople, J., & Frisch, M. (1992). Toward a systematic molecular orbital theory for excited states. *The Journal of Physical Chemistry, 96*, 135–149.

Freddolino, P. L., Dittrich, M., & Schulten, K. (2006). Dynamic switching mechanisms in LOV1 and LOV2 domains of plant phototropins. *Biophysical Journal*, *91*(10), 3630–3639.

Freddolino, P. L., Gardner, K. H., & Schulten, K. (2013). Signaling mechanisms of LOV domains: New insights from molecular dynamics studies. *Photochemical & Photobiological Sciences*, *12*(7), 1158–1170.

Gaci, O. (2010). A topological description of hubs in amino acid interaction networks. *Advances in Bioinformatics*, *2010*, 257512.

Ganguly, A., Thiel, W., & Crane, B. R. (2017). Glutamine amide flip elicits long distance allosteric responses in the LOV protein VIVID. *Journal of the American Chemical Society*, *139*(8), 2972–2980.

Granovsky, A. (2011). Extended multiconfiguration quasi-degenerate perturbation theory: The new approach to multi-state multi-reference perturbation theory. *The Journal of Chemical Physics*, *134*, 214113.

Grimme, S., & Waletzke, M. (1999). A combination of kohn-sham density functional theory and multi-reference confi guration interaction methods. *The Journal of Chemical Physics*, *111*, 5645–5655.

Günther, A., Einwich, A., Sjulstok, E., Feederle, R., Bolte, P., Koch, K.-W., et al. (2018). Double-cone localization and seasonal expression pattern suggest a role in magnetoreception for european robin cryptochrome 4. *Current Biology*, *28*, 211–223.

Halavaty, A. S., & Moffat, K. (2007). N-and c-terminal flanking regions modulate light-induced signal transduction in the LOV2 domain of the blue light sensor phototropin 1 from *Avena sativa*. *Biochemistry*, *46*(49), 14001–14009.

Hamelberg, D., Mongan, J., & McCammon, J. A. (2004). Accelerated molecular dynamics: A promising and efficient simulation method for biomolecules. *The Journal of Chemical Physics*, *120*(24), 11919–11929.

Harper, S. M., Neil, L. C., & Gardner, K. H. (2003). Structural basis of a phototropin light switch. *Science*, *301*(5639), 1541–1544.

Hasegawa, J., Bureekaew, S., & Nakatsuji, H. (2007). SAC-CI theoretical study on the excited states of lumiflavin: Structure, excitation spectrum, and solvation effect. *Journal of Photochemistry and Photobiology, A: Chemistry*, *189*, 205–210.

Heelis, P. F. (1982). The photophysical and photochemical properties of flavins (isoalloxazines). *Chemical Society Reviews*, *11*, 15–39.

Heelis, P., Kim, S.-T., Okamura, T., & Sancar, A. (1993). New trends in photobiology. The photo repair of pyrimidine dimers by DNA photolyase and model systems. *Journal of Photochemistry and Photobiology B: Biology*, *17*, 219–228.

Herrou, J., & Crosson, S. (2011). Function, structure and mechanism of bacterial photosensory LOV proteins. *Nature Reviews Microbiology*, *9*(10), 713–723.

Hohenberg, P., & Kohn, W. (1964). Inhomogeneous electron gas. *Physics Review*, *136*, B864–B871.

Hore, P. J., & Mouritsen, H. (2016). The radical-pair mechanism of magnetoreception. *Annual Review of Biophysics*, *45*, 299–344.

Huang, L., & Roux, B. (2013). Automated force field parameterization for nonpolarizable and polarizable atomic models based on ab initio target data. *Journal of Chemical Theory and Computation*, *9*(8), 3543–3556.

Hünenberger, P., Mark, A., & Vangunsteren, W. (1995). Fluctuation and cross-correlation analysis of protein motions observed in nanosecond molecular-dynamics simulations. *Journal of Molecular Biology*, *252*, 492–503.

Ichiye, T., & Karplus, M. (1991). Collective motions in proteins—A covariance analysis of atomic fluctuations in molecular-dynamics and normal mode simulations. *Proteins*, *11*, 205–217.

Jepsen, K. A., & Solov'yov, I. A. (2017). On binding specificity of (6-4) photolyase to a T(6-4)T DNA photoproduct. *The European Physical Journal D*, *71*, 155–165.

Jorgensen, W. L., Chandrasekhar, J., Madura, J. D., Impey, R. W., & Klein, M. L. (1983). Comparison of simple potential functions for simulating liquid water. *The Journal of Chemical Physics*, *79*(2), 926–935.

Kao, Y.-T., Tan, C., Song, S.-H., Öztürk, N., Li, J., Wang, L., et al. (2008). Ultrafast dynamics and anionic active states of the flavin cofactor in cryptochrome and photolyase. *Journal of the American Chemical Society*, *130*, 7695–7701.

Karplus, M., & Ichiye, T. (1996). Fluctuation and cross correlation analysis of protein motions observed in nanosecond molecular dynamics simulations. *Journal of Molecular Biology*, *263*, 120–122.

Kasahara, K., Fukuda, I., & Nakamura, H. (2014). A novel approach of dynamic cross correlation analysis on molecular dynamics simulations and its application to ETS1 dimer-DNA complex. *PLoS One*, *9*, 0112419.

Kattnig, D. R., Nielsen, C., & Solovýov, I. A. (2018). Molecular dynamics simulations disclose early stages of the photoactivation of cryptochrome 4. *New Journal of Physics*, *20*, 083018.

Kavakli, I. H., & Sancar, A. (2004). Analysis of the role of intraprotein electron transfer in photoreactivation by dna photolyase *in vivo*. *The Biochemist*, *43*, 15103–15110.

Kennis, J. T., Crosson, S., Gauden, M., van Stokkum, I. H., Moffat, K., & van Grondelle, R. (2003). Primary reactions of the lov2 domain of phototropin, a plant blue-light photoreceptor. *Biochemistry*, *42*(12), 3385–3392.

Khrenova, M., Nemukhin, A., Grigorenko, B., Krylov, A., & Domratcheva, T. (2010). Quantum chemistry calculations provide support to the mechanism of the light-induced structural changes in the flavin-binding photoreceptor proteins. *Journal of Chemical Theory and Computation*, *6*, 2293–2302.

Kohn, W., & Sham, L. (1965). Self-consistent equations including exchange and correlation effects. *Physics Review*, *140*, A1133–A1138.

Kottke, T., Heberle, J., Hehn, D., Dick, B., & Hegemann, P. (2003). Phot-LOV1: Photocycle of a blue-light receptor domain from the green alga *Chlamydomonas reinhardtii*. *Biophysical Journal*, *84*(2), 1192–1201.

Kubar, T., & Elstner, M. (2010). Coarse-grained time-dependent density functional simulation of charge transfer in complex systems: Application to hole transfer in DNA. *The Journal of Physical Chemistry. B*, *114*, 11221–11240.

Lamoureux, G., & Roux, B. (2003). Modeling induced polarization with classical Drude oscillators: Theory and molecular dynamics simulation algorithm. *The Journal of Chemical Physics*, *119*(6), 3025–3039.

Lee, E. H., Hsin, J., Sotomayor, M., Comellas, G., & Schulten, K. (2009). Discovery through the computational microscope. *Structure*, *17*(10), 1295–1306.

Lindorff-Larsen, K., Maragakis, P., Piana, S., Eastwood, M. P., Dror, R. O., & Shaw, D. E. (2012). Systematic validation of protein force fields against experimental data. *PLoS One*, 7(2), e32131.

Losi, A., Polverini, E., Quest, B., & Gärtner, W. (2002). First evidence for phototropin-related blue-light receptors in prokaryotes. *Biophysical Journal*, *82*(5), 2627–2634.

Lüdemann, G., Solov'yov, I. A., Kubar, T., & Elstner, M. (2015). Solvent driving force ensures fast formation of a persistent and well-separated radical pair in plant cryptochrome. *Journal of the American Chemical Society*, *137*, 1147–1156.

Lüdemann, G., Woiczikowski, P. B., Kubar, T., Elstner, M., & Steinbrecher, T. B. (2013). Charge transfer in E. coli DNA photolyase: Understanding polarization and stabilization effects via QM/MM simulations. *The Journal of Physical Chemistry. B*, *117*, 10769–10778.

Luo, G., Andricioaei, I., Xie, X. S., & Karplus, M. (2006). Dynamic distance disorder in proteins is caused by trapping. *The Journal of Physical Chemistry B*, *110*(19), 9363–9367.

Marrink, S. J., Risselada, H. J., Yefimov, S., Tieleman, D. P., & De Vries, A. H. (2007). The martini force field: Coarse grained model for biomolecular simulations. *The Journal of Physical Chemistry B*, *111*(27), 7812–7824.

Mayne, C. G., Saam, J., Schulten, K., Tajkhorshid, E., & Gumbart, J. C. (2013). Rapid parameterization of small molecules using the force field toolkit. *Journal of Computational Chemistry*, *34*, 2757–2770.

Möglich, A., & Moffat, K. (2010). Engineered photoreceptors as novel optogenetic tools. *Photochemical & Photobiological Sciences*, *9*(10), 1286–1300.

Nakatsuji, H. (1999). Cluster expansion of the wavefunction. electron correlations in ground and excited states by SAC (symmetry adapted-cluster) and SAC CI theories. *Chemical Physics Letters*, *67*, 329–333.

Nielsen, C., Nørby, M. S., Kongsted, J., & Solov'yov, I. A. (2018). Absorption spectra of FAD embedded in cryptochromes. *Journal of Physical Chemistry Letters*, *9*, 3618–3623.

Olsen, J. M. H. (2012). *Development of quantum chemical methods towards rationalization and optimal design of photoactive proteins.* PhD thesisOdense, Denmark: University of Southern Denmark.

Olsen, J. M., Aidas, K., & Kongsted, J. (2010). Excited states in solution through polarizable embedding. *Journal of Chemical Theory and Computation*, *6*, 3721–3734.

Olsen, J. M. H., & Kongsted, J. (2011). Molecular properties through polarizable embedding. *Advances in Quantum Chemistry*, *61*, 107–143.

Olsen, J. M. H., List, N. H., Kristensen, K., & Kongsted, J. (2015). Accuracy of protein embedding potentials: An analysis in terms of electrostatic potentials. *Journal of Chemical Theory and Computation*, *11*, 1832–1842.

Pande, V. S., Beauchamp, K., & Bowman, G. R. (2010). Everything you wanted to know about Markov state models but were afraid to ask. *Methods*, *52*(1), 99–105.

Paola, L. D., Ruvo, M. D., Paci, P., Santoni, D., & Giuliani, A. (2013). Protein contact networks: An emerging paradigm in chemistry. *Chemical Reviews*, *113*, 1598–1613.

Peter, E., Dick, B., & Baeurle, S. A. (2010). Mechanism of signal transduction of the LOV2-Jα photosensor from Avena sativa. *Nature Communications*, *1*, 122.

Peter, E., Dick, B., Stambolic, I., & Baeurle, S. A. (2014). Exploring the multiscale signaling behavior of phototropin1 from *Chlamydomonas reinhardtii* using a full-residue space kinetic Monte carlo molecular dynamics technique. *Proteins: Structure, Function, and Bioinformatics*, *82*(9), 2018–2040.

Ponder, J. W., & Case, D. A. (2003). Force fields for protein simulations. In *Vol. 66. Advances in protein chemistry* (pp. 27–85): Elsevier.

Pople, J., Binkley, J., & Seeger, R. (1976). Theoretical models incorporating electron correlation. *International Journal of Quantum Chemistry*, *10*, 1–19.

Powell, M. J. D. (1994). A direct search optimization method that models the objective and constraint functions by linear interpolation. In *Advances in Optimization and Numerical Analysis* (pp. 51–67). Dordrecht: Springer Netherlands.

Reese, A., List, N. H., Kongsted, J., & Solov'yov, I. A. (2016). How far does a receptor influence vibrational properties of an odorant? *PLoS One*, *11*, e0152345.

Rhee, Y., & Head-Gordon, M. (2007). Scaled second-order perturbation corrections to configuration interaction singles: Efficient and reliable excitation energy methods. *The Journal of Physical Chemistry. A*, *111*, 5314–5326.

Robertson, M. J., Tirado-Rives, J., & Jorgensen, W. L. (2015). Improved peptide and protein torsional energetics with the OPLS-AA force field. *Journal of Chemical Theory and Computation*, *11*(7), 3499–3509.

Robustelli, P., Piana, S., & Shaw, D. E. (2018). Developing a molecular dynamics force field for both folded and disordered protein states. *Proceedings of the National Academy of Sciences*, *115*(21), E4758–E4766.

Roos, B. (1972). A new method for large-scale Cl calculations. *Chemical Physics Letters, 15*, 153–159.
Roos, B., Andersson, K., Fülscher, M., et al. (1988). Chapter Multiconfigurational perturbation theory: Applications in electronic spectroscopy. In *Advances in chemical physics: New methods in computational quantum mechanics* (pp. 219–331). New York: Wiley.
Runge, E., & Gross, E. (1984). Density functional theory for time-dependent systems. *Chemical Physics Letters, 52*, 997–1000.
Sadeghian, K., Bocola, M., & Schütz, M. (2008). A conclusive mechanism of the photoinduced reaction cascade in blue light using flavin photoreceptors. *Journal of the American Chemical Society, 130*, 12501–12513.
Sadeghian, K., Bocola, M., & Schütz, M. (2010). A QM/MM study on the fast photocycle of blue light using flavin photoreceptors in their light-adapted/active form. *Physical Chemistry Chemical Physics, 12*, 8840–8846.
Sadeghian, K., & Schütz, M. (2007). On the photophysics of artificial blue-light photoreceptors: An ab initio study on a flavin-based dye dyad at the level of coupled-cluster response theory. *Journal of the American Chemical Society, 129*, 4068–4074.
Salzmann, S., Silva, M., Thiel, W., & Marian, C. (2009). Influence of the LOV domain on low-lying excited states of flavin: A combined quantum-mechanics/molecular-mechanics investigation. *The Journal of Physical Chemistry. B, 113*, 15610–15618.
Salzmann, S., Tatchen, J., & Marian, C. (2008). The photophysics of flavins: What makes the difference between gas phase and aqueous solution? *Journal of Photochemistry and Photobiology, A: Chemistry, 198*, 221–231.
Sancar, A. (2003). Structure and function of DNA photolyase and cryptochrome blue-light photoreceptors. *Chemical Reviews, 103*(6), 2203–2237.
Schwabe, T., Olsen, J. M. H., Sneskov, K., Kongsted, J., & Christiansen, O. (2011). Solvation effects on electronic transitions: Exploring the performance of advanced solvent potentials in polarizable embedding calculations. *Journal of Chemical Theory and Computation*, 7(7), 2209–2217.
Seifert, G. (2007). Tight-binding density functional theory: An approximate Kohn – Sham DFT scheme. *The Journal of Physical Chemistry A, 111*(26), 5609–5613.
Shaw, D. E., Grossman, J., Bank, J. A., Batson, B., Butts, J. A., Chao, J. C., et al. (2014). In *Anton 2: Raising the bar for performance and programmability in a special-purpose molecular dynamics supercomputer* (pp. 41–53). *Proceedings of the international conference for high performance computing, networking, storage and analysis*, IEEE Press.
Sjulstok, E., Lüdemann, G., Kubar, T., Elstner, M., & Solov'yov, I. A. (2018). Molecular insights into variable electron transfer in amphibian cryptochrome. *Biophysical Journal, 114*, 2563–2572.
Sjulstok, E., Olsen, J. M. H., & Solov'yov, I. A. (2015). Quantifying electron transfer reactions in biological systems: What interactions play the major role? *Scientific Reports, 5*, 18446.
Sneskov, K., Schwabe, T., Kongsted, J., & Christiansen, O. (2011). The polarizable embedding coupled cluster method. *The Journal of Chemical Physics, 134*(10), 104108.
Solov'yov, I. A., Domratcheva, T., Moughal Shahi, A. R., & Schulten, K. (2012). Decrypting cryptochrome: Revealing the molecular identity of the photoactivation reaction. *Journal of the American Chemical Society, 134*, 18046–18052.
Solov'yov, I. A., Domratcheva, T., & Schulten, K. (2014). Separation of photo-induced radical pair in cryptochrome to a functionally critical distance. *Scientific Reports, 4*, 3845.
Solov'yov, I. A., & Schulten, K. (2012). Reaction kinetics and mechanism of magnetic field effects in cryptochrome. *The Journal of Physical Chemistry B, 116*, 1089–1099.
Song, S.-H., Freddolino, P. L., Nash, A. I., Carroll, E. C., Schulten, K., Gardner, K. H., et al. (2011). Modulating LOV domain photodynamics with a residue alteration outside the chromophore binding site. *Biochemistry, 50*(13), 2411–2423.

Sugita, Y., & Okamoto, Y. (1999). Replica-exchange molecular dynamics method for protein folding. *Chemical Physics Letters*, *314*(1–2), 141–151.

Swartz, T. E., Corchnoy, S. B., Chrisite, J. M., Lewis, J. W., Szundi, I., Briggs, W. R., et al. (2001). The phocycle of a flavin-binding domain of the blue-light photoreceptor phototropin. *The Journal of Biological Chemistry*, *276*, 36493–36500.

Udvarhelyi, A., & Domratcheva, T. (2011). Photoreaction in BLUF receptors: Proton-coupled electron transfer in the Flavin-Gln-Tyr system. *Photochemistry and Photobiology*, *87*, 554–563.

Vanommeslaeghe, K., Hatcher, E., Acharya, C., Kundu, S., Zhong, S., Shim, J., et al. (2010). CHARMM general force field: A force field for drug-like molecules compatible with the CHARMM all-atom additive biological force fields. *Journal of Computational Chemistry*, *31*(4), 671–690.

Vanommeslaeghe, K., & MacKerell, A. D., Jr. (2012). Automation of the CHARMM general force field (cgenff) i: Bond perception and atom typing. *Journal of Chemical Information and Modeling*, *52*(12), 3144–3154.

Vanommeslaeghe, K., Raman, E. P., & MacKerell, A. D., Jr. (2012). Automation of the CHARMM general force field (CGenFF) ii: Assignment of bonded parameters and partial atomic charges. *Journal of Chemical Information and Modeling*, *52*(12), 3155–3168.

Woiczikowski, P. B., Steinbrecher, T., Kubar, T., & Elstner, M. (2011). Nonadiabatic QM/MM simulations of fast charge transfer in *Escherichia coli* DNA photolyase. *The Journal of Physical Chemistry. B*, *115*(32), 9846–9863.

Yoon, H., Lee, S., Park, S., & Wu, S. (2018). Network approach of the conformational change of c-src, a tyrosine kinase, by molecular dynamics simulation. *Scientific Reports*, *8*, 5673.

Zayner, J. P., Antoniou, C., French, A. R., Hause, R. J., & Sosnick, T. R. (2013). Investigating models of protein function and allostery with a widespread mutational analysis of a light-activated protein. *Biophysical Journal*, *105*(4), 1027–1036.

Zayner, J. P., Antoniou, C., & Sosnick, T. R. (2012). The amino-terminal helix modulates light-activated conformational changes in AsLOV2. *Journal of Molecular Biology*, *419*(1–2), 61–74.

Zayner, J. P., & Sosnick, T. R. (2014). Factors that control the chemistry of the lov domain photocycle. *PLoS One*, *9*(1), e87074.

CHAPTER TWELVE

Exploring the sequence, function, and evolutionary space of protein superfamilies using sequence similarity networks and phylogenetic reconstructions

Janine N. Copp[a,*], Dave W. Anderson[a], Eyal Akiva[b,c], Patricia C. Babbitt[b,c], Nobuhiko Tokuriki[a]

[a]Michael Smith Laboratories, University of British Columbia, Vancouver, BC, Canada
[b]Department of Bioengineering and Therapeutic Sciences and Department of Pharmaceutical Chemistry, University of California, San Francisco, CA, United States
[c]Quantitative Biosciences Institute, University of California, San Francisco, CA, United States
*Corresponding author: e-mail address: janine.copp@msl.ubc.ca

Contents

Methods in Enzymology, Volume 620
ISSN 0076-6879
https://doi.org/10.1016/bs.mie.2019.03.015

Abstract

Integrative computational methods can facilitate the discovery of new protein functions and enzymatic reactions by enabling the observation and investigation of complex sequence-structure-function and evolutionary relationships within protein superfamilies. Here, we highlight the use of sequence similarity networks (SSNs) and phylogenetic reconstructions to map the functional divergence and evolutionary history of protein superfamilies. We exemplify this approach using the nitroreductase (NTR) flavoenzyme superfamily, demonstrating that SSN investigations can provide a rapid and effective means to classify groups of proteins, expose sequence similarity relationships across the global scale of a protein superfamily, and efficiently support detailed phylogenetic analyses. Integration of such approaches with systematic experimental characterization will expand our understanding of the functional diversity of enzymes, their evolution, and their associated physiological roles.

Abbreviations

CD-HIT cluster database at high identity with tolerance
EFI-EST enzyme function initiative enzyme similarity networks tool
HMM hidden Markov model
MCL Markov chain clustering
MSA multiple sequence alignment
MUSCLE multiple sequence comparison by log-expectation
NTR nitroreductase
SFLD structure function linkage database
SSN sequence similarity network
TIM triosephosphate isomerase
UPP ultra-large alignments using phylogeny-aware profiles

1. Introduction

The expansive range of protein functions found in Nature is a foundational platform for the development, survival, function and evolution of organisms (Aharoni, Gaidukov, Khersonsky, et al., 2005; Copley, 2003; Jensen, 1976; O'Brien & Herschlag, 1999; Tokuriki & Tawfik, 2009). The breadth of functional diversity and its associated consequences and implications are therefore critical for our understanding of biology.

Researchers have discovered a wide range of protein and enzyme functions during the last century that have been characterized in biochemical and biological terms. However, the full repertoire of functional diversity that exists in nature likely remains concealed. Several functionally diverse superfamilies (a cluster of protein sequences that share a common structural fold, active site and/or mechanistic features, Finn, Attwood, Babbitt, et al., 2017; Gerlt & Babbitt, 2001) and smaller families thereof have been extensively explored by experimental scientists; therefore we have a more comprehensive understanding of their functional repertoire (Baier, Copp, & Tokuriki, 2016; Baier & Tokuriki, 2014; Brown & Babbitt, 2012, 2014; CAZypedia Consortium, 2018; Colin, Kintses, Gielen, et al., 2015; Lukk, Sakai, Kalyanaraman, et al., 2012; Mashiyama, Malabanan, Akiva, et al., 2014; Rao & Oldfield, 2016; Swinehart & Jackman, 2015; Vass, Kooistra, Yang, et al., 2018; Wallrapp, Pan, Ramamoorthy, et al., 2013). For most superfamilies, however, experimental characterization is sparse and it is extremely challenging to assign, annotate and predict function. Conventional computational methods annotate new sequences based on the most similar sequence that has been experimentally studied. However, such approaches are problematic when only a handful of functions have been characterized within the superfamily or sequence set of interest (Hicks, Barber, Giddings, et al., 2011). Enhanced approaches to comprehensively characterize protein superfamilies and guide effective experimental exploration of their sequence space are therefore an emerging need across diverse research fields, including biochemistry, enzymology, and evolutionary biology. Systematic investigation of functional diversity, in molecular and biological terms, will vastly improve our ability to harness and exploit this knowledge for both fundamental and applied sciences (Gerlt & Babbitt, 2009; Glasner, Gerlt, & Babbitt, 2007; Renata, Wang, & Arnold, 2015).

SSN-based approaches provide a framework to assist functional assignment and evolutionary analyses, and can expedite the discovery of new protein functions (Akiva, Copp, Tokuriki, & Babbitt, 2017; Gerlt, Babbitt, Jacobson, & Almo, 2012; Goble, Fan, Sali, & Raushel, 2011; Goble, Feng, Raushel, & Cronan, 2013; Selvadurai, Wang, Seimetz, & Huang, 2014; Wichelecki, Graff, Al-Obaidi, et al., 2014; Zhang, Kumar, Vetting, et al., 2015). SSNs facilitate the inspection of much larger sets of protein sequences compared to conventional multiple sequence alignment (MSA) and phylogenetic tree analyses, and enable the simultaneous assessment of orthogonal information, e.g., functional diversity, when mapped onto the context of sequence similarity (Atkinson, Morris, Ferrin, & Babbitt, 2009). The multiplicity of similarity connections that SSNs provide can enable hypotheses about how superfamily subgroups, i.e., subsets of sequences that share significant sequence similarity within a superfamily, are related (Atkinson et al., 2009). Guided by these SSN analyses, structure-based comparisons and other approaches, the relationships between distant sequence sets within a superfamily can be revealed (Davidson, Baas, Akiva, et al., 2018). SSN-based approaches can also provide a powerful tool to illuminate unexplored sequence space within protein superfamilies, thus guiding biochemists to discover unknown functions within the sequence space (Copp, Akiva, Babbitt, & Tokuriki, 2018). Here, we describe an integrated computational approach to capture a global picture of sequence diversity and evolutionary history of a protein superfamily.

Flavoprotein (proteins that bind flavin cofactors) superfamilies are particularly interesting protein groups for the investigation of sequence and evolutionary relationships (Ahmed, Carr, Lee, et al., 2015; Akiva et al., 2017; Ojha, Meng, & Babbitt, 2007). Flavoprotein superfamilies are typically large (containing >50,000 sequences) and varied in structure, e.g., flavoprotein representatives can be found in the diverse folds such as triosephosphate isomerase (TIM) barrel, flavodoxin-like, nitroreductase (NTR)-like, chorismate synthase-like and Rossmann (Table 1) (Macheroux, Kappes, & Ealick, 2011).

Table 1 Summary of six major flavoprotein oxidase families.

Flavoprotein oxidase superfamilies/families	Pfam identifier(s)[a]	Number of sequences		Convergence ratio[b]
		TrEMBL	Swiss-Prot	
Glucose-methanol-choline oxidoreductases	PF00732	60,814	190	0.49
Vanillyl alcohol oxidases	PF01565	127,896	638	0.222
Amine oxidases	PF01593	77,864	212	0.051
	PF01266	161,620	553	
Sulfhydryl oxidases	PF04777	4226	30	0.31
	PF04137	2247	20	
Acyl-CoA oxidase-type oxidases	PF02770	205,827	200	0.597
	PF00441	219,763	181	
2-Hydroxyacid oxidases	PF03060	38,726	20	0.594

[a]Note that some protein superfamilies involve multiple Pfam domains.
[b]A convergence ratio of 1.0 indicates sequences are very similar or identical, and 0 indicates sequences are very different or unrelated.

Flavoenzyme representatives are found within all of the six major Enzyme Classes (E.C.) and flavoproteins also exist without a demonstrated catalytic activity, e.g., flavin storage proteins (Grininger, Staudt, Johansson, et al., 2009). However, the majority of flavoenzyme superfamilies remain relatively unexplored in terms of sequence and functional diversity. For example, <1% of the six major flavoprotein oxidase superfamilies (Dijkman, de Gonzalo, Mattevi, & Fraaije, 2013) have curated functional annotations (i.e., Swiss-Prot annotation); furthermore, they show a significant range of sequence diversity (as measured by the convergence ratio, Gerlt, 2017). Investigation of flavoprotein sequence and evolutionary relationships is likely to yield novel and interesting hypotheses that can subsequently be explored using functional analyses.

It is particularly important that biochemists and other experimentalists have access to computational tools that are tailored to their questions and needs. In this report, a procedure for "in house" generation and analysis of SSNs is outlined in Sections 2–6. For users with access to advanced computational expertise, an alternative approach is outlined in Section 6. Additionally, webservers such as the Enzyme Function Initiative-Enzyme Similarity Networks Tool (EFI-EST) provide a streamlined framework for the generation of SSNs (Gerlt, Bouvier, Davidson, et al., 2015) and an in-depth manually-curated set of pre-computed networks for several large and diverse enzyme superfamilies is also available from the Structure Function Linkage Database (SFLD) (Akiva, Brown, Almonacid, et al., 2014; Knutson, Westwood, Leuthaeuser, et al., 2017; Pegg, Brown, Ojha, et al., 2006). Many of the latter include SSN networks for specific subgroups and individual reaction families within these superfamilies. Section 7 outlines the use of SSNs to streamline the generation of phylogenetic reconstructions and alternative phylogenetic approaches for advanced users are outlined in Section 8.

2. Curating a sequence dataset

The Pfam database is a large collection of protein families defined by structural regions or "domains" (Finn, Coggill, Eberhardt, et al., 2016) and an appropriate starting point for the generation of sequence datasets. Pfam 31.0 (released March 2017) contains 16,712 families and 604 clans (higher-level groupings, related by similarity of sequence (i.e., Hidden Markov Model (HMM)) and structure.

2.1 Downloading a sequence dataset

2.1.1 Websites/webtools

Pfam webserver, http://pfam.xfam.org
UniProtKB, www.uniprot.org
Galaxy web-tool, http://usegalaxy.org

2.1.2 Procedure

1. Select the Pfam family (or clan) that you wish to investigate and download the sequence dataset from UniProtKB. For example, searching for the *E. coli* NfsA NTR enzyme using keywords or a UniProtKB identifier (P17117) within the Pfam webserver will identify the "nitroreductase" Pfam domain PF00881. In UniProtKB, searching for the PF00881 domain will result in 144 reviewed (Swiss-Prot) sequences and 84,440 unreviewed (TrEMBL) sequences (UniProtKB release 2018_05).
2. To download the resulting set of 84,584 sequences in the appropriate format, click on the "Columns" tab and ensure all options are unchecked except for the "sequence" option. By saving these selections, you will be redirected to a web-page that lists the UniProtKB identifiers and the sequences.
3. Download this dataset in tab-separated format using the "Download" icon.
4. Import the dataset into the Galaxy web-tool server using the "get data" tab. Once loaded, close the pop up window and use the "tabular-to-FASTA" option in the FASTA manipulation section of the left hand panel to convert the tab file into a FASTA file.

2.1.2.1 Notes

(1) The first two lines of the resulting FASTA file will read ">Entry" and "sequence," manually delete these two lines in a text edit program.

(2) Although there is an option to download the sequence dataset in FASTA format from UniProtKB, this is not recommended: The detailed FASTA identifiers that result from the direct download are not compatible with later steps unless they are reformatted (requiring use of specifically written scripts).

(3) Some superfamilies are more complicated and involve multiple domains. For example, the Pfam designations for the cytosolic glutathione transferase superfamily list multiple Pfam identifiers and also breaks the N-terminal and C-terminal domains into two Pfam domains (Mashiyama et al., 2014).

2.2 Reducing the sequence dataset

Due to computational limitations, datasets comprising a large number of sequences (e.g., >10,000) must be condensed in order to enable the visualization and manipulation of SSNs on standard computers. The Cluster Database at High Identity with Tolerance (CD-HIT) webserver (Li & Godzik, 2006) is an effective means to select representative sequences from a group of sequences that share a certain percent identity cutoff, e.g., a single sequence can be selected to represent a set of sequences sharing a particular sequence identity.

2.2.1 Websites and webtools

CD-HIT webserver, http://weizhongli-lab.org/cdhit_suite/cgi-bin/index.cgi

2.2.2 Procedure

1. To condense the dataset above to one containing representative sequences of a certain percent identity threshold, the CD-HIT webserver can be used.
2. Upload the FASTA-formatted file from Section 1 to "h-CD-HIT". This option performs multiple CD-HIT runs: Proteins are first clustered at a high identity (i.e., 90%), then the non-redundant sequences selected from this first clustering process are further clustered at a lower identity (i.e., 70%). For example, after loading the FASTA file above to the webserver, select three CD-HIT runs, at 90%, 70% and 50% ID.
3. Download all files from the CD-HIT calculation and save this folder as "cdhit." The "fas.3" file within this folder contains the reduced dataset, i.e., a representative sequence selected for each 50% ID cluster.

2.2.2.1 Notes

(1) The CD-HIT files are labeled with your unique job number (important for Section 8) and include the results from each CD-HIT run, listing the identifiers for each cluster.

(2) For the PF00881 dataset, performing a multistep CD-HIT (90%, 70% and 50%) reduces the dataset from 84,584 sequences to 6211 "50% ID" representatives. For the same dataset, a lower percent identity cutoff would result in fewer sequences; a higher identity cutoff will result in more sequences.

(3) CD-HIT clusters sequences by first sorting the dataset by length and subsequently, identifying "seed" sequences. Starting from the longest representative of the dataset, each sequence is clustered with seed sequences if they meet the (user defined) sequence identity cut-off. If the sequence identity of any representative sequence is lower than the cut-off, it will become a seed sequence of a new cluster. Iterative cycles of seeding and clustering generate a set of representative sequences with a defined sequence identity threshold. A multi-step run (i.e., sequential CD-HIT runs (h-CD-HIT) with decreasing % identity thresholds) is more efficient and rigorous than a single run (weizhongli-lab.org/cdhit_suite/).

2.3 Filter the dataset by length

2.3.1 Websites/webtools

Galaxy web-based tool, www.usegalaxy.org

2.3.2 Procedure

Galaxy (Afgan, Baker, Batut, et al., 2018) is a helpful platform for filtering sequences by length. This process can be used to eliminate small fragmented sequences (partial gene sequences) and long sequences with multiple-domains.

1. Upload the CD-HIT "fas.3" file created in Section 2.2 and use "filter sequences by length" option found in the "FASTA manipulation" section. The optimal sequence length to analyze the NTR superfamily was manually determined to be >100 aa and <1500 aa (to avoid the inclusion of partial gene fragments and large multidomain enzymes). Thus, set the minimal length to be 100 aa and the maximum length to be 1500 aa.
2. Running this process and downloading the resulting file will create a dataset of 6015 sequences.

2.4 Finalizing your protein database

2.4.1 Websites/webtools

UniProtKB, www.uniprot.org
NCBI, www.ncbi.nlm.nih.gov

2.4.2 Procedure

As the CD-HIT process in Section 2.2 removes many sequences from the dataset, it is important to manually add any specific sequences of interest back into the dataset, e.g., those associated with a biochemically characterized function or with an available crystal structure, thus allowing effective annotation of the sequence space using experimental information.

1. Curated Swiss-Prot sequences can be found by searching for the Pfam domain identifier in UniProtKB. For example, there are 144 Swiss-Prot sequences associated with PF00881. Swiss-Prot sequences have associated functional annotations with "evidence codes" that denote whether the function of a sequence has been determined from a direct experimental assay or by electronic annotation. However, researchers must thoroughly assess all functional information to carefully confirm Swiss-Prot status and perform exhaustive literature reviews. Ensuring that all biochemically-characterized sequences are included is a challenging task due to misannotation and the lack of standardized reporting of accession numbers. For example, although the *M. tuberculosis* Acg enzyme (UniProtKB ID: P9WIZ9) has been physiologically characterized (Hu & Coates, 2011), this enzyme is not associated with the Pfam PF00881 domain; it is instead associated with an alternative definition of the NTR domain, the "nitroreductase-like" InterPro identifier IPR000415. Further complicating evaluation of how function is asserted, "reviewed" Swiss-Prot annotation does not necessarily denote experimental characterization. For example, *M. smegmatis* yasK is a curated Swiss-Prot entry (UniprotKB: P41401); however, closer inspection reveals that this protein is a "predicted protein" with the lowest level of Swiss-Prot review status (www.uniprot.org/uniprot/P41401). Manual curation of "reviewed" Swiss-Prot sequences in the dataset above, and careful assessment of the literature, resulted in a set of 86 NTR sequences which, to the best of our knowledge, are all of the experimentally characterized proteins in the NTR superfamily. These sequences were added to the dataset created in Section 2.3.
2. Structures associated with the Pfam domain can be found, for example, by searching for the Pfam identifier (PF00881) and "pdb" in UniProtKB. Their sequences can then be added to the dataset by downloading and appending to the dataset (as per Section 2.1). For PF00881, this step resulted in 67 additional sequences which should also be added to the dataset created in Section 2.3. Many structures are deposited by structural genomics consortia; therefore, note that some sequences with a PDB structure may not necessarily be assigned with a biochemical function.
3. To ensure this step has not introduced identical (i.e., redundant) sequences to the dataset, a final CD-HIT analysis can be run at a 100% sequence identity cut off. For example, in the dataset above ($6015+86+67=6168$ sequences), there were 40 redundant sequences, removing these created a final dataset of 6128 sequences.

3. Calculation and visualization of sequence similarity relationships

BLAST + can be used for the computation of "all vs all" BLAST, thus obtaining BLAST E-values for every pair of sequences in the dataset (Camacho, Coulouris, Avagyan, et al., 2009).

3.1 Performing "all vs all" BLAST calculations

3.1.1 *Websites/programs/webtools*

BLAST +, www.ncbi.nlm.nih.gov/guide/howto/dwn-software/
Installers/source code, ftp.ncbi.nlm.nih.gov/blast/executables/blast+/LATEST/
EFI-EST, efi.igb.illinois.edu/efi-est/

3.1.2 *Download and install BLAST+*

1. Download BLAST+ and install the version corresponding to your system type.
2. Create a folder labeled "SSN" in your Home directory, and move BLAST+ and your curated database file to this folder. *Important*: the extension of the latter file must be ".txt."

3.1.3 *Procedure (BLAST+)*

1. For Macintosh/linux users, open the Terminal application.
2. Type: "cd SSN."
   (*which changes directory to folder where the file is*)
3. Type: "Makeblastdb -in final.txt -dbtype prot -out finalpro"
   (*creates a protein blast database, named* finalpro, *using the* final.*txt file)*
4. Type: "Blastp -db finalpro -query final.txt -outfmt 6 -out BLASTe22 -num_threads 4 -evalue 10e−22
   (*this action runs "all versus all" BLAST, if the resulting E-value meets the defined threshold, here 10e − 22, it will be listed in the output text file*)

3.1.3.1 Notes

(1) "db" specifies the database; "query" specifies name of the file containing the query sequences; "outfmt" defines the format of the output file; "out" sets that name for the output file; "evalue" defines the threshold value; and "num_threads" defines the number of the processes that will run concurrently for the BLAST calculation.
(2) Four threads are used in this example, for faster machines a higher number can be used and therefore will result in faster calculations.

(3) The BLAST calculations can be run iteratively by repeating the last command line, e.g., to generate an all-by-all BLAST at a threshold of 10e−40, as displayed in Fig. 1C, type (into Terminal): "blastp -db finalpro -query final.txt -outfmt 6 -out **BLASTe40** -num_threads 4 -evalue **10e−40**".

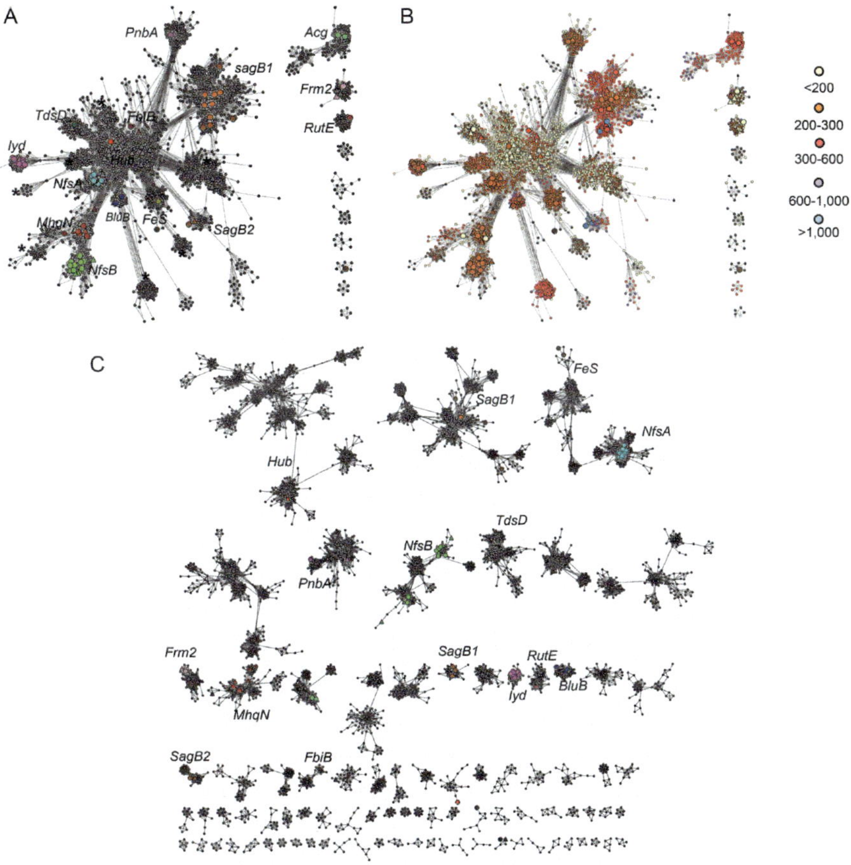

Fig. 1 Sequence similarity networks of the NTR superfamily. (A) A network is shown displaying 5995 nodes (each representing a sequence that shares <50% ID). Edges represent E-values lower than 1×10^{-22}. Sequences that are associated with a PDB structure are depicted as enlarged triangles, nodes with characterized biochemical or physiological function(s) are colored and labeled. Examples of unexplored sequence space (i.e., clusters lacking functionally characterized members) are indicated with asterisks. Clusters containing fewer than five sequences have been removed for clarity. (B) A network is shown as per (A), but colored by sequence length; (C) An SSN of the NTR superfamily is shown, displaying 5207 nodes colored and annotated as per (A). Edges correspond to E-values lower than 1×10^{40}. *Reprinted with permission from Copp, J. N., Akiva, E., Babbitt, P. C., Tokuriki, N. (2018). Revealing unexplored sequence-function space using sequence similarity networks.* Biochemistry, 57, *4651–4662. Copyright 2018 American Chemical Society.*

(4) The output format selected for BLAST calculations (outfmt 6) generates a tab delimited file with 12 columns: **(1) query**, **(2) subject**, (3) percentage of identical matches, (4) alignment length, (5) number of mismatches, (6) number of gap openings, (7) start of alignment in query, (8) end of alignment in query, (9) start of alignment in subject, (10) end of alignment in subject, **(11) E-value**, (12) bit score.

(5) For an alternative method, the EFI-EST webserver (in addition to its utility to generate networks from Pfam domains or InterPro families) can be used to create networks from a user-provided dataset. This can be performed by selecting the "User defined FASTA file" option to enable the upload of a curated dataset, e.g., from Section 4. This facilitates the generation of SSNs using curated datasets without the need for an in-house BLAST+ and command line prompts, which may suit some researchers. If the "read FASTA headers" option is selected when using this method, annotation tables will also be automatically appended to the network (see Section 3.3).

3.2 Visualization of sequence similarities

The Cytoscape platform (Shannon, Markiel, Ozier, et al., 2003) enables the visualization and manipulation of SSNs. Detailed tutorials are available at github.com/cytoscape/cytoscape-tutorials/wiki. In Cytoscape, a "node" (circle) represents a protein sequence and an "edge" (the line connecting nodes) is shown if the BLAST E-value meets the set threshold, e.g., in Fig. 1A, two nodes will be connected by a line if the BLAST E-value (of the respective sequences that they represent) meets the E-value threshold of 1×10^{-22}. In Cytoscape v3.6.1, the "prefuse force-directed (none)" layout algorithm is an effective way to graphically cluster SSNs.

3.2.1 Download and install cytoscape

Cytoscape, *http://www.cytoscape.org/*

3.2.2 Viewing BLAST files

1. To open the file created above, go to File → Import → Network → File, and select the "BLAST22" file. When the preview box opens, ensure "use first line as column names" is unchecked in the Advanced options.
2. Define the column data. First check the "Select None" option to exclude all columns in the table, then change Column 1: Meaning → Source Node, Data type → String; Column 2: Meaning → Target Node, Data

Type → String; Column 11: Meaning → Edge attribute, Data Type → Floating point.

3. Do not select the option to view the network when you first load it on Cytoscape as the file is very large, i.e., click "cancel" on the pop up dialog box.
4. Additional steps before viewing the network: Edit → Remove Self-Loops; Edit → "Remove Duplicated Edges" (ignore edge direction).
5. View the network using Edit → Create view. Ensure the correct layout is displayed by selecting Layout → Prefuse Force Directed (none).

3.3 Appending additional information to the SSN

3.3.1 Websites/webtools

UniProtKB, www.uniprot.org

A wide variety of sequence information (e.g., protein length, taxonomy, E.C. number, etc.) can be found in UniProtKB and can be appended to your network in Cytoscape.

1. For example, after searching for the PF00881 identifier in UniProtKB and selecting the "columns" tab, users can view a list of additional information relating to each sequence. Options can include gene names, protein names, length, sequence, EC number, Swiss-Prot curation (found in the "miscellaneous" section, represented by an icon of a page with a gold star), length, PDB, and superkingdom of the organism from which the gene originated.
2. Save these selections (using the icon at the top right of webpage), and download the resulting dataset in tab-separated format.
3. Append this tab file to your network in Cytoscape by using File → Import → Table → File. In the popup panel that appears, import the Table data using the "To selected networks only" option, import data as "Node Table Columns," and use "shared name" (which is the UniProtKB identifier) for the "key column for networks" option. The additional information will now be found in the Table Panel of Cytoscape.
4. Different graphical attributes such as node size, color and node border width can then be used to display this information in Cytoscape, e.g., sequence length, functional annotation, taxonomy of host organism, or structural information; detailed information can be found on this process on the Cytoscape wiki (https://github.com/cytoscape/cytoscape-tutorials/wiki). For example, from the "Style" option in the Cytoscape control panel, click the "fill color" dropdown option, Column → length,

Mapping Type→Discrete Mapping. The sequences lengths for all nodes will then be displayed. Select all sequences <200 aa and color in pale yellow, then repeat this process with sequences that are 200–300 aa (orange), 300–600 aa (red), 600–1000 aa (purple), and >1000 aa (blue); the resulting network will be colored as per Fig. 1B. Variation in sequence length can guide the investigation of unexplored sequence clusters that may deviate from known/verified enzymes, e.g., groups of sequences that display longer lengths could indicate insertions of loops or domains (Fig. 1B). In particular, sequences harboring large insertions may indicate the fusion of additional Pfam domains (other than the canonical domain of the superfamily of interest), which can guide predictions of metabolic pathways and/or physiological functions. However, SSN clusters that are associated with an additional Pfam domain(s) necessitate careful assessment, as the high sequence similarity relationships may result from segments of the alignment that span beyond the domain of interest. Other attributes, e.g., size, border width, etc., can be similarly selected from the "style" options to visualize various parameters of the dataset. For example, PDB data can be used to determine structural variations from canonical/verified enzyme functions, e.g., structural insertions or deletions, or rearrangements (Fig. 1A; PDB structures indicated by enlarged triangles). However, the structural coverage of superfamilies is typically sparse.

3.3.1.1 Notes

(1) Computational limitations constrain the number of edges that can be displayed in an SSN, e.g., networks with more than one million edges cannot typically be manipulated on a computer with 8GB RAM. Representative sequence sets facilitate the construction of networks that are appropriate for the computational resources that are typically available within a laboratory. At more stringent edge thresholds, a smaller number of edges will meet the threshold and thus, a larger set of sequences (e.g., using a higher CD-HIT cutoff such as 70% or 90%) may be used. In our experience, generating an SSN that displays <6000 nodes with an edge threshold of 1×10^{-15} to 1×10^{-25} is generally a useful starting point to produce computationally manageable networks for the initial analysis of a superfamily. However, it must be emphasized that the broad variation of similarity relationships within superfamilies necessitates case-by-case exploration.

4. Defining appropriate SSN thresholds

Defining the most appropriate threshold to visualize and analyze a protein superfamily is a critically important step in SSN analyses. Thus, for initial analysis of a superfamily, it is recommended that users generate networks using several arbitrary thresholds across a wide range to identify the most suitable threshold for their research aim(s); the optimal E-value threshold can substantially differ depending on the dataset and the purpose of the analysis. For instance, a stringent threshold of $<1 \times 10^{-100}$ may be necessary to delineate highly homologous enzyme sequences that share the same chemical reaction but differ in substrate specificity. For the initial analysis of a superfamily, a threshold "scan" can be performed, generating multiple networks across a broad range of highly permissive (e.g., 1×10^{-10}) to highly stringent (e.g., 1×10^{-120}) E-values. Subsequently, based on these SSNs, the user can then more systematically sample different thresholds on a finer scale. For example, the threshold chosen for Fig. 1A (E-value 1×10^{-22}) was the result of sampling a wide range of E-value scores, from 1×10^{-10} to 1×10^{-100} while looking for consistency between the sequence-similarity based clustering and functional information. As shown in Fig. 1C, at a threshold of 1×10^{-40} sequence clusters are more discernible, however, intercluster relationships are lost. Conversely, the network is highly connected and uninformative at thresholds lower than 1×10^{-15}. "Thresholding," i.e., the process of iteratively analyzing thresholds of increasing stringency, in combination with observation of additional information about the associated proteins (functional assignment, phylogeny, domain architecture, etc.), facilitates the systematic analysis of sequence and functional relationships.

5. Generation of "higher resolution" subgroup-specific SSNs

5.1 Retrieving sequences for the generation of subgroup-specific SSNs

The "cdhit" folder, downloaded in Section 2, contains a master list of all sequences within each cluster, i.e., all sequences attributed (at the respective identity cutoff, e.g., 50% ID or 70% ID) to each representative seed sequence in the reduced dataset. As a multi-step CD-HIT run was used in Section 2.2, the corresponding CD-HIT results folder contains each of the three iterative CD-HIT calculations, 50% ID, 70% ID and 90% ID. For example, the RutE subgroup (Fig. 1A) contains 72 50% ID representative sequences. To expand the RutE subgroup to include more sequences, e.g., 70% ID representatives (instead of 50% ID), a 70% ID representative network of the RutE subgroup can be created.

1. Move the cdhit to the SSN folder created in Section 2.2.
2. A new network can be created in Cytoscape by selecting all nodes of the subgroup and then using the "New Network From Selection" in the Cytoscape Toolbar.
3. Download the "ID2cluster2identifiers.txt" file from the supplementary materials of Copp et al. (2018) and save it to the "cdhit" folder downloaded in Section 2. Manually change the file name to ".pl" (instead of ".txt").
4. For Macintosh/linux users, open the Terminal application.
5. type: "cd cdhit"

(*which changes directory to the cdhit folder*)

6. Type: "perl ID2cluster2identifiers.pl xxxxxxxxxx.fas.3.clstr"

(*where* **xxxxxxxxxx** is the unique 10 digit CD-HIT job number from the clustering performed in Section 2.2, and thus the name of the .fas3.clstr file within the CD-HIT file download)

This will create a file named "ID2clstr2uniprots.tsv," which maps each node name (UniProtKB identifier) to its corresponding CD-HIT cluster number and the list of UniProtKB identifiers represented by this cluster number (i.e., the list of sequences that are represented by a single representative sequence in the 50% ID CD-HIT results).

7. In Cytoscape, select File → Import → Table → File and upload the ID2clstr2uniprots.tsv file. Import the table data "to selected networks only" (selecting the RutE subgroup network created above). Import the data as "Node Table Columns," choose "shared name" as the "Key Column for Networks." Edit the "REPRESENTED ACCs" column (via the menu accessed through the dropdown arrow), to select Meaning → Attribute, Data Type → [ab] (list of strings), List Delimiter → "," (comma).
8. Uploading this list (i.e., identifiers corresponding to the 70% ID clustering of the dataset) will enable its display as a column of the Table panel.
9. Export the resulting table using the "Export Table to File" option. Copying the column "represented ACCs" into a word document will allow the removal of the "|" delimiter and replacement with a space (i.e., " "). Copying the resulting list of identifiers into the "Retrieve/ID mapping" option in the UniProtKB toolbar will result in a list 566 sequences.

10. In order to create the 70% ID representative dataset of the RutE subgroup, the 566 sequences identified above (in addition to the 72 sequences from the 50% ID cluster) can be downloaded as described in Section 1, to create a 70% ID representative dataset of the RutE subgroup. This dataset will contain 638 sequences and can subsequently be used for BLAST calculations and SSN creation as per the process above.

5.1.1 Notes

(1) This process can be run iteratively to create networks for each of the CD-HIT runs, i.e., subgroup-specific networks could be created for the dataset created in Section 2.2 at 50%, 70%, 90% and 100% identity.

6. Advanced generation and analysis of SSNs using expanded sequence sets

Databases such as InterPro (Mitchell, Chang, Daugherty, et al., 2015), Pfam (Finn et al., 2016), and CATH-Gene3D (Sillitoe, Lewis, & Orengo, 2015) are sequence and structure classification systems that provide an excellent resource for associating protein sequences with a superfamily or family of interest. However, each database utilizes different methods with distinctive perspectives, and thus, their datasets can contain subtly different sets of sequences. The vast majority of sequence and domain classifications are not defined by functional features; however, function is paramount for the biological questions of interest. Harnessing data from multiple databases can therefore generate a more comprehensive sequence dataset for a protein superfamily.

For example, in order to collate an all-inclusive set of sequences to represent the NTR superfamily, we collected an exhaustive non-redundant set of all available sequences and structures that can be associated with the NTR superfamily from multiple databases (Akiva et al., 2017). The resulting sequence set was manually filtered based on literature and experimental knowledge, and contained 24,270 sequences that range between 150 and 1580 amino acids in length. This dataset, curated in 2015, contains significantly fewer sequences than the number now associated with the PF00881 domain (>84,000 sequences in the UniProtKB release 2018_5), which can be attributed to redundant (identical) sequences in the database and the recent expansion of sequence space through modern sequencing technologies.

An alternative method to generate "representative SSNs" is to calculate the average pairwise E-value between all the CD-HIT cluster sequences in a given node and the CD-HIT cluster sequences in another node; consequently, the sequence information represented by a single node is maintained. To generate a SSN that retained all sequence information within each representative node of the nitroreductase superfamily, the Pythoscape software (Barber & Babbitt, 2012) was utilized (note that Pythoscape typically requires tailoring for the user's specific hardware environment). A 60% ID cutoff was used to condense the dataset, which resulted in 5604 representative sets of sequences that each contained 1–307 individual members. Average pairwise BLAST E-values were calculated from the individual pairwise BLAST of each member of each representative sequence set.

For an alternative method, network similarity data can also be clustered using algorithms such as Markov chain CLustering (MCL) (Enright, Van Dongen, & Ouzounis, 2002), which is packaged for use within the Cytoscape software, or other approaches such as InfoMap (Rosvall & Bergstrom, 2008). These automated algorithms are designed to achieve a mathematically optimum separation of the network into constituent clusters that are disconnected from each other. Both approaches have been used together to compare classification of sequence groupings using different types of similarity data (Akiva et al., 2017).

A representative SSN created using the Pythoscape platform is shown in Fig. 2A colored by subgrouping classification; for comparative purposes,

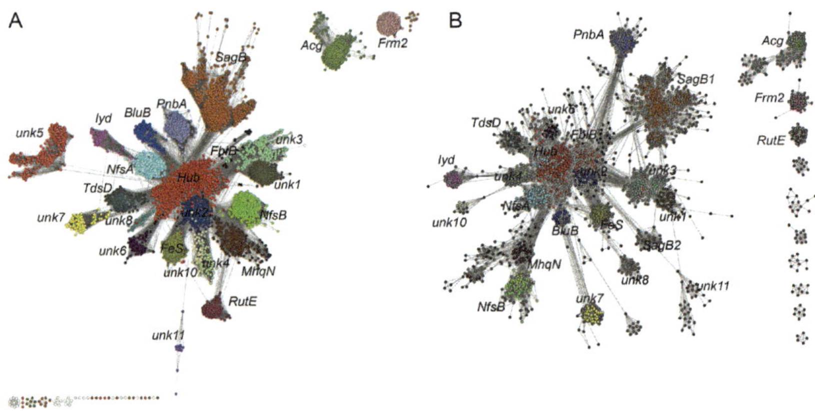

Fig. 2 Sequence similarity networks of the NTR superfamily. (A) A representative sequence similarity network of the NTR superfamily (generated with Pythoscape) at an E-value threshold of 1×10^{-18}. Nodes represent sequence sets that contain 1–307 sequences grouped by 60% identity. Subgroups, as defined previously (Akiva et al., 2017), are colored. (B) A sequence similarity network of 50% identity representatives (as per Fig. 1) is displayed at a threshold of 1×10^{-22}. Subgroups are colored as per (A). *Reprinted with permission from Copp, J. N., Akiva, E., Babbitt, P. C., Tokuriki, N. (2018). Revealing unexplored sequence-function space using sequence similarity networks.* Biochemistry, 57, *4651–4662. Copyright 2018 American Chemical Society.*

the CD-HIT generated SSN in Section 2 is shown in Fig. 2B. Overall, these networks emphasize the robust nature of SSN-based investigations, as although created using different approaches and distinct datasets, they consistently provide similar topologies and sequence similarity relationships. Sequences in the CD-HIT dataset (Fig. 2B) that are not represented in Fig. 2A are displayed as uncolored (gray) nodes and highlight the expansion of the NTR superfamily dataset from that used in earlier investigations (Akiva et al., 2017); SSN clustering is maintained but the coverage of NTR superfamily sequence space is expanding.

Advanced computational approaches can also be used to identify subgroup, or sequence-cluster, specific variation, i.e., using active site profiling or substrate docking approaches (Knutson et al., 2017; Tian, Wallrapp, Holiday, et al., 2014). These variations can be used as graphical features for SSN nodes, thus facilitating global evaluation (i.e., across the entire superfamily) and visualization of their distribution across distinct subgroups (Akiva et al., 2017). Additionally, researchers can guide hypotheses of substrate and catalytic specificities through complementary approaches, such as genomic context analysis (investigations of the neighboring genes that form an operon with the gene of interest and the conservation of these genomic relationships among other members the sequence cluster) or alternatively, profiling abundance in microbiome or metagenome surveys (Levin, Huang, Peck, et al., 2017; Zhang, Carter, Vetting, et al., 2016; Zhao, Kumar, Sakai, et al., 2013).

7. Using SSNs to streamline phylogenetic analyses

SSNs effectively visualize the degree of relatedness between a vast array of sequences; they fall short, however, in their ability to reconstruct the historical relationships between those sequences. The nested hierarchy that is a reflection of the historical evolutionary process (of descent with modification from common ancestors) is the province of phylogenetics. Molecular phylogenetics tools can be used to supplement SSN-based approaches, either by focusing on individual subgroups to enable an enhanced view of evolutionary relationships or for the superfamily as a whole (typically using representative sequences, rather than the full complement of sequences, which may be computationally intractable). In either case, SSNs can guide detailed phylogenetic analyses by presenting potential phylogenetic hypotheses that can be tested, as well as directing the choice of sequences to be used (which is dependent upon the hypothesis in mind).

The common questions that can be resolved using molecular phylogenetics following SSN construction are as follows: (1) What are the detailed evolutionary relationships between the members of a specific subgroup? and (2) What are the relative evolutionary relationships between different subgroups across the protein superfamily?

7.1 Ascertaining the evolutionary relationships within a subgroup

While SSNs provide very good resolution of the relative similarity between the sequences being examined, at thresholds appropriate for observing entire superfamilies, it is common for subgroups to appear as a single "clump" without any clearly defined sub-structure. This can be resolved through the generation of subgroup-specific SSNs (as discussed in Section 5) or alternatively, by reconstructing the historical evolutionary relationships using conventional phylogenetic analyses (Brown & Sjölander, 2006). The latter has the advantage of not only revealing similarity/dissimilarity across the subgroup, but also of revealing the evolutionary mechanisms by which it diversified from a single common ancestor, e.g., via speciation versus duplication events (Szöllősi, Tannier, Daubin, & Boussau, 2015). This information can provide insight into other adaptive processes that could be linked to an enzyme's evolution, such as major ecological shifts that are known to have occurred for particular groups of organisms (Morales-Hojas & Vieira, 2012).

7.2 SSN-guided sequence selection for phylogenetic analysis

Subgroup-specific SSNs, e.g., SSNs displaying sequences within a particular subgroup of a superfamily, can be generated as per Section 5 or downloaded from the SFLD (if available). Increasingly stringent E-value thresholds can then be used to visualize sequence similarity relationships and divergence within the subgroup. At more stringent thresholds, subgroups typically fragment into multiple small clusters (subclusters), therefore representative sequences can be selected from each subcluster for phylogenetic reconstruction. For example, the FASTA sequences of the NfsA subgroup can be extracted from the NTR superfamily SSN (see Section 5) or alternatively, downloaded from the SFLD (http://sfld.rbvi.ucsf.edu/django/subgroup/1228/). SSNs can then be generated at a range of E-value thresholds, e.g., 1×10^{-30} to 1×10^{-90} using the methods described in Section 3 or alternatively, using the EFI-EST webserver. Fig. 3 displays NfsA subgroup SSNs at E-value thresholds of 1×10^{-50}, 1×10^{-60}, and 1×10^{-70}, respectively.

Sequences can then be selected to represent the diversity of the subgroup of interest. As seen in Fig. 3B and C, the NfsA subgroup separates into

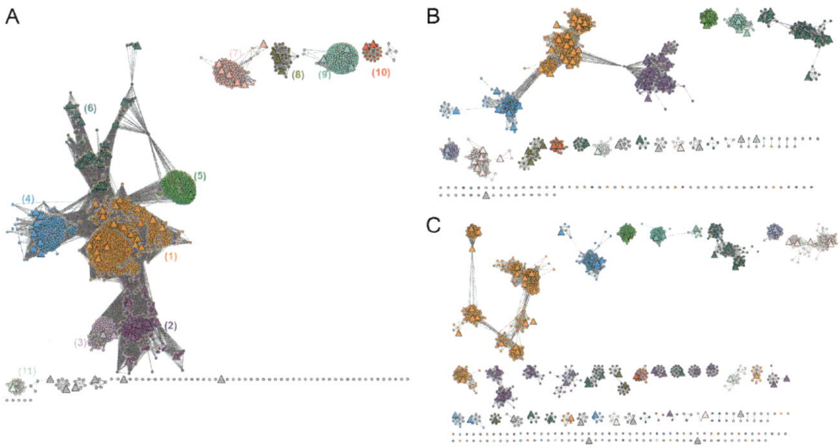

Fig. 3 NfsA subgroup sequence similarity networks. (A) A sequence similarity network shown at an E-value threshold of 1×10^{-50}; sequence clusters containing >10 sequences are colored and numbered. Enlarged triangles indicate sequences selected as representatives for phylogenetic reconstructions (Fig. 4). (B) A sequence similarity network as per (A) is displayed at a threshold of 1×10^{-60}. (C) A sequence similarity network as per (A) is displayed at a threshold of 1×10^{-70}.

>11 subclusters at 1×10^{-50} and >30 subclusters are seen at thresholds lower than 1×10^{-70}. In order to represent the diversity of the NfsA subgroup, 80 sequences were selected to represent the major subclusters of the 1×10^{-70} SSN (i.e., 2–4 sequences per subcluster), which are depicted as enlarged triangles in Fig. 3. Note that the "optimal" number of sequences for analysis varies significantly depending on factors such as the overall diversity of the sequences, gene length, the degree of variation in gene length, and computational limitations that can restrict the size of the MSA. Depending on these factors, successful analysis may be efficacious for hundreds to thousands of distinct sequences.

As previous evolutionary analysis has suggested that the Hub subgroup of the nitroreductase superfamily closely resembles the evolutionary ancestor (Akiva et al., 2017), a representative from the Hub subgroup was included as an outgroup for phylogenetic reconstructions; an "outgroup" is a distantly related gene that can serve as a reference to determine the evolutionary relationships between genes of interest.

7.3 Reconstruction of phylogenetic relationships

There are numerous publicly available software sets that will perform both the phylogenetic reconstruction and the visualization of those results; many of them were extensively reviewed (Akanni, Wilkinson, Creevey, et al., 2015; Mar, Harlow, & Ragan, 2005; Zhou, Shen, Hittinger, & Rokas, 2018). Here, we briefly explore two common approaches to phylogenetic reconstruction using exclusively genetic data: (1) Maximum-likelihood and (2) Bayesian. The former is used by software such as PAML (Yang, 2007), RaxML (Stamatakis, 2014), and PhyML (Guindon, Delsuc, Dufayard, & Gascuel, 2009). Maximum-likelihood has been argued to be as, or more, accurate dependent upon the taxonomic level and genetic diversity of the dataset (Susko, 2008; Wiens & Servedio, 1998). Both RaxML and PhyML can be executed using user-friendly interfaces, such as Phylobot (Hanson-Smith & Johnson, 2016). Bayesian approaches, such as those implemented by MrBayes (Ronquist, Teslenko, van der Mark, et al., 2012) and PhyloBayes (Lartillot, Lepage, & Blanquart, 2009), have the advantage of user-input prior probabilities that can reflect other levels of information (Heaps, Nye, Boys, et al., 2014). In each case, the output from phylogenetic analysis will be a set of nested relationships between all the sequences that were provided in the initial alignment.

Phylogenetic relationships should be carefully and statistically assessed; several different methods are available for this process, the most common of which are bootstrapping, posterior probability, and likelihood ratio statistics. Briefly, these methods differ not just in terms of how they are computed but also in what they are describing. Bootstrapping involves repeatedly re-sampling from a dataset and then calculating the confidence in a particular branching pattern based on how frequently it appears in the re-sampling process (Felsenstein, 1985). Posterior probability indicates the probability that a particular node is true given the data being analyzed (this should reflect the proportion of informative positions that "support" the node versus the proportion of informative positions that "refute" the node). There are a variety of methods of computing posterior probabilities, including estimations from bootstrap results (Efron, Halloran, & Holmes, 1996). Finally, the likelihood ratio test can be used to assess the significance of any node for phylogenetic constructions performed using Maximum Likelihood. This test calculates the likelihood ratio for each node (see Fig. 4), which is the ratio between the likelihood value for the "best" tree and the "second" best topological arrangement surrounding that node (i.e., another tree that contains a node without the same monophyletic descendants) (Anisimova & Gascuel, 2006).

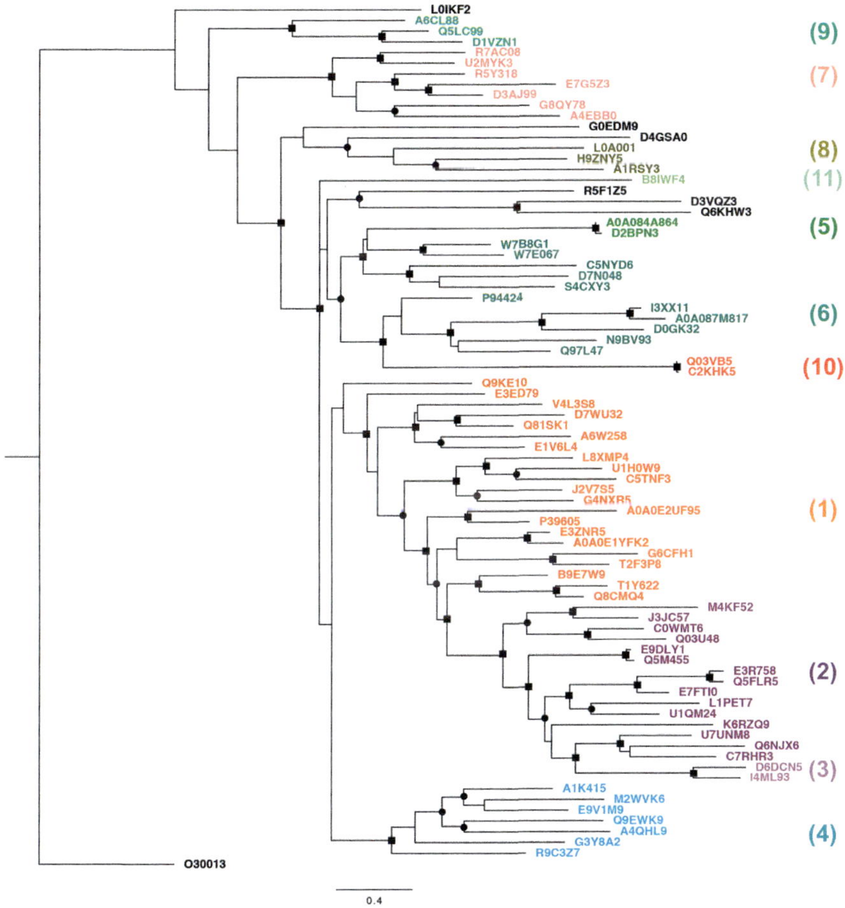

Fig. 4 Phylogenetic reconstruction of representative sequences from the NfsA subcluster. A model depicting the evolutionary history of members of the NfsA subcluster, identified by UniProtKB IDs, were visualized using FigTree (http:/tree.bio.ed.ac.uk/software/figtree/). Nodes are colored by subcluster (as per Fig. 3) and subcluster numbering is indicated in brackets. Tree accuracy is represented by approximate likelihood ratio (aLR); circles represent a likelihood score of 10–1000; squares indicate a score >1000. The scale bar indicates the average number of substitutions per amino acid site.

7.4 Multiple sequence alignments and phylogenetic analysis

7.4.1 Websites/webtools

MUSCLE, https://www.ebi.ac.uk/Tools/msa/muscle/,

Clustal Omega, https://www.ebi.ac.uk/Tools/msa/clustalo/

T-Coffee, http://tcoffee.crg.cat/

RaxML, https://raxml-ng.vital-it.ch/#/

PhyML, http://www.atgc-montpellier.fr/phyml/
IQ-Tree, http://www.iqtree.org/
Phylobot, http://phylobot.com/

7.4.2 Overview of procedure

The first stage of performing a phylogenetic analysis involves the generation of an MSA of all the genes or protein sequences that are to be examined. There are a large number of alternative tools that can perform MSA on both DNA and protein sequences. Three common examples are: MUSCLE (MUltiple Sequence Comparison by Log-Expectation), Clustal Omega, and T-Coffee (which utilizes structural information). MSAs can be used to differentiate important features that determine function, e.g., key catalytic residues. However, protein superfamilies and their respective subgroups are often complex, with multidomain architectures, insertions, deletions, or unstructured regions. Different MSA tools can produce different alignments; thus, detailed assessment of the alignment is critical (Thompson, Plewniak, & Poch, 1999). Manual curation, which may involve removing sequences with insertion or deletions that result in poorly aligned sections of the alignment, is often necessary.

Once a robust and reliable alignment is achieved, the phylogenetic reconstruction can be performed. Again, there are numerous publicly available tools that can be utilized, including (but not limited to): RaxML, PhyML, and IQ-Tree. A useful alternative to the multi-step analysis specified above is to use an integrated pipeline such as Phylobot, which performs the analysis (which may take up to several days) using cloud resources and produces reconstructions for preliminary analyses.

7.5 Visualization of phylogenetic relationships

7.5.1 Websites/webtools

FigTree, http://tree.bio.ed.ac.uk/software/figtree/
TreeView, http://en.bio-soft.net/tree/TreeView.html
Clustal, https://www.ebi.ac.uk/Tools/msa/clustalo/
iTOL, https://itol.embl.de/

7.5.2 Overview of procedure

Appropriate visualization of phylogenetic reconstructions is necessary for their effective analysis (Wilgenbusch, Huang, & Gallivan, 2017). Common software tools include FigTree and TreeView, as well as other visualization options that are included within other analytical packages, e.g., Clustal and iTOL. These visual representations of the evolutionary history, either within or between subgroups of a superfamily, can provide unique hypotheses regarding the processes that have led to their diversification. For example, a PhyloBot phylogenetic reconstruction of the representative sequences selected from the NfsA subgroup is displayed in Fig. 4. This reconstruction reveals the emergence of certain sequence subclusters within the NfsA subgroup, such as subcluster 3 from within the larger subcluster 4, which is congruent with the SSN (Fig. 3). In other respects, the nested hierarchy of evolutionary relationships can reveal details that are missing from the SSN, for example, subcluster 6, which is not a monophyletic clade and instead has members distributed across multiple lineages that led to subcluster 10 and subcluster 5 (Fig. 4). Note that branching support, displayed in Fig. 4 as likelihood ratios, must be carefully considered to assess the reconstruction. For a branching point of interest, if the statistical support is weak, researchers should carefully reselect sequences and manually inspect the MSA; iterative cycles of sequence selection, MSA generation and phylogenetic reconstruction are often necessary to generate stable and robust phylogenetic models. Preliminary evolutionary hypotheses, such as the relationship between NfsA subcluster 3 and 4 discussed above, should be systematically assessed to ensure reliable results.

8. Advanced analysis of evolutionary relationships across a superfamily

Reconstructing the long-range evolution of distantly related genes is a particular challenge in phylogenetics (Lyons-Weiler & Hoelzer, 1997). This is primarily due to two reasons: (1) The genetic distance between extant sequences may be so large that obtaining a phylogenetic tree that has strong statistical support is essentially impossible; or (2) The total number of descendant gene sequences is so great that performing phylogenetic reconstruction is computationally intractable (Nabhan & Sarkar, 2012). The generation of preliminary SSNs, however, can provide critical guidance in overcoming either or both of these obstacles. For the former, the SSN can be adjusted via different similarity cutoffs (e.g., by BLAST E-value—see Section 3) such that overly divergent genes/subclusters are identified and can be excluded from further analysis if necessary. For the latter, the SSN can provide guidance as to how the genotypic relationships between major subgroups are structured; extensively connected SSN subgroups are likely to be more closely evolutionary related to each other and conversely separated clusters are likely to be evolutionary distant. Thus, the total number of gene sequences used in the phylogenetic analysis can be dramatically reduced, either by (a) selecting one or a few sequences to represent subgroups of the network (and reconstructing a phylogeny as outlined in Section 7.4), or (b) by generating an HMM that reflects the sequences of all the members of the subgroup. This can allow one to perform a "higher level" phylogenetic reconstruction that reveals the branching patterns between subgroups while avoiding the computational difficulty that can arise from including all the sequences in the entire superfamily.

One example where the overall evolutionary topology of a superfamily has been successfully analyzed is the nitroreductase superfamily (Akiva et al., 2017). The unique hub-spoke structure of the SSN prompted further examination of the superfamily at the phylogenetic level. Briefly, this was achieved by creating a "backbone" MSA containing 1049 representative sequences using PASTA (Mirarab, Nguyen, Guo, et al., 2015), which was then used to compute a backbone tree using FastTree (Price, Dehal, & Arkin, 2010). The backbone MSA, backbone tree and the complete NTR superfamily dataset (24,270 sequences) were then used as an input to UPP (Ultra-large alignments using phylogeny-aware profiles) (Nguyen, Mirarab, Kumar, & Warnow, 2015). The resulting UPP MSA (of the complete NTR dataset) was used to generate a phylogenetic reconstruction using FastTree and the tree was visualized using FigTree. Combining SSN-based analyses with more detailed phylogenetic reconstructions revealed a pattern of repeated diversification across the superfamily, whereby each of the major subgroups was derived from the same core "Hub" group (Fig. 5). Further structural analysis suggests that "Hub" is composed of genes with a minimal structural scaffold, from which structural adornments were added during the subsequent diversification of descendant subclusters (Akiva et al., 2017). The initial SSN structure provided a critical lens to the generation and interpretation of the reconstructed phylogeny for the superfamily as a whole.

9. Discussion

One of the ultimate goals in biochemistry is to reveal, in comprehensive terms, the relationships between protein sequence, structure, and function. It is therefore essential to elucidate the complete functional capacity of a protein fold. To address this aim with realistic experimental efforts (especially in consideration of the continuous and rapid expansion of sequence data), comprehensive bioinformatic analyses provide an extremely powerful tool

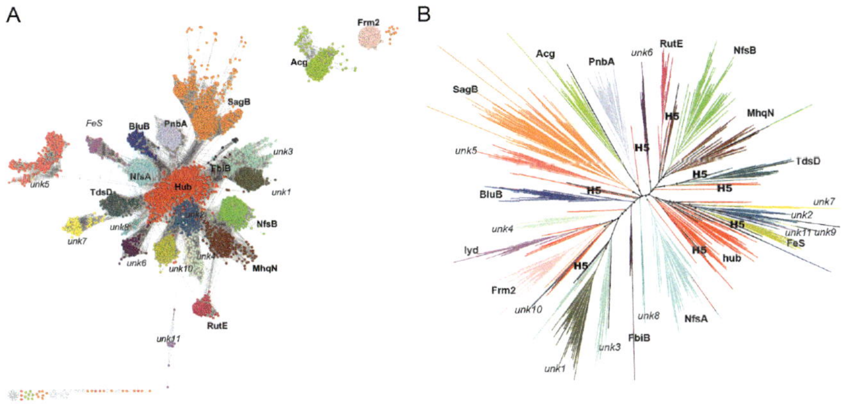

Fig. 5 Sequence similarity and phylogenetic analysis of the NTR superfamily. (A) A representative sequence similarity network of the NTR superfamily at an E-value threshold of 1×10^{-18}. Subgroups, as defined previously (Akiva et al., 2017), are colored and labeled. (B) A phylogenetic reconstruction of the NTR superfamily is displayed, branches are colored and labeled by subgroup. Dispersed red branches represent hub subgroup sequences, e.g., Hub "subcluster 5" branches are labeled (H5). Circles represent branching points with probabilities >0.9; triangles represent probabilities >0.8. *Figure adapted with permission from Akiva, E., Copp, J. N., Tokuriki, N., Babbitt, P. C. (2017). Evolutionary and molecular foundations of multiple contemporary functions of the nitroreductase superfamily.* Proceedings of the National Academy of Sciences of the United States of America, 114, *E9549–E9558.*

to classify sequence diversity and identify unexplored sequence space. A vast amount of untapped potential for novel discoveries of protein and enzyme functions remains hidden in many large superfamilies (Copp et al., 2018).

9.1 Phylogenetic reconstructions can reveal key evolutionary events

A streamlined bioinformatics approach that includes SSNs and phylogenetic reconstructions can effectively establish comprehensive evolutionary models for highly diverse protein superfamilies. SSNs display sequence similarity relationships and correlate strongly with phylogenetic reconstructions (Atkinson et al., 2009). However, SSNs are not constructed using evolutionary models and therefore cannot be used as an alternative to phylogenetic reconstructions or as a basis for inferring evolutionary history. However, SSNs provide a powerful tool to capture a global view of sequence relationships and develop evolutionary hypotheses. Phylogenetic analysis, in comparison, can demonstrate the evolutionary pattern of diversification from common ancestors across a group, supporting proposed evolutionary models and enabling the identification of key periods or points at which functional molecular innovation may have occurred within protein superfamilies.

9.2 Complexities of inferring function from sequence similarity relationships

Within a protein superfamily, diverse functions likely emerged from a common ancestor (Babbitt & Gerlt, 1997; Gerlt & Babbitt, 2001). Divergence, under a given selection pressure to retain a function, can create clusters of sequences that possess the same, or very similar, function (functional families). These sequences are likely to display closer sequence similarity to each other compared to sequences from alternative functional families. In some cases, sequence identity between distinct functional families can be <20%, as they may have diverged billions of years ago and/or their sequences have rapidly evolved due to strong selection forces. Intuitively, sequences that form robust clusters in an SSN may therefore share the same or similar function(s). However, investigations of laboratory and natural protein evolution have demonstrated that a new protein function can emerge via a handful of mutational steps or even a single mutation (Clifton, Kaczmarski, Carr, et al., 2018; Kaltenbach, Burke, Dindo, et al., 2018; McLoughlin & Copley, 2008; Natarajan, Jendroszek, Kumar, et al., 2018; Schmidt, Mundorff, Dojka, et al., 2003; Scott, Jackson, Coppin, et al., 2009; Yang, Hong, Baier, et al., 2016). Thus proteins can exhibit >90% sequence identity but encode distinct functions, as a new function has emerged and evolved only recently (Afriat-Jurnou, Jackson, & Tawfik, 2012; Copley, 2009). Sequence signatures that differentiate distinct functions are thus ambiguous and highly complex, and therefore a single computational approach is unlikely to separate all isofunctional families. Ultimately, comprehensive analyses of a large sample of available sequence, structure and function data, which can be facilitated by sequence analyses (SSNs and phylogenetic reconstructions) at both a global (superfamily) scale and local (subgroup) scale, are necessary to direct and inform hypothesis of functional divergence (Copp et al., 2018).

9.3 Concluding remarks

The methods outlined here can help identify novel sequence space and develop evolutionary hypothesis. However, evolutionary divergence of functions in a superfamily is an extremely complicated process. Separation of sequences into clusters based upon sequence similarity does not necessarily infer that each cluster is isofunctional or that individual clusters may possess different functions at any given threshold; it is essential to combine computational approaches with comprehensive experimental exploration. High throughput methods for activity screening and advances in synthesis technology to dramatically decreasing the cost of gene synthesis (Plesa, Sidore, Lubock, et al., 2018), mean that large scale experimental undertakings (>1000 genes) are now a financially and technically viable option for more laboratories. Integration of the computational approaches we outline in this work with advanced experimental techniques has the potential to greatly expand our understanding of the functional repertoire and evolutionary history of protein superfamilies, enhance our ability to design and modulate protein function for biocatalysis, and ultimately, improve our understanding of the diversity of organisms and cellular processes in nature.

References

Afgan, E., Baker, D., Batut, B., et al. (2018). The galaxy platform for accessible, reproducible and collaborative biomedical analyses: 2018 update. *Nucleic Acids Research*, *46*, W537–W544.

Afriat-Jurnou, L., Jackson, C. J., & Tawfik, D. S. (2012). Reconstructing a missing link in the evolution of a recently diverged phosphotriesterase by active-site loop remodeling. *Biochemistry*, *51*, 6047–6055.

Aharoni, A., Gaidukov, L., Khersonsky, O., et al. (2005). The "evolvability" of promiscuous protein functions. *Nature Genetics*, *37*, 73–76.

Ahmed, F. H., Carr, P. D., Lee, B. M., et al. (2015). Sequence-structure-function classification of a catalytically diverse oxidoreductase superfamily in mycobacteria. *Journal of Molecular Biology*, *427*, 3554–3571.

Akanni, W. A., Wilkinson, M., Creevey, C. J., et al. (2015). Implementing and testing Bayesian and maximum-likelihood supertree methods in phylogenetics. *Royal Society Open Science*, *2*, 140436.

Akiva, E., Brown, S., Almonacid, D. E., et al. (2014). The structure-function linkage database. *Nucleic Acids Research*, *42*, D521–D530.

Akiva, E., Copp, J. N., Tokuriki, N., & Babbitt, P. C. (2017). Evolutionary and molecular foundations of multiple contemporary functions of the nitroreductase superfamily. *Proceedings of the National Academy of Sciences of the United States of America*, *114*, E9549–E9558.

Anisimova, M., & Gascuel, O. (2006). Approximate likelihood-ratio test for branches: A fast, accurate, and powerful alternative. *Systematic Biology*, *55*, 539–552.

Atkinson, H. J., Morris, J. H., Ferrin, T. E., & Babbitt, P. C. (2009). Using sequence similarity networks for visualization of relationships across diverse protein superfamilies. *PLoS One*, *4*, e4345.

Babbitt, P. C., & Gerlt, J. A. (1997). Understanding enzyme superfamilies. Chemistry as the fundamental determinant in the evolution of new catalytic activities. *The Journal of Biological Chemistry*, *272*, 30591–30594.

Baier, F., Copp, J. N., & Tokuriki, N. (2016). Evolution of enzyme superfamilies: Comprehensive exploration of sequence-function relationships. *Biochemistry*, *55*, 6375–6388.

Baier, F., & Tokuriki, N. (2014). Connectivity between catalytic landscapes of the metallo-β-lactamase superfamily. *Journal of Molecular Biology*, *426*, 2442–2456.

Barber, A. E., & Babbitt, P. C. (2012). Pythoscape: A framework for generation of large protein similarity networks. *Bioinformatics*, *28*, 2845–2846.

Brown, S. D., & Babbitt, P. C. (2012). Inference of functional properties from large-scale analysis of enzyme superfamilies. *The Journal of Biological Chemistry*, *287*, 35–42.

Brown, S. D., & Babbitt, P. C. (2014). New insights about enzyme evolution from large scale studies of sequence and structure relationships. *The Journal of Biological Chemistry*, *289*, 30221–30228.

Brown, D., & Sjölander, K. (2006). Functional classification using phylogenomic inference. *PLoS Computational Biology*, *2*, e77.

Camacho, C., Coulouris, G., Avagyan, V., et al. (2009). BLAST+: Architecture and applications. *BMC Bioinformatics*, *10*, 421.

CAZypedia Consortium. (2018). Ten years of CAZypedia: A living encyclopedia of carbohydrate-active enzymes. *Glycobiology*, *28*, 3–8.

Clifton, B. E., Kaczmarski, J. A., Carr, P. D., et al. (2018). Evolution of cyclohexadienyl dehydratase from an ancestral solute-binding protein. *Nature Chemical Biology*, *14*, 542–547.

Colin, P. Y., Kintses, B., Gielen, F., et al. (2015). Ultrahigh-throughput discovery of promiscuous enzymes by picodroplet functional metagenomics. *Nature Communications*, *6*, 10008.

Copley, S. D. (2003). Enzymes with extra talents: Moonlighting functions and catalytic promiscuity. *Current Opinion in Chemical Biology*, *7*, 265–272.

Copley, S. D. (2009). Evolution of efficient pathways for degradation of anthropogenic chemicals. *Nature Chemical Biology*, *5*, 559–566.

Copp, J. N., Akiva, E., Babbitt, P. C., & Tokuriki, N. (2018). Revealing unexplored sequence-function space using sequence similarity networks. *Biochemistry*, *57*, 4651–4662.

Davidson, R., Baas, B.-J., Akiva, E., et al. (2018). A global view of structure-function relationships in the tautomerase superfamily. *The Journal of Biological Chemistry*, *293*, 2342–2357.

Dijkman, W. P., de Gonzalo, G., Mattevi, A., & Fraaije, M. W. (2013). Flavoprotein oxidases: Classification and applications. *Applied Microbiology and Biotechnology*, *97*, 5177–5188.

Efron, B., Halloran, E., & Holmes, S. (1996). Bootstrap confidence levels for phylogenetic trees. *Proceedings of the National Academy of Sciences of the United States of America*, *93*, 13429–13434.

Enright, A. J., Van Dongen, S., & Ouzounis, C. A. (2002). An efficient algorithm for large-scale detection of protein families. *Nucleic Acids Research*, *30*, 1575–1584.

Felsenstein, J. (1985). Confidence limits on phylogenies: An approach using the bootstrap. *Evolution*, *39*, 783–791.

Finn, R. D., Attwood, T. K., Babbitt, P. C., et al. (2017). InterPro in 2017-beyond protein family and domain annotations. *Nucleic Acids Research*, *45*, D190–D199.

Finn, R. D., Coggill, P., Eberhardt, R. Y., et al. (2016). The Pfam protein families database: Towards a more sustainable future. *Nucleic Acids Research*, *44*, D279–D285.

Gerlt, J. A. (2017). Genomic enzymology: Web tools for leveraging protein family sequence-function space and genome context to discover novel functions. *Biochemistry*, *56*, 4293–4308.

Gerlt, J. A., & Babbitt, P. C. (2001). Divergent evolution of enzymatic function: Mechanistically diverse superfamilies and functionally distinct suprafamilies. *Annual Review of Biochemistry*, *70*, 209–246.

Gerlt, J. A., & Babbitt, P. C. (2009). Enzyme (re)design: Lessons from natural evolution and computation. *Current Opinion in Chemical Biology*, *13*, 10–18.

Gerlt, J. A., Babbitt, P. C., Jacobson, M. P., & Almo, S. C. (2012). Divergent evolution in enolase superfamily: Strategies for assigning functions. *The Journal of Biological Chemistry, 287*, 29–34.

Gerlt, J. A., Bouvier, J. T., Davidson, D. B., et al. (2015). Enzyme function initiative-enzyme similarity tool (EFI-EST): A web tool for generating protein sequence similarity networks. *Biochimica et Biophysica Acta, 1854*, 1019–1037.

Glasner, M. E., Gerlt, J. A., & Babbitt, P. C. (2007). Mechanisms of protein evolution and their application to protein engineering. *Advances in Enzymology and Related Areas of Molecular Biology, 75*, 193–239. xii–xiii.

Goble, A. M., Fan, H., Sali, A., & Raushel, F. M. (2011). Discovery of a cytokinin deaminase. *ACS Chemical Biology, 6*, 1036–1040.

Goble, A. M., Feng, Y., Raushel, F. M., & Cronan, J. E. (2013). Discovery of a cAMP deaminase that quenches cyclic AMP-dependent regulation. *ACS Chemical Biology, 8*, 2622–2629.

Grininger, M., Staudt, H., Johansson, P., et al. (2009). Dodecin is the key player in flavin homeostasis of archaea. *The Journal of Biological Chemistry, 284*, 13068–13076.

Guindon, S., Delsuc, F., Dufayard, J.-F., & Gascuel, O. (2009). Estimating maximum likelihood phylogenies with PhyML. *Methods in Molecular Biology, 537*, 113–137.

Hanson-Smith, V., & Johnson, A. (2016). PhyloBot: A web portal for automated phylogenetics, ancestral sequence reconstruction, and exploration of mutational trajectories. *PLoS Computational Biology, 12*, e1004976.

Heaps, S. E., Nye, T. M. W., Boys, R. J., et al. (2014). Bayesian modelling of compositional heterogeneity in molecular phylogenetics. *Statistical Applications in Genetics and Molecular Biology, 13*, 589–609.

Hicks, M. A., Barber, A. E., Giddings, L.-A., et al. (2011). The evolution of function in strictosidine synthase-like proteins. *Proteins, 79*, 3082–3098.

Hu, Y., & Coates, A. R. (2011). Mycobacterium tuberculosis acg gene is required for growth and virulence in vivo. *PLoS One, 6* e20958.

Jensen, R. A. (1976). Enzyme recruitment in evolution of new function. *Annual Review of Microbiology, 30*, 409–425.

Kaltenbach, M., Burke, J. R., Dindo, M., et al. (2018). Evolution of chalcone isomerase from a noncatalytic ancestor. *Nature Chemical Biology, 14*, 988.

Knutson, S. T., Westwood, B. M., Leuthaeuser, J. B., et al. (2017). An approach to functionally relevant clustering of the protein universe: Active site profile-based clustering of protein structures and sequences. *Protein Science, 26*, 677–699.

Lartillot, N., Lepage, T., & Blanquart, S. (2009). PhyloBayes 3: A Bayesian software package for phylogenetic reconstruction and molecular dating. *Bioinformatics, 25*, 2286–2288.

Levin, B. J., Huang, Y. Y., Peck, S. C., et al. (2017). A prominent glycyl radical enzyme in human gut microbiomes metabolizestrans-4-hydroxy-l-proline. *Science, 355*, eaai8386.

Li, W., & Godzik, A. (2006). Cd-hit: A fast program for clustering and comparing large sets of protein or nucleotide sequences. *Bioinformatics, 22*, 1658–1659.

Lukk, T., Sakai, A., Kalyanaraman, C., et al. (2012). Homology models guide discovery of diverse enzyme specificities among dipeptide epimerases in the enolase superfamily. *Proceedings of the National Academy of Sciences of the United States of America, 109*, 4122–4127.

Lyons-Weiler, J., & Hoelzer, G. A. (1997). Escaping from the Felsenstein zone by detecting long branches in phylogenetic data. *Molecular Phylogenetics and Evolution, 8*, 375–384.

Macheroux, P., Kappes, B., & Ealick, S. E. (2011). Flavogenomics—A genomic and structural view of flavin-dependent proteins. *The FEBS Journal, 278*, 2625–2634.

Mar, J. C., Harlow, T. J., & Ragan, M. A. (2005). Bayesian and maximum likelihood phylogenetic analyses of protein sequence data under relative branch-length differences and model violation. *BMC Evolutionary Biology, 5*, 8.

Mashiyama, S. T., Malabanan, M. M., Akiva, E., et al. (2014). Large-scale determination of sequence, structure, and function relationships in cytosolic glutathione transferases across the biosphere. *PLoS Biology*, *12*, e1001843.

McLoughlin, S. Y., & Copley, S. D. (2008). A compromise required by gene sharing enables survival: Implications for evolution of new enzyme activities. *Proceedings of the National Academy of Sciences of the United States of America*, *105*, 13497–13502.

Mirarab, S., Nguyen, N., Guo, S., et al. (2015). PASTA: Ultra-large multiple sequence alignment for nucleotide and amino-acid sequences. *Journal of Computational Biology*, *22*, 377–386.

Mitchell, A., Chang, H.-Y., Daugherty, L., et al. (2015). The InterPro protein families database: The classification resource after 15 years. *Nucleic Acids Research*, *43*, D213–D221.

Morales-Hojas, R., & Vieira, J. (2012). Phylogenetic patterns of geographical and ecological diversification in the subgenus Drosophila. *PLoS One*, *7*, e49552.

Nabhan, A. R., & Sarkar, I. N. (2012). The impact of taxon sampling on phylogenetic inference: A review of two decades of controversy. *Briefings in Bioinformatics*, *13*, 122–134.

Natarajan, C., Jendroszek, A., Kumar, A., et al. (2018). Molecular basis of hemoglobin adaptation in the high-flying bar-headed goose. *PLoS Genetics*, *14*, e1007331.

Nguyen, N.-P. D., Mirarab, S., Kumar, K., & Warnow, T. (2015). Ultra-large alignments using phylogeny-aware profiles. *Genome Biology*, *16*, 124.

O'Brien, P. J., & Herschlag, D. (1999). Catalytic promiscuity and the evolution of new enzymatic activities. *Chemistry & Biology*, *6*, R91–R105.

Ojha, S., Meng, E. C., & Babbitt, P. C. (2007). Evolution of function in the "two dinucleotide binding domains" flavoproteins. *PLoS Computational Biology*, *3*, e121.

Pegg, S. C. H., Brown, S. D., Ojha, S., et al. (2006). Leveraging enzyme structure – function relationships for functional inference and experimental design: The structure – function linkage database. *Biochemistry*, *45*, 2545–2555.

Plesa, C., Sidore, A. M., Lubock, N. B., et al. (2018). Multiplexed gene synthesis in emulsions for exploring protein functional landscapes. *Science*, *359*, 343–347.

Price, M. N., Dehal, P. S., & Arkin, A. P. (2010). FastTree 2-approximately maximum-likelihood trees for large alignments. *PLoS One*, *5*, e9490.

Rao, G., & Oldfield, E. (2016). Structure and function of four classes of the 4Fe-4S protein, IspH. *Biochemistry*, *55*, 4119–4129.

Renata, H., Wang, Z. J., & Arnold, F. H. (2015). Expanding the enzyme universe: Accessing non-natural reactions by mechanism-guided directed evolution. *Angewandte Chemie (International Ed. in English)*, *54*, 3351–3367.

Ronquist, F., Teslenko, M., van der Mark, P., et al. (2012). MrBayes 3.2: Efficient Bayesian phylogenetic inference and model choice across a large model space. *Systematic Biology*, *61*, 539–542.

Rosvall, M., & Bergstrom, C. T. (2008). Maps of random walks on complex networks reveal community structure. *Proceedings of the National Academy of Sciences of the United States of America*, *105*, 1118–1123.

Schmidt, D. M. Z., Mundorff, E. C., Dojka, M., et al. (2003). Evolutionary potential of (β/α) 8-barrels: Functional promiscuity produced by single substitutions in the enolase superfamily. *Biochemistry*, *42*, 8387–8393.

Scott, C., Jackson, C. J., Coppin, C. W., et al. (2009). Catalytic improvement and evolution of atrazine chlorohydrolase. *Applied and Environmental Microbiology*, *75*, 2184–2191.

Selvadurai, K., Wang, P., Seimetz, J., & Huang, R. H. (2014). Archaeal Elp3 catalyzes tRNA wobble uridine modification at C5 via a radical mechanism. *Nature Chemical Biology*, *10*, 810–812.

Shannon, P., Markiel, A., Ozier, O., et al. (2003). Cytoscape: A software environment for integrated models of biomolecular interaction networks. *Genome Research*, *13*, 2498–2504.

Sillitoe, I., Lewis, T., & Orengo, C. (2015). Using CATH-Gene3D to analyze the sequence, structure, and function of proteins. *Current Protocols in Bioinformatics*, *50*, 1.28.1–21.

Stamatakis, A. (2014). RAxML version 8: A tool for phylogenetic analysis and post-analysis of large phylogenies. *Bioinformatics*, *30*, 1312–1313.

Susko, E. (2008). On the distributions of bootstrap support and posterior distributions for a star tree. *Systematic Biology*, *57*, 602–612.

Swinehart, W. E., & Jackman, J. E. (2015). Diversity in mechanism and function of tRNA methyltransferases. *RNA Biology*, *12*, 398–411.

Szöllősi, G. J., Tannier, E., Daubin, V., & Boussau, B. (2015). The inference of gene trees with species trees. *Systematic Biology*, *64*, e42–e62.

Thompson, J. D., Plewniak, F., & Poch, O. (1999). A comprehensive comparison of multiple sequence alignment programs. *Nucleic Acids Research*, *27*, 2682–2690.

Tian, B.-X., Wallrapp, F. H., Holiday, G. L., et al. (2014). Predicting the functions and specificity of triterpenoid synthases: A mechanism-based multi-intermediate docking approach. *PLoS Computational Biology*, *10*, e1003874.

Tokuriki, N., & Tawfik, D. S. (2009). Protein dynamism and evolvability. *Science*, *324*, 203–207.

Vass, M., Kooistra, A. J., Yang, D., et al. (2018). Chemical diversity in the G protein-coupled receptor superfamily. *Trends in Pharmacological Sciences*, *39*, 494–512.

Wallrapp, F. H., Pan, J.-J., Ramamoorthy, G., et al. (2013). Prediction of function for the polyprenyl transferase subgroup in the isoprenoid synthase superfamily. *Proceedings of the National Academy of Sciences of the United States of America*, *110*, E1196–E1202.

Wichelecki, D. J., Graff, D. C., Al-Obaidi, N., et al. (2014). Identification of the in vivo function of the high-efficiency D-mannonate dehydratase in Caulobacter crescentus NA1000 from the enolase superfamily. *Biochemistry*, *53*, 4087–4089.

Wiens, J. J., & Servedio, M. R. (1998). Phylogenetic analysis and intraspecific variation: Performance of parsimony, likelihood, and distance methods. *Systematic Biology*, *47*, 228–253.

Wilgenbusch, J. C., Huang, W., & Gallivan, K. A. (2017). Visualizing phylogenetic tree landscapes. *BMC Bioinformatics*, *18*, 85.

Yang, Z. (2007). PAML 4: Phylogenetic analysis by maximum likelihood. *Molecular Biology and Evolution*.

Yang, G., Hong, N., Baier, F., et al. (2016). Conformational tinkering drives evolution of a promiscuous activity through indirect mutational effects. *Biochemistry*, *55*, 4583–4593.

Zhang, X., Carter, M. S., Vetting, M. W., et al. (2016). Assignment of function to a domain of unknown function: DUF1537 is a new kinase family in catabolic pathways for acid sugars. *Proceedings of the National Academy of Sciences of the United States of America*, *113*, E4161–E4169.

Zhang, X., Kumar, R., Vetting, M. W., et al. (2015). A unique cis-3-hydroxy-l-proline dehydratase in the enolase superfamily. *Journal of the American Chemical Society*, *137*, 1388–1391.

Zhao, S., Kumar, R., Sakai, A., et al. (2013). Discovery of new enzymes and metabolic pathways by using structure and genome context. *Nature*, *502*, 698–702.

Zhou, X., Shen, X.-X., Hittinger, C. T., & Rokas, A. (2018). Evaluating fast maximum likelihood-based phylogenetic programs using empirical phylogenomic data sets. *Molecular Biology and Evolution*, *35*, 486–503.

CHAPTER THIRTEEN

Structural methods for probing the interaction of flavoenzymes with dioxygen and its surrogates

Raspudin Saleem-Batcha[a,b], Robin Teufel[a,b,*]

[a]ZBSA, Center for Biological Systems Analysis, University of Freiburg, Freiburg, Germany
[b]Faculty of Biology, University of Freiburg, Freiburg, Germany
*Corresponding author: e-mail address: robin.teufel@zbsa.uni-freiburg.de

Contents

Abstract

As a rare feature among organic cofactors, reduced flavins (Fl_{red}) can efficiently react with dioxygen (O_2). As a consequence, many flavin-dependent enzymes may serve as either oxidases that use O_2 as an electron acceptor or as monooxygenases that transfer one oxygen atom derived from O_2 to an organic substrate. For the latter functionality, covalent flavin: oxygen adducts are formed that function as oxygenating species. Remarkably, despite intensive research, many open questions remain how flavoenzymes control the reaction with O_2. Here, we describe O_2-pressurized protein crystallography in detail as a structural approach to gain insight into the interactions between the protein scaffold, the flavin cofactor and O_2. This may allow to further our understanding of how flavoenzymes can steer the formation of different oxygenating species and thus provide missing puzzle pieces for rational flavoenzyme design.

Methods in Enzymology, Volume 620
ISSN 0076-6879
https://doi.org/10.1016/bs.mie.2019.03.016

1. Introduction

Flavin-dependent enzymes contain either flavin mononucleotide (FMN) or flavin adenine dinucleotide (FAD) derived from vitamin B2 (riboflavin) as cofactor and are ubiquitous redox catalysts with diverse functionalities (Teufel, Agarwal, & Moore, 2016; Walsh & Wencewicz, 2013). Many oxidoreductases that employ a flavin cofactor utilize O_2 as a cosubstrate, either as terminal electron acceptor (oxidases, which produce H_2O_2) or for the generation of covalently bound oxygenating species (monooxygenases) in the form of the flavin-C4a-(hydro)peroxide (Fl_{C4aOOH}) (Huijbers, Montersino, Westphal, Tischler, & van Berkel, 2014; Palfey & McDonald, 2010) and the flavin-N5-oxide (Fl_{N5O}) (Fig. 1) (Teufel, 2017; Teufel et al., 2013). On the other hand, flavin dehydrogenases generally react slowly with O_2, which also applies to free Fl_{red} in solution. Commonly, the protein microenvironment around the Fl_{red} fine-tunes O_2 reactivity, thereby promoting the acceleration or deceleration of the reaction of Fl_{red} with O_2 over six orders of magnitude (Mattevi, 2006). Interestingly, there are examples where only one amino acid, i.e. the "gatekeeper" residue,

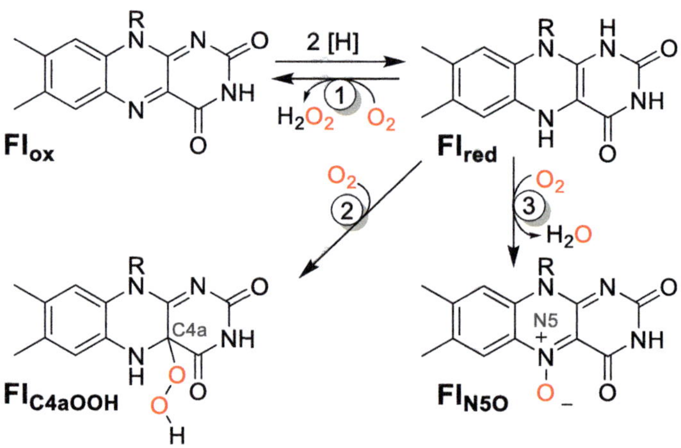

Fig. 1 Overview of selected flavin redox states and oxygenating species. Flavoproteins tightly control the reaction of Fl_{red} with O_2. Some react sluggishly with O_2 (dehydrogenases, not shown). (1) Oxidases rapidly reduce O_2 to H_2O_2. (2) Common monooxygenases generate a transient covalent adduct, the Fl_{C4aOOH}. (3) In contrast, the non-canonical monooxygenase EncM produces a stable Fl_{N5O}. The protein microenvironment of the flavin cofactor fine tunes the reaction with O_2, yet how exactly is still a matter of debate.

controls access of O_2 to the flavin cofactor, e.g., in the vanillyl alcohol oxidase/p-cresol methylhydroxylase (VAO/PCMH) structural superfamily that mostly function as dehydrogenases and oxidases (Ewing, Fraaije, Mattevi, & van Berkel, 2017). Here, a gatekeeper residue was identified in a conserved 18-residue long "oxygen reactivity motif" (Leferink et al., 2009). Later, it was found that other members of this family such as the grass pollen allergen Phl p 4 feature an alternative gatekeeper reside (I153) (Zafred et al., 2015). When I153 is replaced by the less bulky valine in Phl p 4, the enzyme is transformed from a dehydrogenase into an oxidase. Contrariwise, the structurally related berberine bride enzyme (BBE) naturally functions as an oxidase with a valine at the gatekeeper position (V169) and can be converted to a dehydrogenase through substitution with the bulkier isoleucine that blocks access of O_2 to the C4a-N5 locus (Zafred et al., 2015).

Remarkably, the bacterial monooxygenase EncM, another structural homolog of Phl p 4 and BBE, harbors a threonine at this position (T139) (Saleem-Batcha et al., 2018; Saleem-Batcha & Teufel, 2018; Teufel et al., 2013). EncM-Fl_{red} reacts with O_2, but in contrast to BBE, EncM does not produce H_2O_2 but rather a covalent oxygen adduct, the flavin-N5-oxide (Fl_{N5O}), which serves as a novel type of oxygenating species and is stably maintained in the enzyme's resting state. Mutagenesis indicates that T139 indeed adopts a pivotal role in Fl_{N5O} formation. An isosteric T139V variant produced only minor amounts of Fl_{N5O} and was transformed into an H_2O_2-producing oxidase (Saleem-Batcha et al., 2018). It was proposed that T139 promotes the generation of Fl_{N5O} by hydrogen bonding to the transient hydroperoxide radical intermediate, retaining it in the O_2 reaction site until spin-inversion and radical coupling can occur. The resulting N5-bound peroxy species then eliminates OH^-, thereby giving rise to the Fl_{N5O}. These cases highlight that drastic changes in O_2 reactivity of flavoenzymes can sometimes be achieved by minimal changes in the protein surroundings of the cofactor. However, it also poses challenges with respect to the prediction of O_2 reactivity based on sequence information that does not currently allow the prediction of fine structural details in the flavin vicinity. EncM serves as prime example, which appeared as an inconspicuous member of the VAO/PCMH family, before detailed mechanistic studies revealed the presence of the Fl_{N5O}. Hence, experimental approaches aimed at gaining more insight into key factors that govern O_2 reactivity in flavoenzymes are crucial for future functional predictions and rational enzyme design. In the following sections, structural methods and in particular the methodology for O_2-pressurized X-ray crystallography will be introduced, which we successfully applied to precisely determine the spatial positioning of O_2 required for Fl_{N5O} formation in EncM. Meanwhile, we successfully applied the method for a structurally unrelated flavoenzyme (unpublished data) and it thus appears that the method may be useful for a broader range of flavoenzymes. Aside from the obvious advantages, we will furthermore discuss some challenges and caveats of the method.

2. Terminology of the interaction between O_2 and proteins

In the research field, terms such as O_2 binding site, O_2 reaction site, or O_2 surrogate are commonly used in the literature, but sometimes in a confusing and inconsistent manner. In general, a distinction should be made between O_2-binding and O_2-consuming proteins (Baron, McCammon, & Mattevi, 2009) and it is still an open question, whether some flavoproteins can truly bind O_2. Commonly, however, it seems that the reactions of O_2 with most flavoproteins are diffusion controlled, show no saturation with O_2, and thereby follow second-order kinetics (K_d for $O_2 \gg$ max. O_2 concentration in solution) (Baron, McCammon, et al., 2009; Baron, Riley, et al., 2009; Klinman, 2007). Hence, the typical flavin monooxygenases and oxidases appear to have very low affinity for O_2 and rather directly reduce it without prior binding. EncM may be a true exception and stopped-flow kinetics as well as O_2-pressurized X-ray crystallography were consistent with O_2 binding, although further studies are required to unambiguously solve this question.

To study the O_2 interaction with proteins, X-ray crystallography is poised to become one of the most important tools. In the past, chemical surrogates were commonly used to probe for sites in which no reaction occurs, for example in protein cavities that may serve as gateways for O_2 funneling. On the other hand, similar methodology was applied for O_2 reaction sites that primarily promote the reduction of O_2 and formation of superoxide. O_2 and superoxide have markedly different chemical properties. While O_2 is hydrophobic and apolar, superoxide is hydrophilic and negatively charged and can thus provide more possibilities for interaction with the protein via ionic and H-bonding. Therefore, inert and hydrophobic xenon gas may in many cases be a suitable chemical surrogate to probe for O_2 binding sites but not for O_2 reaction sites close to the flavin cofactor of monooxygenases and oxidases. The reason may be that O_2 reaction sites are presumably optimized to interact with superoxide rather than O_2 and are thus amphiphilic in nature rather than hydrophobic. The escape of superoxide is highly detrimental to the cell and flavoenzymes are likely optimized to retain superoxide for subsequent spin inversion and radical coupling (monooxygenases) or for a second electron transfer step to afford H_2O_2 (oxidases).

Therefore, polar and negatively charged halides (chloride and bromide) are likely more suitable to find O_2 reaction sites (Kommoju, Chen, Bruckner, Mathews, & Jorns, 2011) but should more accurately be referred to as superoxide surrogates. Such surrogates, however, are not necessarily reliable, as showcased by studies with EncM. Here, neither xenon, bromide nor chloride could be detected in the dedicated O_2 reaction site at the re-side of the FAD cofactor in EncM, which could only be identified via O_2-pressurized X-ray crystallography as outlined below (Saleem-Batcha et al., 2018) (Fig. 2).

3. Brief overview of strategies to study the interaction of O_2 with flavoenzymes

Various methods have been successfully applied to elucidate diffusion of O_2 into proteins as well as the interaction with cofactors. Computational methods such as enhanced-statistics molecular dynamics simulations can be applied and combined with experimental approaches (e.g., site-directed mutagenesis, rapid kinetics experiments, structural studies) to study diffusion of O_2, e.g., into flavin monooxygenases and oxidases (Baron, Riley, et al., 2009). Accordingly, dedicated cavities in proteins were identified that seem crucial for directing and funneling O_2 from the bulk solvent into the protein along several possible diffusion pathways to the flavin cofactor.

As a structural experimental approach, X-ray crystallography is most commonly applied with chemical surrogates. The noble gas xenon is often used as O_2 surrogate due to similar chemical properties and a strong anomalous signal in the range of 1.5–3.0 Å that allows derivatives to be obtained with high isomorphism (Duff et al., 2004). Due to these features, xenon (or krypton) pressurization has been traditionally used to solve the phase problem with the isomorphous replacement method during structural elucidation of proteins. Xenon derivatives can be easily prepared and do not require modification of the buffer or crystal mother liquor in contrast to many other derivatives (e.g., heavy metal soaks), thus requiring only crystal pressurization and freezing. Specialized xenon-pressurization equipment is therefore widely available at synchrotron facilities. In the case of EncM, e.g., a Xe-derivatized EncM crystal structure (pressurized with bar 15 bars xenon)

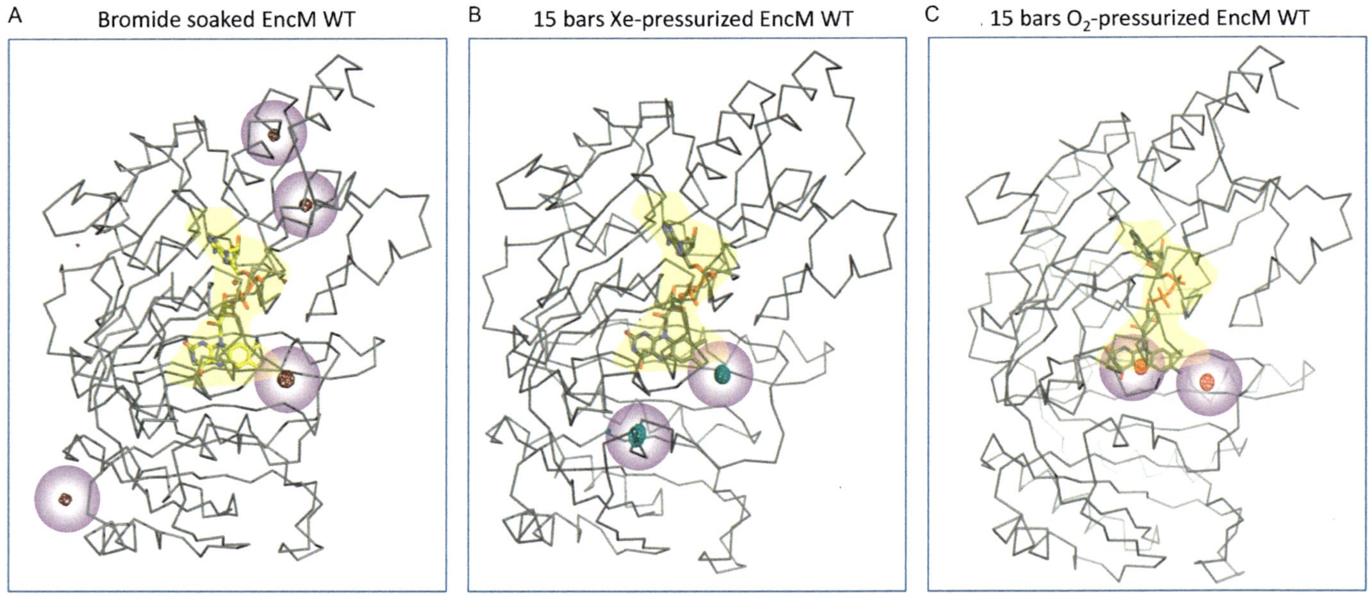

Fig. 2 (A) Br-derivatized EncM wild-type crystal structure (anaerobically grown crystals were soaked in crystallization solution containing 1.5 M sodium bromide) displaying Br anomalous maps with electron density distribution at 5σ above the mean. Four distinct bromide ions can be observed, but not at the O_2 reaction site. (B) Xe-derivatized EncM wt crystal structure displaying Xe anomalous maps with electron density distribution at 5σ above the mean. Two distinct xenon atoms can be observed, but again not at the O_2 reaction site. Protein crystals were pressurized with 15 bars xenon gas using a Xe-pressurization cell (Oxford Cryosystem Xcell) before data collection. (C) O_2-derivatized EncM wt crystal structure displaying O_2 Polder OMIT maps with electron density distribution at 5σ above the mean. Two distinct O_2 molecules can be observed, one of them at the O_2 reaction site. Protein crystals were pressurized with 15 bar O_2 gas using a Xe-pressurization cell before data collection. *Data were similarly previously published in Saleem-Batcha (2018).*

could be obtained that showed xenon anomalous maps with electron density distribution at 8σ above the mean. Two distinct xenon binding sites were identified, but not close to the flavin cofactor. For pressurization, a Xe-pressurization cell (Oxford Cryosystem Xcell) was used before data collection. The same device was furthermore modified for use in O_2-pressurization experiments, as shown below. As an alternative method to address the phase problem, short soaks with bromide generate anomalous signals and are traditionally used for isomorphous replacement, but are also applicable as method for the detection of O_2 reaction sites with bromide serving as superoxide surrogate. In the case of EncM, two distinct bromide binding sites could be identified. Once more, however, they were not in the vicinity of the flavin cofactor. In contrast to that, using O_2-pressurized X-ray crystallography, two binding sites for O_2 could be identified in EncM, one of which corresponds to a bromide binding site, the other one representing the sought-for amphiphilic site of O_2 activation adjacent to the N5 of the flavin. The important role of this O_2 reaction sites for Fl_{N5O} formation and thus catalysis was furthermore confirmed via site-directed mutagenesis. As an alternative structural experimental approach, sensitive O_2 gas-pressure NMR measurements have been used to detect O_2 in hydrophobic protein cavities, e.g., by observing changes in ^{1}H, ^{15}N, and ^{13}C chemical shifts for surrounding backbone amide and side chain methyl groups (Kitahara, Yoshimura, Xue, Kameda, & Mulder, 2016). In the following sections, however, we will focus on the methodology and data analysis required for finding O_2 binding and reaction sites in (flavo)proteins using O_2-pressurized X-ray crystallography.

4. O_2-pressurized X-ray crystallography: Experimental setup

The methodology for O_2-pressurized X-ray crystallography was pioneered by Colloc'h and colleagues (Colloc'h et al., 2008) with cofactorless urate oxidase and extended by us by implementing additional controls (anaerobic crystals, varying O_2 pressure levels) and improved data analysis (see below) (Saleem-Batcha et al., 2018). We think that this method is applicable for a wider range of flavoproteins, although this may possibly depend on whether true binding of O_2 occurs as opposed to direct consumption (as discussed above). Below we summarized the required equipment, reagents, and experimental procedures that were used to study the interaction of EncM and other flavoenzymes in our laboratory. The complete setup used for O_2 pressurization is shown in Fig. 3.

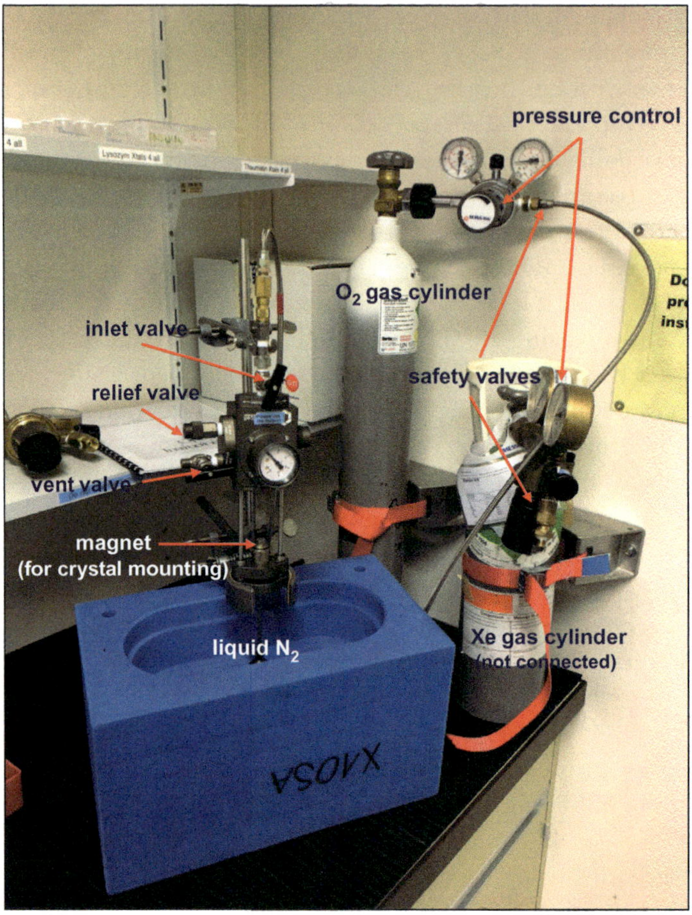

Fig. 3 Overview of the gas pressurization setup using a xenon derivatization chamber (Xcell) manufactured by Oxford Cryosystems, Oxford, UK at the PXII X10SA beamline, PSI-SLS, Villigen, Switzerland.

4.1 Freeze-trapping O_2 and Xe in protein crystals

Equipments

- Cryo-tools for crystallography (e.g., Magnetic Cryo Wand, Magnetic CryoVials, CryoVial tong from Jena Bioscience, Germany)
- Optical Microscope to fish protein crystals
- CrystalCap™ with CryoLoop (Hampton Research)
- Liquid nitrogen dewar
- Whatman filter paper discs
- Xe derivatization apparatus XCell (Oxford Cryosystems, Oxford, UK)

Reagents

- Compressed O_2 gas
- Compressed Xe gas
- Liquid nitrogen
- Cryoprotectants

Procedure (as shown in Fig. 4)

- Crystallize the protein to be pressurized following standard sitting-drop vapor diffusion technique.
- Wet a Whatman filter paper disc with the cryoprotectant solution and fix it to the filterpaper cap in the Xcell apparatus (Fig. 4(1)) (The filter paper disc with cryoprotectant prevents the CryoLoop from drying out during the gas pressurization process.)

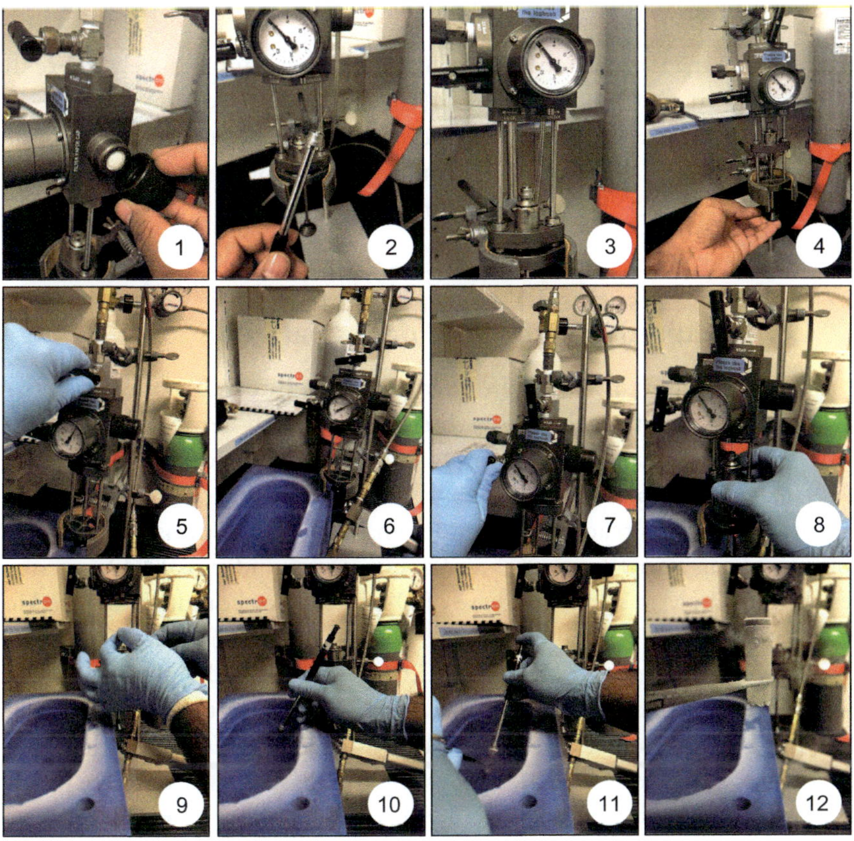

Fig. 4 Required steps for the gas pressurization experiments (individual steps are explained in detail in the main text).

- Transfer the protein crystal to a cryoprotectant solution.
- Collect the cryoprotected crystal in a CryoLoop attached to CrystalCap™ (Fig. 4(2)).
- Place the CrystalCap™ onto the magnetic base of the Xcell chamber plunger (Fig. 4(3)).
- Close the chamber (Fig. 4(4)) and pressurize with, e.g., 15 bars Xe or 15 bars O_2 (Fig. 4(5)). If the gas pressure within the chamber exceeds the limit, the pressure will vent through the safety valve. Tolerance of the present protein crystals to higher pressure levels should be tested first, some may crack and lower pressure levels should then be tested accordingly (in the case of EncM, even at only 5 bars O_2, distinct electron density became evident).
- Incubate the protein crystal under pressure for 10–15 min (Fig. 4(6)).
- Slowly release gas from the chamber so as not to disturb the protein crystal (Fig. 4(7)).
- Remove the CrystalCap™ (Fig. 4(8) and (9)) and immediately freeze the pressurized protein crystal (Fig. 4(10)) in liquid nitrogen.
- Transfer the CrystalCap™ to a vial submerged in liquid nitrogen within the dewar (Fig. 4(11) and (12)). Shock freezing of the crystal will immobilize Xe and O_2 bound to the protein.

5. O_2-pressurized X-ray crystallography: Required controls and data analysis

As an important control to avoid misinterpretation of the electron density maps, additional datasets for unpressurized aerobic crystals and, most importantly, anaerobically prepared crystals should be acquired. This allows a comparison between each distinct condition and should result in an unoccupied O_2 reaction site under anaerobic conditions, while showing increasing signal intensities and occupancies under rising O_2 pressure levels. For EncM, a steady increase of O_2 occupancy could be observed for 0.21 (atmospheric partial pressure), 5, 10, and 15 bars of O_2 pressure, respectively (Saleem-Batcha et al., 2018). Importantly, in our datasets obtained from anaerobic crystals, the omit maps calculated from the anaerobically grown control crystals revealed no significant electron density in the dedicated O_2 reaction site, suggesting the absence of alternative ligands (such as water) and underlining the high selectivity for O_2. However, it cannot be generally excluded that water or chloride (present in the crystallization buffer for example) may bind at O_2 reaction sites in other flavoenzymes. In fact, chloride was used as another suitable analog in some flavoprotein oxidases to probe for O_2 reaction sites that are otherwise occupied by water (Kommoju et al., 2011). Notably, electron density maps around the flavin cofactor should be thoroughly scrutinized to avoid erronous assignment, e.g., by automatic (mis)placement of a water molecule during computational processing of the acquired data, as originally occurred for EncM (PDB code: 4XLO), where the electron density stemming from the presence of O_2 was falsely assigned as a water molecule (Saleem-Batcha et al., 2018; Teufel et al., 2013). Importantly, such mistakes will likely go unnoticed in the fully refined (but biased) 2Fo–Fc maps. To avoid this, we strongly recommend the use of unbiased polder Fo–Fc omit maps to verify whether the electron density supports the presence of O_2 rather than water.

Note: If protein crystals crack under pressurization and are thus unsuitable for the above-described methodology, the simplest approach would be to compare anaerobic and aerobic (unpressurized) crystals. In case of EncM crystals, a clear difference between both conditions could be observed.

5.1 Polder OMIT map calculations

For enhanced accuracy of our electron density maps in the putative O_2 reaction site, we used PHENIX.polder (Liebschner et al., 2017) from the PHENIX software suite (Adams et al., 2010). PHENIX.polder is a recently established tool to provide more accurate omit maps, allowing to exclude the bulk solvent during calculation of the omit map around the O_2 reaction site, which facilitates the detection and interpretation of weak densities otherwise obscured by the bulk solvent and therefore improves detection of O_2 (or any other ligand). In short, a polder map is in essence an omit map, which assists to improve weak features in electron-density maps by holding bulk solvent out of the area. Also, the density in the omitted region is therefore not biased by the bulk solvent.

5.2 Diffraction data collection and analysis

Software

- XDS (Kabsch, 2010) and SCALA (Evans, 2011) were used for processing diffraction data.
- Phaser (McCoy et al., 2007) and PHENIX (Adams et al., 2010; Afonine et al., 2012) were used for phasing and model refinement.
- COOT (Emsley, Lohkamp, Scott, & Cowtan, 2010) enabled visualization of electron density maps and manual model manipulation.

Procedure

- Collect diffraction data of the O_2 pressurized protein crystals at the native wavelength of the synchrotron beamline. In case of Xe-pressurized crystals, anomalous scattering can be maximized by collecting diffraction data at longer wavelengths closer to the absorption edges of Xe: 5.452keV (L_I), 5.103keV (L_{II}), or 4.782keV (L_{III}) (Roose, Zemerov, & Dmochowski, 2018).
- Calculate molecular replacement-phased anomalous difference Fourier maps (ΔF_{ano}) for the locations of Xe using Phaser (McCoy et al., 2007). As an alternative, evaluate isomorphous difference maps for Xe-derivative using $|F_{Xe}|-|F_{native}|$ and as amplitudes and phase using the atomic coordinates of the native protein structure (Roose et al., 2018).
- In case of O_2-pressurized structure evaluations, O_2 lacks significant anomalous signal to locate the sites. Therefore, calculate molecular replacement phasing of aerobic and anaerobic crystal datasets (isomorphous) and compare the density at the expected O_2 reaction site. Anaerobic crystals are grown using sitting-drop vapor diffusion technique in an anaerobic and air-sealed environment. Since anaerobic crystals have to be fished for measurement and air, one has to minimize the exposure to air by
  (i) performing the crystal transfer using anaerobic cryoprotectant and
  (ii) flash-freezing of the crystals in liquid N_2 as fast as possible.
- Use COOT (Emsley et al., 2010) and add Xe atoms and the O_2 molecule into the respective models for large spheres of positive electron density in the difference map.
- As final step, perform refinement of the Xe or O_2 modeled structures. The occupancies and B-factors of Xe or O_2 within the respective protein model should converge to their final values (Roose et al., 2018).

With this in mind, revisiting of previously published flavoprotein structures for the presence of O_2 may be worthwhile. Notably, the determination of O_2 occupancies at various pressure levels does not allow the determination of a K_d due to the fact that following pressurization, crystals are exposed to atmospheric O_2 partial pressure for several seconds when removed from the pressurization device, before being frozen in liquid N_2. Moreover, a quantitative comparison with K_d values derived from, e.g., stopped-flow data should not be made due to differences in solvent composition (buffers, additives, viscosity, etc.) and protein dynamics of crystallized protein compared to protein in solution. However, the methodology could be improved in the future, e.g., by ensuring a more rapid freezing. In addition, different redox states of the flavin cofactor should be taken into account, e.g., EncM was crystallized and pressurized in the oxidized form, whereas Fl_{red} is used in stopped-flow experiments. In addition, the reductive properties of X-ray radiation should be considered. For example, it is well known that flavins and other cofactors often become reduced during crystal measurement, which may cause for example the "butterfly"-shaped electron density of Fl_{red}. This was also the case for EncM and the Fl_{N5O} (that is susceptible to reduction) therefore eluded detection by X-ray crystallography (Teufel et al., 2013, 2015). Similarly, O_2 reduction may conceivably occur, giving rise to reactive oxygen species such as superoxide radicals (that cannot be distinguished from O_2 by the herein presented methodology), which in turn may react with the flavin cofactor. In our datasets, however, we found no indication of bond formation between O_2 and the flavin. To trap and visualize possible intermediates such as a flavin-N5-peroxy species by crystallography, freeze-quenching of reactions between reduced flavoenzyme crystals and O_2 should be considered in the future.

6. Summary and conclusion

This chapter describes the methodology underlying O_2-pressurized X-ray crystallography that we adapted for flavoproteins. To our knowledge, the recent study with EncM was the first successful example of O_2-pressurized X-ray crystallography with a flavoprotein, which revealed the precise positioning of O_2 within the active site (Saleem-Batcha et al., 2018). However, EncM is a non-canonical flavin monooxygenase that shows markedly different interaction with O_2 than enzymes utilizing the prevalent Fl_{C4aOOH}. It remains to be seen whether this methodology is applicable for a wide range of flavoproteins. We anticipate, however, that the method should work for enzymes employing a similar route for O_2 activation than EncM, i.e., enzymes with N5-functionalized flavin cofactors that may be discovered in the future. An experimental approach to screen for such Fl_{N5O}-forming enzymes has been described before (Teufel, 2018). Future crystallographic studies are required to find out if highly ordered electron density for O_2 molecules is only detectable in flavoenzymes that truly bind O_2.

Acknowledgments

We thank the DFG for funding (grant no. TE 931/2-1) and the Swiss Light Source (Villigen, CH) as well as the staff, in particular Takashi Tomizaki, for helping us establishing the herein described methodology.

References

Adams, P. D., Afonine, P. V., Bunkóczi, G., Chen, V. B., Davis, I. W., Echols, N., et al. (2010). PHENIX: A comprehensive python-based system for macromolecular structure solution. *Acta Crystallographica, Section D: Biological Crystallography, 66*(2), 213–221.

Afonine, P. V., Grosse-Kunstleve, R. W., Echols, N., Headd, J. J., Moriarty, N. W., et al. (2012). Towards automated crystallographic structure refinement with phenix.refine. *Acta Crystallographica, Section D: Biological Crystallography, 68*(4), 352–367.

Baron, R., McCammon, J. A., & Mattevi, A. (2009). The oxygen-binding vs. oxygen-consuming paradigm in biocatalysis: Structural biology and biomolecular simulation. *Current Opinion in Structural Biology, 19*(6), 672–679.

Baron, R., Riley, C., Chenprakhon, P., Thotsaporn, K., Winter, R. T., et al. (2009). Multiple pathways guide oxygen diffusion into flavoenzyme active sites. *Proceedings of the National Academy of Sciences of the United States of America, 106*(26), 10603–10608.

Colloc'h, N., Gabison, L., Monard, G., Altarsha, M., Chiadmi, M., et al. (2008). Oxygen pressurized X-ray crystallography: Probing the dioxygen binding site in cofactorless urate oxidase and implications for its catalytic mechanism. *Biophysical Journal, 95*(5), 2415–2422.

Duff, A. P., Trambaiolo, D. M., Cohen, A. E., Ellis, P. J., Juda, G. A., et al. (2004). Using xenon as a probe for dioxygen-binding sites in copper amine oxidases. *Journal of Molecular Biology, 344*(3), 599–607.

Emsley, P., Lohkamp, B., Scott, W. G., & Cowtan, K. (2010). Features and development of coot. *Acta Crystallographica, Section D: Biological Crystallography, 66*, 486–501.

Evans, P. R. (2011). An introduction to data reduction: Space-group determination, scaling and intensity statistics. *Acta Crystallographica, Section D: Biological Crystallography, 67*(4), 282–292.

Ewing, T. A., Fraaije, M. W., Mattevi, A., & van Berkel, W. J. H. (2017). The VAO/PCMH flavoprotein family. *Archives of Biochemistry and Biophysics, 632*, 104–117.

Huijbers, M. M., Montersino, S., Westphal, A. H., Tischler, D., & van Berkel, W. J. (2014). Flavin dependent monooxygenases. *Archives of Biochemistry and Biophysics, 544*, 2–17.

Kabsch, W. (2010). XDS. *Acta Crystallographica Section D: Biological Crystallography, 66*(2), 125–132.

Kitahara, R., Yoshimura, Y., Xue, M., Kameda, T., & Mulder, F. A. A. (2016). Detecting O_2 binding sites in protein cavities. *Scientific Reports, 6*, 20534.

Klinman, J. P. (2007). How do enzymes activate oxygen without inactivating themselves? *Accounts of Chemical Research, 40*, 325–333.

Kommoju, P. R., Chen, Z. W., Bruckner, R. C., Mathews, F. S., & Jorns, M. S. (2011). Probing oxygen activation sites in two flavoprotein oxidases using chloride as an oxygen surrogate. *Biochemistry, 50*, 5521–5534.

Leferink, N. G., Fraaije, M. W., Joosten, H. J., Schaap, P. J., Mattevi, A., & van Berkel, W. J. (2009). Identification of a gatekeeper residue that prevents dehydrogenases from acting as oxidases. *Journal of Biological Chemistry, 284*(7), 4392–4397.

Liebschner, D., Afonine, P. V., Moriarty, N. W., Poon, B. K., Sobolev, O. V., et al. (2017). Polder maps: Improving OMIT maps by excluding bulk solvent. *Acta Crystallographica, Section D: Structural Biology, 73*, 148–157.

Mattevi, A. (2006). To be or not to be an oxidase: Challenging the oxygen reactivity of flavoenzymes. *Trends in Biochemical Sciences, 31*(5), 276–283.

McCoy, A. J., Grosse-Kunstleve, R. W., Adams, P. D., Winn, M. D., Storoni, L. C., & Read, R. J. (2007). Phaser crystallographic software. *Journal of Applied Crystallography, 40*(4), 658–674.

Palfey, B. A., & McDonald, C. A. (2010). Control of catalysis in flavin-dependent monooxygenases. *Archives of Biochemistry and Biophysics, 493*(1), 26–36.

Roose, B. W., Zemerov, S. D., & Dmochowski, I. J. (2018). Xenon–protein interactions: Characterization by X-ray crystallography and hyper-CEST NMR. *Methods in Enzymology*, *602*, 249–272.

Saleem-Batcha, R., Stull, F., Sanders, J. N., Moore, B. S., Palfey, B. A., et al. (2018). Enzymatic control of dioxygen binding and functionalization of the flavin cofactor. *Proceedings of the National Academy of Sciences of the United States of America*, *115*(19), 4909–4914.

Saleem-Batcha, R., & Teufel, R. (2018). Insights into the enzymatic formation, chemical features, and biological role of the flavin-N5-oxide. *Current Opinion in Chemical Biology*, *47*, 47–53.

Teufel, R. (2017). Flavin-catalyzed redox tailoring reactions in natural product biosynthesis. *Archives of Biochemistry and Biophysics*, *9861*(17), 30270–30279.

Teufel, R. (2018). Preparation and characterization of the favorskiiase flavoprotein EncM and its distinctive favin-N5-oxide cofactor. *Methods in Enzymology*, *604*, 523–540.

Teufel, R., Agarwal, V., & Moore, B. S. (2016). Unusual flavoenzyme catalysis in marine bacteria. *Current Opinion in Chemical Biology*, *31*, 31–39.

Teufel, R., Miyanaga, A., Michaudel, Q., Stull, F., Louie, G., Noel, J. P., et al. (2013). Flavin-mediated dual oxidation controls an enzymatic Favorskii-type rearrangement. *Nature*, *503*, 552–556.

Teufel, R., Stull, F., Meehan, M. J., Michaudel, Q., Dorrestein, P. C., et al. (2015). Biochemical establishment and characterization of EncM's flavin-N5-oxide cofactor. *Journal of the American Chemical Society*, *137*, 8078–8085.

Walsh, C. T., & Wencewicz, T. A. (2013). Flavoenzymes: Versatile catalysts in biosynthetic pathways. *Natural Product Reports*, *30*(1), 175–200.

Zafred, D., Steiner, B., Teufelberger, A. R., Hromic, A., Karplus, P. A., et al. (2015). Rationally engineered flavin-dependent oxidase reveals steric control of dioxygen reduction. *FEBS Journal*, *282*(16), 3060–3074.

CHAPTER FOURTEEN

Reduction midpoint potentials of bifurcating electron transfer flavoproteins

A.-F. Miller*, H.D. Duan, T.A. Varner, N. Mohamed Raseek
Department of Chemistry, University of Kentucky, Lexington, KY, United States
*Corresponding author: e-mail address: afmill3r2@gmail.com

Contents

Methods in Enzymology, Volume 620
ISSN 0076-6879
https://doi.org/10.1016/bs.mie.2019.03.017

Abstract

Recently, a variety of enzymes have been found to accept electrons from NAD(P)H yet reduce lower-potential carriers such as ferredoxin and flavodoxin semiquinone, in apparent violation of thermodynamics. The reaction is favorable overall, however, because these enzymes couple the foregoing endergonic one-electron transfer to exergonic transfer of the other electron from each NAD(P)H, in a process called "flavin-based electron bifurcation." The reduction midpoint potentials (E°s) of the multiple flavins in these enzymes are critical to their mechanisms. We describe methods we have found to be useful for measuring each of the E°s of each of the flavins in bifurcating electron transfer flavoproteins.

1. Introduction to flavin-based bifurcation

1.1 Redox considerations

Some of life's most demanding reactions require extremely potent reductants, for example reduction of N_2 to NH_3. Indeed, a plethora of reactions at the core of anaerobic metabolism occur at very low reduction midpoint potentials (E°s) corresponding to strong reductants. However a ubiquitous currency for redox biochemistry is NAD(P)H, which is less reducing (higher-E°) and therefore apparently incapable of supplying electrons for these reactions. In addition, several anaerobic processes display higher metabolic efficiency than can be accounted for by consideration of only low-potential electron donors for low-potential reactions (Buckel & Thauer, 2013). Enter electron transfer bifurcation or "bifurcation."

Bifurcation employs a strongly exergonic transfer of one electron from NAD(P)H to drive endergonic transfer of the other electron to yield a more potent reductant (Herrmann, Jayamani, Mai, & Buckel, 2008). These enzymes thus trade quantity (two electrons from NAD(P)H) for quality (one more potent reducing equivalent, Fig. 1). Conversely, a variety of

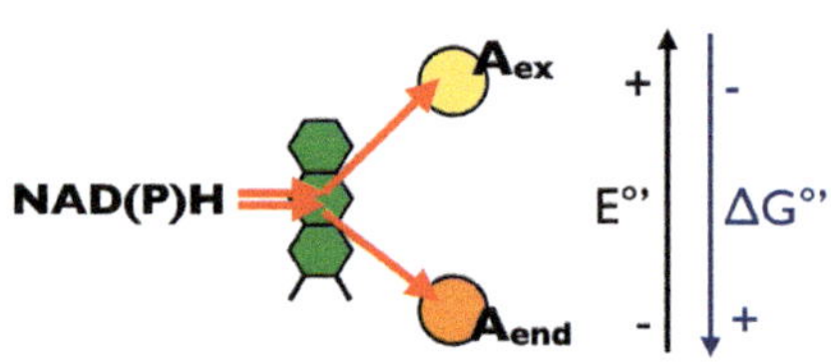

Fig. 1 Cartoon of a minimal flavin-based bifurcating system, in which the flavin is represented by the green cartoon, a higher-E° cofactor able to accept one electron in an exergonic transfer is in yellow, whereas a lower-E° acceptor whose reduction by the flavin represents an endergonic, energy conserving step, is in orange.

different overall reactions have been found to incorporate bifurcation in reverse to conserve excess energy in the form of potent reductants. This process: confurcation, has been found to maximize the energy economy of multi-electron reactions by obtaining half the electrons needed from less potent reductants while drawing on powerful reducing equivalents only as needed to achieve overall thermodynamic favorability.

A particularly well-understood example is that of NADPH-dependent NAD reductase, Nfn, in which a combination of crystallography and spectroscopy have revealed that a pair of electrons are transferred from NADPH to one of two flavins (Demmer et al., 2015; Lubner et al., 2017). This flavin transfers one electron to a Fe_2S_2 cluster which in turn reduces a second flavin (A_{ex} path in Fig. 1), leaving the first one in a highly-energetic anionic semiquinone state (ASQ). The latter possesses sufficient driving force to reduce the first of a pair of Fe_4S_4 clusters that rapidly convey the electron to ferredoxin (Fd), at ≈ -400 mV (A_{end} path in Fig. 1). A second round of similar electron transfer results in a second electron being passed to the second flavin, which then reduces NAD^+ at a higher-potential due to its status as the majority species in the NAD^+/NADH pool, in contrast to NADPH which acts as a low-E° donor due to a 10-fold excess of NADPH over $NADP^+$ (Buckel & Thauer, 2013). Thus, consumption of 2 NADPH effects reduction of one NAD^+ at higher E° and two Fd at lower E°. Although the first flavin acquires electrons as a pair, it donates them individually to two different acceptors, such that the path of electron transfer bifurcates at this flavin, the so-called bifurcation site.

More and more enzymes are being found that perform bifurcating electron transfer from NAD(P)H. So far, all have been found to contain at least one flavin. Thus, use of a flavin as the site of bifurcation upon electron acceptance from NAD(P)H may be a unifying theme, and Buckel and Thauer have popularized the term "flavin-based electron bifurcation," that explicitly distinguishes it from the earlier example of quinone-based electron bifurcation (Buckel & Thauer, 2013).

Electron bifurcation was first recognized as the Q-cycle in respiratory complex III (cytochrome bc_1 complex) where pairs of electrons are acquired by a quinone and bifurcated to the Rieske Fe_2S_2 cluster and cytochrome b_L. Energy is stored by augmentation of the transmembrane proton gradient and one of the two electrons retains a lower E° than the initial pair and therefore can contribute half the reducing equivalents needed for another round (Buckel & Thauer, 2013; Zhu, Egawa, Yeh, Yu, & Yu, 2007). Quinones are well suited to this role by their ready acceptance of pairs of electrons, but formation of a high-energy one-electron reduced intermediate which retains much of the energy from the exergonic electron transfer to the Rieske Fe_2S_2 cluster. From an electrochemical standpoint, we say that the semiquinone state is thermodynamically unfavorable or "thermodynamically suppressed." Suppression of the semiquinone state has been proposed to be essential for bifurcation activity (Nitschke & Russell, 2012).

The extent to which a semiquinone state is suppressed is a continuum. Because it is intermediate in degree of reduction between the oxidized (OX) and two-electron reduced hydroquinone (HQ) states, the semiquinone state (SQ) will be maximally populated at a potential half-way between the $E°$s of reduction of OX to SQ ($E°_{OX/SQ}$) and of reduction of SQ to HQ ($E°_{SQ/HQ}$). At this potential, one can show that the formation constant of SQ, $K = [SQ]^2/[OX][HQ]$ is related to the two $1e^-$ $E°$s by $\log(K) = (E°_{OX/SQ} - E°_{SQ/HQ})/59\,mV$ (Clark, 1960 pg. 186). Thus, when $E°_{SQ/HQ}$ is more positive than $E°_{OX/SQ}$ we say that the two potentials are crossed, and less than half the population can adopt the SQ state in equilibrium with OX and HQ. For the example of $1e^-$ E°s crossed by 120 mV, $(E°_{OX/SQ} - E°_{SQ/HQ}) = -120\,mV$, SQ will constitute only 5% of the population when $[OX] = [HQ]$ at $E°_{OX/HQ}$, and the free energy of the SQ will be 5.7 kJ/mol higher than either OX or HQ.

As noted, all enzymes known so far to bifurcate electrons from NAD(P)H possess at least one flavin, which therefore has been suggested to be the site of bifurcation. The flavin is inherently well-suited to bifurcation, poised by its intrinsic $E°$s at a cusp between one-electron ($1e^-$) and two-electron ($2e^-$) redox reactivity. The $E°$s of free flavin mononucleotide (FMN) are $E°_{OX/SQ} = -313$ mV and $E°_{SQ/HQ} = -101$ mV (Anderson, 1983; Mayhew, 1999a), crossed by 212 mV, causing population of SQ to be a maximum of ~1% at neutral pH (Ehrenberg, Müller, & Hemmerich, 1967). The SQ is nonetheless energetically accessible via mechanisms employed in proteins to tune flavin $E°$s, as evidenced by the stable SQ of flavodoxin (Mayhew, Foust, & Massey, 1969). However in other enzymes analogous non-covalent interactions suffice to diminish the maximum SQ population to well below 0.01% of the total (Koder, Haynes, Rodgers, Rodgers, & Miller, 2002) and give this state the status of a transient high-energy intermediate, akin to a transition state. Thus, electrochemical characterization of flavin sites can shed light on whether they may be competent to execute bifurcation by virtue of suppressed SQ states, provided supportive kinetics also apply (Zhang et al., 2017).

1.2 Multiplicity of pathways and carriers in bifurcation

Electron bifurcation can be understood to couple two electron transfer events such that the exergonic transfer in-essence pays for the endergonic one (Fig. 1, Buckel & Thauer, 2018). A second implication is that bifurcation-competent enzymes must contain at least three redox-active cofactors: the bifurcation site, the higher-E° acceptor and the lower-E° acceptor (green, yellow and orange, respectively, in Fig. 1). Titrations aiming to measure E°s of such systems will therefore have to parse the contributions of multiple sites. This is not only a spectral deconvolution challenge, but the fact that an E° measurement by definition addresses a system at equilibrium limits such measurements to descriptions of systems in which each cofactor's E° is measured in the presence of the reduced states of all higher-E° cofactors. Hence the system may not necessarily be in the same state as that prevailing under turnover, wherein the paths and kinetics of electron transfer can produce intermediates in which a lower-E° cofactor is reduced in the presence of oxidized states of higher-E° cofactors (Hirst, 2010). Thus, a full mechanistic description of bifurcating enzymes must include elucidation of the paths and kinetics of electron transfer in addition to the E°s of the carriers, and it is desirable to measure E°s for individual cofactors in the presence of different mechanistically relevant states of the others (Sevrioukova, Shaffer, Ballou, & Peterson, 1996). The work is advancing, but these and other critical measurements remain open opportunities to be addressed.

A consequence of bifurcation's employment of multiple paths of electron transfer is that there is in-principle the possibility that the proposed high-energy (strongly-reducing) SQ state could lose its electron to the high-E° acceptor instead of the energy-conserving low-E° acceptor. Thus, a gating mechanism is believed to be required, that would limit the exergonic path to acceptance of only one electron per pair from NADH, essentially shutting this path to the second and thus forcing it to employ the other, less favorable one.

Kinetics can play a critical role, for example if the path to the low-E° acceptor is very efficient (short) it can compete effectively even with a more favorable path that is slower. Moreover bifurcating electron transfer flavoproteins have been found to adopt multiple conformations that produce very different inter-cofactor distances, so it is proposed that conformational changes gate electron transfer (Demmer, Chowdhury, Selmer, Ermler, & Buckel, 2017). Since conformational change can also affect E°s, one would like to assess the extent to which the E°s of each cofactor depend on the status of the others including redox state, conformational state and presence of bound substrate or product. Overall, measurement of $E°$s, in conjunction with kinetics and course of electron transfer, are needed for a basic description of the mechanism of bifurcation, and knowledge of the $E°$s' dependence on events throughout the system will provide an invaluable picture of how the steps involved in bifurcation are coordinated and coupled. In what follows, we describe experiments and strategies we have employed to measure $1e^-$ and $2e^-$ $E°$s in an emerging group of bifurcating enzymes with diverse metabolic roles: the bifurcating electron transfer flavoproteins (Bf-ETFs).

1.3 Bifurcating ETFs

ETFs in general have been known for a long time as the heterodimeric proteins that convey $1e^-$ equivalents between fatty acyl-CoA dehydrogenases and the respiratory quinone pool in mitochondria (Watmough & Frerman, 2010). These group-1 ETFs employ a single bound FAD cycling between its OX and ASQ states to execute the redox role, but the protein also contains a bound AMP which is required for protein folding, FAD incorporation and stability (Sato, Nishina, & Shiga, 1993). Recently members of group-2 of the ETF family have been found to possess a second FAD in place of the AMP (Sato, Nishina, & Shiga, 2003). Members of this family have also been shown to have bifurcating activity in combination with partner enzymes able to pass electrons to a higher-$E°$ acceptor (Chowdhury et al., 2014; Chowdhury, Kahnt, & Buckel, 2015; Ledbetter et al., 2017). The Bf-ETF associated with nitrogen fixation in *Azotobacter vinelandii* acquired the name FixAB due to its metabolic role, but we refer to it as ETF here in accordance with its phylogenetics in the ETF family (note, however, that the A,B nomenclature of the two constituent proteins is reversed in Fix notation from its use among ETFs where "A" denotes the larger protein of the two) (Garcia Costas et al., 2017). The nitrogen fixation-associated ETFs display genetic context that is common beyond diazotrophs, with a gene for a putative ETF-quinone oxido<u>r</u>eductase (ETF-QR, = FixC) and a Fe_4S_4 protein (FixX) annotated as a *fixABCX* cluster, even in non-diazotrophs (Fig. 2).

In FixABCX of *A. vinelandii*, NADH reduces a Bf-FAD bound in the AB dimer that is the ETF portion of the complex (Fig. 2). One electron is believed to pass via the ETF's ET-FAD to the higher-$E°$ ETF-QR component of the complex (product of *fixC*) while the other electron passes via

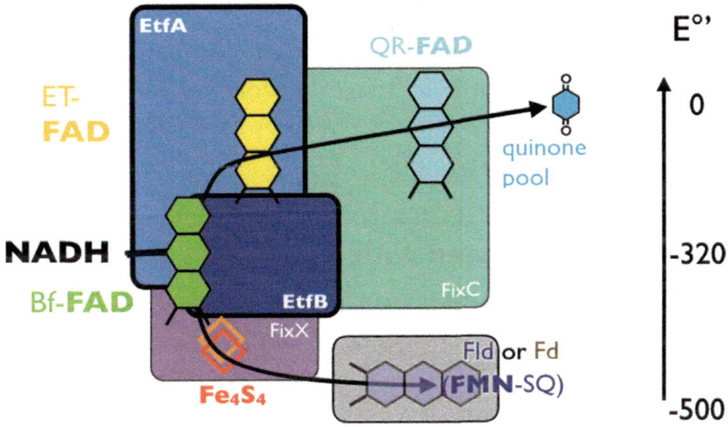

Fig. 2 Cartoon of Bf-ETF associated with nitrogen fixation, and its partner proteins the ETF/quinone oxidoreductase (FixC) and the FixX component believed to mediate electron transfer to Fd or Fld. Bold borders identify the components of the ETF (light and dark blue).

FixX' Fe_4S_4 to Fd or flavodoxin (Fld) SQ (Ledbetter et al., 2017). An analogous mechanism is proposed for the Bf-ETF of *Acidaminococcus fermentans* which forms a complex with a co-expressed butyryl-CoA dehydrogenase that functions in the direction of reducing crotonyl-CoA to butyryl-CoA using a bound FAD (Chowdhury et al., 2014). After two iterations, the net reaction is 2 NADH + 2 Fd_{ox} + crotonyl-CoA → 2 NAD^+ + 2 Fd_{red} + butyryl-CoA (Herrmann et al., 2008).

Among the Bf-ETF-associated complexes, E°s have been measured for the ETF of *Megasphaera elsdenii*, that displays bifurcating activity in conjunction with butyryl-CoA dehydrogenase (Sato, Nishina, & Shiga, 2013). More recently, we have measured E°s for the ETF of FixABCX of *Rhodopseudomonas palustris* (Duan et al., 2018). In both cases, one flavin executes sequential $1e^-$ acquisitions at potentials of $E^{\circ\prime}_{ASQ/HQ} = -136$ mV and -83 mV (*M. elsdenii* and *R. palustris*, respectively) whereas the other flavin displays $2e^-$ reactivity at a lower potential compatible with oxidation of NADH ($E^{\circ\prime}_{OX/HQ} = -279$ mV and -223 mV *M. elsdenii* and *R. palustris*, respectively, Duan et al., 2018; Sato et al., 2013). These values are consistent with the model, and the former higher-E° flavin with $1e^-$ activity is understood to be the ET-FAD while the lower-E° $2e^-$ flavin is suited to the role of Bf-FAD. Crucially, the high $E^{\circ}_{OX/ASQ}$s of +81 and −47 mV for the ET-flavin provide the insight that this FAD most likely rests in the ASQ state in vivo, and therefore would only be capable of accepting one electron from the Bf-FAD. Thus the protein has tuned this FAD's E°s to preclude passage of both electrons via the exergonic path, and the redox tuning thus enforces bifurcation, in these systems (although conformational gating may also occur). This illustrates the insights provided by E° values in understanding bifurcation.

2. Theory associated with measurement of E°s

2.1 One- and two-electron redox reactions

The reduction midpoint potential E° is a statement of the free energy change per electron, for a reductive reaction:

$$Ox + \mathrm{n\,e}^- \rightarrow Red^{-n}, E^\circ = \frac{-1}{nF}\Delta G^\circ \tag{1}$$

where n is the number of electrons involved in the reductive reaction, Ox is the oxidized species (could be OX or one of the SQ states, for flavins), Red is the reduced species to which it is converted (could be a SQ or HQ state for flavins) and F is Faraday's constant.

Because ΔG° is in turn related to the equilibrium constant K_{eq}

$$E^\circ = \frac{RT}{nF}\ln\left(K_{eq}\right) \tag{2}$$

and we have that at an arbitrary potential E (vs. NHE), the concentrations of the Ox and Red species are related as follows:

$$E = E^\circ + \frac{RT}{nF}\ln\left(\frac{[Ox]}{[Red]}\right) = E^\circ + \frac{2.303RT}{nF}log\left(\frac{[Ox]}{[Red]}\right) \tag{3}$$

which is the Nernst equation.

At "room temperature" (25 °C) $RT/F = 25.7\,\mathrm{mV}$. Plots of fraction reduced vs. E-E° show that 90% of the population will be reduced at a potential 60 mV below the E° whereas 90% of the population will be oxidized at a potential 60 mV above E° (Fig. 3A). Thus measurements of the concentrations of Ox and Red species within the range of $E^\circ \pm 60\,\mathrm{mV}$ avoid the difficulty of evaluating minute concentrations or small deviations from 100%. Since E° is generally not known in advance, this means that the first order of business will be some exploratory titrations designed to survey a promising range of E, and most determinations involve multiple attempts with different indicator dyes as part of identifying an optimal range and a dye that does not interfere or interact with the enzyme. These preliminary

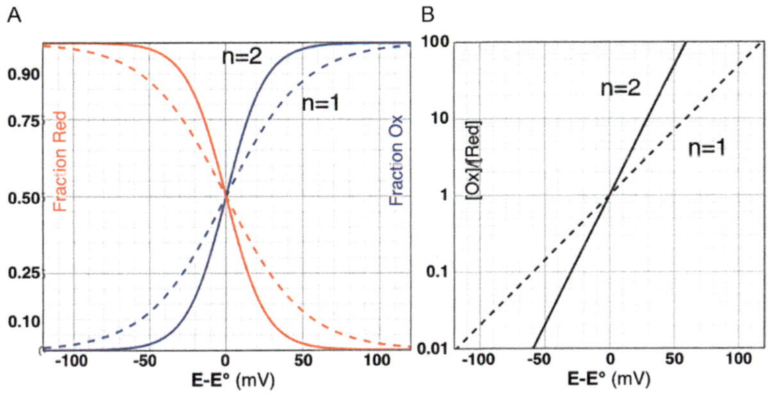

Fig. 3 Dependence on ambient potential of the concentrations of Reduced and Oxidized states for reactions of *Ox* +1e$^-$ *Red*→ (n = 1, dashed lines) and *Ox* +2e$^-$ →*Red*$^{-2}$ (n = 2, solid lines). Panel A provides the response of one or other state as a fraction of the total, since it is often the case that only one of the two states is detected directly. Panel B provides the response of the ratio of populations or concentrations after taking the $\log_{10}$, which provides the benefit of a linear plot in which the slope provides a test of the nature of the reaction via the value of n. The slope of log([*Ox*]/[*Red*]) vs. $E - E°$ is *nF/2.303RT*.

experiments also serve to establish the nature of the reaction, as the slope of the plot of $\log([Ox]/[Red])$ vs. E depends on n, the number of electrons, as n^* $(F/2.303RT)$ (Eq. 3 and Fig. 3B).

2.2 Equilibrium between the flavin and an indicator dye

For cases where a redox indicator dye (D) is present at equilibrium with the flavin (F), the Nernst equation holds for both of them, and one can write that

$$E = E°_D + \frac{RT}{n_D F} ln\left(\frac{[D_{Ox}]}{[D_{Red}]}\right) = E°_F + \frac{RT}{n_F F} ln\left(\frac{[F_{Ox}]}{[F_{Red}]}\right) \tag{4}$$

thus

$$\ln\left(\frac{[F_{Ox}]}{[F_{Red}]}\right) = \frac{n_F}{n_D} ln\left(\frac{[D_{Ox}]}{[D_{Red}]}\right) + \frac{F n_F}{RT}(E°_D - E°_F) \tag{5}$$

and a plot of $\ln([F_{Ox}]/[F_{Red}])$ vs. $\ln([D_{Ox}]/[D_{Red}])$ is expected to produce a straight line of slope n_F/n_D of 2, 1 or 0.5 depending on the reactions undergone by the flavin and the dye (F in a subscript or with a subscript stands for flavin and D stands for indicator dye). The $E°_F$ is determined from the value of $\ln([F_{Ox}]/[F_{Red}])$ and the known $E°$ of the dye, $E°_D$, at the point at which

$\ln([D_{Ox}]/[D_{Red}]) = 0$ (Massey, 1991). This point is in-essence the Y-intercept of the plot and equal to $n_F\ (F/RT)(E°_D - E°_F)$. Such a plot appears later, in Section 6.

2.3 Effect of pH when proton acquisition is coupled to reduction

While reduction of OX to ASQ is described by Eq. 1 with $n = 1$, reduction of OX to NSQ (neutral semiquinone), as in flavodoxin or FixC, requires consideration of uptake of a proton as well. In this case the reaction becomes.

$$Ox + e^- + H^+ \rightarrow Red\text{H}$$

and one can show (Dutton, 1978) that

$$E°_{(pH)} = E°_{lim\ low\ pH} - 59\,mV \log \frac{([H^+] + K_{Ox})}{([H^+] + K_{Red})} \tag{6}$$

where K_{Ox} and K_{Red} are the acid dissociation constants of the *Ox* and *Red* species, and $E°_{lim\ low\ pH}$ indicates the $E°$ relating the protonated version of the *Ox* state to the protonated version of the *Red* state (the NSQ that is protonated at N5). In practice, OX is not protonated at pHs above 2, and even below pH 2, model flavins are protonated at N1, so this limiting value is not measured under physiological circumstances. However given that the *pK* of NSQ of flavodoxin is above 10 (O'Farrel et al., 1998), but the *pK* of OX is much lower than 4, most experimental determinations occur at pKs between the two where $K_{Ox} > [H^+] > K_{Red}$

$$E°_{(pH)} \approx E°_{lim\ low\ pH} - 59\,mV \log \frac{(K_{Ox})}{[H^+]} \tag{7}$$

and the observed $E°$ is expected to decrease by 59 mV for every unit increase in pH (Fig. 4).

For the case of $2e^-$ reduction of OX to anionic HQ^- that is associated with one H^+ per $2e^-$, analogous algebra shows that a 30 mV decrease in $E°$ is expected per unit increase in measurement pH. Since many enzymes are not maximally stable or soluble at pH 7.0, the same equations can be used to calculate the potentials of the dyes to be used at the pH of use from tabulated values (Duan et al., 2018). Before doing so, one must look up the *pKs* of the dye in both its reduced and oxidized form in order to learn how many protons, if any, must be accounted for. Excellent reference data can be found in (Clark, 1960; Ottaway, 1972).

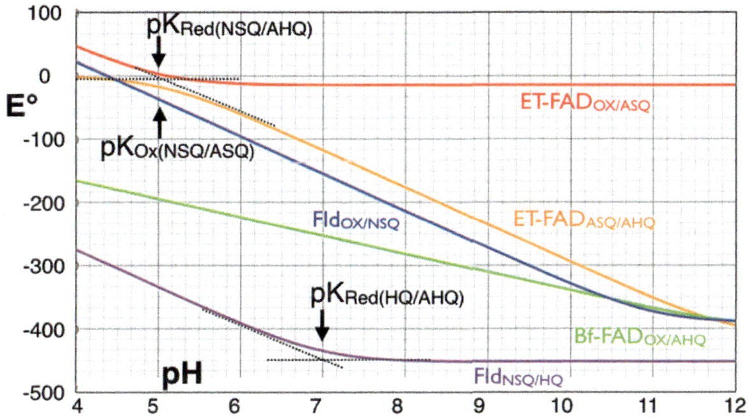

Fig. 4 Generic pH dependence expected for flavins that undergo reduction from OX to ASQ (e.g., ET-FAD of Bf-ETF, red), and then reduction to anionic HQ (AHQ, orange), flavins such as that of Fld that undergo $1e^- + 1H^+$ reduction to NSQ (blue) and then $1e^-$ reduction to AHQ or $1e^- + 1H^+$ reduction to neutral HQ (purple), contrasted with expectations for a Bf-flavin undergoing $2e^- + 1H^+$ reduction to AHQ (green, where the active site is assumed to depress the pK for protonation of anionic HQ). Applicable values for $E^o{}_{\text{lim low pH}}$ and pKs are inspired by literature when possible, but otherwise provided as plausible guesses for illustrative purposes. They are chosen as 7 for the pK of HQ/AHQ, 11 for the pK of Fld NSQ/ASQ and 5 for the pK of ET-FAD NSQ/ASQ.

3. Basic overview of a measurement

In general, an electrode is used to measure the ambient reductive potential *E* relative to the normal hydrogen electrode (NHE), and the natures and populations of the oxidized (*Ox*) and reduced (*Red*) states related by the E^o are revealed by spectroscopy after each of a series of stepwise partial reductions (Dutton, 1978). However, it is also convenient to employ a dye with an E^o near that of the sample as a mediator cum *E* indicator (within ± 30 mV is recommended, Massey, 1991). This can be combined with slow continuous enzymatic delivery of reducing equivalents, to obtain a very stable self-contained experiment that yields excellent results (Massey, 1991). The flavin community has a history of both methods, and electrode-monitored measurements have been reviewed (Byron, Stankovich, Husain, & Davidson, 1989; Mayhew, 1999b). Therefore this treatise will concentrate on use of enzyme-delivered reducing equivalents and dye-based monitoring of the ambient E, because this method provides a very stable experiment that runs on its own, even overnight, once launched.

The challenge is to identify a dye whose optical signal does not overly complicate determinations of the population of flavin in its *Ox* and *Red* states. Some of our favorites are listed in Table 1. We highly recommend conducting the experiment in an inert atmosphere (recall that some glove boxes include H_2 in the atmosphere, which is not necessarily inert, depending on the enzymes and cofactors present). If a glove box or anaerobic chamber is available that is ideal, however skilled use of a gas train with O_2-depleted Ar ($\leq$2 ppm) and high-quality gas-tight glassware can also yield excellent data sets.

Ideally, "pure spectra" corresponding to the fully *Ox* and fully *Red* sample should be in hand (along with analogous spectra of the indicator dye), obtained at potentials >120 mV above and below the $E°$, respectively. These make it straightforward to deconvolute spectra containing mixtures of all four species, to extract their amplitudes at each point in the titration. Measurements of the populations (or concentrations) of *Ox* and *Red* should be made when these are at equilibrium with one-another, at each of several different values of E within ±60 mV of $E°$. Emphasizing data obtained within 60 mV of $E°$ optimizes the accuracy with which populations are quantified, as neither *Ox* nor *Red* represents less than 10% of the total concentration within this range. It also allows more rapid attainment of equilibrium. (Analogous considerations apply for the dye, producing the recommendation that the $E°$ of the dye be within 30 mV of that of the sample, ideally.) Samples should be sufficiently concentrated that at least one of *Ox* or *Red* can be quantified to better than ±10% when present at only 10% of the population. Determinations should be made only after the populations have stabilized after each addition of titrant, in order to avoid confounding the measurement with kinetic effects. Moreover, the enzyme's cofactors must have reached equilibrium with the indicator dye (or eletrode) reporting the solution E. To aid in this, a redox mediator should be included in the solution (Dutton, 1978). The indicator dye plays the same role as buffers do for pH, and the mediator catalyzes any needed interconversion of $1e^-$ and $2e^-$ reactions that may be necessitated by the use of a dye or enzyme with $2e^-$ reactivity, in conjunction with a $1e^-$ reductant. The goal is to establish a single redox equilibrium encompassing the solution, the sample and the electrode.

Ox and *Red* populations determined at several values of E are analyzed collectively by plotting ln([*Ox*]/[*Red*]) vs. E, or equivalently ln(fraction *Ox*/fraction *Red*) vs. E. When only one of the two is measured directly

Table 1 Some useful reference dyes and mediators.[a]

Name	$E°$/mV (vs. NHE)	Number of electrons/n
Potassium Ferricyanide (Guerrero et al., 2011)	430	1
p-benzoquinone (Guerrero et al., 2011)	280	2
N,*N*,*N′*,*N′*,-tetramethyl phenylene diamine (Zhang et al., 1996)	260	1
Diaminodurol (Guerrero et al., 2011)	220	2
2,6-dichlorophenolindophenol (Kolaj-Robin et al., 2011)	217	2
2,5-dimethyl benzoquinone (Brugna, Albouy, & Nitschke, 1998)	180	2
Toluylene blue (Tyree & Webster, 1978)	115	2
Phenazine methosulfate (Zhu & Davidson, 1998)	80	2
Thionine (Jung et al., 1996)	56	2
Phenazine ethosulfate (Covian, Zwicker, Rotsaert, & Trumpower, 2007)	55	2
Toluidine blue (Winkler et al., 2009)	34	2
Ascorbate (Matsui, Kitagawa, Okumura, & Shigeta, 2015)	30	2
Gallocyanine (Nakajima et al., 2001)	21	2
Methylene blue (Vollmer, Switzer, & Debrunner, 1983)	11	2
Duroquinone (Nedbal, Samson, & Whitmarsh, 1992)	5	2
Menadione (Alric, Pierre, Picot, Lavergne, & Rappaport, 2005)	0	2
Pyocyanine (Ugulava & Crofts, 1998)	−34	2
5,5′,7,7′-indigotetrasulphonate (Chen et al., 1993)	−46	2
5,5′,7-indigotrisulphonate (Chen et al., 1993)	−81	2
5,5′-indigodisulphonate (Cheng et al., 2013)	−125	2
Nile Blue (Hu, Chuenchor, & Rokita, 2015)	−116	2
2-hydroxyl-1,4-naphthoquinone (Kay, Solomonson, & Barber, 1986)	−145	2

Continued

Table 1 Some useful reference dyes and mediators.[a]—cont'd

Name	$E°$/mV (vs. NHE)	Number of electrons/n
Anthraquinone-2,6-disulphonate (Millner, Widger, Abbott, Cramer, & Dilley, 1982)	−185	2
Anthraquinone-2-sulfonate (Guerrero et al., 2011)	−225	2
Phenosafranine (Arents, Perez, Hendriks, & Hellingwerf, 2011)	−252	2
Safranine T (Arents et al., 2011)	−289	2
Neutral red (Pueyo, Gomez-Moreno, & Mayhew, 1991)	−325	
−323	2	
Benzyl viologen (Stankovich, Schopfer, & Massey, 1978)	−311	1
Diquat (Chen et al., 1993)	−361	1
Dithionite[b] (Jang, Scrutton, & Hille, 2000)	−386	1
Methyl viologen (Corrado, Aliverti, Zanetti, & Mayhew, 1996)	−430	1

[a]Also see (Clark, 1960; Ottaway, 1972).
[b]Note: dithionite reduction midpoint potential depends strongly on pH and concentration.

(for example, *Ox*) the other is calculated from the total concentration or (fraction *Red*) = 1-(fraction *Ox*). According to the Nernst Eq. (3) a straight line is predicted providing an excellent test of applicability of the model. An integer value is expected for n, and may be predicted from the spectra obtained. For the example of a flavin site in the OX state, observation of the spectral signature of ASQ upon reduction indicates that $n = 1$ so a slope $\neq RT/F$ would reveal a need to improve the experiment. Assuming a straight line with a slope within 10% of that expected for the value of n, the value of E when $\ln([Ox]/[Red]) = 0$ is the experimental value of $E°$ $(E - E°) = 0$. (Note that when the log is used instead of the natural ln, an additional factor of 2.303 is present, as in Fig. 3B.)

4. Protocol for $E°$ determination using xanthine/xanthine oxidase to deliver reducing equivalents

When a dye is used instead of an electrode to report on E, the analysis described in Sections 5.3 and 5.4 can be used to obtain the $E°$ of the flavin directly from that of the dye (below). This approach is commonly employed in conjunction with xanthine and xanthine oxidase to deliver reducing equivalents. The latter obviate the need to add titrants and allows the reaction proceed without intervention producing high quality data sets that may be acquired over night.

4.1 Small apparatus and reagents: Have these in the anaerobic chamber

- 2.0 mL cryovials
- 20 mL scintillation vials
- Precision Seal rubber septa
- Precision Glide needles
- Syringe filters (optional)
- Aluminum foil
- 500 μL self-masking cuvettes
- Micro stirring bars and a larger one to aid in retrieving them.
- Milli-Q water and Kimwipes
- Gel filtration mini-columns for switching buffers/removing dyes/testing for flavin dissociation.
- Set of pipetmen
- Buffer of your choice. Note that if your normal buffers contain redox-supplements such as dithiothreitol, β-mercaptoethanol or TCEP (*Tris*(2-carboxyethyl)phosphine hydrochloride) you should determine in advance whether these interact with the indicator dyes, or react in the course of the titration. It is best to avoid possible complications. It is wise to place some extra buffer and deionized water into the chamber the night before use, and allow them to equilibrate with the inert atmosphere. These can permit rinsing of small apparatus and mid-course corrections.

4.2 Major instrumentation present in the anaerobic chamber

- Ecotherm chilling dry bath with lid, for keeping samples refrigerated.
- Spectrophotometer (associated computer is outside the chamber).

4.3 Reagents to prepare on the day of the titration, and use when fresh

- 1 μM stock solution of xanthine oxidase in the same buffer of your choice.

- 500 μM xanthine in the same buffer of your choice. Xanthine does not dissolve easily at neutral pH. When preparing a more concentrated stock solution, 1–2 μL of 10 M NaOH may be added to 1 mL buffer to aid dissolution. After dilution to the desired working concentration, check the pH.
- 100 μM methyl viologen in the same buffer of your choice.
- Dithionite for a stock solution. Weigh out powder into a vial and degas, prior to transfer into chamber, but dissolve by addition of desired volume of alkaline buffer only at time of use, inside the chamber (e.g., to confirm a reduction end-point).
- Reference dyes in the same buffer of your choice. (Note that some dyes are sparingly soluble in water, and may need to be prepared as stock solutions in ethanol)
- The enzyme to be characterized, in the buffer of choice.

4.4 Procedure

Set-up (on the day of the titration):

1. Pierce the cap of the 2.0 mL cryovial containing your protein sample with a needle to allow degassing (see below) without protein foaming. Multiple holes can be made to facilitate faster purging.
2. Place the capped cryovial into a 20 mL scintillation vial and then seal the scintillation vial with a Precision Seal rubber septum.
3. Pierce the septum of the scintillation vial with two Precision Glide needles far apart and then leave the needles in place. One of the needles will be used as an input for purging with high-purity N_2 gas and the other will serve as the vent during purging.
4. Attach one of the needles to a syringe filter that is hooked up to the nitrogen gas line. The syringe filter is optional.
5. Purge the scintillation vial with nitrogen for 1–2 h while keeping the vial on ice and shielded from light using the aluminum foil.
6. Remove both needles and transfer the scintillation vial into the anaerobic chamber.
7. Store your protein sample in the Ecotherm chilling dry bath at 3 °C, shielded from light, inside the anaerobic chamber.
8. Make all solutions anaerobic by multiple cycles of alternating pumping and refilling with purified N_2 or Ar gas, on a gas train. Similarly, all dry reagents should be weighed out into septum vials, then their head spaces should be made fully anaerobic.

9. Transfer stable reagents into the anaerobic chamber and allow them to equilibrate with the inert atmosphere. Note that some protein samples undergo slow auto-reduction under anaerobic conditions or may be even unstable vs. overnight storage at 3 °C. In these cases, the protein samples should be thawed from −80 °C, made anaerobic on the day of the experiment and used right away, for consistent titration data.
10. Shield methyl viologen and reference dyes from light with aluminum foil.
11. Keep xanthine oxidase and any other labile reagents in the Ecotherm chilling dry bath at 3 °C.

To initiate the titration

12. If needed, pretreat the enzyme sample with 0.5 mM ferricyanide, to obtain a starting state of 100% in the oxidized state. Then pass the enzyme over a gel filtration minicolumn to remove oxidant.
13. Place 385 µL of 520 µM xanthine in a 500 µL self-masking cuvette and collect a spectrum to serve as the blank reference.
14. Add 5 µL of 100 µM methyl viologen, mix well with a micro stirring bar and then record another spectrum. There should be no apparent absorbance in the 300–800 nm region, indicating that the methyl viologen is in the oxidized state. (This is your mediator blank spectrum.)
15. Add 100 µL of your enzyme to produce a flavin concentration of at minimum 20 µM. Mix gently and well, then record another spectrum. This is your starting OX spectrum.
16. Add up to 5 µL of the reference dye and mix gently but thoroughly. Record the spectrum of oxidized protein plus oxidized dye. (The amount of reference dye used in the titration depends on its extinction coefficient. Generally, keep the absorbance of the dye comparable to that of the 450 nm band of the enzyme.)

The components added are summarized as follows:

Volume (µL)	Component	Final concentration (µM)
385	520 µM xanthine	400
5	100 µM methyl viologen	1
100	≥50 µM enzyme	≥10
5	reference dye	–

17. To initiate the titration, add 5 μL of 1 μM xanthine oxidase (10 nM final concentration). Mix thoroughly and collect a spectrum every 1 min from 200 to 800 nm (or further if possible). The rate of reduction is controlled by the concentration of xanthine oxidase with a larger amount of enzyme producing more rapid reduction. Therefore, trials run are usually necessary to find a good concentration of xanthine oxidase which allows the titration to last 1–2 h. Too rapid a titration can preclude maintenance of equilibrium, however, higher concentrations of xanthine and xanthine oxidase may be needed to reach lower reduction potentials (e.g., for the $E^{\circ}_{NSQ/AHQ}$ of Fld).

5. Ingredients of success: Tips

Every enzyme is different so the advice that follows will need to be adapted for the specific system of interest.

Self-masking quartz cuvettes permit use of as little as 0.5 mL of sample, at flavin concentrations of no less than 10 μM. Even when used in an anaerobic chamber, an airtight screw-cap cuvette is helpful.

In our work, reducing equivalents have been provided continuously by including xanthine and xanthine oxidase in the reaction following the procedure of (Massey, 1991). We use 0.5 mL reactions containing 400 μM xanthine, 1 μM methyl viologen (mediator), ≥10 μM *Rpal*ETF and a concentration of the reference dye that yields comparable absorbance to that of the flavins in the buffer of choice (Efimov et al., 2014; Mishanina, Yadav, Ballou, & Banerjee, 2015). The reaction is initiated by adding a catalytic amount of xanthine oxidase (0.5–20 nM) with a larger amount used for titrations aiming for more reduced states of the ETF. Spectral changes are monitored every 1 min. Analyses exploit spectra demonstrating concurrent reduction of the reference dye and progress of the flavin reduction of interest.

5.1 Identification of the species formed, isosbestic and characteristic wavelengths

It is helpful to begin the investigation by performing reductive titrations in the absence of dyes, to learn what oxidation states are formed with different reductants and oxidants, and to determine the absorbance maxima and isosbetic wavelengths. Difference spectra characterizing the different stages of the titration will identify the wavelengths at which measurements are maximally sensitive to the oxidation state change for each of the flavin and the dye. It is ideal if a wavelength that is an isosbestic for the flavin can be used to monitor the oxidation state of the dye, and vice versa. Alternatively, if a sufficient portion of the dye's signal is non-overlapping with the flavin, then this region can be used both to determine the fraction of dye that is oxidized and to determine the weighting factor needed to subtract the contribution of the dye from the rest of the spectrum, using the previously obtained spectrum of dye alone.

Thus, a titration with dithionite revealed that one of the FADs in Bf-ETF undergoes sequential $1e^-$ reduction via an ASQ state whereas the other undergoes concerted $2e^-$ reduction from OX to HQ, for *R. palustris* ETF (*Rpal*ETF). Titration with NADH as the reductant revealed a qualitatively identical result for *Rpal*ETF, as for the Bf-ETFs from *Acetaminococcus fermentans* and *Megasphaera elsdenii* (Chowdhury et al., 2014; Chowdhury et al., 2015; Sato et al., 2013). However mutant *Rpal*ETF lacking the Bf-FAD was not susceptible to reduction by NADH, whereas Bf-ETF from *Megasphaera elsdenii* containing only the Bf-FAD underwent only $2e^-$ reduction without accumulating any SQ (Sato et al., 2013). Titrations in which partner proteins serve to deliver reducing or oxidizing equivalents are also invaluable (Chowdhury et al., 2014). (Note that long-lived kinetic intermediates can occur, which may require that more mediators be used, longer times be allowed or partner proteins be provided, to hasten attainment of equilibrium (Talfournier et al., 2001).)

Fig. 5A shows the sequence of spectra collected for *Rpal*ETF in the course of a titration using dithionite as the reductant. They show an initial increase at 374 nm indicative of ASQ formation while a decrease at 456 nm reveals diminution of OX. Later in the titration intensity decreases at all wavelengths producing a typical OX spectrum roughly half way through that then gives way to the weak signal of HQ. Plots of the absorbance at the absorption maxima of the different species chart their appearance and disappearance (Fig. 5B, Byron et al., 1989; Duan et al., 2018; Mayhew et al., 1969; Sevrioukova et al., 1996; Swenson & Krey, 1994). Three phases are readily distinguished (Fig. 5B).

5.2 Nature of the reaction in each phase of the titration and calculation of flavin populations, oxidized and reduced

The nature of the reaction characterizing each phase can be deduced by difference spectra calculated for each phase of the titration, as the first spectrum

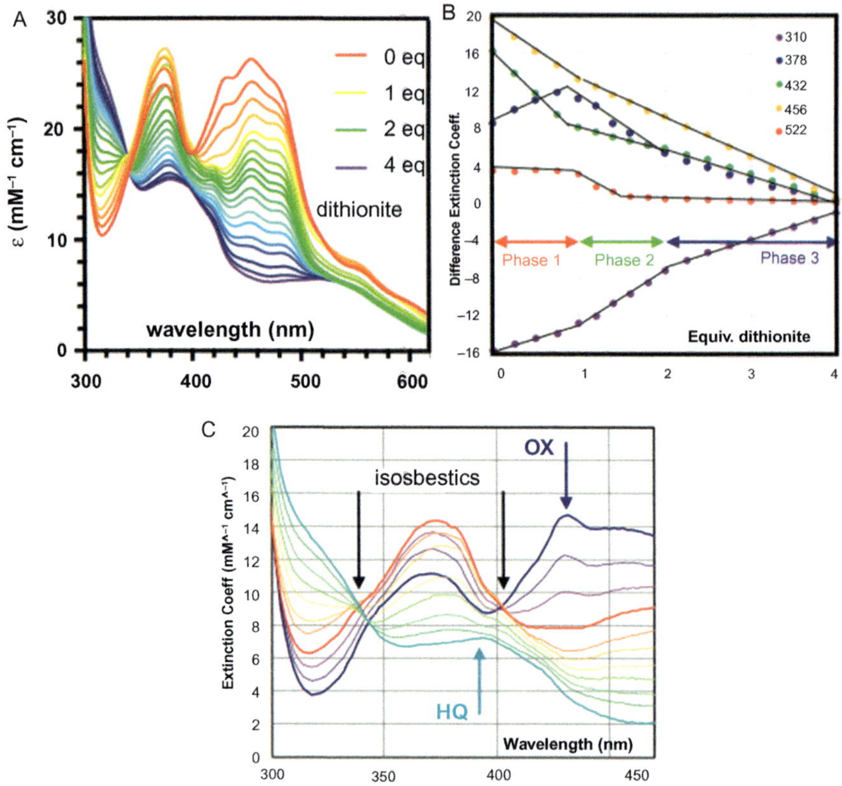

Fig. 5 (A) Absorption spectra of *Rpal*ETF documenting the sequence of species present as the enzyme is reduced by substoichiometric dithionite additions, reproduced from (Duan et al., 2018). A dithionite stock solution producing 14.5 μM increases in concentration with each addition was used. (The concentration of the dithionite stock solution can be estimated via its absorbance at 315 nm, $\varepsilon_{315} = 7.05\,\text{mM}^{-1}\,\text{cm}^{-1}$ (Mayhew, 1978).) If NADH is the titrant, the concentration of the stock solution can be assessed via $\varepsilon_{340} = 6.22\,\text{mM}^{-1}\,\text{cm}^{-1}$ (Haid, Lehmann, & Ziegenhorn, 1975). The *Rpal*ETF concentration was 65 μM. (B) Plots of extinction coefficient change at select wavelengths, showing distinct behaviors in different phases of the reduction, and wavelengths at which different trends are detected. A series of difference spectra were calculated from those in panel A by subtracting the final spectrum of the series from every other spectrum. In the resulting difference spectra, the extinction coefficients at different wavelengths were plotted vs. equivalents of dithionite added as a way of monitoring the formation and decay of different species. (C) difference spectra derived from the first two phases whence the last spectrum of phase 2 has been subtracted but a HQ signature derived from the last spectrum has been restored, to depict spectra from the higher-$E°$ flavin without interference from the lower-$E°$ flavin. *This data set was originally published in the Journal of Biological Chemistry, in Duan, H.D., Lubner, C.E., Tokmina-Lukaszewska, M., Gauss, G.H., Bothner, B., King, P.W., et al. (2018). Distinct flavin properties underlie flavin-based electron bifurcation within a novel electron-transferring flavoprotein FixAB from* Rhodopseudomonas palustris. Journal of Biological Chemistry 293, *4688–4701 © the American Society of Biochemistry and Molecular Biology or © the authors.*

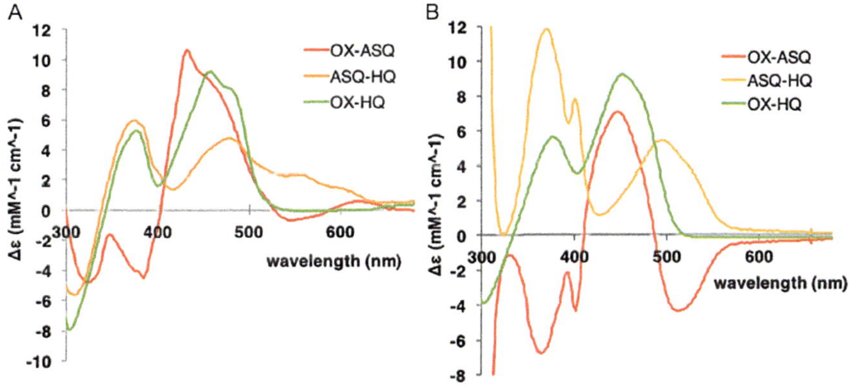

Fig. 6 Difference spectra corresponding to each phase of reduction of *Rpal*ETF by dithionite (A), compared to synthetic "model" spectra produced from spectra of OX, ASQ and HQ obtained from systems in which these states can be fully populated: FAD (pH 7.0), glucose oxidase (pH 10) and FAD (pH 7.0), respectively. *Data in panel A were originally published in the Journal of Biological Chemistry, in Duan, H.D., Lubner, C.E., Tokmina-Lukaszewska, M., Gauss, G.H., Bothner, B., King, P.W., et al. (2018). Distinct flavin properties underlie flavin-based electron bifurcation within a novel electron-transferring flavoprotein FixAB from* Rhodopseudomonas palustris. Journal of Biological Chemistry *293, 4688–4701 with permission from the Journal of Biological Chemistry © the American Society of Biochemistry and Molecular Biology or © the authors.*

for the phase minus the last one (Fig. 6A). Comparison with synthetic difference spectra constructed from spectra that typify each of the flavin oxidation states in well-studied systems confirms that the dominant process in the first phase is conversion of OX to ASQ (Fig. 6B), the second phase is best described by conversion of ASQ to HQ and the third phase is OX to HQ. The difference spectra in Fig. 6A also reveal which wavelengths are maximally sensitive to the change being monitored, which may not be the same as the wavelengths with the maximum absorbance (Fig. 7A). Division of the difference absorbance (ΔA) at the chosen wavelength by the concentration change yields a difference extinction coefficient.

It is equally easy to note that the full difference for the phase represents a population change of 1. Then, subtraction of the last spectrum of a phase from each of the preceding spectra yields the same difference but with a decreasing amplitude representing the extent to which the population is approaching the final state. The amplitude of difference spectrum "i" divided by that of the complete phase ($\Delta_{(phase)}$) yields the population change it represents permitting calculation of the population or concentrations $[Ox]$ and $[Red]$ present (Scheme 1).

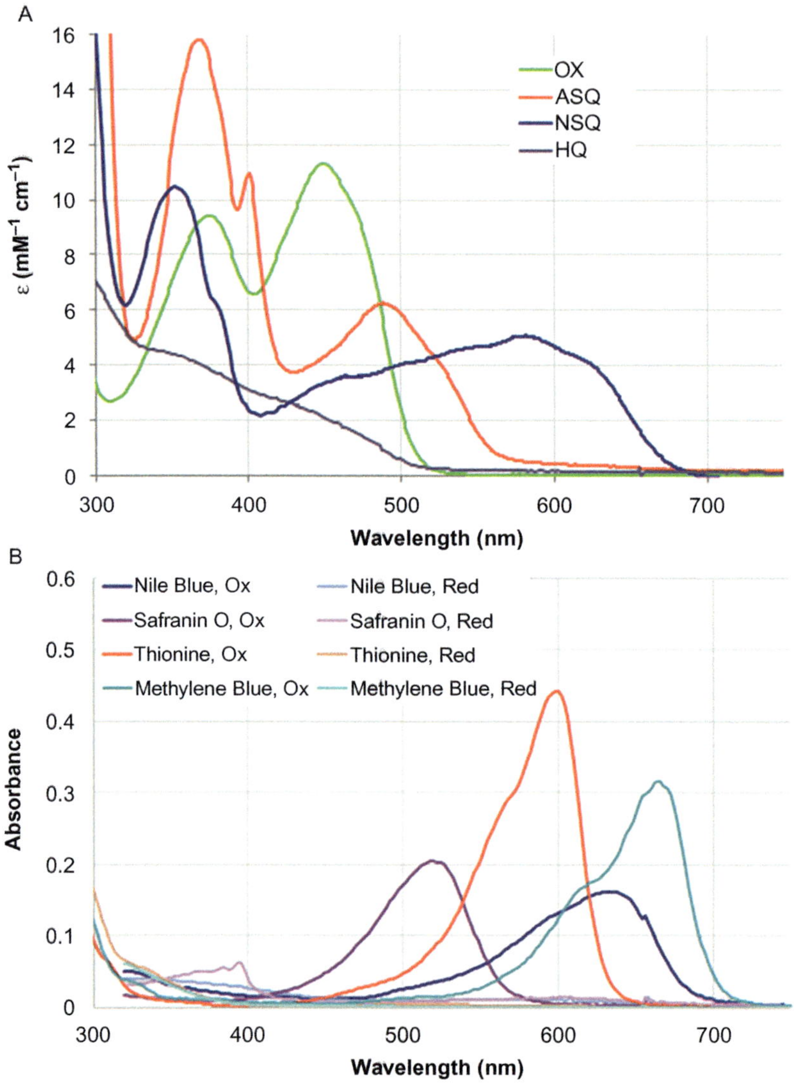

Fig. 7 Optical signatures of each of the common oxidation states of flavin (A) compared with the oxidized-state (*Ox*) and reduced-state (*Red*) spectra of four commonly-used indicator dyes (B). Flavin OX, ASQ, NSQ and HQ spectra were obtained from systems in which these states can be fully populated: FAD (pH 7.0), glucose oxidase (pH 10), flavodoxin from *Desulfovibrio vulgaris* (pH 7.5) and FAD (pH 7.0), respectively. *Three spectra in panel A were originally published in the Journal of Biological Chemistry, in Hoben, J.P., Lubner, C.E., Ratzloff, M.W., Schut, G.J., Nguyen, D.M.N., Hempel, K.W., et al. (2017). Equilibrium and ultrafast kinetic studies manipulating electron transfer: A short-lived flavin semiquinone is not sufficient for electron bifurcation. Journal of Biological Chemistry 292, 14039–14049 with permission from the Journal of Biological Chemistry © the American Society of Biochemistry and Molecular Biology or © the authors.*

Calculation of populations from difference spectra

Spectra	Diff. spectra = Current - End	=
Start = 1.0 Ox + 0.0 Red	Δ Start=1.0 Ox - 1.0 Red	1.0 $\Delta_{(Phase)}$
Step 1 = 0.8 Ox +0.2 Red	Δ 1 = 0.8 Ox -0.8 Red	0.8 $\Delta_{(Phase)}$
Step 2 = 0.6 Ox +0.4 Red	Δ 2 = 0.6 Ox -0.6 Red	0.6 $\Delta_{(Phase)}$
Step 3 = 0.4 Ox +0.6 Red	Δ 3 = 0.4 Ox -0.4 Red	0.4 $\Delta_{(Phase)}$
Step 4 = 0.2 Ox +0.8 Red	Δ 4 = 0.2 Ox -0.2 Red	0.2 $\Delta_{(Phase)}$
End (5) = 0.0 Ox +1.0 Red	$\Delta_{(Phase)}$ = 0.0 Ox -0.0 Red	0

For a difference spectrum of amplitude 0.6, add 0.6 Ox and subtract 0.6 Red from the ending population of 0.0 Ox and 1.0 Red to yield populations of OX=0.6, Red=0.4.

Start = 0.8 Ox + 0.2 Red	Δ Start=0.8 Ox - 0.8 Red	$\Delta_{(Phase)}$
Step 1 = 0.6 Ox +0.4 Red	Δ 1 = 0.6 Ox - 0.6 Red	0.75 $\Delta_{(Phase)}$
Step 2 = 0.4 Ox +0.6 Red	Δ 2 = 0.4 Ox - 0.4 Red	0.5 $\Delta_{(Phase)}$
Step 3 = 0.2 Ox +0.8 Red	Δ 3 = 0.2 Ox - 0.2 Red	0.25 $\Delta_{(Phase)}$
End (4) = 0.0 Ox +1.0 Red	$\Delta_{(Phase)}$ = 0.0 Ox - 0.0 Red	0.0

For a difference spectrum of amplitude 0.5, add 0.4 Ox and subtract 0.4 Red from the ending population of 0.0 Ox and 1.0 Red to yield populations of OX=0.4, Red=0.6.

Scheme 1 Recipies for calculating populations from difference spectrum amplitudes, for two scenarios. Top: the case where the high-$E°$ flavin is fully oxidized at the outset, or, below: the case where the high-$E°$ flavin is partially reduced at the start, as a consequence of its high $E°$.

5.3 Calculation of oxidized and reduced populations of indicator dye

The spectra obtained in the course of an $E°$ measurement are more complicated because they include the indicator dye in addition to the flavin. The dyes are chosen to provide a reference $E°$ near that of the flavin (usually within $\pm 30\,mV$) but also to not interfere with the flavin's optical signals. Fig. 7 shows spectra of four of our favorites in panel B, aligned beneath with the signatures of different flavin oxidation states in panel A. A concentration of indicator dye yielding an optical absorbance comparable to that of the flavin is usually chosen. While it may not be possible to find a dye that does not overlap with the flavin signals, the dye's oxidized-state amplitude can nonetheless be determined at a wavelength that is an isosbestic for the flavin oxidation state under study. Thus, at the wavelength at which the flavin OX and ASQ spectra cross, any change in absorbance will be attributable to the dye's change in oxidation state as the titration proceeds. Knowledge of the progress of the dye's reduction then permits subtraction of the dye's contribution to the optical spectrum, for each step in the phase, and analysis for flavin populations as above. Alternately, a set of simultaneous equations can be used to account for conversion to the reduced states of flavin and dye in each reduction phase (Trimmer et al., 2005).

5.4 Ln/ln plots of population ratios

For the example of OX→ASQ (Phase 1) 4.2 μM methylene blue was included and its oxidized population monitored at 664 nm using the extinction coefficient $\varepsilon_{664} = 76\,\text{mM}^{-1}\,\text{cm}^{-1}$, $E^{o\prime} = +11\,\text{mV}$ (Norden & Tjerneld, 1982) while the population of ET-FAD converted to ASQ was determined from difference spectra and $\Delta\varepsilon_{456}(ET\text{-}FAD_{OX\text{-}SQ}) = 6.57\,\text{mM}^{-1}\,\text{cm}^{-1}$ (Duan et al., 2018). Panel A of Fig. 8 shows the difference spectra obtained for this phase as per Scheme 1, panel B shows the result of deconvolution of the spectrum of methylene blue, accounting for its increasing extent of reduction and facilitating analysis of the flavin's absorbance changes. At the wavelength used to evaluate the extent of flavin reduction, the contribution of this dye is very small, and moreover corrected out.

Using the determined populations of *Ox* and *Red* flavin and dye, a plot of $\ln([FAD_{Ox}]/[FAD_{Red}])$ vs. $\ln([Dye_{Ox}]/[Dye_{Red}])$ is fit with a linear regression model corresponding to Eq. (8a) and (8b) (see Eq. 5).

$$\ln\left(\frac{[F_{Ox}]}{[F_{Red}]}\right) = b + \frac{n_F}{n_D}\ln\left(\frac{[D_{Ox}]}{[D_{Red}]}\right) \tag{8a}$$

where b is the intercept produced by the fit. For methylene blue, $n_D = 2$. Thus the obtained slope of 0.51 confirms the spectral signatures in assigning the reaction to $1e^-$ reduction of OX to ASQ (Fig. 8C).

The value obtained for b was then used with the midpoint potential of the dye at the pH used to calculate the E°.

$$b = \frac{n_F}{25.7\,mV}\left(E_D^0 - E_F^0\right) \tag{8b}$$

where 25.7 mV replaces RT/F. In this case, because the measurement was made at pH 9, we calculated the dye $E^\circ{}_D$ at that pH to be −48 mV from the published $E^{o\prime}$ value of 11 mV and a correction of −59 mV/2 per pH unit increase consistent with acquisition of one proton per two electrons in the range of pH 7–9 for methylene blue (Norden & Tjerneld, 1982). Thus for the example of the intercept in Fig. 8C we have that $-0.13 = (-48\,\text{mV} - E_F^0)/25.7\,\text{mV}$ so $-45\,\text{mV} = E_F^0$. Since the OX/ASQ reduction does not involve any protons, the value of E° is pH independent.

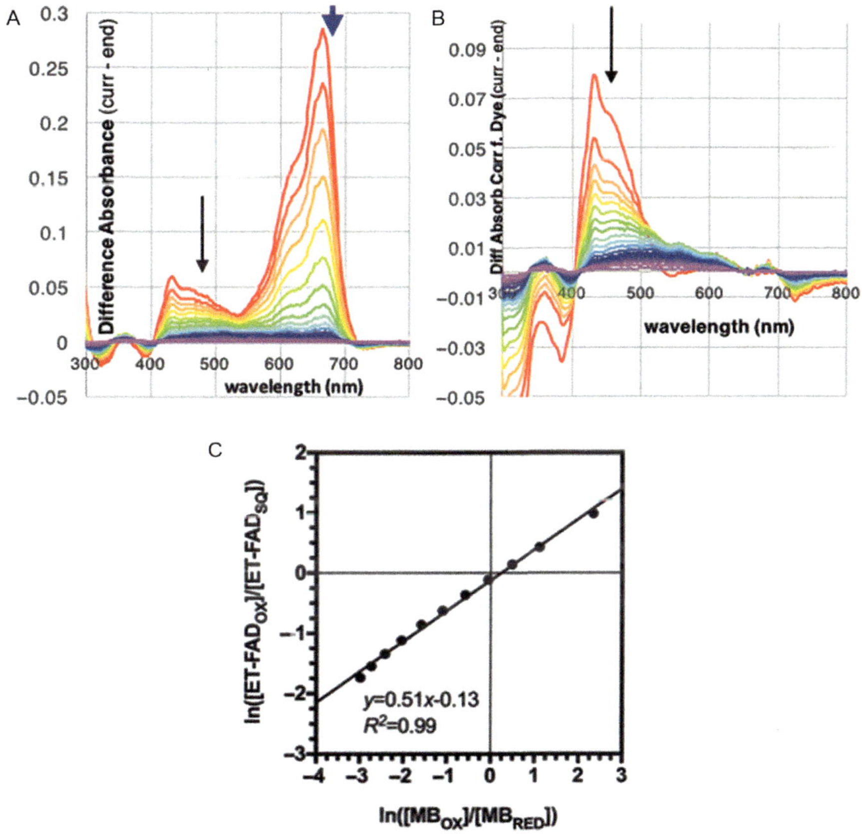

Fig. 8 Difference spectra from Phase 1 of reduction of 10 μM *Rpal*ETF in the presence of 4.2 μM of the indicator dye methylene blue (panel (A) upon subtraction of the last spectrum in the phase, and panel (B) upon further subtraction of the contribution of MB). The blue arrow indicates λ_{max} for the dye, used to determine the *Ox* population, and the black arrow marks the position at which the difference in *Rpal*ETF population was measured, to be converted to populations OX and ASQ as in Scheme 1. Panel C shows the resuling "ln/ln" plot where the natural logarithms of the population ratios are plotted for flavin *Ox* = OX and *Red* = ASQ, and dye = methylene blue (MB). *This data set was originally published in the Journal of Biological Chemistry, in Duan, H.D., Lubner, C.E., Tokmina-Lukaszewska, M., Gauss, G.H., Bothner, B., King, P.W., et al. (2018). Distinct flavin properties underlie flavin-based electron bifurcation within a novel electron-transferring flavoprotein FixAB from* Rhodopseudomonas palustris. Journal of Biological Chemistry 293, *4688–4701 with permission from the Journal of Biological Chemistry © the American Society of Biochemistry and Molecular Biology or © the authors.*

Three independent repetitions yielded an average $E^{\circ}{}_{OX/ASQ}$ of -47 ± 4 mV for this system (Duan et al., 2018). Sample data for conversion of OX to NSQ, and NSQ to HQ are provided in Fig. 9, to show the signature of NSQ which often overlaps with that of the dye.

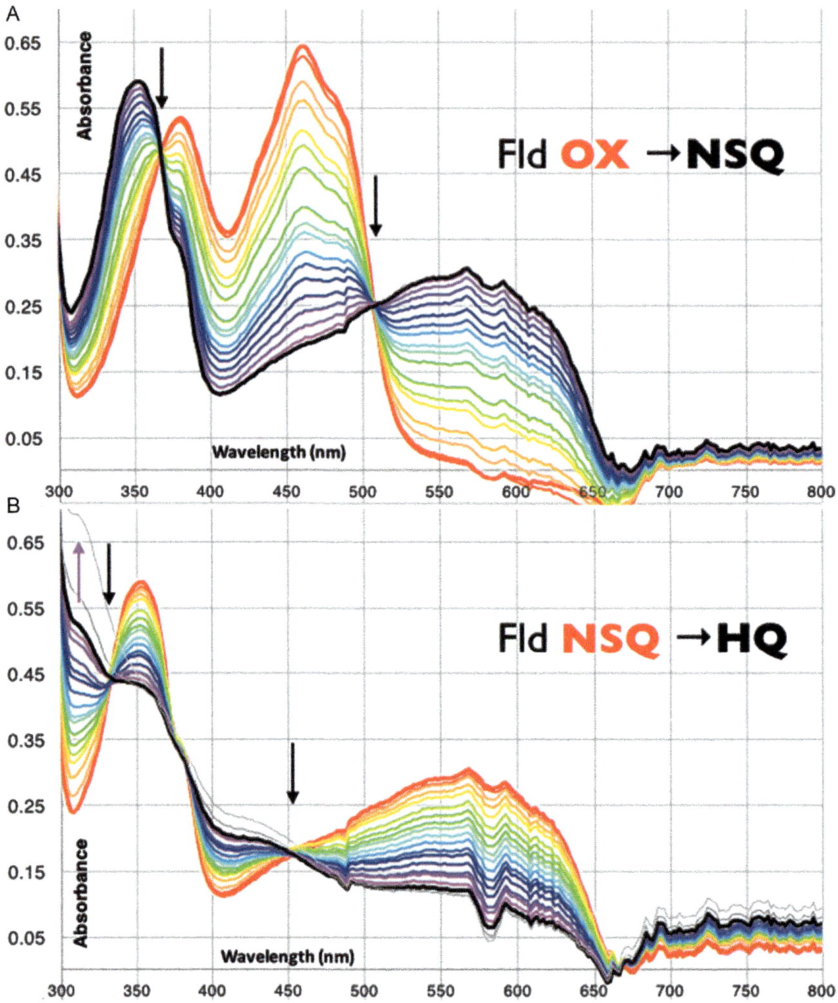

Fig. 9 A text-book titration of flavodoxin with dithionite as the reductant, showing desirable features such as clean isosbestics for each successive reaction (black arrows), as well as the increase at 315 nm at the end (pink arrow) indicating that the Fld is fully reduced and excess dithionite is no longer reacting, but accumulating instead. Panel (A) displays spectra collected as the OX state (red) reacts to form NSQ (black). Panel (B) displays spectra collected as the NSQ state (now in red) reacts to form HQ (black). *D. vulgaris* Fld (provided by J. Hoben) was at pH 7.5, and divets due to spectrophotometer problems are evident at 486, 582 and 656 nm.

6. Controls and precautions

6.1 Enzyme integrity

It is crucial that the enzyme retain its integrity throughout collection of the data set, so that all the data can be considered to represent the same condition of the enzyme, albeit at different ratios of oxidation states. The most important consideration is that no cofactors should be released, as free flavin (or other cofactor) will contribute to the spectra but introduce additional species that will be essentially impossible to deconvolute from the species of interest. At the end of a titration, it is good practice to assay enzyme activity when an assay exists, for comparison with activity beforehand. Passage of the sample over a gel filtration column (in the anaerobic chamber) is recommended to separate the enzyme from all mediators, dyes and titrants to permit assessment of any flavin release via an optical spectrum of the small-molecule fraction after reoxidation (Fig. 10). Similarly, the recovered enzyme upon reoxidation should display the same optical spectrum as it did beforehand.

If the enzyme does not tolerate a full titration (typically taking 1–2 h, but in some cases requiring days, Byron et al., 1989), then it is better to make a series of short measurements by setting up several individual equilibrations with different amounts of reductant (oxidant) and monitoring each for

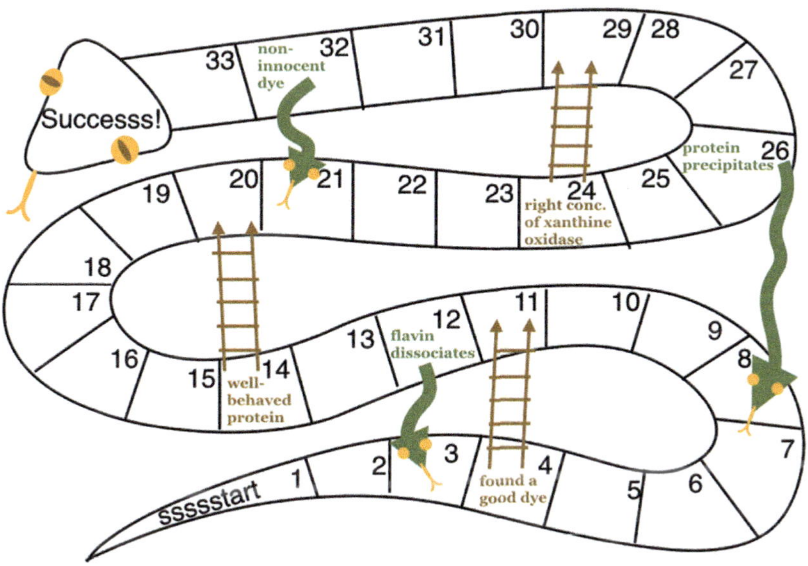

Fig. 10 Cartoon guide to obtaining excellent data quality, in geek game format.

approach to equilibrium. Single exponential kinetics are anticipated and the equilibrium point can often be predicted with good accuracy without waiting for it to be achieved (an asymptotic process that becomes increasingly slow) (Anderson, 1983).

6.2 Non-interference by dyes

Mediators and dyes that are intended to accelerate equilibration of the electrode with the rest of the system need to populate their SQ states or at least have thermal access to them in order to be effective (Dutton, 1978). They should ideally also have E°s within 30 mV of the target value of the experiment. However if they are additionally to serve as the means of measuring E, they must be present at a concentration sufficient that spectral changes corresponding to $\approx$2 mV can be measured reproducibly. Since spectra of the flavin are also desired (unless EPR will be used to measure flavin SQ concentrations), the best dyes have optical spectra that overlap minimally with the signatures of the flavin. This must be true for both oxidation states of the dye. A dye that interacts preferentially with one or other oxidation state of the enzyme can shift the enzyme's apparent E° via the binding equilibrium. In cases where a different dye cannot be found, measurements at a series of dye concentrations permit extrapolation of the measured $E^{\circ}_{apparent}([D])$ to $[D] = 0$ to obtain the value of E° corresponding to no binding of dye (for example, Koder & Miller, 1998). As a compromise between provision of enough dye to permit rapid equilibration, but not so much as to obscure the optical signals of the flavin, many authors employ a concentration that produces a similar maximum absorbance to that of the flavin (Mishanina et al., 2015), or comparable concentration (Pitsawong et al., 2010).

Obviously the dyes should be pure. In many cases this will require that they be recrystallized prior to use, as some are photochemically labile and a bottle may be decades old and still in use. Storage in the dark with an Ar head space or in Ar-filled secondary containment is recommended for some. Since some of the mediator/dyes are toxic and carcinogenic, self-education and due precautions are essential. The optical spectrum of the dye should not be altered by addition of the enzyme, as this could indicate reaction between the two, and the dyes should not succumb to reaction in the course of the titration, as confirmed by the optical spectrum of the small molecules separated from the enzyme, after the titration (above). The titration should be protected from light to minimize photochemically driven reactions of photosensitive dyes as well as the flavin.

6.3 Anaerobicity

Prepare all solutions, apparatus and reagents to be fully anaerobic before beginning the titration, and keep everything that way. We degas large volumes on a gas-train and then move them into an anaerobic chamber the night before use. To do this, bubble scrubbed Ar through large volumes of buffer while stirring. Recall that the solubility of O_2 is higher in cold aqueous solutions than in warm ones so allow refrigerated solutions to come to room temperature before degassing. For small volumes, five cycles of evacuation and refilling with scrubbed Ar while stirring is generally sufficient. We have sets of pipettors and syringes that reside permanently in the anaerobic chamber. Protein samples are brought into the chamber on the day of use and are kept in a chilling dry bath shielded from light, along with stock solutions of the mediators and short-lived reactants such as NADH or dithionite. A chilling dry bath is very helpful. We also have the luxury of a spectrophotometer inside the anaerobic chamber. This has improved the quality of our results and diminishes the stress involved in performance of the experiments.

6.4 Chemical titrants

Sodium dithionite is able to produce *E* values as low as −420 mV and is essentially colorless (Dutton, 1978), but has a signal at 315 nm which is a useful indication at the end of a reductive titration that the sample is no longer reacting with added reductant. However the products of dithionite oxidation can react with some flavoenzymes and are acidic so the stock solution should be well buffered and adequate controls performed. In addition, dithionite tends to be impure as purchased and the solutions have a limited shelf life. We prefer to make them up in degassed buffer immediately before use, keep them cold under inert atmosphere and use them within 6 h. A solution buffered at high pH (e.g., 12) will last longer. Some researchers prefer to use Ti citrate as the reductant. This too must be used promptly upon preparation. Best practice is to test the solution regularly by depositing a drop onto a piece of filter paper that has been wet with methyl viologen and allowed to dry. A solution retaining reducing activity will create a purple spot which will fade quickly in air, but persist in a good inert atmosphere. Methyl viologen is toxic and must be handled with due precaution, but it is helpful to prepare a small stash of filter papers dipped in methyl violagen, dried and then cut into 1 cm squares. These can be stored conveniently in small stoppered vials for use at gas trains or in glove boxes. They should be handled with tweezers or gloves and disposed of as toxic, but provide immediate diagnosis of solutions.

Ferricyanide can produce a solution E up to +400 mV, using only a few μL of 100 mM stock solution. This suffices to produce a spectrum of fully oxidized enzyme.

As with any other measurement, results should be repeatable for different preparations of the enzyme. Ideally it should be possible to confirm reversibility and independence from the rate at which the titration is performed (confirms absence of a kinetic impediment). The value of $E°$ should not depend on the identity of the mediator/dye.

At the end of a reductive titration, it is valuable to confirm the end-point via full reduction using dithionite or Ti citrate, and for oxidative titrations ferricyanide can be used (but obtain spectra of ferri- and ferrocyanide as the former contributes a yellow color of its own, Dutton, 1978).

7. Concluding remarks

Careful study of metabolic energetics was crucial in revealing the importance of bifurcation to life (Buckel & Thauer, 2013; Nitschke & Russell, 2012). Similarly, determination of the $E°$s of individual sites contributing to this activity can educate us as to the mechanisms used to execute bifurcation in enzymes. Since these provide means of optimizing energy conservation and versatility, we hope that increased understanding of enzymatic bifurcation can inspire more efficient man-made materials and devices, allowing us to lighten the load we place on our planet's ecosystems.

References

Alric, J., Pierre, Y., Picot, D., Lavergne, J., & Rappaport, F. (2005). Spectral and redox characterization of the heme ci of the cytochrome b6f complex. *Proceedings of the National Academy of Sciences of the United States of America*, *102*, 15860–15865.

Anderson, R. F. (1983). Energetics of the one-electron reduction steps of riboflavin, FMN and FAD to their fully reduced forms. *Biochimica et Biophysica Acta*, *722*, 158–162.

Arents, J. C., Perez, M. A., Hendriks, J., & Hellingwerf, K. J. (2011). On the midpoint potential of the FAD chromophore in a BLUF-domain containing photoreceptor protein. *FEBS Letters*, *585*, 167–172.

Brugna, M., Albouy, D., & Nitschke, W. (1998). Diversity of cytochrome bc complexes: Example of the Rieske protein in green sulfur bacteria. *Journal of Bacteriology*, *180*, 3719–3723.

Buckel, W., & Thauer, R. K. (2013). Energy conservation via electron bifurcating ferredoxin reduction and proton/Na+ translocating ferredoxin oxidation. *Biochimica et Biophysica Acta*, *1827*, 94–113.

Buckel, W., & Thauer, R. K. (2018). Flavin-based electron bifurcation, a new mechanism of biological energy coupling. *Chemical Reviews*, *118*, 3862–3886.

Byron, C. M., Stankovich, M. T., Husain, M., & Davidson, V. L. (1989). Unusual redox properties of electron-transfer flavoprotein from *Methylophilus methylotrophus*. *Biochemistry*, *28*, 8582–8587.

Chen, L., Liu, M. Y., Legall, J., Fareleira, P., Santos, H., & Xavier, A. V. (1993). Purification and characterization of an NADH-rubredoxin oxidoreductase involved in the utilization of oxygen by Desulfovibrio gigas. *European Journal of Biochemistry*, *216*, 443–448.

Cheng, V. W., Tran, Q. M., Boroumand, N., Rothery, R. A., Maklashina, E., Cecchini, G., et al. (2013). A conserved lysine residue controls iron-sulfur cluster redox chemistry in Escherichia coli fumarate reductase. *Biochimica et Biophysica Acta*, *1827*, 1141–1147.

Chowdhury, N. P., Kahnt, J., & Buckel, W. (2015). Reduction of ferredoxin or oxygen by flavin-based electron bifurcation in Megasphaera elsdenii. *The FEBS Journal*, *282*, 3149–3160.

Chowdhury, N. P., Mowafy, A. M., Demmer, J. K., Upadhyay, V., Koelzer, S., Jayamani, E., et al. (2014). Studies on the mechanism of electron bifurcation catalyzed by electron transferring flavoprotein (Etf) and butyryl-CoA dehydrogenase (Bcd) of *Acidaminococcus fermentans*. *The Journal of Biological Chemistry*, *289*, 5145–5157.

Clark, W. M. (1960). *Oxidation-reduction potentials of organic systems*. Baltimore MD: Williams and Wilkins.

Corrado, M. E., Aliverti, A., Zanetti, G., & Mayhew, S. G. (1996). Analysis of the oxidation-reduction potentials of recombinant ferredoxin-NADP+ reductase from spinach chloroplasts. *European Journal of Biochemistry*, *239*, 662–667.

Covian, R., Zwicker, K., Rotsaert, F. A., & Trumpower, B. L. (2007). Asymmetric and redox-specific binding of quinone and quinol at center N of the dimeric yeast cytochrome bc1 complex. Consequences for semiquinone stabilization. *The Journal of Biological Chemistry*, *282*, 24198–24208.

Demmer, J. K., Chowdhury, N. P., Selmer, T., Ermler, U., & Buckel, W. (2017). The semiquinone swing in the bifurcating electron transferring flavoprotein/butyryl-coA dehydrogenase complex from Clostridium difficile. *Nature Communications*, *8*.

Demmer, J. K., Huang, H., Wang, S., Demmer, U., Thauer, R. K., & Ermler, U. (2015). Insights into flavin-based electron bifurcation via the NADH-dependent reduced ferredoxin:NADP oxidoreductase structure. *The Journal of Biological Chemistry*, *290*, 21985–21995.

Duan, H. D., Lubner, C. E., Tokmina-Lukaszewska, M., Gauss, G. H., Bothner, B., King, P. W., et al. (2018). Distinct flavin properties underlie flavin-based electron bifurcation within a novel electron-transferring flavoprotein FixAB from *Rhodopseudomonas palustris*. *The Journal of Biological Chemistry*, *293*, 4688–4701.

Dutton, P. L. (1978). Redox potentiometry: Determination of midpoint potentials of oxidation-reduction components of biological electron-transfer systems. *Methods in Enzymology*, *54*, 411–435.

Efimov, I., Parkin, G., Millett, E. S., Glenday, J., Chan, C. K., Weedon, H., et al. (2014). A simple method for the determination of reduction potentials in heme proteins. *FEBS Letters*, *588*, 701–704.

Ehrenberg, A., Müller, F., & Hemmerich, P. (1967). Basicity, visible spectra an delectron spin resonance of flavosemiquinone anions. *European Journal of Biochemistry*, *2*, 282–293.

Garcia Costas, A. M., Poudel, S., Miller, A.-F., Schut, J. G., Ledbetter, R. N., Fixen, K., et al. (2017). Defining electron bifurcation in the electron transferring flavoprotein family. *Journal of Bacteriology*, *199*, e00440-17.

Guerrero, F., Sedoud, A., Kirilovsky, D., Rutherford, A. W., Ortega, J. M., & Roncel, M. (2011). A high redox potential form of cytochrome c550 in photosystem II from Thermosynechococcus elongatus. *The Journal of Biological Chemistry*, *286*, 5985–5994.

Haid, E., Lehmann, P., & Ziegenhorn, J. (1975). Molar absorptivities of beta-NADH and beta-NAD at 260 nm. *Clinical Chemistry*, *21*, 884–887.

Herrmann, G., Jayamani, E., Mai, G., & Buckel, W. (2008). Energy conservation via electron-transferring flavoprotein in anaerobic bacteria. *Journal of Bacteriology*, *190*, 784–791.

Hirst, J. (2010). Towards the molecular mechanism of respiratory complex I. *The Biochemical Journal*, *425*, 327–339.

Hu, J., Chuenchor, W., & Rokita, S. E. (2015). A switch between one- and two-electron chemistry of the human flavoprotein iodotyrosine deiodinase is controlled by substrate. *The Journal of Biological Chemistry*, *290*, 590–600.

Jang, M. H., Scrutton, N. S., & Hille, R. (2000). Formation of W(3)A(1) electron-transferring flavoprotein (ETF) hydroquinone in the trimethylamine dehydrogenase x ETF protein complex. *The Journal of Biological Chemistry*, *275*, 12546–12552.

Jung, Y. S., Vassiliev, I. R., Qiao, F., Yang, F., Bryant, D. A., & Golbeck, J. H. (1996). Modified ligands to FA and FB in photosystem I. Proposed chemical rescue of a [4Fe-4S] cluster with an external thiolate in alanine, glycine, and serine mutants of PsaC. *The Journal of Biological Chemistry*, *271*, 31135–31144.

Kay, C. J., Solomonson, L. P., & Barber, M. J. (1986). Thermodynamic properties of the heme prosthetic group in assimilatory nitrate reductase. *The Journal of Biological Chemistry*, *261*, 5799–5802.

Koder, R. L., Haynes, C. A., Rodgers, M. E., Rodgers, D. W., & Miller, A.-F. (2002). Flavin thermodynamics explain the oxygen insensitivity of enteric nitroreductases. *Biochemistry*, *41*, 14197–14205.

Koder, R. L., Jr., & Miller, A.-F. (1998). Steady state kinetic mechanism, stereospecificity, substrate and inhibitor specificity of Enterobacter cloacae nitroreductase. *Biochimica et Biophysica Acta*, *1387*, 394–405.

Kolaj-Robin, O., O'Kane, S. R., Nitschke, W., Leger, C., Baymann, F., & Soulimane, T. (2011). Biochemical and biophysical characterization of succinate: Quinone reductase from Thermus thermophilus. *Biochimica et Biophysica Acta*, *1807*, 68–79.

Ledbetter, R. N., Garcia Costas, A. M., Lubner, C. E., Mulder, D. E., Tokmina-Lukaszewska, M., Artz, J. H., et al. (2017). The electron bifurcating FixABCX protein complex from Azotobacter vinelandii: Generation of low-potential reducing equivalents for nitrogenase catalysis. *Biochemistry*, *56*, 4177–4190.

Lubner, C. E., Jennings, D. P., Mulder, D. W., Schut, G. J., Zadvornyy, O. A., Hoben, J. P., et al. (2017). Mechanistic insights into energy conservation by flavin-based electron bifurcation. *Nature Chemical Biology*, *13*, 655–659.

Massey, V. (1991). A simple method for determination of redox potentials. In B. Curti, S. Ronchi, & G. Zanetti (Eds.), *Flavins and flavoproteins* (pp. 59–66). Walter de Gruyer: Berlin.

Matsui, T., Kitagawa, Y., Okumura, M., & Shigeta, Y. (2015). Accurate standard hydrogen electrode potential and applications to the redox potentials of vitamin C and NAD/NADH. *The Journal of Physical Chemistry. A*, *119*, 369–376.

Mayhew, S. G. (1978). The redox potential of dithionite and SO_2^- from equilibrium reactions with flavodoxins, methyl viologen and hydrogen plus hydrogenase. *European Journal of Biochemistry*, *85*.

Mayhew, S. G. (1999a). The effects of pH and semiquinone formation on the oxidation-reduction potentials of flavin mononucleotide: A reappraisal. *European Journal of Biochemistry*, *265*, 698–702.

Mayhew, S. G. (1999b). Potentiometric measurements of oxidation-reduction potentials. In S. K. Chapman & G. A. Reid (Eds.), *Flavoprotein protocols* (pp. 49–59). Humana Press: Totowa, NJ.

Mayhew, S. G., Foust, G. P., & Massey, V. (1969). Oxidation-reduction properties of flavodoxin from *Peptostreptococcus elsdenii*. *The Journal of Biological Chemistry*, *244*, 803–810.

Millner, P. A., Widger, W. R., Abbott, M. S., Cramer, W. A., & Dilley, R. A. (1982). The effect of adenine nucleotides on inhibition of the thylakoid protein kinase by sulfhydryl-directed reagents. *The Journal of Biological Chemistry*, *257*, 1736–1742.

Mishanina, T. V., Yadav, P. K., Ballou, D. P., & Banerjee, R. (2015). Transient kinetic analysis of hydrogen sulfide oxidation catalyzed by human sulfide Quinone oxidoreductase. *The Journal of Biological Chemistry*, *290*, 25072–25080.

Nakajima, H., Honma, Y., Tawara, T., Kato, T., Park, S. Y., Miyatake, H., et al. (2001). Redox properties and coordination structure of the heme in the co-sensing transcriptional activator CooA. *The Journal of Biological Chemistry*, *276*, 7055–7061.

Nedbal, L., Samson, G., & Whitmarsh, J. (1992). Redox state of a one-electron component controls the rate of photoinhibition of photosystem II. *Proceedings of the National Academy of Sciences of the United States of America*, *89*, 7929–7933.

Nitschke, W., & Russell, M. J. (2012). Redox bifurcations: Mechanisms and importance to life now, and at its origin. *BioEssays*, *34*, 106–109.

Norden, B., & Tjerneld, F. (1982). Structure of methylene blue-DNA complexes studied by linear and circular dichroism spectroscopy. *Biopolymers*, *21*, 1713–1734.

O'Farrel, P. A., Walsh, M. A., McCarthy, A. A., Higgins, T. M., Voordouw, G., & Mayhew, S. G. (1998). Modulation of the redox potentials of FMN in Desulfovibrio vulgaris flavodoxin: Thermodynamic properties and crystal structures of glycine-61 mutants. *Biochemistry*, *37*, 8405–8416.

Ottaway, J. M. (1972). Oxidation-reduction indicators of E'2 < 0.76 volt. In E. Bishop (Ed.), *Indicators* (pp. 469–528). New York: Pergamon.

Pitsawong, W., Jeerus, S., Methinee, P., Tan, T.-C., Spadiut, O., Haltrich, D., et al. (2010). A conserved active-site threonine is important for both sugar and flavin oxidations of pyranose 2-oxidase. *The Journal of Biological Chemistry*, *285*, 9697–9705.

Pueyo, J. J., Gomez-Moreno, C., & Mayhew, S. G. (1991). Oxidation-reduction potentials of ferredoxin-NADP+ reductase and flavodoxin from Anabaena PCC 7119 and their electrostatic and covalent complexes. *European Journal of Biochemistry*, *202*, 1065–1071.

Sato, K., Nishina, Y., & Shiga, K. (1993). Electron-transferring flavoprotein has an AMP-binding site in addition to the FAD-binding site. *Journal of Biochemistry*, *114*, 215–222.

Sato, K., Nishina, Y., & Shiga, K. (2003). Purification of electron-transferring flavoprotein from Megasphaera elsdenii and binding of additional FAD with an unusual absorption spectrum. *Journal of Biochemistry*, *134*, 719–729.

Sato, K., Nishina, Y., & Shiga, K. (2013). Interaction between NADH and electron-transferring flavoprotein from *Megasphaera elsdenii*. *Journal of Biochemistry*, *153*, 565–572.

Sevrioukova, I., Shaffer, C., Ballou, D. P., & Peterson, J. A. (1996). Equilibrium and transient state spectrophotometric studies of the mechanism of reduction of the flavoprotein domain of P450BM-3. *The Biochemist*, *35*, 7058–7068.

Stankovich, M. T., Schopfer, L. M., & Massey, V. (1978). Determination of glucose oxidase oxidation-reduction potentials and the oxygen reactivity of fully reduced and semiquinoid forms. *The Journal of Biological Chemistry*, *253*, 4971–4979.

Swenson, R. P., & Krey, G. D. (1994). Site-directed mutagenesis of Tyrosine-98 in the flavodoxin from *Desulfovibrio vulgaris* (Hildenborough): Regulation of the oxidation-reduction properties of the bound FMN cofactor by aromatic, solvent and electrostatic interactions. *Biochemistry*, *33*, 8505–8514.

Talfournier, F., Munro, A. W., Basran, J., Sutcliffe, M. J., Daff, S., Chapman, S. K., et al. (2001). Alpha Arg-237 in Methylophilus methylotrophus (sp. W3A1) electron-transferring flavoprotein affords approximately 200-millivolt stabilization of the FAD anionic semiquinone and a kinetic block on full reduction to the dihydroquinone. *The Journal of Biological Chemistry*, *276*, 20190–20196.

Trimmer, E. E., Ballou, D. P., Galloway, L. J., Scannell, S. A., Brinker, D. R., & Casas, K. R. (2005). Aspartate 120 of Escherichia coli methylenetetrahydrofolate reductase: Evidence for major roles in folate binding and catalysis and a minor role in flavin reactivity. *The Biochemist, 44*, 6809–6822.

Tyree, B., & Webster, D. A. (1978). Electron-accepting properties of cytochrome o purified from Vitreoscilla. *The Journal of Biological Chemistry, 253*, 7635–7637.

Ugulava, N. B., & Crofts, A. R. (1998). CD-monitored redox titration of the Rieske Fe-S protein of Rhodobacter sphaeroides: pH dependence of the midpoint potential in isolated bc1 complex and in membranes. *FEBS Letters, 440*, 409–413.

Vollmer, S. J., Switzer, R. L., & Debrunner, P. G. (1983). Oxidation-reduction properties of the iron-sulfur cluster in Bacillus subtilis glutamine phosphoribosylpyrophosphate amidotransferase. *The Journal of Biological Chemistry, 258*, 14284–14293.

Watmough, N. J., & Frerman, F. E. (2010). The electron transfer flavoprotein: Ubiquinone oxidoreductases. *Biochimica et Biophysica Acta, 1797*, 1910–1916.

Winkler, A., Motz, K., Riedl, S., Puhl, M., Macheroux, P., & Gruber, K. (2009). Structural and mechanistic studies reveal the functional role of bicovalent flavinylation in berberine bridge enzyme. *The Journal of Biological Chemistry, 284*, 19993–20001.

Zhang, H., Carrell, C. J., Huang, D., Sled, V., Ohnishi, T., Smith, J. L., et al. (1996). Characterization and crystallization of the lumen side domain of the chloroplast Rieske iron-sulfur protein. *The Journal of Biological Chemistry, 271*, 31360–31366.

Zhang, P., Yuly, J. L., Lubner, C. E., Mulder, D. W., King, P. W., Peters, J. W., et al. (2017). Electron bifurcation: Thermodynamics and kinetics of two-electron brokering in biological redox chemistry. *Accounts of Chemical Research, 50*, 2410–2417.

Zhu, Z., & Davidson, V. L. (1998). Redox properties of tryptophan tryptophylquinone enzymes. Correlation with structure and reactivity. *The Journal of Biological Chemistry, 273*, 14254–14260.

Zhu, J., Egawa, T., Yeh, S. R., Yu, L., & Yu, C. A. (2007). Simultaneous reduction of iron-sulfur protein and cytochrome b(L) during ubiquinol oxidation in cytochrome bc(1) complex. *Proceedings of the National Academy of Sciences of the United States of America, 104*, 4864–4869.

CHAPTER FIFTEEN

Investigations of two-component flavin-dependent monooxygenase systems

John M. Robbins[a], Holly R. Ellis[b,*]
[a]School of Chemical & Biomolecular Engineering, Georgia Institute of Technology, Atlanta, GA, United States
[b]Department of Chemistry and Biochemistry, Auburn University, Auburn, AL, United States
*Corresponding author: e-mail address: ellishr@auburn.edu

Contents

Abstract

Bacterial two-component flavin-dependent monooxygenase systems catalyze the oxidation of diverse metabolic reactions. There are several shared mechanistic features in the two-component monooxygenase systems that differ from canonical monooxygenase enzymes. The flavin reductases catalyze the reductive half-reaction, and

Methods in Enzymology, Volume 620
ISSN 0076-6879
https://doi.org/10.1016/bs.mie.2019.03.018

the reduced flavin is transferred to the monooxygenase enzyme. The oxidative half-reaction catalyzed by the monooxygenase enzyme has been proposed to occur through the formation of a (hydro)peroxyflavin intermediate. In some two-component flavin-dependent systems the mechanism of flavin transfer involves protein-protein interactions between the flavin reductase and monooxygenase enzyme. Methods are presented that provide an alternative approach from flavin-bound monooxygenases to evaluate the kinetic properties and flavin transfer mechanism of the two-component flavin-dependent monooxygenase systems.

Abbreviations

EmoB/EmoA EDTA monooxygenase
FAD oxidized flavin adenine dinucleotide
FMN oxidized flavin mononucleotide
$FMNH_2$ reduced flavin mononucleotide
HDX-MS hydrogen/deuterium exchange mass spectrometry
MsuE/MsuD methanesulfonate monooxygenase
NAD(P)H reduced nicotinamide adenine dinucleotide (phosphate)
SfnF/SfnG dimethylsulfone monooxygenase
SsuE/SsuD alkanesulfonate monooxygenase
TIM triose phosphate isomerase

1. Introduction

Flavoprotein monooxygenases catalyze the cleavage of the oxygen-oxygen bond of dioxygen, inserting one oxygen atom into a substrate and reducing the other oxygen to water (Ghisla & Massey, 1989; Massey, 1994; Palfey & McDonald, 2010; van Berkel, Kamerbeek, & Fraaije, 2006). The catalytic mechanism of many flavoprotein monooxygenase enzymes involves the activation of oxygen to form a C4a-(hydro) peroxyflavin intermediate (Massey, 1994; Palfey & McDonald, 2010). While flavins are generally prosthetic groups that are covalently or tightly bound to the enzyme, a group of flavin-dependent enzymes has been identified that utilize the flavin as a substrate in the reaction. These two-component flavin-dependent systems are composed of a flavin reductase and a monooxygenase that together catalyze the oxidation of various substrates (Fig. 1). The role of the flavin reductase is to supply reduced flavin to the monooxygenase, while the monooxygenase utilizes the reduced flavin to activate molecular oxygen. Unlike flavoproteins with a tightly or covalently bound prosthetic group, the reductive and oxidative half-reactions occur on two separate enzymes.

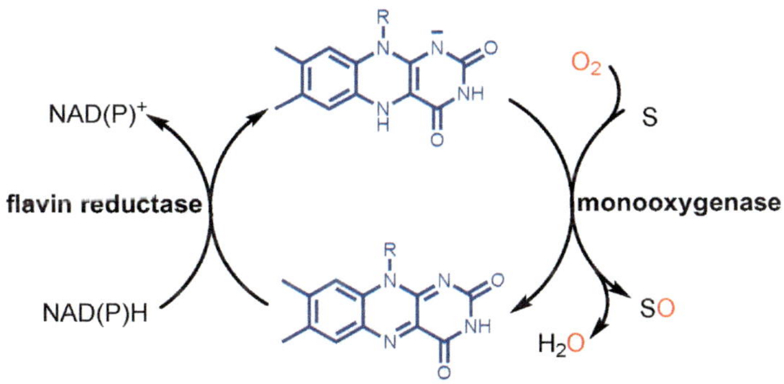

Fig. 1 General mechanism for two-component systems. The flavin is reduced by the flavin reductase with reducing equivalents provided by NAD(P)H. The reduced flavin is transferred to the monooxygenase enzyme, which activates dioxygen for substrate oxidation. Oxidized flavin is released from the monooxygenase enzyme and can bind to the FMN reductase to continue the catalytic cycle. In most two-component systems, the flavin reductase and monooxygenase show a specificity for a particular redox form of the flavin.

Two-component flavin-dependent systems have only been identified in bacteria, and are involved in diverse reactions that provide the organism with alternative metabolic pathways for survival (Ellis, 2010). Many of the flavin reductases are expressed on the same operon as their monooxygenase partner to ensure that both enzymes are available to maintain the metabolic needs of the organism. Because the enzymes are purified in the absence of flavin, the experimental design to evaluate the mechanism of two-component flavin-dependent systems differs from canonical flavoproteins. Kinetic assays can monitor each enzyme independently, or in a coupled assay that includes both enzymes. Typically, the flavin reductase has a higher affinity for the oxidized flavin, while the monooxygenase has a higher affinity for reduced flavin. The affinity of each enzyme for a specific redox form of the flavin ensures that the reduced flavin is transferred and rapidly bound by the monooxygenase enzyme. The mechanism of reduced flavin transfer between the flavin reductase and monooxygenase can occur through a channeling or diffusion mechanism, and is dependent on the system under investigation. There are several two-component flavin-dependent systems with a specificity for FMN that are grouped into the FMN:NAD(P)H reductase family based on a similar flavodoxin fold (Fig. 2A). Interestingly, the two-component FMN reductases in the FMN:NAD(P)H family structurally diverge from their flavin-bound counterparts (Driggers et al., 2014). Enzymes within the subgroup possess a

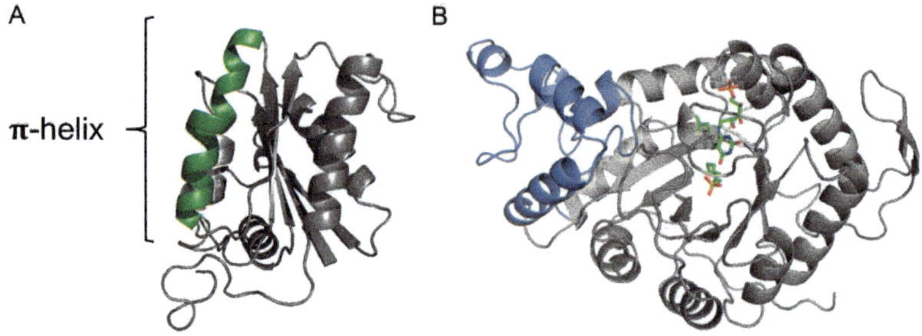

Fig. 2 Structures of the two-component FMN-dependent monooxygenase enzymes. (A) Monomeric structure of SsuE enzyme from *E. coli*. FMN reductases that belong to the NAD(P)H:FMN family have a flavodoxin type fold, and a π-helix structure (highlighted in green) that participates in promoting flavin release (PDB: 4PTY). (B) Structure of SsuD from *E. coli*. Enzymes belonging to the luciferase family contain an insertion region (highlighted in blue) that closes over the active site following the binding of $FMNH_2$ and the oxidized substrate. The figure was rendered with Pymol using PDB ID 1M41. The unstructured loop region was modeled with PDB ID 1NQK and the structure was superpositioned with the bound flavin of LadA (PDB ID: 3RAO) (Driggers, Dayal, Ellis, & Karplus, 2014; Eichhorn, Davey, Sargent, Leisinger, & Richmond, 2002; Li et al., 2008).

π-helix located at the tetramer interface that has been proposed to provide the unique function of flavin transfer (Driggers et al., 2014; Musila & Ellis, 2016; Musila, Forbes, & Ellis, 2018 Nissen et al., 2008). The monooxygenase partners belong to the luciferase family, and are characterized by a classic TIM-barrel fold (Fig. 2B) (Eichhorn et al., 2002; Jun, Lewis, Youn, Xun, & Kang, 2016; Li et al., 2008). The two-component FMN-dependent systems with a defined FMN:NAD(P)H reductase and monooxygenase characterized by a TIM-barrel fold are involved in sulfur acquisition and EDTA degradation (Fig. 3). Evidence for a channeling mechanism involving protein-protein interactions in two-component FMN-dependent systems has been supported through kinetic studies to evaluate flavin transfer and the identification of protein-protein interaction sites (Abdurachim & Ellis, 2006; Dayal, Singh, Busenlehner, & Ellis, 2015; Jun et al., 2016; Musila et al., 2018). Therefore, these two-component FMN-dependent systems invoke similar mechanisms for effective flavin transfer while still retaining their defined substrate specificity. This chapter will focus on methods to evaluate the kinetic properties and substrate specificity of two-component flavin-dependent systems, and evaluate the formation of protein-protein interactions between the flavin reductase and monooxygenase partner. The studies described have been optimized for

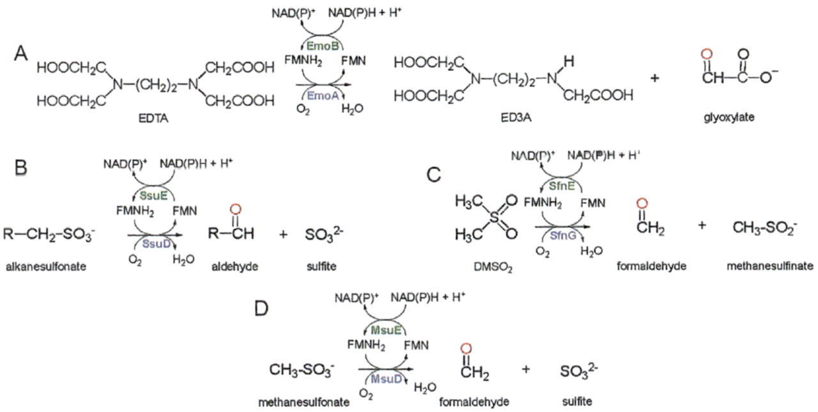

Fig. 3 Reactions catalyzed by the two-component FMN monooxygenase enzymes. Certain two-component FMN reductases have similar structural properties but catalyze distinct reactions. (A) EDTA monooxygenase (EmoB/EmoA). (B) Alkanesulfonate monooxygenase (SsuE/SsuD). (C) Dimethylsulfone monooxygenase (SfnF/SfnG). (D) Methanesulfonate monooxygenase (MsuE/MsuD).

the FMN-dependent alkanesulfonate monooxygenase system, but the experimental techniques can be applied to most two-component systems that utilize flavin as a substrate.

2. Expression and purification of the alkanesulfonate monooxygenase enzymes

Protocols for the expression and purification of two-component FMN-dependent monooxygenase enzymes typically describe the individual isolation of the enzymes (Bohuslavek, Payne, Liu, Bolton, & Xun, 2001; Endoh et al., 2003; Gao & Ellis, 2005; Kantz, Chin, Nallamothu, Nguyen, & Gassner, 2005; Li et al., 2008; Wicht, 2016). The native enzymes are purified to >90% homogeneity using a combination of chromatography techniques. Affinity chromatography using a 6 × His tag was not initially used for purification of the alkanesulfonate monooxygenase enzymes to eliminate potential complications when evaluating protein-protein interactions. Because the flavin is a substrate for each enzyme, the native enzymes often purify without the bound flavin cofactor. Therefore, standard protocols to quantify and initially characterize canonical flavoproteins are not a viable option for most two-component flavin-dependent systems.

2.1 Gene-cloning of the alkanesulfonate monooxygenase enzymes

Expression of two-component FMN-dependent systems utilize pET vectors that are dependent on gene induction from a T7 promoter (Bohuslavek et al., 2001; Endoh et al., 2005; Gao & Ellis, 2005; Wicht, 2016). The alkanesulfonate monooxygenase genes were cloned from *Escherichia coli* K12 chromosomal DNA, and the genes were PCR-amplified with primers containing engineered *Nde*I and *Xho*I cut sites for cloning into the pET21A expression vector (Gao & Ellis, 2005). Representative pET21a constructs containing the *ssuE* and *ssuD* genes were transformed into XL-1 to house the constructs, or transformed into Bl21(DE3) for gene expression. The constructs in each cell line are stored as 40% glycerol stocks at −80 °C. Expression vectors will need to be retransformed into Bl21(DE3) cells if decreased enzyme expression is observed, but are typically stable for 2 years.

2.2 Expression and purification of the alkanesulfonate monooxygenase enzymes

Reported expression and purification of two-component FMN-dependent systems are varied and are based on procedures utilized by different groups (Bohuslavek et al., 2001; Chen & Baldwin, 1989; Gao & Ellis, 2005; Kantz et al., 2005; Li et al., 2008). Protocols for the expression and purification of the alkanesulfonate monooxygenase have been previously reported (Gao & Ellis, 2005). Details noted here highlight subtleties observed during the growth and purification of each enzyme. Cultures are grown at 37 °C until an A_{600} between 0.4 and 0.6 is reached. Following the addition of isopropyl-β-D-thiogalactoside, the temperature is lowered to 18 °C to prevent the formation of inclusion bodies. An optimal expression temperature can be determined prior to initiating large-scale expression using small-scale expression analysis (100 mL). Standard buffers used for purification consist of 25 mM phosphate, pH 7.5, 10% glycerol, with different salts added when needed. The proteins are stored at −80 °C in 25 mM potassium phosphate, pH 7.5, 100 mM sodium chloride, and 10% glycerol. Initial studies used potassium phosphate buffer for kinetic studies, but it was determined that phosphate can inhibit the reactions of each enzyme. Good buffers containing a sulfonate group should also be avoided for two-component monooxygenases involved in desulfonation because the buffers are substrates for the monooxygenases. A working concentration of ~60 μM of SsuE or SsuD is typically obtained from 4 to 6 L cultures. The SsuD monooxygenase is usually stable at −80 °C for 6 months, while the SsuE reductase demonstrates diminished activity after 3 months.

3. Steady-state kinetic analyses of the FMN-dependent monooxygenase enzymes

The steady-state kinetic analyses of the flavin reductases and monooxygenases can be studied independently or as a complex monitoring the monooxygenase reaction. There is some inherent difficulty in evaluating only the monooxygenase reaction because it requires reduced flavin for activity, but steady-state kinetic studies have been performed to determine the kinetics parameters of the two-component monooxygenase enzymes without the flavin reductase (Chen & Baldwin, 1989; Hastings & Presswood, 1978). Initial steady-state kinetic studies should focus on the flavin reductase enzyme to determine the flavin and pyridine nucleotide preference based on the kinetic parameters, as these preferences are also important in coupled assays that include the monooxygenase. Some flavin-dependent reductases show a preference for one or both flavin (FMN/FAD) or pyridine nucleotide (NADH/NADPH) substrates, so both flavin and pyridine nucleotide substrates will need to be kinetically evaluated (Ellis, 2010).

3.1 Kinetic analyses of the flavin reductases

The two-component flavin reductase assays are performed with 0.001–0.5 μM enzyme (Bohuslavek et al., 2001; Gao & Ellis, 2005; Payne, Bolton, Campbell, & Xun, 1998; Wicht, 2016). Initial flavin reductase assays vary the concentrations of the pyridine nucleotide or flavin while keeping the other substrate at a fixed concentration. At least five different concentrations of the pyridine nucleotide or flavin are selected to provide a complete data set, and the substrate that is not varied is kept at saturating levels. Each set will need to be performed in triplicate for accurate statistical analyses when fitting the data. Different buffers and pH ranges (pH 7.0–7.5) have been used to evaluate the activities of the two-component flavin reductases. However, phosphate buffer used for the purification protocol is not used in the assay because it can affect flavin binding. Pyridine nucleotide and flavin stock solutions are standardized prior to beginning the reaction using the molar extinction coefficient for each substrate. The concentration of the enzyme stock is determined from the calculated molar extinction coefficient. Standardizing the substrates and enzymes should be performed prior to starting any of the kinetic experiments. The steady-state kinetic parameters of SsuE (0.04 μM) for NADH or NADPH are measured with 1.0 μM FMN, and the NADH or NADPH concentrations varied from 10 to 200 μM in 25 mM Tris-HCl, pH 7.5, and 100 mM NaCl at 25 °C. It is important that 10 mM Tris-HCl, pH 8.5, is used to make the pyridine nucleotide stock solution to prevent decomposition. The flavin reductases in the two-component systems involved in sulfur metabolism show an absolute preference for FMN, with no activity observed with FAD. Steady-state kinetic parameters of SsuE (0.04 μM) for FMN are obtained with 200 μM NADPH or NADH, and varying concentrations of FMN (0.04–1 μM) in 25 mM Tris-HCl, pH 7.5, and 100 mM NaCl at 25 °C. It is often necessary to modify the substrate concentrations based on the preliminary findings. All reaction components except SsuE are added to the cuvette and mixed, and the initial rates for pyridine nucleotide oxidation monitored at 340 nm. The SsuE enzyme is then added with a cuvette-mixer once the 340 nm absorbance has stabilized. The background rate due to autooxidation of NAD(P)H can be taken into account when determining the initial rate. The initial rate is obtained from the linear portion of the plot in the first 10–20 s. The $v_o/[E_t]$ for each substrate concentration is obtained by dividing the initial rate of change of absorbance by the molar extinction coefficient 6220 M^{-1} cm^{-1} and the concentration of enzyme used. A higher k_{cat}/K_m value for a specific pyridine nucleotide or flavin suggests a preference for that particular substrate. The substrate preference can also be evaluated through coupled reactions that include both the reductase and monooxygenase enzymes. The flavin reductases are not as stable as the monooxygenases enzymes and show diminished activity if refrozen for future assays. Freezing aliquots of appropriate amounts of the enzyme after purification will eliminate protein waste.

3.2 Kinetic analyses of the monooxygenase enzymes

The monooxygenase reactions are usually studied in a coupled assay with the flavin reductase supplying reduced flavin. Endpoint assays are often used to measure product formation in two-component monooxygenase reactions. The reactions are quenched after incubating the enzymes for a specified time and the products analyzed through various approaches. The assay described for the alkanesulfonate monooxygenase system uses a colorimetric assay, but each monooxygenase assay will need to be modified for the reaction under investigation (Zhan, Carpenter, & Ellis, 2008). Other groups working with two-component FMN-dependent enzymes have utilized HPLC, mass spectrometry, NMR, and UV–visible spectroscopy to identify monooxygenase products (Bohuslavek et al., 2001; Endoh et al., 2005; Tu & Hastings, 1975; Wicht, 2016; Xin, Xi, & Tu, 1991). It is important that flavin reduction not be rate-limiting. To optimize the ratio of flavin reductase to monooxygenase before starting kinetic analyses, assays are performed with various ratios of flavin reductase to monooxygenase (Eichhorn, van der Ploeg, & Leisinger, 1999; Payne et al., 1998). Initial assays for the alkanesulfonate monooxygenase system utilized saturating substrate concentrations with the SsuE and SsuD ratio ranging from 0.5:1 to 4:1 (Eichhorn et al., 1999). Based on these studies a 3:1 ratio of FMN reductase to monooxygenase is used in studies to evaluate the monooxygenase reaction.

Steady-state kinetic assays for SsuE and SsuD are performed by measuring the reaction of the sulfite product with DTNB. A stock solution of DTNB (~20 mM) is prepared in DMSO, and either covered or placed in a dark glass container due to the light-sensitivity of DTNB. The DTNB stock solution is made fresh weekly to avoid a decrease in sensitivity. A working solution of DTNB (1 mM) for the assays can be prepared by diluting the stock solution into the assay reaction buffer. Reactions are initiated by the addition of NADPH (500 μM) into a reaction mixture containing SsuD (0.2 μM), SsuE (0.6 μM), FMN (1 μM), and a range of octanesulfonate concentrations (5–1000 μM) in 25 mM Tris-HCl, pH 7.5, and 100 mM NaCl at 25 °C. After a 3 min incubation, the reaction is stopped by the addition of urea (2.0 M), and 50 μL of the DTNB solution (1 mM) is added to a 200 μL aliquot of the reaction solution. The colorimetric reaction is developed at room temperature for 2 min, and the absorbance measured at 412 nm. The $v_o/[E_t]$ is determined using the molar extinction coefficient of the TNB anion of 14.1 mM^{-1} cm^{-1}. The reactions are performed as a set of varied octanesulfonate concentrations, and each reaction set is performed in triplicate.

4. Measuring the substrate affinity of the alkanesulfonate monooxygenase enzymes

The K_d value can be determined for oxidized and reduced flavin based on the substrate preference obtained in the steady-state kinetic assays. Determining the K_d value for different redox forms of the flavin can provide information regarding the propensity for flavin transfer. The SsuE enzyme has a 1000-fold higher affinity for oxidized flavin than SsuD, and a 50-fold lower affinity for reduced flavin to ensure that the flavin is not transferred to SsuD until it is reduced (Zhan et al., 2008). Fluorescence spectroscopy has been used successfully with several flavin reductases and monooxygenases to determine the substrate affinity, but are dependent on the presence of an aromatic amino acids within the active site to observe fluorescent quenching. For most two-component FMN reductases the enzyme does not contain a bound flavin, so there is no need to remove the flavin prior to initiating the titrations. Alternative methods have also been used to monitor flavin affinity in the two-component FMN-dependent monooxygenase systems (Baldwin, 1974; Jun et al., 2016; Niviere, Fieschi, Decout, & Fontecave, 1996; Valton, Filisetti, Fontecave, & Niviere, 2004).

4.1 Affinity of the alkanesulfonate monooxygenase enzymes for oxidized flavin

The affinity of the flavin reductase and monooxygenase for oxidized flavin is often determined by measuring the decrease in the intrinsic fluorescence with the addition of oxidized flavin (Baldwin, Nicoli, Becvar, & Hastings, 1975; Bohuslavek et al., 2001; Gao & Ellis, 2005; Liu, Lei, Ding, Lee, & Tu, 1997; Tang, Jeffers, Nichols, & Tu, 2001). Flavin binding to SsuE is monitored on a fluorescence spectrophotometer with a fluorescent cuvette using an excitation wavelength of 280 nm and emission measurements at 344 nm. A 1.0 mL solution of flavin-free SsuE (0.1 μM) in 25 mM potassium phosphate, pH 7.5, and 0.1 M NaCl is titrated with FMN (0.022–0.44 μM), and the fluorescence spectrum recorded following a 2 min incubation after each addition of FMN in order for the binding to reach equilibrium. For the titration of SsuD with FMN, a 1 mL solution of SsuD (0.5 μM) in 25 mM potassium phosphate, pH 7.5, and 0.1 M NaCl is titrated with FMN (0.16–1.99 μM) and the fluorescence spectra recorded following a 2 min incubation after each addition. The incubation times may need to be adjusted based on the time it takes the bound substrate to reach equilibrium. An important consideration in planning the titration is that the total amount of titrant added should be only 1–2% of the total volume to avoid significant dilution effects. Titrations are performed in triplicate, and the K_d value is obtained by determining the amount of bound FMN. Bound FMN is determined with Eq. (1) (Gao & Ellis, 2005):

$$[\text{flavin}]_{bound} = [\text{Enzyme}]\left[\frac{I_0 - I_c}{I_0 - I_f}\right] \tag{1}$$

where $[\text{flavin}]_{bound}$ represents the concentration of FMN bound to the enzyme, [Enzyme] represents the initial concentration of enzyme, I_0 is the initial fluorescence intensity of enzyme prior to the addition of substrate, I_c is the fluorescence intensity of enzyme following each addition, and I_f is the final fluorescence intensity. The concentration of FMN bound is then plotted against the total substrate to obtain the dissociation constant (K_d) according to Eq. (2) (Gao & Ellis, 2005):

$$y = \frac{(K_d + x + n) - \sqrt{(K_d + x + n)^2 - 4xn}}{2} \tag{2}$$

where y and x represent the concentration of bound and total substrate, respectively, following each addition, and n represents the total enzyme concentration. The affinity of the FMN reductases for oxidized FAD can also be determined using the same experimental protocol. The range of flavin concentrations may need to be adjusted to ensure that the binding of flavin is saturating under the experimental conditions used.

4.2 Affinity of the alkanesulfonate monooxygenase enzymes for reduced flavin

Reduced flavin binding can also be measured by spectrofluorimetric titration; however, the titrations must be performed anaerobically to avoid oxidation of the reduced flavin. The protocol described is specific for $FMNH_2$, but similar methods are used to evaluate binding of $FADH_2$. SsuE or SsuD (0.5 μM) in 25 mM potassium phosphate, pH 7.5, is prepared in a glass titration cuvette by at least 15 cycles of evacuation followed by equilibration with ultrahigh purity argon gas. FMN (100 μM prepared in 25 mM potassium phosphate, pH 7.5) is bubbled with ultrahigh purity argon gas for 15 min. Both protein and flavin solution are then incubated in an anaerobic glovebox for 20 min after adding glucose (20 mM) and glucose oxidase (10 U) to remove trace amounts of dioxygen. The anaerobic FMN solution is photoreduced by two consecutive 15 min irradiations in the presence of EDTA (10 mM). The reduced flavin is added to an air-tight titration syringe, and the anaerobic cuvette is assembled in the anaerobic glove-box. For the titration of SsuD with reduced flavin, a 1 mL solution of SsuD (0.5 μM) in standard buffer is titrated with reduced flavin (0.16–7.8 μM), and the fluorescence spectra recorded following a 2 min incubation after each addition. The titration of SsuE (0.5 μM) with reduced flavin was performed as described for SsuD; however the reduced flavin concentration used for the titration ranged from 0.5 to 97 μM because of the lower affinity of two-component FMN reductases for reduced flavin. Bound $FMNH_2$ is determined using Eq. (1) for the titration of SsuE and SsuD with oxidized flavin (Section 4.1). The concentration of $FMNH_2$ bound is plotted against the free substrate to obtain the dissociation constant (K_d) according to Eq. (2) (Section 4.1).

4.3 Alkanesulfonate monooxygenase affinity for the octanesulfonate

The affinity of the octanesulfonate for the monooxygenase enzyme can also be monitored through fluorescent titrations. These experiments will provide the K_d for the substrate, but can also provide valuable information regarding the substrate binding order. Octanesulfonate binding to SsuD or the SsuD-$FMNH_2$ complex is performed by spectrofluorimetric methods similar to those used in the reduced flavin binding experiments. A 1 mL solution of SsuD (1 μM), either alone or with $FMNH_2$ (2 μM), is made anaerobic in a glass titration cuvette as described in Section 4.2. Aliquots of octanesulfonate (2.7–54 μM or 2.7–108 μM) in an air-tight titrating syringe are added to SsuD alone or to the SsuD-$FMNH_2$ complex. The fluorescence spectra are recorded with an excitation wavelength at 280 nm and emission intensity measurements at 344 nm following a 2-min incubation. The concentration of bound octanesulfonate is determined by Eq. (1), and plotted against the concentration of total octanesulfonate to determine the dissociation constant (K_d) according to Eq. (2) (Section 4.1). With SsuD, there was no decrease in fluorescence observed with the addition of octanesulfonate unless reduced FMN was included with the enzyme in the anaerobic titration cuvette. Experiments can also be performed with $FADH_2$ based on the flavin specificity of the monooxygenase determined in Section 3.1. The results supported ordered binding with reduced flavin binding first followed by octanesulfonate.

5. Evaluating protein-protein interactions

The two-component flavin-dependent monooxygenase systems rely on reduced flavin transfer between the reductase and monooxygenase enzymes. Flavin transfer can either occur through a channeling mechanism involving direct protein-protein interactions, or through a diffusion mechanism (Fig. 4). Both mechanisms have been observed in different two-component monooxygenase systems, but several of the two-component

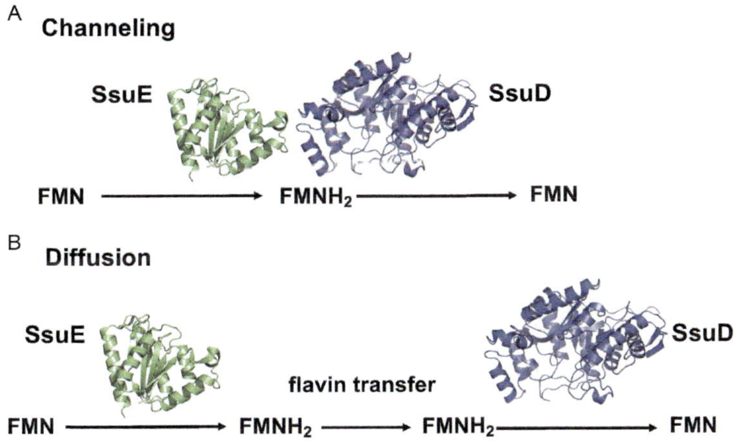

Fig. 4 Mechanism of flavin transfer in the two-component flavin-dependent systems. Both a channeling and diffusion mechanism for flavin transfer have been identified in two-component flavin-dependent systems. (A) Channeling mechanism: Protein-protein interactions form between the FMN reductase and monooxygenase, and reduced flavin is transferred between interacting proteins. (B) Diffusion mechanism: the reduced flavin is released from the flavin reductase and diffuses to the monooxygenase enzyme.

FMN-dependent monooxygenase systems are dependent on protein-protein interactions for effective flavin transfer (Dayal et al., 2015; Jeffers, Nichols, & Tu, 2003; Jun et al., 2016). Different methods exist to evaluate static protein-protein interactions in the two-component FMN-dependent systems. The methods described utilize a comprehensive approach to identify the affinity of the protein-protein interactions, and specific regions of each enzyme involved in forming the interactions.

5.1 Pull-down assays

A 6 × His-tagged reductase or monooxygenase can be used to evaluate static protein-protein interactions. The 6 × His-tagged protein, referred to as the bait protein, is bound to metal affinity column, and is used to capture a potential binding partner referred, to as the prey, that lacks a fusion tag (Fig. 5). Because the 6 × His-tag can block interaction sites, it may be necessary to test each enzyme independently as the bait protein. The experiments can be performed with purified protein, but the assays described use cell lysate containing expressed SsuE and 6 × His-tagged SsuD. Other types of affinity columns and tags may also be used for these experiments. Before beginning the experiment it is important that the imidazole elution

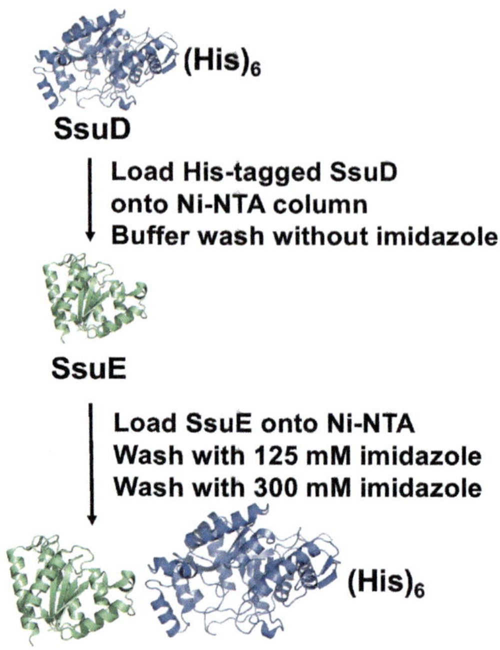

Fig. 5 Pull-down assays to evaluate protein-protein interactions. Lysate containing the expressed monooxygenase enzyme with a 6 × His tag are loaded onto a Ni-NTA column. The column is washed with buffer to remove any adventitiously bound protein, and lysate containing the expressed native flavin reductase is loaded onto the column. Following a wash with low imidazole concentrations (125 mM), the proteins are eluted with 300 mM imidazole. Interacting proteins will elute with the higher imidazole wash.

concentration be determined for the 6 × His-tagged enzyme. This allows an estimation of the imidazole concentration needed for the wash and subsequent elution from the column. 6 × His-tagged SsuD is expressed in a 1.0 L culture as noted in Section 2.2 and harvested by centrifugation at 5000 × *g* for 15 min. The cell pellet is then resuspended in 100 mL 25 mM potassium phosphate, pH 7.5, and 10% glycerol containing 4 μg/mL lysozyme. Sonication is used to lyse the resuspended pellet followed by centrifugation (20 min, 15,000 × *g*). Solid streptomycin sulfate (1.5%, w/v) is added to the cell supernatant and stirred slowly for 1 h at 4 °C to precipitate nucleic acids. The treated supernatant is centrifuged (20 min, 15,000 × *g*), and the clarified supernatant loaded onto a Ni-NTA column and washed with of 25 mM potassium phosphate, pH 7.5, and 10% glycerol (100 mL). A 1.0 L cell culture of expressed SsuE is treated similarly to SsuD and the resulting clarified supernatant loaded onto the Ni-NTA column, followed by an additional wash with 100 mL of 25 mM potassium phosphate, pH 7.5, and 10% glycerol to remove unbound proteins. Glycerol is included in the reaction because of the relative instability of the FMN reductase. Salts may need to be added to promote protein-protein interactions, and the buffer pH adjusted based on protein stability. The 6 × His tagged SsuD enzyme elutes from the Ni-NTA column with 250 mM imidazole. Therefore, adventitiously bound proteins are removed from the column with 200 mL of 25 mM potassium phosphate, pH 7.5, 10% glycerol, and 125 mM imidazole. 6 × His-tagged SsuD and any interacting proteins are eluted with 100 mL of 25 mM potassium phosphate, pH 7.5, 10% glycerol, and 300 mM imidazole. The fractions from the 125 and 300 mM imidazole wash are collected and analyzed by SDS-PAGE. Two sets of control experiments should also be performed to ensure that complex formation between SsuE and 6 × His-tagged SsuD protein in the Ni-NTA column is due to specific interactions. The first control experiment evaluates the ability of the bait protein (SsuE) to bind to the Ni-NTA column in the absence of the 6 × His-tagged SsuD protein. In the second control experiment, bovine serum albumin (or an expressed protein unrelated to the system) is used as the bait protein to ensure that the binding is specific. Each control reaction is performed as described for SsuE and SsuD. Both SsuE and SsuD eluted with the 300 mM wash supporting the proposal of protein-protein interactions. There was some unbound SsuE that eluted with the 125 mM wash, but the unbound protein was removed in the early fractions of the 125 mM isocratic gradient.

5.2 Fluorimetric titration of FMN-bound SsuE with SsuD

Further studies can be performed to determine the K_d and stoichiometric ratio of subunit binding through fluorescence spectroscopy. For these experiments an increase in flavin fluorescence is observed with the addition of the monooxygenase enzyme due to changes in the flavin environment for a flavin-bound reductase. The FMN-bound SsuE protein is prepared by titrating 0.4 μM FMN with 1 μL aliquots of SsuE for a total of 10–15 additions (0.04–0.4 μM). Titration of FMN ensures that the majority of SsuE is flavin-bound. The 0.40 μM FMN-bound SsuE sample in 25 mM potassium phosphate buffer, pH 7.5, (1.0 mL total volume) is titrated with 1 μL aliquots of SsuD for a total of 20 additions (0.02–0.95 μM). Emission intensities from 470 to 650 nm are measured using an excitation wavelength of 450 nm. Protein-protein interactions result in an increase in the fluorescence intensity until a binding equilibrium is reached. If the interactions are due to aggregation, then a continual decrease in flavin fluorescence is observed. The concentration of SsuD bound to SsuE is determined by applying Eq. (3):

$$[\mathrm{SsuD}]_{bound} = [\mathrm{SsuE}]\left[\frac{\mathrm{I}_0 - \mathrm{I}_c}{\mathrm{I}_0 - \mathrm{I}_f}\right] \tag{3}$$

where [SsuE] represents the initial concentration of enzyme, I_0 is the initial fluorescence intensity of FMN prior to addition of SsuD, I_c is the fluorescence intensity of FMN following each addition of SsuD, and I_f is the final fluorescence intensity. The concentration of SsuD bound ($[\mathrm{SsuD}]_{\mathrm{bound}}$, y) is plotted against the total SsuD ($[\mathrm{SsuD}]_{\mathrm{total}}$, x) to obtain the dissociation constant (K_d) for the binding between SsuE and SsuD according to Eq. (2).

It is important to verify that the observed fluorescent changes are not due to release of the flavin from the flavin reductase with the addition of the monooxygenase. The flavin-bound spectrum is obtained before ultrafiltration with 20 μM FMN in 25 mM potassium phosphate, pH 7.5, 10% glycerol, and 100 mM NaCl at 25 °C in the presence of 20 μM SsuE and SsuD in a 1.0 mL total volume. The solution is added to a 10 kDa-molecular-weight-cutoff centrifugation filter and the reaction is brought up to 4 mL with 25 mM potassium phosphate, pH 7.5, 10% glycerol, and 100 mM NaCl and centrifuged at 3000 rpm. It is important that the final volume is 1.0 mL to compare the concentration of flavin and enzyme before and after centrifugation. Spectra of the retentate show absorbance peaks at 280 and 450 nm, and the concentration of enzyme and flavin can be determined using the molar extinction for SsuE (20.3 $\mathrm{mM^{-1}\,cm^{-1}}$ at 280 nm) and FMN (12.2 $\mathrm{mM^{-1}\,cm^{-1}}$ at 450 nm). Control experiments with FMN only are performed as a control under similar conditions described for FMN-bound SsuE and SsuD.

5.3 Hydrogen-deuterium exchange mass spectrometry

Hydrogen-deuterium exchange mass spectrometry (HDX MS) can be used to identify protein-protein interaction sites in the presence or absence of substrates. If proteins form a complex, the interacting sites are protected from the exchange of hydrogen with solvent-deuterium. The resulting peptides will have a lower mass compared to comparable peptides obtained from deuterium-exchange of the individual peptides (Fig. 6). It is often necessary to collaborate for these studies with groups specialized in HDX MS analysis

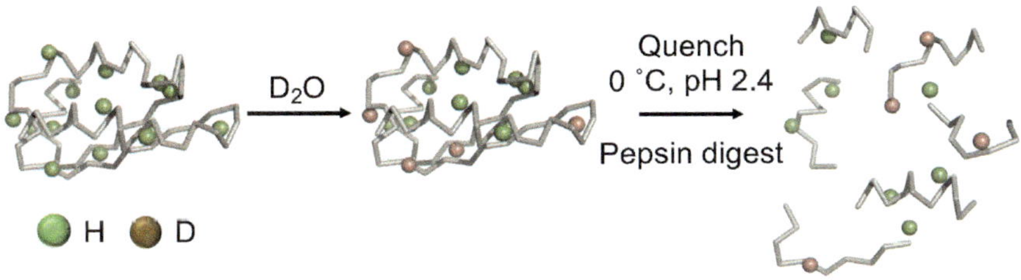

Fig. 6 HDX MS evaluation of protein-protein interactions. Backbone hydrogens of the flavin reductases and monooxygenases are exchanged with deuterium in separate reactions. The deuterium exchange reaction is quenched and digested with pepsin for 5 min. The separated peptides are sequenced by MS/MS to obtain the peptide map of each enzyme. Both the flavin reductases and monooxygenase enzymes are incubated together in deuterium to identify protein-protein interactions sites. Interacting sites will be protected from solvent exchange with deuterium, and the peptides identified in that region will have a lower mass when compared to the peptides obtained independently.

for the experimental design and implementation. However, the experiments can be performed on any mass spectrometer capable of LC-MS/MS analysis. Principle difficulties can arise when evaluating the data to determine the masses of each peptide. Peptide mapping of each enzyme will need to be performed initially to confirm that optimal coverage is obtained for each enzyme. The SsuE or SsuD enzymes (60 μM) are digested with 5 mg/mL pepsin for 5 min, and the digested peptides separated on a C18 reverse-phase column using an HPLC inline to a mass spectrometer. Alternatively an in-line pepsin column can be used to digest the peptide prior to separation by HPLC and subsequent analysis by mass spectrometry. For SsuE and SsuD, the peptides were sequenced over 15 min at 0.1 mL/min using a linear gradient of 0–50% solvent B (98% acetonitrile, 2% H_2O, 0.4% formic acid) with the mass spectrometer in positive ion mode by data dependent MS/MS, and identified by PEAKS Client 6 (Dayal et al., 2015). There are several programs available to evaluate the MS spectra, and most of these programs are specific to the instrumentation used. If poor peptide coverage is obtained, the experiments can be repeated using alternative proteases, incubation times, and inclusion of a denaturant or reductant.

Individual HDX MS experiments are performed with 60 μM (2 μL total volume) of SsuE or SsuD in 25 mM potassium phosphate, pH 7.5, and 5 mM DTT. The SsuE-SsuD complex is generated by combining equal volumes (1 μL each) of 120 μM SsuE and SsuD. The HDX reaction is initiated with the addition of 23 μL of D_2O (99.9%) (Dayal et al., 2015). All the samples are incubated at 25 °C for 30 s, after which the reaction is stopped with 25 μL quench buffer (water, pH 2.3, 0 °C). The samples are then immediately digested on ice with 2 μL of 5 mg/mL pepsin for 5 min. To maintain consistency, the digested peptides are separated as described in initial experiments to evaluate peptide coverage. All samples for HDX are prepared individually and evaluated on the same day as the corresponding water and 100% deuterated control samples. The spectra from each HDX-MS sample were analyzed using HDExaminer (Sierra Analytics). The percent deuterium incorporated for each peptide can be determined with Eq. (4):

$$\%D = \frac{(m_t - m_0)}{(m_{100} - m_0)} \times 100 \tag{4}$$

where m_t, m_0, m_{100} are the average molecular masses of the same peptide in the partially deuterated at 30 s, non-deuterated, and the fully-deuterated control samples, respectively. Deuterium-protected regions in SsuE and SsuD are located at sites adjacent to the active site of each enzyme. The SsuE and SsuD interaction region is conserved in other two-component FMN-dependent monooxygenase enzymes (Dayal et al., 2015). If protein-protein interactions sites are identified, then further studies can be performed to introduce substitutions in these regions to determine if protein-protein interactions are affected in these variants.

6. Monitoring flavin transfer between SsuE and SsuD

The identification of protein-protein interactions between the FMN reductase and monooxygenase supports a channeling mechanism for flavin transfer. However, the studies do not monitor the actual flavin transfer event. For the two-component FMN-dependent systems the reductive and oxidative half-reactions occur on separate enzymes. Rapid-reaction kinetic analyses can be performed with both enzymes to monitor each half-reaction independently. There is a clear delineation between the reductive and oxidative half-reaction that can provide information regarding reduced flavin transfer from the flavin reductase and subsequent transfer to the monooxygenase. In addition, there are several reviews describing methods for evaluating channeling mechanisms in enzyme systems (Huang, Holden, & Raushel, 2001; Spivey & Ovadi, 1999). Competition assays with the alkanesulfonate monooxygenase system have also been developed to evaluate reduced flavin channeling (Geck & Kirsch, 1999). Methods used to evaluate channeling between the flavin reductase and monooxygenase have provided valuable insights on the role of protein-protein interactions in effective flavin transfer. Both rapid-reaction kinetic studies and competition assays can be used together to differentiate between a channeling or diffusion mechanism for flavin transfer. Additional protocols to evaluate the flavin-transfer mechanism have also been reported by other groups working on two-component flavin-dependent systems (Jawanda, Ahmed, & Tu, 2008; Jeffers et al., 2003; Jun et al., 2016; Lei & Tu, 1998; Tu, 2001).

6.1 Rapid reaction kinetic analysis of flavin transfer

Single-turnover experiments can be performed to evaluate the reduced flavin transfer step from SsuE to SsuD. In stopped-flow kinetic traces at 450 nm, a lag phase is observed between the reductive (SsuE) and oxidative (SsuD) half-reaction. The lag phase has been proposed to represent the transfer step and/or conformational changes that occur with the binding of substrates to SsuD. In the absence of SsuD, the flavin is released and is oxidized quickly because it is not protected by SsuD. It is not necessary to maintain anaerobiosis in these studies because the FMN reductase supplies reduced flavin to the monooxygenase.

The stopped-flow kinetic experiments monitoring both the reductive and oxidative half-reactions are performed with aerobic solutions of SsuD (35 μM), SsuE (35 μM), and FMN (25 μM) in one drive-syringe mixed against NADPH (250 μM) in air-saturated 25 mM potassium phosphate, pH 7.5, and 10% glycerol at 4 °C. Glass syringes can be used to add the solutions to the drive-syringes of the stopped-flow. It is important that the flavin-reductase and monooxygenase are at a 1:1 molar equivalence to maintain single-turnover conditions. The NADPH is in excess, but the flavin is released by SsuE following reduction because of the 50-fold lower affinity of SsuE for reduced flavin. Therefore, the flavin-bound SsuE is not continually reduced by NADPH. Saturating concentrations of the monooxygenase substrate can also be included to determine if the flavin transfer step is altered. When included in the reaction, octanesulfonate (250 μM) was present in the drive syringe with NADPH. All experiments are carried out in single-mixing mode by mixing equal volumes of the solutions, and monitoring the reactions by absorbance at 450 nm. The reaction is also repeated in the absence of the monooxygenase to evaluate the release of flavin by the flavin reductase. Variants that interrupt protein-protein interactions can also be tested to determine if flavin transfer is affected.

If protein-protein interactions are disrupted, then there will be a decrease in the steady-state kinetic activity of the monooxygenase in coupled assays, and the flavin is immediately released from the flavin reductase following the reductive half-reaction. It is often difficult to quantitatively fit these kinetic plots, but these experiments provide a qualitative assessment of flavin transfer.

6.2 Competition assays

There are several variations of competition assays to evaluate protein-protein interactions during catalysis. An approach that gives a definitive evaluation of protein-protein interactions employs a variant of the substrate donor enzyme in coupled assays (Geck & Kirsch, 1999). The variant will not be able to participate in catalysis, but will still effectively bind to the recipient enzyme. A variant of SsuE (Y118A SsuE) was identified that was unable to support flavin transfer to SsuD, but was still able to form stable protein-protein interactions with SsuD (Musila et al., 2018). Because both Y118A and wild-type SsuE enzymes were shown to interact with SsuD, the two enzymes will compete for the interaction sites in SsuD resulting in a decline in catalytic activity when both enzymes are included in the assay. The observed desulfonation activity would only be possible from SsuD interacting wild-type SsuE because the Y118A SsuE cannot transfer reduced flavin. If the transfer of reduced flavin between SsuE and SsuD occurs through diffusion, the catalytic activity would not change with increasing amounts of Y118A SsuE because reduced flavin produced by wild-type would diffuse to SsuD (Fig. 7A). Conversely, a decrease in kinetic activity is observed when the SsuE deletion variant is included in the reaction with wild-type SsuE and SsuD if reduced flavin transfer occurs through channeling (Fig. 7B).

An experiment to evaluate the dependence of the SsuD activity on the concentration of wild-type SsuE was initially performed in the absence of Y118A SsuE by varying wild-type SsuE (0.01–0.12 μM) at constant concentrations of substrates and SsuD as outlined above. This preliminary analysis is critical because the experimental premise is based on the ability of both wild-type and Y118A SsuE to compete for SsuD interaction sites. A concentration of SsuE was selected based on these findings. The reaction mixture contained SsuE (0.06 μM), FMN (2 μM), SsuD (0.06 μM), octanesulfonate (1 mM), NADPH (0.25 mM), and varying concentrations of Y118A SsuE (0.01–0.3 μM) in 25 mM Tris, pH 7.5, and 0.1 M NaCl in

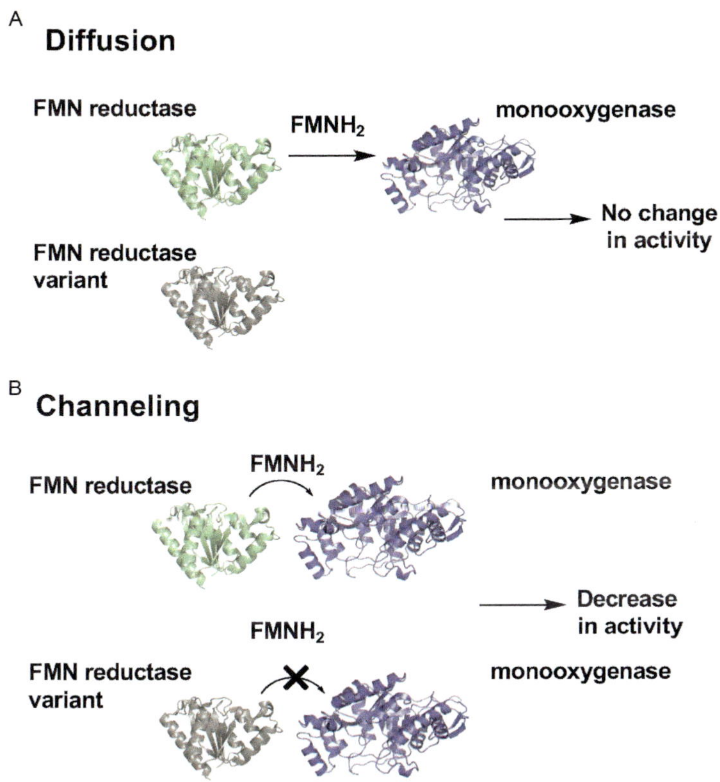

Fig. 7 Competition assays to evaluate flavin transfer in the FMN-dependent two-component systems. Competition assays to evaluate flavin transfer utilize a variant of the FMN reductase that is defective in flavin transfer, but is still able to bind to the monooxygenase. For the alkanesulfonate monooxygenase system a variant was generated that bound flavin, but was not able to release the flavin once reduced. There are two possible outcomes for competition assays: (A) Diffusion mechanism: Increasing concentrations of the variant enzyme will have no effect on the activity of the monooxygenase. (B) Channeling mechanism: Increasing concentrations of the variants enzyme will decrease the activity of the monooxygenase enzyme.

a total volume of 0.5 mL. The endpoint-assay is initiated by the addition of NADPH, incubated for 3 min at 25 °C, and quenched with the addition of 8 M urea (167 μL). The sulfite product is measured as previously described (Zhan et al., 2008). The rates obtained are plotted against the concentration of Y118A SsuE included in the reaction. For the alkanesulfonate monooxygenase system, there was a decline in activity with increasing concentrations of Y118A SsuE, supporting a channeling mechanism for flavin transfer.

7. Conclusions and future directions

Two-component flavin-dependent systems often catalyze reactions that enable bacterial survival when standard metabolites are limiting. Operons containing putative two-component systems have been identified, but the functional assignments for many of these systems have not been evaluated. The two-component systems offer subtle mechanistic differences from canonical flavoproteins that need to be considered when evaluating these systems. Methods have been developed that provide an alternative analyses of two-component systems compared to their canonical counterparts. An important aspect of two-component flavin-dependent systems is the mechanism of flavin transfer. There have been different mechanisms reported for flavin transfer for different systems, and these differences may be inherent in the structural properties of the enzyme within a specific family. Determining the structural features that promote flavin-transfer in two-component systems will provide important information on how these enzymes have evolved to accommodate this mechanistic variation, and what advantage it provides over classic canonical flavoproteins.

Acknowledgment

This work was supported by Grant 1244320 (to H.R.E.) from the National Science Foundation.

References

Abdurachim, K., & Ellis, H. R. (2006). Detection of protein-protein interactions in the alkanesulfonate monooxygenase system from *Escherichia coli*. *Journal of Bacteriology*, *188*(23), 8153–8159.

Baldwin, T. O. (1974). The binding and spectral alterations of oxidized flavin mononucleotide by bacterial luciferase. *Biochemical and Biophysical Research Communications*, *57*(4), 1000–1005.

Baldwin, T. O., Nicoli, M. Z., Becvar, J. E., & Hastings, J. W. (1975). Bacterial luciferase. Binding of oxidized flavin mononucleotide. *The Journal of Biological Chemistry*, *250*(8), 2763–2768.

Bohuslavek, J., Payne, J. W., Liu, Y., Bolton, H. J., & Xun, L. (2001). Cloning, sequencing, and characterization of a gene cluster involved in EDTA degradation from the bacterium BNC1. *Applied and Environmental Microbiology*, *67*(2), 688–695.

Chen, L. H., & Baldwin, T. O. (1989). Random and site-directed mutagenesis of bacterial luciferase: Investigation of the aldehyde binding site. *Biochemistry*, *28*(6), 2684–2689.

Dayal, P. V., Singh, H., Busenlehner, L. S., & Ellis, H. R. (2015). Exposing the alkanesulfonate monooxygenase protein-protein interaction sites. *Biochemistry*, *54*(51), 7531–7538.

Driggers, C. M., Dayal, P. V., Ellis, H. R., & Karplus, P. A. (2014). Crystal structure of *Escherichia coli* SsuE: Defining a general catalytic cycle for FMN reductases of the flavodoxin-like superfamily. *Biochemistry*, *53*(21), 3509–3519.

Eichhorn, E., Davey, C. A., Sargent, D. F., Leisinger, T., & Richmond, T. J. (2002). Crystal structure of *Escherichia coli* alkanesulfonate monooxygenase SsuD. *Journal of Molecular Biology*, *324*(3), 457–468.

Eichhorn, E., van der Ploeg, J. R., & Leisinger, T. (1999). Characterization of a two-component alkanesulfonate monooxygenase from *Escherichia coli*. *The Journal of Biological Chemistry*, *274*(38), 26639–26646.

Ellis, H. R. (2010). The FMN-dependent two-component monooxygenase systems. *Archives of Biochemistry and Biophysics*, *497*(1–2), 1–12.

Endoh, T., Habe, H., Nojiri, H., Yamane, H., Omori, T., Habe, H., et al. (2005). The sigma54-dependent transcriptional activator SfnR regulates the expression of the *Pseudomonas putida* sfnFG operon responsible for dimethyl sulphone utilization. *Molecular Microbiology*, *55*(3), 897–911.

Endoh, T., Kasuga, K., Horinouchi, M., Yoshida, T., Habe, H., Nojiri, H., et al. (2003). Characterization and identification of genes essential for dimethyl sulfide utilization in *Pseudomonas putida* strain DS1. *Applied Microbiology and Biotechnology*, *62*(1), 83–91.

Gao, B., & Ellis, H. R. (2005). Altered mechanism of the alkanesulfonate FMN reductase with the monooxygenase enzyme. *Biochemical and Biophysical Research Communications*, *331*(4), 1137–1145.

Geck, M. K., & Kirsch, J. F. (1999). A novel, definitive test for substrate channeling illustrated with the aspartate aminotransferase/malate dehydrogenase system. *Biochemistry*, *38*(25), 8032–8037.

Ghisla, S., & Massey, V. (1989). Mechanisms of flavoprotein-catalyzed reactions. *European Journal of Biochemistry*, *181*(1), 1–17.

Hastings, J. W., & Presswood, R. P. (1978). Bacterial luciferase: $FMNH_2$-aldehyde oxidase. *Methods in Enzymology*, *53*, 558–570.

Huang, X., Holden, H. M., & Raushel, F. M. (2001). Channeling of substrates and intermediates in enzyme-catalyzed reactions. *Annual Review of Biochemistry*, *70*, 149–180.

Jawanda, N., Ahmed, K., & Tu, S. C. (2008). *Vibrio harveyi* flavin reductase-luciferase fusion protein mimics a single-component bifunctional monooxygenase. *Biochemistry*, *47*(1), 368–377.

Jeffers, C. E., Nichols, J. C., & Tu, S. C. (2003). Complex formation between *Vibrio harveyi* luciferase and monomeric NADPH:FMN oxidoreductase. *Biochemistry*, *42*(2), 529–534.

Jun, S. Y., Lewis, K. M., Youn, B., Xun, L., & Kang, C. (2016). Structural and biochemical characterization of EDTA monooxygenase and its physical interaction with a partner flavin reductase. *Molecular Microbiology*, *100*(6), 989–1003.

Kantz, A., Chin, F., Nallamothu, N., Nguyen, T., & Gassner, G. T. (2005). Mechanism of flavin transfer and oxygen activation by the two-component flavoenzyme styrene monooxygenase. *Archives of Biochemistry and Biophysics*, *442*(1), 102–116.

Lei, B., & Tu, S. C. (1998). Mechanism of reduced flavin transfer from *Vibrio harveyi* NADPH-FMN oxidoreductase to luciferase. *Biochemistry*, *37*(41), 14623–14629.

Li, L., Liu, X., Yang, W., Xu, F., Wang, W., Feng, L., et al. (2008). Crystal structure of long-chain alkane monooxygenase (LadA) in complex with coenzyme FMN: Unveiling the long-chain alkane hydroxylase. *Journal of Molecular Biology*, *376*(2), 453–465.

Liu, M., Lei, B., Ding, Q., Lee, J. C., & Tu, S. C. (1997). *Vibrio harveyi* NADPH:FMN oxidoreductase: Preparation and characterization of the apoenzyme and monomer-dimer equilibrium. *Archives of Biochemistry and Biophysics*, *337*(1), 89–95.

Massey, V. (1994). Activation of molecular oxygen by flavins and flavoproteins. *The Journal of Biological Chemistry*, *269*(36), 22459–22462.

Musila, J. M., & Ellis, H. R. (2016). Transformation of a flavin-free FMN reductase to a canonical flavoprotein through modification of the π-helix. *Biochemistry*, *55*(46), 6389–6394.

Musila, J. M., Forbes, D., & Ellis, H. R. (2018). Functional evaluation of the π-helix in the NAD(P)H:FMN reductase of the alkanesulfonate monooxygenase system. *Biochemistry*, *57*(30), 4469–4477.

Nissen, M. S., Youn, B., Knowles, B. D., Ballinger, J. W., Jun, S. Y., Belchik, S. M., et al. (2008). Crystal structures of NADH:FMN oxidoreductase (EmoB) at different stages of catalysis. *The Journal of Biological Chemistry*, *283*(42), 28710–28720.

Niviere, V., Fieschi, F., Decout, J. L., & Fontecave, M. (1996). Is the NAD(P)H:flavin oxidoreductase from *Escherichia coli* a member of the ferredoxin-$NADP^+$ reductase family? Evidence for the catalytic role of serine 49 residue. *The Journal of Biological Chemistry*, *271*(28), 16656–16661.

Palfey, B. A., & McDonald, C. A. (2010). Control of catalysis in flavin-dependent monooxygenases. *Archives of Biochemistry and Biophysics*, *493*(1), 26–36.

Payne, J. W., Bolton, H. J., Campbell, J. A., & Xun, L. (1998). Purification and characterization of EDTA monooxygenase from the EDTA-degrading bacterium BNC1. *Journal of Bacteriology*, *180*(15), 3823–3827.

Spivey, H. O., & Ovadi, J. (1999). Substrate channeling. *Methods*, *19*(2), 306–321.

Tang, C. K., Jeffers, C. E., Nichols, J. C., & Tu, S. C. (2001). Flavin specificity and subunit interaction of *Vibrio fischeri* general NAD(P)H-flavin oxidoreductase FRG/FRase I. *Archives of Biochemistry and Biophysics*, *392*(1), 110–116.

Tu, S. C. (2001). Reduced flavin: Donor and acceptor enzymes and mechanisms of channeling. *Antioxidants & Redox Signaling*, *3*(5), 881–897.

Tu, S., & Hastings, J. W. (1975). Differential effects of 8-anilino-1-naphthalenesulfonate upon binding of oxidized and reduced flavines by bacterial luciferase. *Biochemistry*, *14*(19), 4310–4316.

Valton, J., Filisetti, L., Fontecave, M., & Niviere, V. (2004). A two-component flavin-dependent monooxygenase involved in actinorhodin biosynthesis in *Streptomyces coelicolor*. *The Journal of Biological Chemistry*, *279*(43), 44362–44369.

van Berkel, W. J., Kamerbeek, N. M., & Fraaije, M. W. (2006). Flavoprotein monooxygenases, a diverse class of oxidative biocatalysts. *Journal of Biotechnology*, *124*(4), 670–689.

Wicht, D. K. (2016). The reduced flavin-dependent monooxygenase SfnG converts dimethylsulfone to methanesulfinate. *Archives of Biochemistry and Biophysics*, *604*, 159–166.

Xin, X., Xi, L., & Tu, S. C. (1991). Functional consequences of site-directed mutation of conserved histidyl residues of the bacterial luciferase alpha subunit. *Biochemistry*, *30*(47), 11255–11262.

Zhan, X., Carpenter, R. A., & Ellis, H. R. (2008). Catalytic importance of the substrate binding order for the $FMNH_2$-dependent alkanesulfonate monooxygenase enzyme. *Biochemistry*, *47*(7), 2221–2230.

CHAPTER SIXTEEN

The styrene monooxygenase system

George T. Gassner*
Department of Chemistry and Biochemistry, San Francisco State University, San Francisco, CA, United States
*Corresponding author: e-mail address: gassner@sfsu.edu

Contents

Abstract

Styrene monooxygenases are soluble two-component flavoproteins that catalyze the NADH and FAD-dependent enantioselective epoxidation of styrene to styrene oxide in the aqueous phase. These enzymes present interesting mechanistic features and potential as catalysts in biotechnological applications ranging from green chemical synthesis to bioremediation.

This chapter presents approaches for the expression of the reductase (SMOB, StyB) and epoxidase (SMOA, StyA) components of SMO from pET-vectors as native or N-terminally histidine-tagged proteins in commercial strains of *E. coli*.

The two-component structure of SMO and hydrophobic nature of styrene substrate requires some special consideration in evaluating the mechanism of this enzyme. The

Methods in Enzymology, Volume 620
ISSN 0076-6879
https://doi.org/10.1016/bs.mie.2019.03.019

modular composition of the enzyme allows the flavin-reduction reaction of SMOB and styrene epoxidation reaction of SMOA to be evaluated both independently and as a composite catalytic system. The freedom to independently study the reductase and epoxidase components of SMO significantly simplifies studies of equilibrium-binding and the coupling of the free energy of ligand binding to the electrochemical potential of bound FAD.

In this chapter, methods of steady-state and pre-steady-state kinetic assay, experimental approaches to equilibrium-binding reactions of flavin and substrate, and determination of the electrochemical midpoint potential of FAD bound to the reductase and epoxidase components of SMO are presented. This presentation focuses on approaches that have been successfully used in the study of the wild-type styrene monooxygenase system recovered from *Pseudomonas putida* (S12), but similar approaches may be effective in the characterization of related two-component enzyme systems.

Abbreviations

SMO styrene monooxygenase
StyA, (N)SMOA (N-terminally histidine-tagged) styrene monooxygenase component A
StyB, (N)SMOB (N-terminally histidine-tagged) styrene monooxygenase component B
NADH β-nicotinamide adenine dinucleotide
FAD flavin adenine dinucleotide

1. Introduction

Microorganisms employ catabolic pathways that allow them to access energy and carbon captured in naturally occurring repositories of biochemical, geochemical, and synthetic origins. This complex menu of reduced carbon demands a diverse inventory of biochemical pathways to allow microbes to access and transform available carbon energy and biomass (Dagley, 1987).

The broad range of substrates encountered at the entry points of microbial catabolism requires enzymes of diverse structure and catalytic capability. Research focused on the elucidation of the structure and mechanisms of these enzymes has provided a rich source of information that contributes regularly to the advancement of understanding of fundamental principles of enzyme catalysis and such enzymes have been increasingly targeted for their potential as biocatalysts in the synthesis of products of biotechnological value (Turner & Kumar, 2018).

In aerobic metabolism, mechanistically versatile flavin monooxygenases occur frequently at the initial step of biochemical pathways targeting the transformation of activated aromatic compounds to metabolic intermediates. These enzymes react as NAD(P)H-dependent mixed function oxygenases to reductively activate and insert oxygen into target substrates (Huijbers, Montersino, Westphal, Tischler, & van Berkel, 2014; Palfey & McDonald, 2010; Romero, Gomez Castellanos, Gadda, Fraaije, & Mattevi, 2018).

The mechanistic diversity of these enzymes is dependent on structural and electronic features of their substrate and flavin-binding active sites and the molecular assembly of subunits that defines the catalytic unit. Studies targeting both the active site structure and thermodynamic linkage of substrate and flavin-binding and further provide insight into the principles guiding flavoprotein catalysis. This is of significance both in our fundamental understanding of flavoprotein monooxygenase mechanisms of action and in the advancement of new technologies and applications of flavoprotein in biocatalysis.

Intrinsic properties of flavins and their pyridine nucleotide substrates provide spectroscopic signatures that make many steps in the flavin reduction and subsequent reactions with oxygen amenable to study by UV-vis and fluorescence spectroscopy. Notably, the oxidized and reduced forms of flavin and pyridine nucleotide, the peroxyflavin, and hydroxyflavin are each characterized by unique spectral features that have been used extensively in the evaluation of flavin monooxygenase mechanisms. In addition, unique spectra of substrates and products of flavin monooxygenases in free or enzyme-bound form may provide additional information about the state of reaction.

In microbial catabolic pathways, flavin monooxygenase occurs both as single-component systems, which use a single polypeptide to catalyze flavin-reduction and substrate oxygenation, and two-component flavin monooxygenases, which employ separate flavin reductase and monooxygenase components in catalysis (Heine, van Berkel, Gassner, van Pee, & Tischler, 2018; Huijbers et al., 2014). Many aspects of the one- and two-component flavoproteins are conserved and for this reason similar experimental approaches have proven valuable in the study of these enzymes.

These enzymes have gathered interest both as model systems for structural and mechanistic studies of catalysis by two-component flavoproteins and for their uncommon ability to react efficiently with hydrophobic hydrocarbons solvated in the aqueous phase, and their stability and flexibility in biocatalytic applications. This chapter focuses on experimental approaches for the characterization of styrene monooxygenases.

2. Recovery of recombinant SMOs

Depending on the experimental objectives it can be convenient to express the component parts of SMOs from a single expression system or as separate protein components. In general, the reductase (SMOB or StyB) and epoxidase (SMOA or StyA) and their N-terminally histidine-tagged derivatives are readily expressed in *E. coli* by using commercially available vectors with high output promoters (Kantz, Chin, Nallamothu, Nguyen, & Gassner, 2005).

The N-terminally histidine-tagged epoxidase (NSMOA) can be conveniently recovered by nickel affinity chromatography in higher yield and purity than the native enzyme. It is possible to recover both native SMOB and the N-terminally histidine-tagged protein (NSMOB) in similar yield and purity (Kantz et al., 2005). However, NSMOB has a significantly higher binding affinity for oxidized FAD than the native reductase, SMOB, and proceeds through a double displacement mechanism in place of the sequential ternary mechanism observed by the wild-type enzyme (Heine et al., 2017; Kantz et al., 2005; Otto, Hofstetter, Rothlisberger, Witholt, & Schmid, 2004).

3. Biochemical characterization of SMOs

3.1 Practical considerations

Several features of styrene monooxygenases require special consideration in the biochemical evaluation of styrene monooxygenases. These enzymes function as soluble, cytosolic, two-component flavoproteins, but target styrene, a hydrophobic oil, as their preferred substrate. This disparity of physical properties poses a challenge in studies of the substrate specificity and reaction mechanisms of these enzymes.

The two-component structure of SMOs poses a further experimental consideration for studies of the catalytic mechanism of SMOs. The modular structure of SMOs greatly simplifies studies of the FAD-reduction reaction catalyzed by SMOB and styrene-epoxidation reaction catalyzed by NSMOA. Each SMO component can be expressed, purified, and mechanistically characterized independent of the other. Alternatively the reductase and epoxidase activities can be reconstituted and joined in a single catalytic system to provide a comprehensive view into the operation of the entire system

(Corrado, Knaus, & Mutti, 2018; Heine et al., 2017; Kantz et al., 2005; Kantz & Gassner, 2011; Tischler et al., 2010).

Under conditions of steady-state turnover, the reductase and epoxidase activities collectively react with three consumable substrates, NADH, dioxygen, and shared FAD that shuttles NADH-derived electrons from reductase to the epoxidase where it subsequently engages in the reductive activation of molecular oxygen and the enantioselective insertion of an oxygen atom into vinyl side chain of styrene. Oxidized FAD is then returned to the reductase completing the catalytic cycle. Thus, the efficiency with which the reduction reaction of SMOB and oxygen reaction of SMOA are coupled during catalysis best characterized by simultaneous monitoring and comparison of the rates of NADH and styrene consumption under conditions of steady-state turnover (Kantz et al., 2005).

3.2 Single-turnover reaction of SMOA with oxygen and styrene

3.2.1 Preparation of SMOA with reduced FAD bound

Styrene monooxygenase binds oxidized FAD with low affinity (millimolar range) and to reduced FAD with sub-micromolar affinity (Ukaegbu, Kantz, Beaton, Gassner, & Rosenzweig, 2010).This makes possible the preparation enzyme with reduced flavin bound in near 1:1 stoichiometry. To prepare the enzyme in this format, NSMOA is first exchanged into the desired reaction buffer from storage buffer by gel filtration through a Biogel DG-10 column. The concentration of protein recovered in this step is determined by measurement of the absorbance at 280 nm using the molar extinction coefficient of the apo protein at this wavelength. Typically, a protein concentration of 30–60 μM is targeted. The apo protein is then reconstituted with a stoichiometric amount of oxidized FAD. Due to the low affinity of NSMOA for oxidized FAD, only a small fraction of the epoxidase is bound with FAD at this point. The concentration of FAD binding sites is verified by measurement of absorbance at 450 nm where FAD has a molar extinction coefficient of 11,300 M^{-1} cm^{-1}.

To prepare NSMOA with reduced FAD bound, the solution of epoxidase and oxidized FAD is transferred to an anaerobic cuvette similar to that previously described (Maehly, Smith, & Graham, 1955) (Fig. 1). Dissolved oxygen is replaced with prepurified nitrogen, by attaching the anaerobic cuvette to a Schlenk line and completing 10–12 vacuum-gas exchange cycles (Kantz et al., 2005).

The Schlenk line should preferably include a manifold of three-way valves to allow several activities to be accomplished in parallel. The manifold

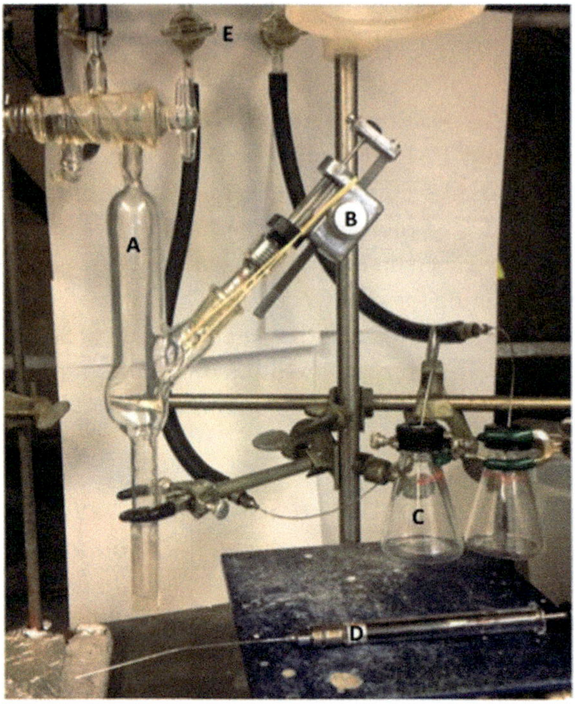

Fig. 1 Apparatus used for the preparation of reduced SMO components. (A) Anaerobic Cuvette; (B) Repeating Dispenser fitted with a 250 μL gas-tight Hamilton Syringe; (C) Erlenmeyer flask fitted with septum top for the preparation of dithionite solutions; (D) Hamilton Gastight syringe fitted with 22-gage needle for transfer of air sensitive solutions; (E) Schlenk line fitted with butyl rubber tubing.

can be inexpensively and effectively plumbed with ¼ in. copper pipe and butyl rubber tubing. A basic vacuum pump capable of decreasing the pressure to about 100 Torr is attached to one inlet of the manifold. A source of prepurified nitrogen gas (grade 4.8) regulated at 5 psi is attached to the other inlet. It is essential that the gas delivered to experimental samples is essentially oxygen-free. This is accomplished by including an oxygen-scrubbing column in the line leading from the regulated gas source to the gas manifold. Commercially available columns such as the Restek Oxygen Scrubber (Cat. # 20600) are well suited for this purpose.

The enzyme is then reduced by incremental addition of dithionite from a 250 μL Hamilton syringe equipped with a PB600-1 Repeating Dispenser. The syringe is attached to the cuvette by a #7 Ace Thred mini adapter with a 14/20 inner joint bottom (Fig. 1).

Sodium hydrosulfite solutions for reductive titrations are prepared by using two 25 mL Chemglass Erlenmeyer flasks with GPI 20–400 threaded caps with Supelco PTFE-backed silicone septa (Fig. 1B). Each flask is filled with 10 mL of buffer, attached to a Schlenk line and bubbled with purified nitrogen for 10 min. A stock solution of approximately 10 mM sodium hydrosulfite is prepared by addition of 20 mg of 85% solid sodium hydrosulfite to one of the flasks and after mixing, 1–3 mL of the stock solution is transferred to the second flask depending in the desired final concentration. The concentration of the diluted sodium hydrosulfite solution is determined by transfer to a low volume 1 mL quartz cuvette and reading the absorbance at 360 nm where sodium hydrosulfite has a molar extinction coefficient of 500 M^{-1} cm^{-1}. For example, an absorbance of 1 at 360 nm indicates a dithionite concentration of 2 mM.

The desired concentration of sodium hydrosulfite sub-stock is drawn into the titrating syringe. The loaded syringe is attached to the anaerobic cuvette under positive gas flow via a 14/20 adaptor seated with Apiezon N grease. After mounting the syringe, reductant is added incrementally (5 μL increments if using a 250 μL Hamilton gas tight syringe with the Repeating Dispenser) to the enzyme solution in the anaerobic cuvette while monitoring the extent of FAD reduction spectrophotometrically. This is done conveniently with a USB 4000 or USB 2000+ Ocean Optics Diode Array spectrophotometer by monitoring the absorbance decrease at 450 nm, the absorbance maximum for oxidized FAD and 340 nm, an isosbestic point for the oxidized and reduced forms of FAD. The reduction is complete when the absorbance ceases to change at 450 nm or the absorbance is observed to increase slightly at 314 nm due to the accumulation of a slight excess of sodium hydrosulfite after the titration endpoint.

The reduced FAD complex of NSMOA can be transferred to a stopped-flow spectrophotometer after first making one of the stopped-flow drive syringes anaerobic. Not all stopped-flow instruments are well suited for anaerobic work. Best results will be obtained with an instrument equipped with PEEK plastic valves and PEEK or stainless steel tubing leading from the loading and drive syringes to the flow cell. Stopped-flows that are plumbed in this way can be made anaerobic by simply loading the drive syringe with 10 mM dithionite and then replacing the dithionite with anaerobic buffer prepared by sparging with purified nitrogen and subsequently loading the reduced enzyme. Reduced enzyme can be easily transferred to the stopped-flow instrument by using a gas tight Hamilton syringe fitted with a 22-gage needle of 8 in. in length (Fig. 1). For the transfer, the syringe and needle are washed with 10 mM sodium hydrosulfite and then anaerobic buffer. The syringe now scrubbed free of oxygen contamination is then used to transfer reduced enzyme solution to the stopped-flow instrument. This accomplished by first drawing reduced enzyme solution into the anaerobic, gas-tight syringe through the 22-gauge needle under positive nitrogen pressure on the Schlenk line. The reduced enzyme solution is then transferred from the Hamilton syringe to an oxygen-free drive syringe of the stopped-flow instrument.

3.2.2 Preparation of styrene solutions for enzyme studies

An aqueous stock solution of 1 mM styrene can be prepared conveniently by adding 10 μL of styrene to 100 mL of water or buffer while stirring vigorously with a Teflon stir bar in a volumetric flask. It is critical that the solution is stirred continuously after addition of styrene and the volumetric flask should remain capped except when removing samples for concentration determination or other experimental use. The styrene stock solution can be transferred by using standard polypropylene pipette tips, but these must be first familiarized with the styrene stock solution by washing the stock solution in and out of the tip several times prior to the final filling and delivery and the time the aqueous styrene solution remains in the pipette tip should be kept to a minimum. The concentration of the styrene stock can be verified based on its molar extinction coefficient of 8880 M^{-1} cm^{-1} at 245 nm. This is typically accomplished by transferring 100 μL of styrene solution to a 1 mL quartz cuvette containing 900 μL of water and mixing with a PEEK or Teflon mixing rod or by inversion if the cuvette is equipped with a Teflon cap. The stock solution is stable for several hours if capped and stirred continuously. The stock solution can be diluted conveniently into a disposable 13 × 100 mm borosilicate glass tubes followed by vortexing with a Teflon-backed septum covering the top of the tube. It is recommended that a total volume of 5 mL be prepared just prior to experimental use. Preparation of solutions of this volume will help to minimize error associated with the transfer of small volumes and generally eliminate the need for serial dilutions.

3.2.3 Single-turnover reactions of SMOs with oxygen and styrene

The kinetics of the reaction of the epoxidase component of SMO with oxygen and styrene substrates can be thoroughly evaluated by using a single-mixing stopped-flow instrument equipped with absorbance and fluorescence detection capability. Studies will generally involve loading one drive syringe with an anaerobic complex of the epoxidase with reduced

FAD bound and the other drive syringe with buffer containing a defined amount of oxygen or a solution containing both oxygen and styrene or a styrene analog of interest. Unless there is a particular need to vary the oxygen concentration, single turnover experiments can readily be completed using air-saturated buffer. At atmospheric pressure and 25 °C, the dissolved oxygen concentration is approximately 260 μM. The exact concentration of dissolved oxygen under a range of experimental conditions oxygen can be conveniently calculated by using the USGS dissolve oxygen calculator at user defined ionic strength, temperature and pressure (https://water.usgs.gov/software/DOTABLES). It is often convenient to prepare solutions at room temperature and then cool then use the thermostatted water bath to adjust the temperature to the desired reaction conditions. Since oxygen solubility increases with decreasing temperature in the 0–40 °C range, an oxygen-saturated solutions prepared at room temperature can be cooled to desired reaction temperature in the stopped-flow drive without desolvation of the dissolved oxygen.

The reduced FAD complex of styrene monooxygenases reacts rapidly with dissolved oxygen to generate a C(4a)-FAD hydroperoxide intermediate (Kantz & Gassner, 2011). The kinetics of the formation and elimination of this intermediate can be observed by both time resolved absorbance and fluorescence measurements to establish its stability under a variety of reaction conditions.

The kinetics of the reaction of reduced FAD with oxygen in the active site of SMOA can be observed by monitoring the absorbance 382 nm and at 450 nm where the FAD C(4a)-hydroperoxide intermediate and oxidized FAD have absorbance maxima, respectively (Kantz & Gassner, 2011). The kinetics of peroxide formation is observed as a rapid increase in 382 nm absorbance (Fig. 2). The kinetics of hydrogen peroxide elimination can be observed as a decrease at 382 nm and corresponding increase at 450 nm associated with the elimination of hydrogen peroxide and reoxidation of FAD. Giving consideration to the molar absorbtivities involved, it is recommended that an enzyme concentration no <15 μM after mixing be used in this type of study for a stopped-flow instrument with a 1 cm pathlength for best quality data.

The reduced FAD-bound enzyme and C(4a)-FAD hydroperoxide intermediates occurring in this reaction are essentially non-fluorescent, whereas oxidized FAD is highly fluorescent. Thus, the kinetics of peroxide elimination can also observed by monitoring the accumulation of oxidized FAD by fluorescence. This can be done by setting the stopped-flow excitation

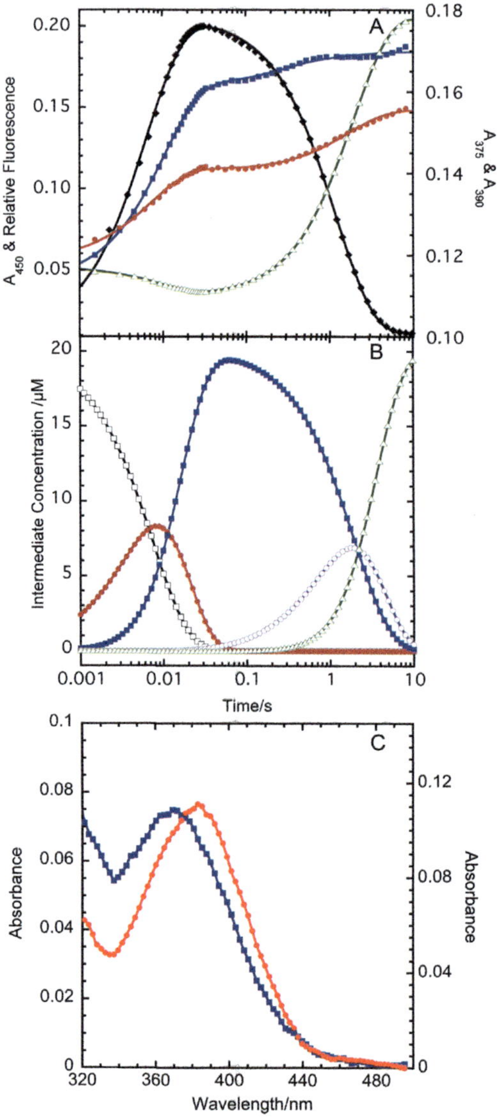

Fig. 2 Fluorescence and absorbance data from single turnover reaction of NSMOA with oxygen and styrene. (A) Reaction kinetics monitored by absorbance at 375 nm (■), 390 nm (●), 450 nm (Δ) and fluorescence emission at 520 nm (◆). (B) Reaction intermediate concentrations calculated from sequential exponential fitting: SMOA(FAD_{red}) (□), SMOA(FAD_{OOH}) (●), SMOA($FAD_{OH\text{-}SO}$) (■), SMOA(FAD_{OH}) (❍), and SMOA(FAD_{ox})(Δ). (C) Time-Resolved Absorbance Spectra representative of SMOA(FAD_{OOH}) (●) and SMOA($FAD_{OH\text{-}SO}$) (■). *Figure adapted from Kantz, A., & Gassner, G. T. (2011). Nature of the reaction intermediates in the flavin adenine dinucleotide-dependent epoxidation mechanism of styrene monooxygenase.* Biochemistry, *50 (4), 523–532. https://doi.org/10.1021/bi101328r.*

monochromator to an excitation wavelength of 450 nm and placing an Edmund Optics VG-6 optical bandpass filter in front of the emission photomultiplier. The reaction monitored in this way will occur with a lag phase corresponding to the formation of the non-fluorescent FAD C(4a)-FAD hydroperoxide followed by an increase in fluorescence that occurs with an observed rate constant corresponding to elimination of hydrogen peroxide and the regeneration of oxidized FAD.

If a solution of styrene and oxygen is included in the drive syringe, stopped-flow absorbance and fluorescence experiments can be used evaluate the kinetics of the styrene epoxidation reaction (Fig. 2). Upon reacting with oxygen, the epoxidase rapidly forms a FAD C(4a)-hydroperoxide intermediate. Styrene binds rapidly and with high affinity to this intermediate (Kantz et al., 2005; Ukaegbu et al., 2010), which in turn reacts with styrene to form styrene oxide and a highly fluorescent C(4a)-hydroxyFAD intermediate (Kantz & Gassner, 2011). The hydoxyflavin then eliminates water to regenerate oxidized FAD.

The kinetics of the epoxidation reaction can be monitored by absorbance at 382, 368, and 450 nm corresponding to peak absorbance values of the C (4a)-FAD hydroperoxide, C(4a)-FAD hydroxide, and oxidized FAD, respectively (Kantz & Gassner, 2011). Intermediate extinction coefficients are once again in the 10,000 M^{-1} cm^{-1} range, so an enzyme concentration $\leq$15 μM after mixing is recommended for a good quality of data.

The highly fluorescent hydroxyflavin intermediate proves to be and extremely valuable resource for studying the kinetics of the styrene epoxidation reaction. The accumulation and decay of this intermediate can be monitored selectively in a stopped-flow equipped with fluorescence detection by using an excitation wavelengths of 368 nm and monitoring emission with the VG6 band pass filter in front of the emission photomultiplier. The observed epoxidation kinetics occur with a lag phase corresponding to the formation of the flavin-hydroperoxide intermediate followed by a sharp increase in fluorescence corresponding to the formation of the hydroxy-FAD intermediate produced in the styrene epoxidation reaction (Fig. 2). In the last phase of the reaction the fluorescence decays as water is eliminated from the hydroxyFAD with the corresponding regeneration of oxidized FAD, which has relatively weak fluorescence compared to the hydroxyFAD intermediate (Fig. 2). Exponential fitting of the fluorescence data provides rate constants corresponding to the kinetics of the styrene epoxidation and hydroxyFAD dehydration reactions. The relative ease of execution and analysis coupled to the intrinsic sensitivity of the fluorescence stopped-flow experiment identify this approach as a target method for high-throughput screening reactivity of libraries of styrene analogs and active site directed mutants of the epoxidase component of styrene monooxygenases.

3.3 Flavin reduction kinetics of styrene monooxygenase

The reduction reaction of SMOB involves steps of FAD and pyridine nucleotide binding, flavin reduction by hydride-transfer from NADH to FAD, and the dissociation of reduced FAD and NAD^+. The kinetics of the single-turnover reaction of styrene monooxygenase provides insight into the kinetics of each or these reaction steps and the nature of the flavin and pyridine nucleotide complexes formed in catalysis (Morrison, Kantz, Gassner, & Sazinsky, 2013).

Styrene monooxygenase binds FAD with dissociation equilibrium constants in the micromolar range and demonstrates only a weak linkage of flavin redox and binding equilibria such that oxidized FAD binds to SMOB with only slightly higher affinity than reduced FAD (Morrison et al., 2013). FAD plays a central role in nucleating the correct folding of SMOB into an active reductase and only the FAD-bound version of SMOB is stable at the high concentrations required for single turnover studies. Reaction steps leading to and including the reduction of FAD are unaffected by dissolved oxygen and can be studied under anaerobic or aerobic conditions by mixing the complex of SMOB and oxidized FAD with NADH in a stopped-flow spectrophotometer and monitoring the kinetics of FAD reduction at 450nm. Under anaerobic conditions, the reaction ceases once the FAD bound in the active site is reduced. Under aerobic conditions, reduced FAD generated in the hydride transfer reaction is reoxidized by an equivalent amount molecular oxygen to generate superoxide and hydrogen peroxide. If NADH is present in molar excess over dissolved oxygen, the reductase will make the solution anaerobic by the catalytic transformation of oxygen to hydrogen peroxide and superoxide.

3.4 Establishing the rate limiting step in the epoxidation reaction

To establish conditions that are most suitable for efficient enzyme turnover, it is helpful to identify rate-limiting steps in the kinetics of flavin-reduction, transport, and styrene epoxidation reactions. In addition the possibility that protein-protein interactions between the reductase components of SMO may influence reactivity should be evaluated.

Comparison of the kinetics of the reduction and epoxidation reactions helps to establish rate-limiting steps in the reaction mechanisms of two-component SMOs. The kinetics of the binding of reduced FAD to SMOA and the subsequent reaction of the reduced flavin are rapid compared with the kinetics of reduction of FAD by SMOB, so if SMOB is used as the source of the reduced flavin the rate of epoxidation will be limited by the kinetics of FAD reduction. This set of reactions can be readily evaluated by stopped-flow absorbance or fluorescence measurements by rapidly mixing a solution containing equimolar SMOB with oxidized FAD bound and apoSMOA with a solution containing styrene, NADH under aerobic conditions. In this type of experiment it is convenient to monitor the time-dependent absorbance change at 340 nm tracking the consumption of NADH and 450 nm corresponding to the reduction of FAD. The kinetics of the epoxidation reaction limited by the flavin-reduction rate can be monitored by inducing fluorescence of the hydroxy FAD intermediate at 400 nm (longer wavelength selected to avoid overlap with the NADH absorbance band) and monitoring fluorescence emission at 520 nm as described above for studies of the epoxidase as an isolated entity (Fig. 3).

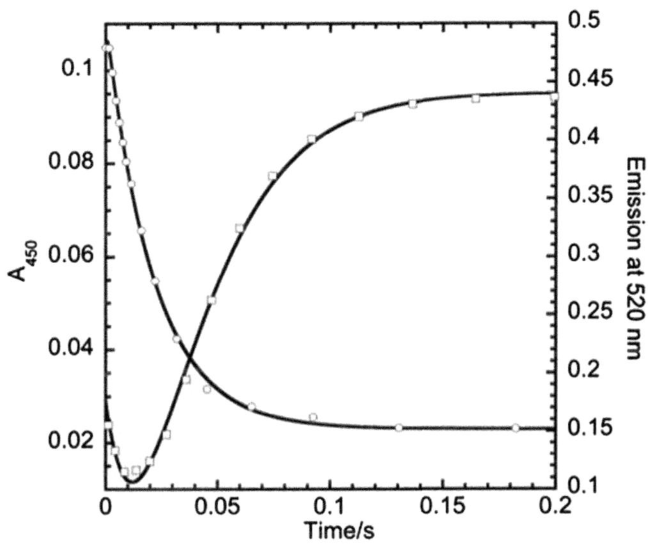

Fig. 3 Pre-steady-state reaction of SMO with NADH, styrene and molecular oxygen. Exponential fits through time-dependent absorbance (○) and fluorescence (□) changes corresponding to the SMOB-dependent FAD reduction and SMOA-dependent styrene epoxidation reactions of SMO indicate that FAD reduction is the rate-limiting step in the styrene epoxidation reaction. *Figure adapted from Morrison, E., Kantz, A., Gassner, G. T., & Sazinsky, M. H. (2013). Structure and mechanism of styrene monooxygenase reductase: New insight into the FAD-transfer reaction.* Biochemistry, *52(35), 6063–6075. https://doi.org/10.1021/bi400763h.*

4. Steady-state kinetic assay of two-component SMOs

The steady-state reaction of styrene monooxygenase can be recorded by using UV-vis spectroscopy to monitor simultaneously the oxidation of NADH to NAD^+ by the reductase and styrene to styrene oxide by the epoxidase. In addition to providing initial rate data that can be used to establish steady-state parameters, the stoichiometry of NADH and styrene consumption can be readily measured from the ratio of these reaction rates and used to establish the efficiency with which the reduction and epoxidation reactions are coupled (Kantz et al., 2005). Under ideal conditions, NADH and styrene are consumed in a 1:1 ratio, but if the reductase activity begins to outpace the epoxidase, excess reduced FAD will accumulate and react with dissolved oxygen outside of the epoxidase active site resulting in flavin-reoxidation and the accumulation of reactive oxygen species in solution.

4.1 Steady-state assay of reductase activity

In steady-state studies using apo reductase, the stock protein should be kept at ice temperature and diluted as needed with ice-cold buffer. Best results are obtained when the apo protein is diluted directly from storage buffer into reaction buffer containing the desired amount of FAD at the target reaction temperature. If serial dilution of the apo protein is required this should be done with ice-cold buffer to avoid activity losses. Care should be taken when handling the apo protein at higher temperatures. At 30 °C the apo protein has a half-life of about 2 min. In the presence of 30 μM FAD the protein half-life increases to about 5 h. A convenient standard activity assay of 1 mL reaction volume can be done in a 1 cm pathlength quartz cuvette at room temperature by the following procedure: Combine 200 μL of 100 mM MOPSO buffer, 30 μL of 1 mM FAD, 650 μL of water and mix. Add 20 μL of 1 μM apo reductase and 100 μL of 1 mM NADH, mix and monitor oxidation of NADH at 340 nm. The initial rate of NADH oxidation can be computed by dividing the slope of the time-dependent absorbance change recorded at 340 nm by the molar extinction coefficient of NADH and the cuvette pathlength.

4.2 Steady-state assay of epoxidase activity

A convenient assay has been developed that can be used to study simultaneously the kinetic activity of the reductase and epoxidase activity of SMOs

(Kantz et al., 2005). This assay takes advantage of the changes in absorbance that occur as NADH is oxidized to NAD^+ and as styrene is oxidized to styrene oxide.

The steady-state rates are computed by inputting apparent initial rate and pathlength data from the time-dependent absorbance changes monitored simultaneously at 245 and 340 nm into Eqs. (1) and (2) below in which NOR and SOR represent the NADH and Styrene Oxidation Rates in units of $\mu M s^{-1}$

$$NOR = \left(\frac{160.77(\mu M \cdot cm)}{b(cm)}\right) \cdot \left(\frac{\Delta A_{340}}{s}\right) \tag{1}$$

$$SOR = \left(\frac{41.207(\mu M \cdot cm)}{b(cm)}\right) \cdot \left(\frac{\Delta A_{340}}{s}\right) - \left(\frac{112.61(\mu M \cdot cm)}{b(cm)}\right) \cdot \left(\frac{\Delta A_{245}}{s}\right) \tag{2}$$

The stoichiometry coupling the NADH and styrene oxidation reactions is computed by from the ratio of these reaction rates with the most efficient coupling giving a ratio of 1.0. The constant terms in Eqs. (1) and (2) are calculated from ratios of the micromolar extinction coefficients and optical pathlength (b) (Kantz et al., 2005). To maximize the range of this assay, an optical cell is generally selected with a pathlength of 0.5 cm or less.

As NADH is oxidized in this assay, the absorbance increases at 245 nm ($\Delta\varepsilon_1{}^M = (+)\ 0.002276\,\mu M^{-1}\ cm^{-1}$) and decreases at 340 nm ($\Delta\varepsilon_2{}^M = (-)\ 0.00622\,\mu M^{-1}\ cm^{-1}$). The oxidation of styrene occurs with a decrease in absorbance at 245 nm ($\Delta\varepsilon_3{}^M = (-)\ 0.00888\,M^{-1}\ cm^{-1}$) and no change in absorbance at 340 nm. Thus, the absorbance at 340 nm will always decrease, but the absorbance at 245 nm may increase, remain nearly constant, or decrease depending on the relative rates of styrene and NADH consumption (see Fig. 4, for example).

For steady-state initial rate measurements, the reductase concentration is typically in the 20–200 nM concentration range and epoxidase in the 1–5 μM concentration range. The reaction is typically monitored over a period of 1–3 min. Under these conditions, there is no net reduction of FAD as the rate of FAD re-oxidation significantly outpaces the rate of FAD reduction. These reaction conditions can be verified by establishing that no significant decrease in absorbance occurs at 450 nm where the difference in molar absorbtivity $\Delta\varepsilon^M$(oxidized-reduced) is $0.0113\,\mu M^{-1}\ cm^{-1}$ for FAD. Once reaction conditions have been established, the initial time-dependent absorbance changes

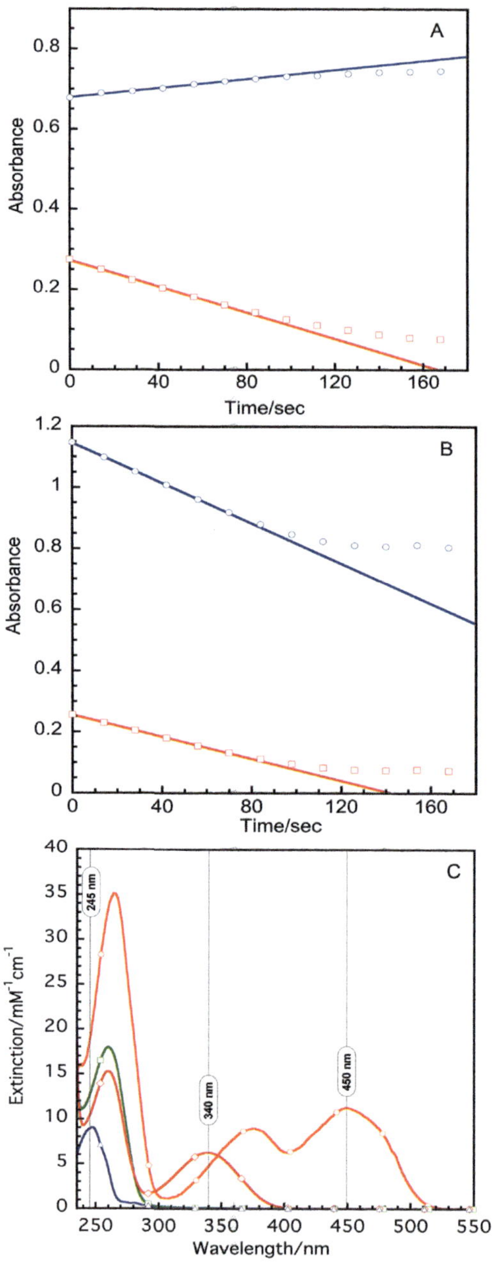

Fig. 4 Assay for evaluating the steady-state kinetics of the reaction of SMO. Reaction of 100 μM NADH, and 270 μM oxygen with (A) 20 nM SMOB in the absence of styrene and (B) 20 nM SMOB and 1 μM SMOA in the presence of 100 μM styrene. Time-dependent absorbance changes monitored at 245 nm (❍) and 340 nm (❑) in a 0.5 cm quartz cuvette. Observed initial reaction rates were calculated from slopes of the best-fitting lines passing through the initial part of the data. Initial Reaction rates and stoichiometries were then computed by using Eqs. (1) and (2) in the text. (C) Absorbance spectra of the components present in the steady-state assay: Flavin adenine dinucleotide (❍), Nicotinamide adenine dinucleotide, oxidized form (□), Nicotinamide adenine dinucleotide, reduced form (◊), and styrene (Δ).

recorded at 245 and 340 nm can be fit with line functions and the slopes of these fits used to compute observed rates of absorbance change at these wavelengths (Fig. 4).

5. Thermodynamic equilibrium properties of SMOs

5.1 Linkage of binding and redox equilibria

The equilibria of FAD binding to the reductase and epoxidase components of SMOs are linked to the redox state of the flavin with the reduced flavin generally binding with highest affinity to the epoxidase and the oxidized flavin binding with highest affinity to the reductase. This arrangement contributes to the efficient transport of reduced flavin from the reductase to the epoxidase and the return of oxidized flavin from the epoxidase to the reductase. The FAD binding equilibria of the epoxidase are further linked to the styrene-binding equilibria (Kantz et al., 2005; Ukaegbu et al., 2010).

The linkage of these binding equilibria illustrated in Fig. 5 can be discussed graphically in terms of thermodynamic cycles and mathematically

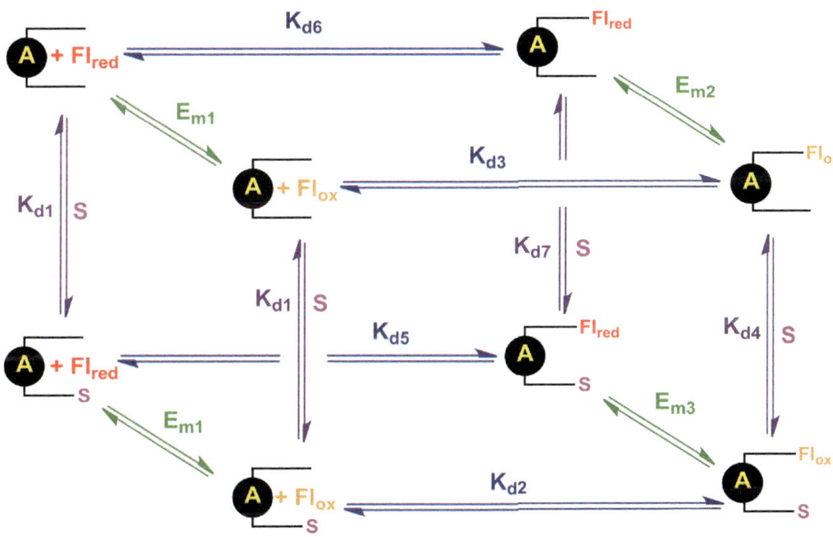

Fig. 5 Representation of the binding and redox equilibria of the epoxidase component of styrene monooxygenase adapted from Ukaegbu et al. (2010). The upper face of the thermodynamic cube represents the redox and FAD-binding equilibria of the substrate-free form of SMOA. The lower face of the cube represents redox and FAD-binding equilibria of the substrate-bound form of SMOA. The four equilibria joining the upper and lower planes represent substrate binding to the apo and FAD-bound forms of SMOA.

by joining the Nernst equations describing the electronic state of the FAD and ligand-binding equilibrium equations describing the interaction of FAD and styrene with SMO (Ukaegbu et al., 2010).

Establishing the linkage electronic and substrate binding equilibria provides insight into extent to which the shuttling of flavin between the reductase and epoxidase components is influenced by the redox state of the flavin and the occupancy of the styrene-binding active site. Approaches to measuring FAD binding equilibria are discussed below.

5.2 Estimation of FAD binding affinity to SMO components

Changes in the electronic state of FAD occur when it binds to the reductase and epoxidase components of SMO. This interaction can be detected experimentally by measuring the intrinsic steady-state fluorescence emission of FAD, which is greater in the bound state than the free state of the coenzyme. This common feature allows the same general quadratic equation (Eq. 3) to be used to estimate the binding affinity of FAD to the reductase and epoxidase components (Morrison et al., 2013; Ukaegbu et al., 2010).

$$F_T = \left(\frac{\left(K_d + [SMOx]_{tot} + [FAD]_{tot}\right) - \sqrt{\left(K_d + [SMOx]_{tot} + [FAD]_{tot}\right)^2 - 4[SMOx]_{tot} \cdot [FAD]_{tot}}}{2}\right) \cdot \left(\epsilon_b - \epsilon_f\right) + [FAD]_{tot} \cdot \epsilon_f \quad (3)$$

In this equation, F_T represents the total fluorescence emission of the free and bound forms of FAD, $[SMOx]_{tot}$ represents the total concentration of reductase or epoxidases monomer present in the titration and ε_f and ε_b represent extinction coefficients for fluorescence emission from the free and bound forms of the flavin. K_d values are computed by using a program such as Kaleidagraph to complete non-linear least squares fitting of the data according to this model. The disparate equilibrium dissociation constants of oxidized FAD for the epoxidase (K_d in the millimolar range) and reductase (K_d in the micromolar range) require different experimental approaches to obtain data suitable for the fitting procedure.

To determine the equilibrium dissociation constant of FAD of styrene monooxygenase reductase, apo enzyme prepared in the hundreds of nanomolar to micromolar range will provide a good result (Morrison et al., 2013). A sensitive research-grade steady-state fluorescence instrument equipped with a temperature controller and magnetic stirrer such as the

Horiba-Jobin-Yvonne Fluorlog-3 will be needed to obtain high quality data. Excitation and emission slits are set to 1 nm and fluorescence is induced in the sample by using light of 450 nm excitation wavelength. The optimum emission wavelength typically $\geq$520 nm is established by emission scanning. Fluorescence emission at this wavelength is used will initially increase hyperbolically as FAD binds to the apo enzyme in the range of K_d and then linearly as the coenzyme concentration is increased in excess of the K_d. These regions of the plot are used to obtain estimates of the K_d (FAD concentration giving half saturation of the enzyme in the hyperbolic phase of the titration), ε_f (estimated from the slope of the linear phase of the titration), and ε_b (estimated from the y-intercept of a line fit through the linear phase of the titration). An example of experimental data fit in this way is shown in Fig. 6.

Much higher experimental ligand concentrations are required to estimate the binding affinity of oxidized FAD to the epoxidase component of SMO. The top reading fluorescence configuration of the Molecular

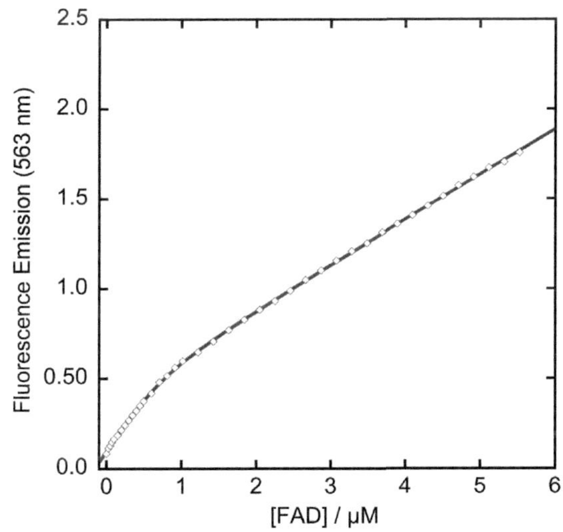

Fig. 6 Equilibrium-titration of styrene monooxygenase reductase with FAD. Increases in the intrinsic steady-state fluorescence emission of FAD observed upon binding to SMOB at 4 °C. The fluorescence increases hyperbolically as the FAD-binding sites approach saturation in the beginning of the titration and then linearly as excess free FAD accumulates later in the titration. Estimates of the K_d of the SMOB-FAD_{ox} complex and fluorescence extinction coefficients corresponding to the bound and free states of FAD are computed by using Eq. (3). *Figure adapted from Morrison, E., Kantz, A., Gassner, G. T., & Sazinsky, M. H. (2013). Structure and mechanism of styrene monooxygenase reductase: New insight into the FAD-transfer reaction.* Biochemistry, *52(35), 6063–6075. doi:10.1021/bi400763h.*

Devices SpectraMax M5 spectrofluorimetric plate reader is well suited to this application (Ukaegbu et al., 2010). By using a black masked 96-fluorescene microplate and a 100 μL sample volume a linear fluorescence response is obtained up to a concentration of 50 μM FAD without the need to introduce a correction for inner-filter effect (Puchalski, Morra, & von Wandruszka, 1991). To complete the titration, apo SMO is first exchanged into 20 mM pH 7 MOPSO buffer containing 5% glycerol and concentrated by centrifugal ultrafiltration by using a Millipore CentriPrep 30 to concentration in the millimolar range. The protein is very apt to foam and bubble due the high concentration so great care in handling is required from this point onward. After recovering the protein from the concentrator, the sample is transferred to an Eppendorf centrifuge tube of known mass. The exact sample concentration is determined spectrophotometrically using the molar extinction coefficient at 280 nm. The sample volume is then estimated by weight by using a precision milligram balance. A small aliquot of a concentrated FAD is then added to the enzyme to give a final total flavin concentration of 50 μM. Appropriate volumes of the protein stock are then diluted in with 20 mM MOPSO buffer containing 5% glycerol and 50 μM FAD in Eppendorf centrifuge tubes. For best result each tube should contain at least 110 μL of sample. The samples are then mixed by spinning at 13,000 × *g* for a minute in an Eppendorf centrifuge. One hundred microliters of each sample is then transferred to the fluorescence microplate. This last step must be done with great finesse to avoid the production of bubbles in the microplate well. Once the transfer is complete the microplate is read top read with fluorescence excitation at 450 nm and emission monitored at 520 nm. The fluorescence data are then fit using Eq. (3) to obtain estimates of the K_d of FAD and extinction coefficients of the bound and free forms of the flavin.

5.3 Estimation of styrene binding affinity

Styrene is rapidly epoxidized upon binding to the (C4a)-hydroperoxyFAD intermediate of SMO ($k_{epoxidation} > 100\,s^{-1}$) (Kantz & Gassner, 2011). For this reason, the apparent K_M of styrene recorded under steady-state conditions (~4.6 μM) contains information about both the kinetics of styrene binding and reacting with the C(4a)-peroxyFAD intermediate of the epoxidase. The apparent K_M therefore does not provide a good estimate of the equilibrium dissociation constant of styrene and the epoxidase.

Isothermal titration calorimetry has been used to obtain an estimate of the equilibrium dissociation constant of the substrate analog, benzene binding to the apo epoxidase (Ukaegbu et al., 2010). Benzene is a competitive inhibitor of styrene ($K_i = 173\,\mu M$), which is unable to react with the C (4a)-peroxyFAD intermediate of SMO (Kantz et al., 2005; Kantz & Gassner, 2011). Benzene is intrinsically more water-soluble than styrene, which allows a significantly higher concentration of benzene to be titrated into the aqueous phase than is possible with styrene.

To obtain an estimate of the equilibrium dissociation constant of benzene and the epoxidase component of SMO a sensitive isothermal titration calorimeter such as the MicroCal VP-ITC model is required. To complete the experiment a solution of 20 μM epoxidase is degassed in 20 mM pH 7 MOPSO buffer at 25 °C and titrated with a solution of 5 mM benzene in the same buffer. Under these conditions the binding reaction is exothermic and a monotonic binding isotherm is obtained. Non-linear least squares analysis provides a K_d value of 4.2 mM (Ukaegbu et al., 2010).

5.4 Linkage of oxidized FAD and substrate binding equilibria

Substrate and FAD bind to the epoxidase with heterotrophic positive cooperativity. The coupling of the oxidized FAD and substrate binding reactions can be established by evaluating the result of fluorescence-monitored titration of the epoxidase with oxidized FAD in buffered solutions containing defined amounts of substrate. As oxidized FAD is titrated, it binds to both the styrene free and bound forms of the epoxidase (see front face of Fig. 5).

Quadratic fits through the FAD-binding data according to Eq. (3) (Section 5.2) provide the apparent equilibrium dissociation constants for FAD at each styrene concentration. The apparent K_d of FAD includes all forms of the FAD-free and FAD-bound enzyme is given in Eq. (4) in which S represent substrate and F represents FAD.

$$K_{d(app)} = \left[\frac{([E] + [E_S])[F]}{[E^F] + [E_S^F]}\right] \tag{4}$$

Substitution of $[E_S]$, $[E^F]$, *and* $[E_S^F]$ with expressions for K_{d1}, K_{d3}, *and* K_{d4} (see Fig. 5) and simplifying gives Eq. (5):

$$K_{d(app)} = \left[\frac{\left(1 + \frac{[S]}{K_{d1}}\right)K_{d3}}{1 + \frac{[S]}{K_{d4}}}\right] \tag{5}$$

Limiting values of this equation at zero substrate concentration and as substrate concentration approaches infinity are K_{d3} and K_{d2}, respectively. Apparent equilibrium dissociation constants of FAD plotted as a function of benzene concentration can then be fit with Eq. (5) after substituting in experimentally determined values of K_{d1} *and* K_{d3} to obtain estimates of K_{d2}, the equilibrium dissociation constant of oxidized FAD for the substrate saturated enzyme. The remaining of substrate binding to the FAD-saturated enzyme (K_{d4}) calculated from the best fitting values of K_{d1}, $K_{d2,}$ *and* K_{d3} by using Eq. (6).

$$K_{d4} = \frac{K_{d1}K_{d2}}{K_{d3}} \tag{6}$$

5.5 Apparent redox potential of FAD bound to SMO

In catalysis, FAD serves as a shuttle for the transport of electrons from the reductase to the epoxidase components of SMO. This mechanism is facilitated by linkage FAD-binding and redox equilibria such that reduced FAD is bound more tightly by the epoxidase and oxidized FAD is bound more tightly by the reductase (Fig. 7).

The apparent midpoint potential of FAD in the presence of each protein component can be estimated by reductive titration. To complete the titration, a buffered solution of the protein component of interest is prepared by gel filtration to replace the sample storage buffer with the desired buffer system for the experiment.

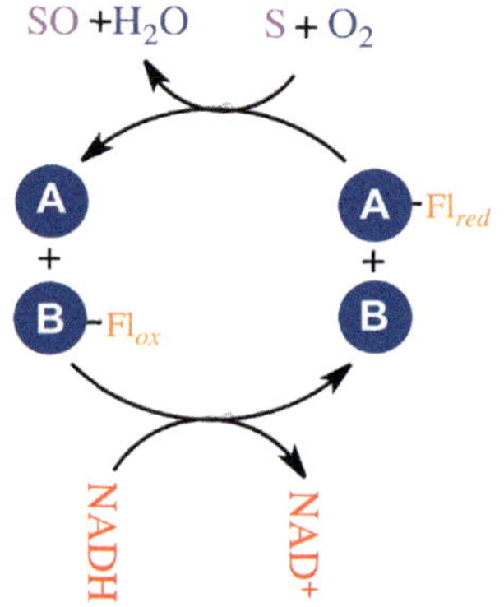

Fig. 7 Flavin exchange in the catalytic cycle of SMO. In the reaction of SMO oxidized FAD is associated primarily with SMOB and reduced FAD is associated primarily with SMOA. The steady-state reaction rate is limited by the kinetics of dehydration of the C4a-hydroxy FAD intermediate of SMOA (Morrison et al., 2013).

Different results are obtained with the epoxidase and reductase components due to their disparate binding affinities for oxidized FAD. Exchange of the holoreductase into the experimental buffer system results in an enzyme sample with oxidized FAD tightly bound. The concentration of FAD bound to the reductase in the recovered stock solution is calculated by using a molar extinction coefficient at 450 nm of 10.8 mM^{-1} cm^{-1}. The epoxidase is stored in the absence of FAD and recovered in experimental buffer in the apo form. The concentration of the epoxidase stock solution recovered from the gel filtration is computed by using its molar extinction coefficient at 280 nm.

A variety of solution potential indicators are commercially available that span a suitable range for the determination of the midpoint potential of FAD bound to the two-component SMOs (Fig. 8). These compounds are soluble in the millimolar range and can be prepared as aqueous stock solutions as needed.

To complete the titration, a buffered 1 mL solution 25–50 μM of the enzyme of interest and a similar concentration of a selected solution potential indicator was prepared. The sample was then transferred to the titration cuvette, made anaerobic by vacuum-gas exchange on the Schlenk line, and fitted with a titrating syringe containing a stock solution containing 1–2 mM sodium hydrosulfite. Data are conveniently acquired with a diode array spectrophotometer.

The relatively high affinity of the reductase for both the oxidized and reduced forms FAD makes it possible to gain a good estimate of the bound-flavin midpoint potential by titration of a solution containing equimolar FAD and protein active sites and a solution potential indicator (Morrison et al., 2013) (Fig. 8).

The concentration of oxidized and reduced forms of flavin and indicator can be computed by deconvolution of experimental spectra acquired in the reductive titration by using a matched set of basis spectra. Experimental spectra recorded during the titration are first transformed to (Reduced-Oxidized) difference spectra by subtracting each of the experimental spectra from the final reduced spectrum (Fig. 8).

It is important that experimental spectra and basis spectra are recorded with the same spectral resolution and signal-to-noise ratio. (In the example provided, samples used for the acquisition of experimental and basis spectra were illuminated with an Ocean optics deuterium lamp and spectra were recorded using an Ocean Optics USB2000+ diode array instrument set to average ten 200 ms spectra with approximately 5 nm spectral resolution).

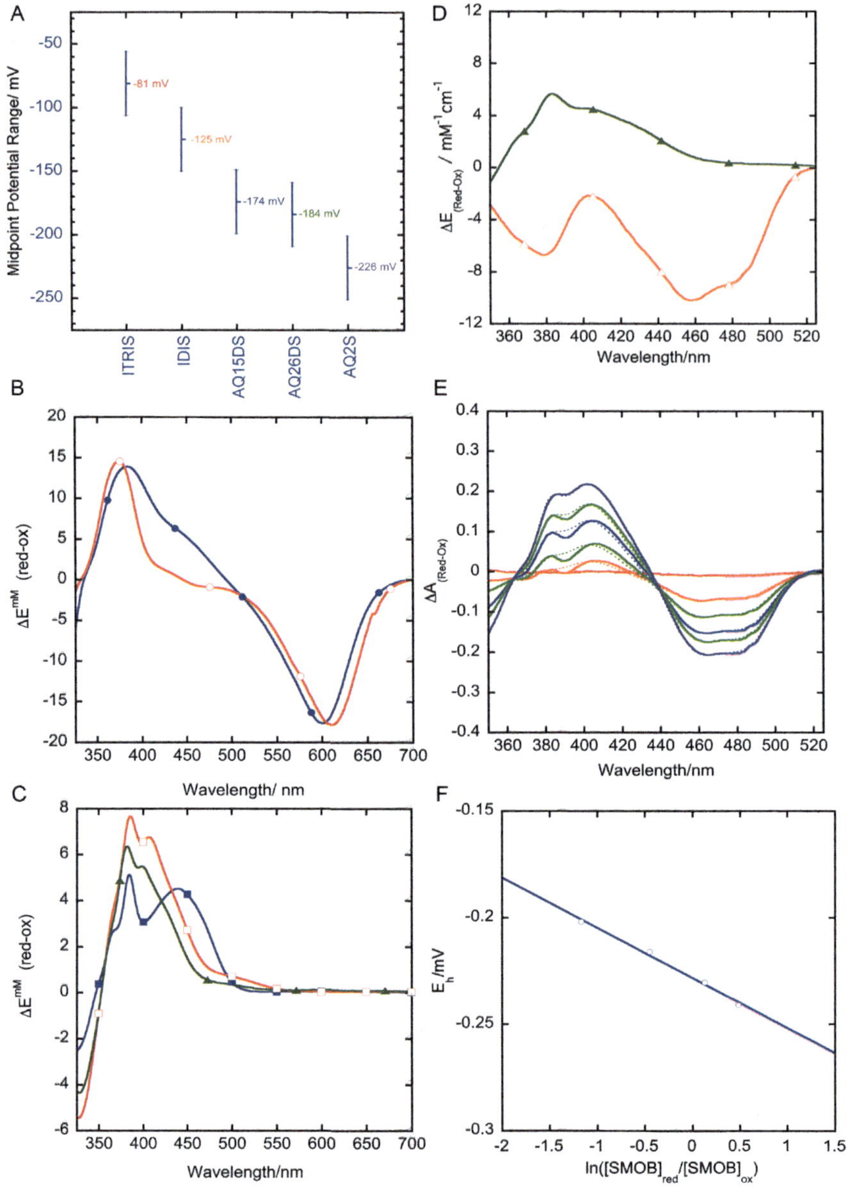

Fig. 8 Estimation of apparent equilibrium midpoint potentials of SMO Components. (A) Effective ranges for solution potential indicators: ITRIS (Indigo 5,5′,7-trisulfonic acid), IDIS (Indigo 5,5′-disulfonic acid), AQ15DS (Anthraquinone 1,5-disulfonic acid), AQ26DS (Anthraquinone 2,6-disulfonic acid), AQ2S (Anthraquinone 2-sulfonic acid); Difference spectra of (B) ITRIS (●) and IDIS (❍), (C) AQ15DS (■), AQ26DS (□), and AQ2S (▲) (D) AQ2S (▲) and SMOB (Δ); (E) Fit of absorbance difference spectra from the reductive titration of SMOB and AQ2S; (F) Plot of solution potentials computed from titration data used to estimate the equilibrium midpoint potential of SMOB. *Adapted from Morrison, E., Kantz, A., Gassner, G. T., & Sazinsky, M. H. (2013). Structure and mechanism of styrene monooxygenase reductase: New insight into the FAD-transfer reaction.* Biochemistry, *52(35), 6063–6075. doi:10.1021/bi400763h.*

The concentration of reduced solution potential indicator and flavin present is first computed by finding the best fitting linear combination of basis spectra for each experimental spectrum recorded in the reductive titration. This is readily achieved by using curve fitting software such as Kaleidagraph or Igor. In Kaleidagraph, this is achieved by first arranging the data to be fit in table configured as follows:

C0	C1	C2	C3-Cn
Wavelength (nm)	(Oxidized-Reduced) Flavin Basis Spectrum (M^{-1} cm^{-1})	(Oxidized-Reduced) Indicator Basis Spectrum (M^{-1} cm^{-1})	Experimental (Oxidized-Reduced) Absorbance Difference Spectra

After plotting the experimental difference spectra, the Kaleidagraph table function is used to find the best linear combination basis spectra to fit the experimental data. A table function is entered into the General Equation Editor window in Kaleidagraph (Eq. 7).

$$M1^{*}table(MO, c0, c1) + M2^{*}table(M0, c0, c2) \tag{7}$$

In this equation, the coefficients M1 and M2 represent the concentrations of reduced solution potential indicator and flavin, respectively. The term M0 represents the experimental absorbance difference spectrum from the titration. The term c0 represents the column containing wavelength data common to the experimental and basis spectra. The terms c1 and c2 represent the columns of molar absorbtivities corresponding to the solution potential indicator and flavin basis spectra.

Once the concentrations of reduced flavin and indicator are determined at each point in the titration, the corresponding concentrations of oxidized species are computed by subtracting from the total concentration of flavin and indicator included in the experiment. The computed indicator concentrations are then substituted into the Nernst equation to determine the solution potential at each point in the data set (Eq. 8).

$$E_h = E_{m7}^{Indicator} - \frac{RT}{nF} ln\left(\frac{[Indicator_{red}]}{[Indicator_{oxidized}]}\right) \tag{8}$$

The natural logarithm of the [reduced flavin]/[oxidized flavin] ratio is then plotted as a function of the computed solution potentials at each titration point (Fig. 8). In accord with the Nernst equation (Eq. 7) the y-intercept of this plot provides the apparent midpoint potential of FAD.

$$E_h = E_{m7}^{app} - \frac{RT}{nF} ln\left(\frac{[SMOx_{FAD(red)}]}{[SMOx_{FAD(ox)}]}\right) \tag{9}$$

In the case that the FAD remains tightly bound in the oxidized and reduced state as is the case for SMOB, the apparent midpoint potential is the same as the bound-FAD midpoint potential. In the case of more weakly bound FAD, both free and bound states of the flavin will contribute to the value of E_{m7}^{app}. In this case additional titrations will be needed before the bound-FAD midpoint potential can be determined as is discussed in Section 5.6.

5.6 Linkage of FAD redox and ligand binding equilibria

The epoxidase binds to reduced FAD with a much higher affinity than it binds to oxidized FAD. For this reason, bound and free forms of FAD are present throughout the course of the titration with free, oxidized FAD predominating the beginning of the titration and bound, reduced FAD predominating toward the end of the titration (Ukaegbu et al., 2010).

The extent to which the epoxidase is saturated with substrate further influences the apparent midpoint potential of bound FAD. The bound and free states involved are summarized in Fig. 7. A general strategy for obtaining estimates of the bound flavin midpoint potential in the presence or absence of substrates is outlined below.

In the absence of substrate, the apparent midpoint potential is a function of flavin and enzyme concentration. The Nernst equation that describes this system includes two free and bound forms of FAD as indicated in Eq. (10).

$$\boldsymbol{E_h = E_{m7(app)} - \frac{RT}{nF} ln\left(\frac{[FAD_{red}] + \frac{[FAD_{red}][SMOA]}{K_{d6}}}{[FAD_{ox}] + \frac{[FAD_{ox}][SMOA]}{K_{d3}}}\right)} \tag{10}$$

Comparison of $Em_{7(app)}$ to the midpoint potential of free FAD provides information about the linkage of the FAD-binding and redox equilibria.

For example: If E_{m7app} is more positive than E_{m7}(FAD) this implies that reduced FAD binds to the enzyme with higher affinity than oxidized FAD.

The E_{m7} of FAD in the protein-bound state can be computed after completing a series of solution potential measurements over a range of total protein concentrations by using the approach described in Section 5.4. These data sets are then fit to determine the apparent E_{m7} value at each protein concentration by using Eq. (9) (Fig. 9).

Substitution of the Nernst expression for free FAD into Eq. (9) gives Eq. (11), which describes $E_{m7}^{FAD(app)}$ as a function of protein concentration and the midpoint potential of free FAD (Ukaegbu et al., 2010).

$$E_{m7}^{FAD(app)} = E_{m7}^{FAD} + \frac{RT}{nF} \ln \left(\frac{1 + \frac{[SMOA]}{K_{d6}}}{1 + \frac{[SMOA]}{K_{d3}}} \right) \tag{11}$$

The limit of this expression as protein concentration approaches zero gives the midpoint potential of free FAD. The limiting value of the apparent midpoint potential of FAD as the protein concentration approaches infinity is the midpoint potential of bound FAD (Eqs. 12 and 13) (Fig. 9A).

$$\lim_{[SMOA]\to 0} \left(E_{m7(app)} \right) = E_{m7}^{FAD} \tag{12}$$

$$\lim_{[SMOA]\to \infty} \left(E_{m7(app)} \right) = E_{m2} = E_{m7}^{FAD} + \frac{RT}{nF} \ln \left(\frac{K_{d6}}{K_{d3}} \right) \tag{13}$$

The linkage of substrate and FAD binding and electronic equilibria can be investigated similarly by fitting data from a series apparent redox potential measurements in solutions containing a defined amount of substrate over a range of protein concentrations (Ukaegbu et al., 2010). In this case, the relationship between the apparent midpoint potential of bound FAD relative to that of free FAD is given by Eq. (14).

$$E_{m7}^{app(FAD)} = E_{m7}^{FAD} + \frac{RT}{nF} \ln \left(\frac{1 + \frac{[SMOA][S]}{K_{d1}K_{d5}}}{1 + \frac{[SMOA][S]}{K_{d1}K_{d2}}} \right) \tag{14}$$

To obtain estimates of the midpoint potential of FAD bound to the SMOA-substrate complex (E_{m3}) and the equilibrium dissociation constant of reduced FAD for the NSMOA-substrate complex (K_{d5}), a series of equilibrium redox titrations are first completed over a range of SMOA concentrations in the presence of substrate and spectroelectrochemical solution potential indicator. Apparent equilibrium midpoint potentials corresponding to each protein concentration are then computed as above (Fig. 9B).

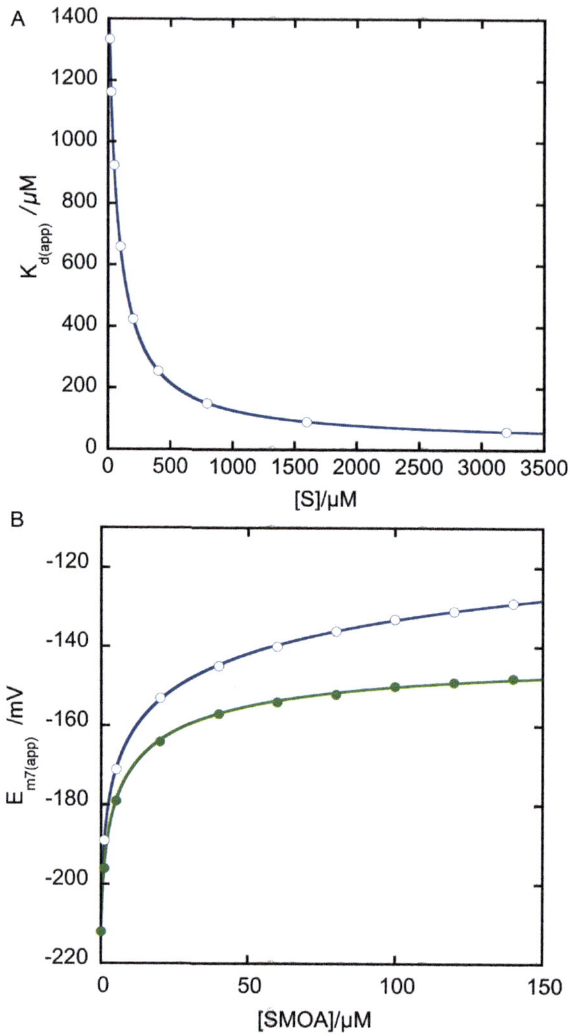

Fig. 9 Equilibrium dissociation constant and midpoint potentials of FAD. (A) Plot of the apparent K_d FAD as a function of substrate analog concentration used to compute the K_d of oxidized FAD in the presence of saturating substrate analog; (B) Simulated plot of the apparent FAD midpoint potential as a function SMOA concentration used to compute the midpoint potential of bound FAD in the presence (●) or absence (○) of saturating substrate analog. *Adapted from Ukaegbu, U. E., Kantz, A., Beaton, M., Gassner, G. T., & Rosenzweig, A. C. (2010). Structure and ligand binding properties of the epoxidase component of styrene monooxygenase.* Biochemistry, 49*(8), 1678–1688. doi:10.1021/bi901693u.*

The apparent midpoint potentials determined in this way are then plotted as a function of [SMOA] and fit with Eq. (14) after substituting in known values of E_{m1}, K_{d1}, K_{d2}, and $[S]$ to obtain an estimate of K_{d5}. Limiting values of the apparent midpoint potential as protein concentration approaches zero and infinity are the solution potentials of free FAD and FAD bound to the complex of NSMOA and substrate. The value of $\boldsymbol{E_{m3}}$ is computed by using Eq. (15) after substituting in experimentally determined values K_{d1}, K_{d2}, K_{d5}, and E_{m1} (Ukaegbu et al., 2010).

$$\lim_{[SMOA \to \infty]} E_{m7}^{app} = E_{m3} = E_{m1} + \frac{RT}{nF} ln\left(\frac{K_{d2}}{K_{d5}}\right) \tag{15}$$

6. Summary and conclusion

The modularity structure of the two-component flavin monooxygenases offers significant flexibility both from the standpoint of enzymological study and as a resource for practical development in the field of biocatalysis (Heine et al., 2018; Oelschlagel, Zimmerling, & Tischler, 2018). Styrene monooxygenases represent a novel members of the two-component enzyme family with respect to their solvent tolerance and unique ability catalyze enantioselective epoxidation of styrene and an array of styrene analogs suspended the aqueous phase (Oelschlagel et al., 2018).

The modular structure of SMOs allows the flavin reduction and styrene oxygenation reactions to be studied as orchestrated by both the isolated components and in the context of the fully assembled SMO system. The modularity of this system is a further asset in the field of biocatalysis where the isolated epoxidase component can be readily modified by site directed mutagenesis without impacting the reductase activity, implemented as a standalone biocatalyst (Hollmann, Hofstetter, & Schmid, 2006), or as an engineered reductase-epoxidase fusion protein (Corrado et al., 2018; Heine et al., 2017).

This chapter addresses approaches to the experimental study and characterization of two-component SMOs that have proven to be of value in the biochemical characterization of wild-type two-component systems, site-directed mutants, as well as genetically engineered, and naturally occurring SMO fusion proteins.

It is anticipated that the approaches presented here will be of value to researchers engaged in the purification, mechanistic characterization, and development of existing newly discovered SMOs.

References

Corrado, M. L., Knaus, T., & Mutti, F. G. (2018). A chimeric styrene monooxygenase with increased efficiency in asymmetric biocatalytic epoxidation. *Chembiochem*, *19*(7), 679–686. https://doi.org/10.1002/cbic.201700653.

Dagley, S. (1987). Lessons from biodegradation. *Annual Review of Microbiology*, *41*, 1–23. https://doi.org/10.1146/annurev.mi.41.100187.000245.

Heine, T., Tucker, K., Okonkwo, N., Assefa, B., Conrad, C., Scholtissek, A., et al. (2017). Engineering styrene monooxygenase for biocatalysis: Reductase-epoxidase fusion proteins. *Applied Biochemistry and Biotechnology*, *181*(4), 1590–1610. https://doi.org/10.1007/s12010-016-2304-4.

Heine, T., van Berkel, W. J. H., Gassner, G., van Pee, K. H., & Tischler, D. (2018). Two-component FAD-dependent monooxygenases: Current knowledge and biotechnological opportunities. *Biology*, 7(3), 1–35. https://doi.org/10.3390/biology7030042.

Hollmann, F., Hofstetter, K., & Schmid, A. (2006). Non-enzymatic regeneration of nicotinamide and flavin cofactors for monooxygenase catalysis. *Trends in Biotechnology*, *24*(4), 163–171. https://doi.org/10.1016/j.tibtech.2006.02.003.

Huijbers, M. M., Montersino, S., Westphal, A. H., Tischler, D., & van Berkel, W. J. (2014). Flavin dependent monooxygenases. *Archives of Biochemistry and Biophysics*, *544*, 2–17. https://doi.org/10.1016/j.abb.2013.12.005.

Kantz, A., Chin, F., Nallamothu, N., Nguyen, T., & Gassner, G. T. (2005). Mechanism of flavin transfer and oxygen activation by the two-component flavoenzyme styrene monooxygenase. *Archives of Biochemistry and Biophysics*, *442*(1), 102–116. https://doi.org/10.1016/j.abb.2005.07.020.

Kantz, A., & Gassner, G. T. (2011). Nature of the reaction intermediates in the flavin adenine dinucleotide-dependent epoxidation mechanism of styrene monooxygenase. *Biochemistry*, *50*(4), 523–532. https://doi.org/10.1021/bi101328r.

Maehly, A. C., Smith, L., & Graham, J. D. (1955). Anaerobic cuvette. *Science*, *122*(3173), 767.

Morrison, E., Kantz, A., Gassner, G. T., & Sazinsky, M. H. (2013). Structure and mechanism of styrene monooxygenase reductase: New insight into the FAD-transfer reaction. *Biochemistry*, *52*(35), 6063–6075. https://doi.org/10.1021/bi400763h.

Oelschlagel, M., Zimmerling, J., & Tischler, D. (2018). A review: The styrene metabolizing cascade of side-chain oxygenation as biotechnological basis to gain various valuable compounds. *Frontiers in Microbiology*, *9*, 490. https://doi.org/10.3389/fmicb.2018.00490.

Otto, K., Hofstetter, K., Rothlisberger, M., Witholt, B., & Schmid, A. (2004). Biochemical characterization of StyAB from *Pseudomonas* sp. strain VLB120 as a two-component flavin-diffusible monooxygenase. *Journal of Bacteriology*, *186*(16), 5292–5302. https://doi.org/10.1128/JB.186.16.5292-5302.2004.

Palfey, B. A., & McDonald, C. A. (2010). Control of catalysis in flavin-dependent monooxygenases. *Archives of Biochemistry and Biophysics*, *493*(1), 26–36. https://doi.org/10.1016/j.abb.2009.11.028.

Puchalski, M. M., Morra, M. J., & von Wandruszka, R. (1991). Assessment of inner filter effect corrections in fluorimetry. *Fresenius' Journal of Analytical Chemistry*, *340*, 341–344. https://doi.org/10.1007/BF00321578.

Romero, E., Gomez Castellanos, J. R., Gadda, G., Fraaije, M. W., & Mattevi, A. (2018). Same substrate, many reactions: Oxygen activation in flavoenzymes. *Chemical Reviews*, *118*(4), 1742–1769. https://doi.org/10.1021/acs.chemrev.7b00650.

Tischler, D., Kermer, R., Groning, J. A., Kaschabek, S. R., van Berkel, W. J., & Schlomann, M. (2010). StyA1 and StyA2B from *Rhodococcus opacus* 1CP: A multifunctional styrene monooxygenase system. *Journal of Bacteriology*, *192*(19), 5220–5227. https://doi.org/10.1128/JB.00723-10.

Turner, N. J., & Kumar, R. (2018). Editorial overview: Biocatalysis and biotransformation: The golden age of biocatalysis. *Current Opinion in Chemical Biology*, *43*, A1–A3. https://doi.org/10.1016/j.cbpa.2018.02.012.

Ukaegbu, U. E., Kantz, A., Beaton, M., Gassner, G. T., & Rosenzweig, A. C. (2010). Structure and ligand binding properties of the epoxidase component of styrene monooxygenase. *Biochemistry*, *49*(8), 1678–1688. https://doi.org/10.1021/bi901693u.

CHAPTER SEVENTEEN

Flavin-N5-oxide intermediates in dibenzothiophene, uracil, and hexachlorobenzene catabolism

Sanjoy Adak, Tadhg P. Begley*
Department of Chemistry, Texas A&M University, College Station, TX, United States
*Corresponding author: e-mail address: begley@tamu.edu

Contents

Abstract

Flavin-N5-oxide is a new intermediate in flavoenzymology. Here we describe the identification of DszA (dibenzothiophene catabolism), RutA (uracil catabolism) and HcbA1 (hexachlorobenzene catabolism) as flavin-N5-oxide-utilizing enzymes. Mechanistic analysis of these reactions suggests a model for the identification of other examples of this catalytic motif.

Methods in Enzymology, Volume 620
ISSN 0076-6879
https://doi.org/10.1016/bs.mie.2019.03.020

1. Introduction

The long-held idea that all flavin-dependent oxygenases use flavin-C4a-hydroperoxide as the oxygenating species was challenged when Bradley Moore and Bruce Palfey identified EncM on the enterocin biosynthetic pathway as a flavin-N5-oxide-utilizing-enzyme (Fig. 1) (Saleem-Batcha et al., 2018; Teufel, 2017, 2018; Teufel, Agarwal, & Moore, 2016; Teufel et al., 2013, 2015). The identification of this intermediate raised the question as to its generality in flavin-dependent oxygenases. In this chapter, we describe the identification of flavin-N5-oxide intermediates in dibenzothiophene, uracil, and hexachlorobenzene catabolism.

2. General protocols for enzyme overexpression and purification

The DszA gene from *Rhodococcus erythropolis* was cloned into the pTHT vector (Adak & Begley, 2016) (derivative of pET28b with a TEV protease cleavage site after the N-terminal His-tag). The DszA-pTHT plasmid was transformed into electrocompetent *Escherichia coli* BL21 (DE3) by electroporation. A single colony was inoculated in 90 mL of LB medium containing 40 μg/mL of kanamycin. This starter culture was grown at 37 °C overnight with agitation (220 rpm). LB medium (9 L, Lennox, 20 g/L) containing 40 μg/mL of kanamycin was inoculated with 90 mL of starter culture. The cells were grown at 37 °C with shaking (220 rpm) until the culture reached an OD_{600} of 0.6 (3–3.5 h). After incubating the flasks

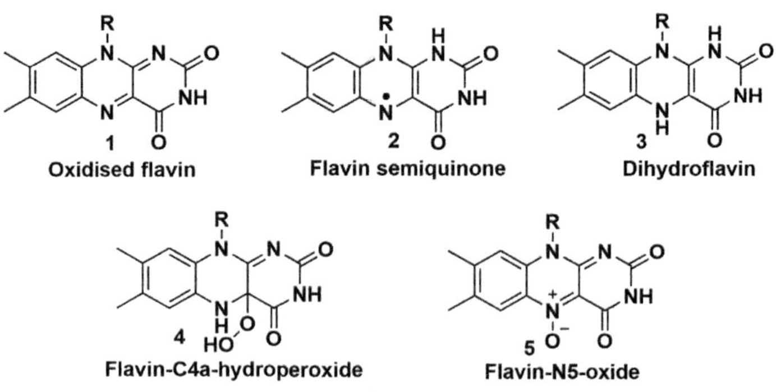

Fig. 1 The structures of relevant flavins discussed in the chapter.

at 4 °C for ~1 h, without shaking, the cultures were induced by adding isopropyl-β-D-thiogalactoside (IPTG) to a final concentration of 0.4 mM. For overproduction of the protein, the cultures were further incubated at 15 °C for 15 h with shaking (180 rpm). The cells were then harvested by centrifugation at 10,000 *g* for 10 min at 4 °C. Typical yields were 20–25 g cell pellet (wet weight) from the 9 L of culture.

Cells from 9 L of culture were re-suspended in 30 mL of binding buffer (50 mM KH_2PO_4, 150 mM NaCl, 10 mM imidazole, pH 8.0) and approximately 5 mg of lysozyme was added. The cells were then lysed by five cycles of sonication on ice (each cycle consisted of a 1 s pulse followed by a 1 s delay repeated 15 times with 5 min of cooling between cycles.

The cell debris was removed by centrifugation at 39,000*g* for 40 min at 4 °C. The clarified supernatant was loaded onto a 5 mL Ni-NTA-affinity column pre-equilibrated with binding buffer kept at 4 °C. The Ni-NTA-affinity column was then washed with 50 mL wash buffer (50 mM KH_2PO_4, 150 mM NaCl, 20 mM imidazole, pH 8.0). The protein was eluted from the column with elution buffer (50 mM KH_2PO_4, 150 mM NaCl, 200 mM imidazole, pH 8.0) at 4 °C. The fractions containing protein were pooled and concentrated to 3 mL using YM-10 Amicon centrifugal ultrafilters at 5000*g*. The concentrated sample was buffer exchanged into 100 mM phosphate buffer at pH 7.5 containing 100 mM NaCl and 15% glycerol using an Econo-Pac 10DG desalting column. Protein concentration was determined from its absorbance at 280 nm (A_{280}) with an extinction coefficient calculated using the ProtParam tool of the ExPASy proteomics Server. A typical yield was 6 mg/L.

The RutA gene encoded in the pCA24N overexpression plasmid (Mukherjee, Zhang, Abdelwahed, Ealick, & Begley, 2010) was transformed into *Escherichia coli* BL21 (DE3) by electroporation. For protein overexpression, a starter culture was grown overnight in 90 mL of LB medium containing 20 μg/mL of chloramphenicol at 37 °C. LB medium (9 L, 20 g/L) containing 20 μg/mL of chloramphenicol was inoculated with this starter culture. RutA overexpression and purification were carried out as described above for DszA. A typical yield was 20 mg/L.

The HcbA1 gene from *Nocardioides* sp. strain PD653 was cloned into the pTHT vector. Electrocompetent *Escherichia coli* BL21 (DE3) were transformed with HcbA1-pTHT by electroporation. HcbA1 overexpression and purification were carried out as described above for DszA. A typical yield was 10 mg/L.

3. Analytical methods

3.1 HPLC parameters

An Agilent 1260 HPLC, equipped with a quaternary pump, was used. The system included a diode array UV–Vis detector and products were detected using absorbance at 254, 280, and 450 nm. Analysis was performed either on a ZORBAX Eclipse XDB-C18 column (15 cm × 4.6 mm, 5 μm particles, Agilent Technologies) or a PFP column (150 × 4.6 mm, 2.6 μm particles, Kinetex). Data were processed using Agilent ChemStation ver. B.04.01.

HPLC conditions for the analysis of the DszA reaction mixture (using a C18 column):

A—Water

B—10 mM ammonium acetate buffer, pH 6.6

C—Methanol

HPLC method (flow rate 1 mL/min):

0 min—90% A 10% B, 2 min—90% A 10% B, 22 min—20% A 10% B 70% C, 27 min—20% A 10% B 70% C, 28 min—90% A 10% B, 36 min—90% A 10% B

HPLC conditions for the analysis of the RutA reaction mixture (using a C18 column):

A—Water

B—100 mM potassium phosphate buffer, pH 6.6

C—Methanol

HPLC method (flow rate 1 mL/min):

0 min—100% B, 5 min—10% A 90% B, 12 min—48% A 40% B 12% C, 14 min—50% A 30% B 20% C, 18 min—30% A 10% B 60% C, 20 min—100% B, 25 min—100% B

HPLC conditions for the analysis of the HcbA1 reaction mixture (using a C18 column):

A—Water with 0.1% phosphoric acid

B—Acetonitrile

HPLC method (flow rate 1 mL/min):

0 min—10% A 90% B, 30 min—10% A 90% B

HPLC conditions for separation of FMN and FMN-N5-oxide (using a PFP column):

A—Water

B—100 mM potassium phosphate buffer, pH 6.6

C—Methanol

HPLC method (flow rate 0.4 mL/min):

0 min—90% A 10% B, 2 min—90% A 10% B, 22 min—20% A 10% B 70% C, 27 min—20% A 10% B 70% C, 28 min—90% A 10% B, 36 min—90% A 10% B

3.2 LC-MS parameters

LC-ESI-TOF-MS was performed using an Agilent 1260 HPLC system equipped with a binary pump and a 1200 series diode array detector followed by a MicroToF-Q II mass spectrometer (Bruker Daltonics) using an ESI source either in negative mode or positive mode. Analysis was performed on an LC-18-T column (15 cm × 3 mm, 3 μm particles, Supelco). Data were processed with DataAnalysis ver. 4.0 (Bruker Daltonics).

LC conditions:

A—5 mM Ammonium acetate buffer, pH 6.6

B—75% Methanol and 25% Water.

LC method (for positive and negative mode on MS):

0 min—100% A, 2 min—100% A, 12 min—30% A 70% B, 17 min—30% A 70% B, 18 min—100% A, 30 min—100% A

The mass spectrometer parameters: Capillary, 4500 V; end plate offset, 500 V; nebulizer gas, 3.0 bar; Dry gas, 10.0 L/min; dry gas temperature, 200 °C; funnel 1 RF, 200.0 Vpp; funnel 2 RF, 200.0 Vpp; ISCID, 0.0 eV; hexapole RF, 100 Vpp; quadrupole, Ion energy, 3.0 eV; low mass, 100 *m/z*; collision cell, collision energy, 8.0 eV; collision RF, 150.0 Vpp; transfer time, 80.0 μs; prepulse storage, 5.0 μs.

4. Chemical methods

4.1 Synthesis of FMN-N5-oxide standard

The synthesis of FMN-N5-oxide has been previously described (Mansurova, Koay, & Gaertner, 2008; Teufel et al., 2013). One can therefore easily compare the putative flavin-N5-oxide formed in the enzymatic reaction mixture with a chemically synthesized authentic sample. The scheme for the synthesis is shown in Fig. 2. In this route, 2,3 dimethylaniline **6** is converted to riboaniline **7** which is then reacted sequentially with chlorouracil and acetic anhydride to give compound **8**. Treatment of **8** with sodium nitrite in glacial acetic acid yields compound **9**. Deprotection using ammonia in methanol gives riboflavin-N5-oxide **10**. Finally, enzymatic phosphorylation, using riboflavin kinase, gives the desired FMN-N5-oxide **5** (Adak & Begley, 2016).

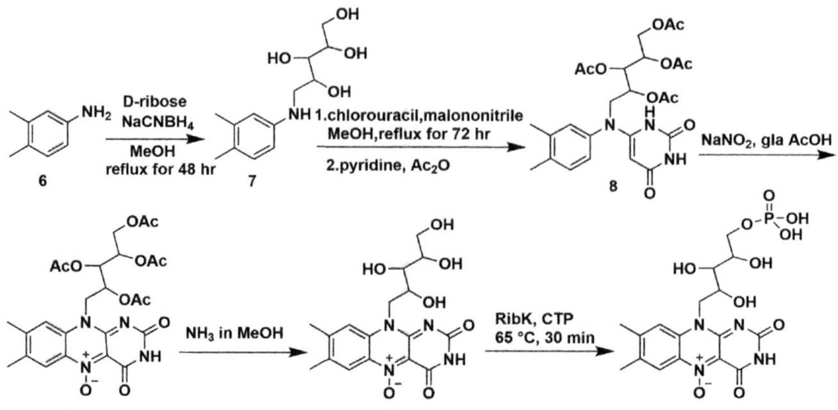

Fig. 2 The chemical synthesis of FMN-N5-oxide.

4.2 Photochemical reduction of flavin

Photoreduction of flavin is an effective way to produce reduced flavin that is free from excess reducing agent. The flavin is reduced by irradiating for 5 min under anaerobic conditions in a glove box in the presence of EDTA as the reducing agent (A21 LED white light bulb, 100-W) (Enns & Burgess, 1965; Heelis, 1982; McCormick, Koster, & Veeger, 1967).

4.3 Reaction procedure for flavin-N5-oxide detection

A typical reaction procedure for the detection of flavin-N5-oxide is as follows. A solution containing FMN (500 μM) and EDTA (10 mM) is irradiated with white light (A21 LED light bulb, 100-W) for 5 min under anaerobic conditions. Once flavin is photo-reduced, substrate and enzyme are added. Typical final concentrations of reduced flavin, substrate and enzyme are 100 μM. The reaction mixture is incubated under anaerobic conditions for 30 min and then exposed to atmospheric oxygen for an additional 30 min. The reaction is quenched by heat-denaturation of the protein and the solution is filtered (10 kDa cut-off). The samples are analyzed by HPLC (PEP column is effective for the separation of flavin and flavin-N5-oxide) and LC-MS.

5. Enzyme studies

5.1 Mechanistic studies on the DszA-catalyzed reaction

Sulfur dioxide, a known environment pollutant, is released during the combustion of sulfur-containing petroleum. While there are several methods for the catalytic desulfurization of petroleum, these are not effective for the desulfurization of compounds like dibenzothiophene, the most abundant sulfur-containing heterocycle found in crude petroleum (Soleimani, Bassi, & Margaritis, 2007). The search for a microbial desulfurization process has resulted in the identification of a dibenzothiophene catabolic pathway in *Rhodococcus erythropolis* (Fig. 3) (Denome, Oldfield, Nash, & Young, 1994; Denome, Olson, & Young, 1993; Gray et al., 1996; Piddington, Kovacevich, & Rambosek, 1995). The pathway involves thioether oxidation catalyzed by the DszC (Barbosa, Neves, Sousa, Ramos, & Fernandes, 2018; Gonzalez-Osorio, Luong, Jirde, Palfey, & Vey, 2016; Guan et al., 2015) flavoenzyme, oxidative C–S bond cleavage catalyzed by the DszA flavoenzyme and a final C–S bond cleavage catalyzed by DszB (Lee, Ohshiro, Matsubara, Izumi, & Tanokura, 2006). The DszA catalyzed C–S bond cleavage is a novel reaction meriting further mechanistic exploration.

The current mechanism for the DszA-catalyzed reaction is shown in Fig. 4 (Adak & Begley, 2016, 2017a). In this mechanism, the flavin-C4a-peroxide **4** is added to **13** to give **16**; release of oxidized flavin gives hydroperoxide **17**, reduction of **17** gives **18**, which upon C–S bond cleavage and tautomerization gives the final product **14**. This mechanism is consistent with the requirement of the reaction for molecular oxygen, and the incorporation of oxygen from molecular oxygen and not from water into the product **14**.

Since there are no cysteine residues in the DszA sequence, the reduction of peroxide **17** using peroxiredoxin type chemistry (Hall, Karplus, & Poole, 2009) is not possible. The possibility of flavin functioning as the reducing agent was therefore considered (Saleem-Batcha et al., 2018; Teufel, 2017, 2018; Teufel et al., 2016, 2013, 2015). Since flavin-N5-oxide **5**, the

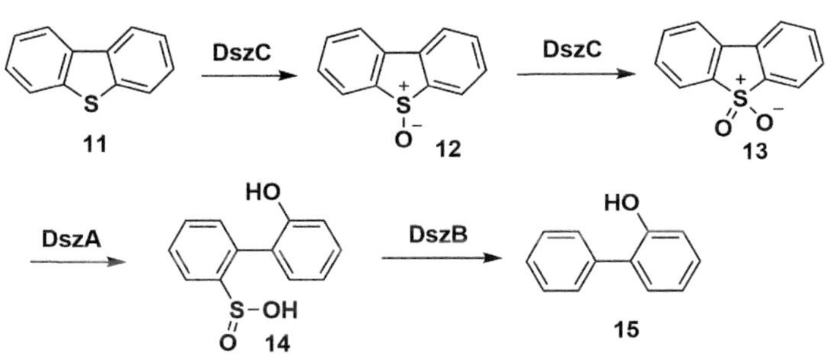

Fig. 3 The dibenzothiophene catabolic pathway in *Rhodococcus erythropolis*.

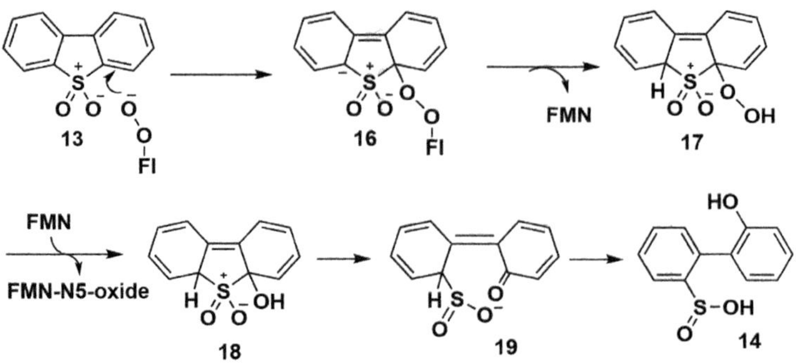

Fig. 4 Mechanistic proposal for the DszA-catalyzed reaction involving flavin-N5-oxide formation.

expected flavin product, is readily converted to flavin by reducing agents, it was essential to develop reaction conditions free of reducing agent. Thus, instead of using Fre and NADH, reduced flavin was generated in the glove box by photoreduction using EDTA as the electron source. Then enzyme and substrate were added and the reaction mixture was incubated in the glove box for 30 min before exposure to oxygen to complete the reaction. Under these conditions it was possible to detect the formation of a new flavin-derived species which was identified as FMN-N5-oxide **5** based on LC-MS analysis and coelution with chemically synthesized FMN-N5-oxide. When the reaction was carried out in the presence of $^{18}O_2$, the predicted 2 Da increase in the FMN-N5-oxide mass was observed.

5.2 Mechanistic studies on the RutA catalyzed reaction

In 2006 Sydney Kustu's group identified an operon in *Escherichia coli* K12 responsible for uracil catabolism (Loh et al., 2006). In this catabolic pathway, RutA catalyzes the conversion of uracil to 3-ureidoacrylic acid **21** (Fig. 5A). Although this reaction seems like a simple amide hydrolysis, a labeling experiment showed that molecular oxygen is the source of one of the carboxylate oxygens of **21**; thus, RutA catalyzes an oxidative amide bond cleavage reaction instead of an amide hydrolysis (Mukherjee et al., 2010). Our current mechanistic proposal for this reaction is shown in Fig. 5B (Adak & Begley, 2017a, 2017b). The flavin-C4a-peroxide **4** is added to uracil to give the tetrahedral intermediate **23**; C–N bond cleavage and release of oxidized flavin give the peracid intermediate **25**; reduction of this peracid, by flavin, gives the final product **21**. The absence of any active site cysteine residue suggests that oxidized flavin reduces peracid **25** to form **21** and a

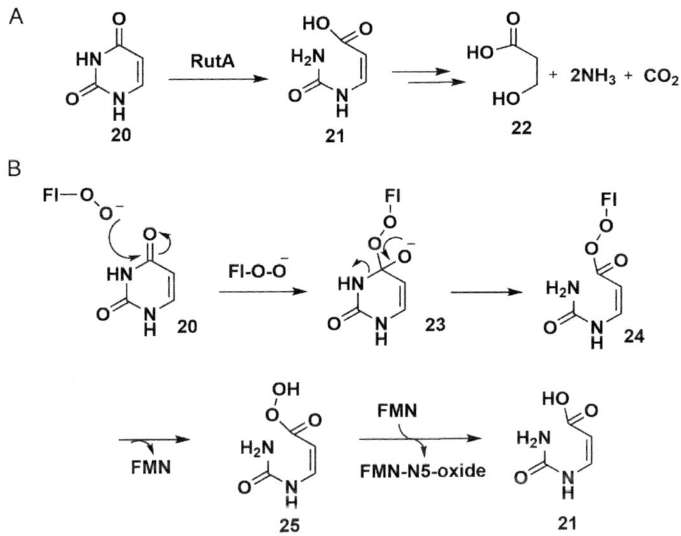

Fig. 5 (A) The uracil catabolic pathway. (B) The mechanistic proposal for the RutA-catalyzed reaction.

flavin-N5-oxide **5**. Similar strategies to those used for the DszA system were followed to characterize this putative flavin-N5-oxide. To avoid a reducing environment in which the flavin-N5-oxide is unstable, the RutA-catalyzed reaction was carried out in the presence of photo-reduced FMN. Under these conditions the formation of FMN-N5-oxide was observed and confirmed by LC-MS analysis and coelution with an authentic standard. When the reaction was carried out in the presence of $^{18}O_2$, the predicted 2 Da increase in the FMN-N5-oxide mass was observed.

5.3 Mechanistic studies on the HcbA1-catalyzed reaction

Hexachlorobenzene (HCB, **34**) is a well-known fungicide that has been extensively used in agriculture since the 1940s (Barber, Sweetman, van Wijk, & Jones, 2005). Although HCB is banned under the Stockholm convention on persistent organic pollutants, it is still widely distributed across the globe. As part of a bioremediation program, a gene cluster, responsible for the conversion of HCB to pentachlorophenol (PCP, **26**), has been identified in *Nocardioides* sp. strain PD653 (Ito et al., 2017, 2018). This gene cluster consists of a putative flavin monooxygenase HcbA1, and two putative flavin reductases HcbA2 and HcbA3. Experiments with cell lysate provided evidence that HcbA1 and HcbA3 together can convert HCB to PCP and we have demonstrated that purified HcbA1 alone can convert HCB to PCP in the presence of a heterologous flavin reductase (Fig. 7A).

The best-studied flavin-dependent dehalogenases are PcpB and iodotyrosine deiodinase (Hu, Chuenchor, & Rokita, 2015; Phatarphekar, Buss, & Rokita, 2014; Sun, Su, & Rokita, 2017; Thomas, McTamney, Adler, LaRonde-LeBlanc, & Rokita, 2009) (Fig. 6A and B). PcpB catalyzes the conversion of pentachlorophenol **26** to tetrachlorobenzoquinone **28**. In this reaction, a flavin hydroperoxide functions as an electrophile mediating the hydroxylation of the *para* position of **26**. Iodotyrosine deiodinase catalyzes the conversion of **29** to **33**. In this reaction, the substrate undergoes a keto-enol tautomerization, followed by a single electron transfer from the reduced flavin. Loss of iodide followed by reduction of the resulting phenoxy radical completes the reaction.

The current mechanism for the HcbA1-catalyzed reaction is shown in Fig. 7B. In this mechanism, the flavin peroxide is added to the benzene ring of **34** to give the Meisenheimer intermediate **35**. Elimination of chloride to give **36** followed by elimination of flavin gives peroxide **37**. Flavin-mediated peroxide reduction gives the flavin-N5-oxide **5** and the product **26**. This mechanism is consistent with the requirement of the reaction for molecular oxygen, and the incorporation of oxygen from molecular oxygen and not from water into the product **26**.

A similar strategy to that used for DszA and RutA was used to test for flavin-N5-oxide **5** formation. Under anaerobic conditions, stoichiometric amounts of **34**, HcbA1 and $FMNH_2$ were mixed, incubated for 30min and then exposed to atmospheric oxygen for flavin-C4a-peroxide

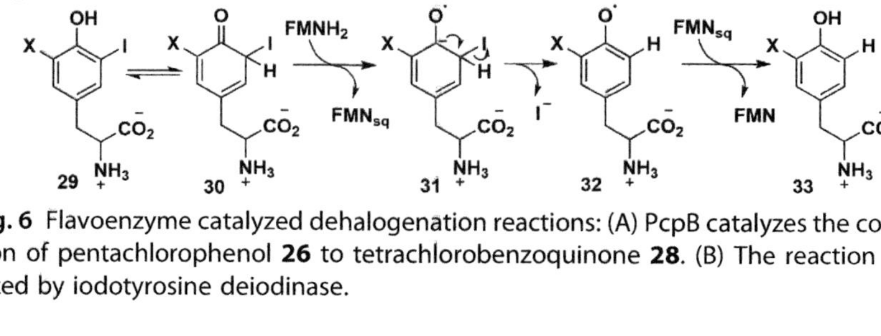

Fig. 6 Flavoenzyme catalyzed dehalogenation reactions: (A) PcpB catalyzes the conversion of pentachlorophenol **26** to tetrachlorobenzoquinone **28**. (B) The reaction catalyzed by iodotyrosine deiodinase.

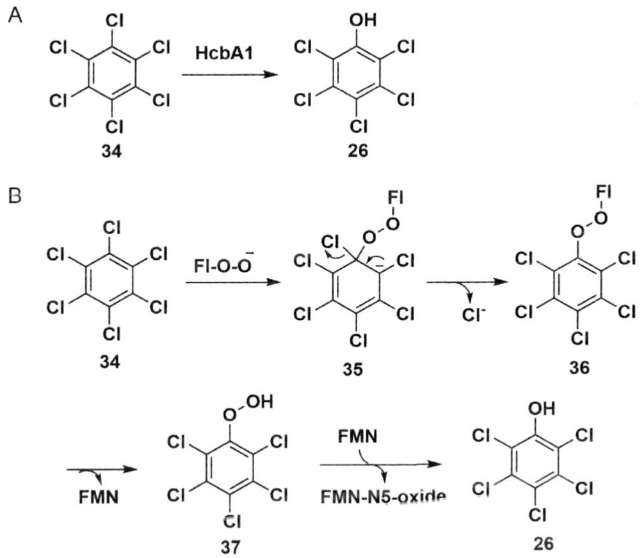

Fig. 7 (A) The reaction catalyzed by HcbA1. (B) Mechanistic proposal for the HcbA1-catalyzed reaction involving flavin-N5-oxide 5 formation.

4 formation and completion of the reaction. LC-MS analysis of the resulting reaction mixture demonstrated the formation of the flavin-N5-oxide **5**. When the reaction was carried out in the presence of $^{18}O_2$, the predicted 2 Da increase in its mass was also observed (Adak & Begley, 2019).

5.4 General guidelines for the detection of flavin-N5-oxide intermediates

1. Flavin-N5-oxide **5** is readily reduced to flavin. It is therefore essential to exclude reducing agents such as DTT, TCEP and reduced nicotinamide from the reaction mixture. Reduced flavin is prepared by photoreduction in the presence of EDTA rather than by enzymatic reduction using NAD(P)H as the reducing agent.
2. Although the UV–Visible spectra of oxidized flavin **1** and flavin-N5-oxide **5** are similar, there is a clear distinction between the two species in the 450 nm region (λ_{max} of flavin is 447 nm whereas λ_{max} of flavin-N5-oxide is 462 nm). This can be used as an initial indicator of the formation of a flavin-N5-oxide **5** intermediate.
3. Flavin-N5-oxide **5** can be detected by HPLC analysis of the enzymatic reaction mixture (single turnover) using synthesized FMN-N5-oxide **5** as an authentic standard.

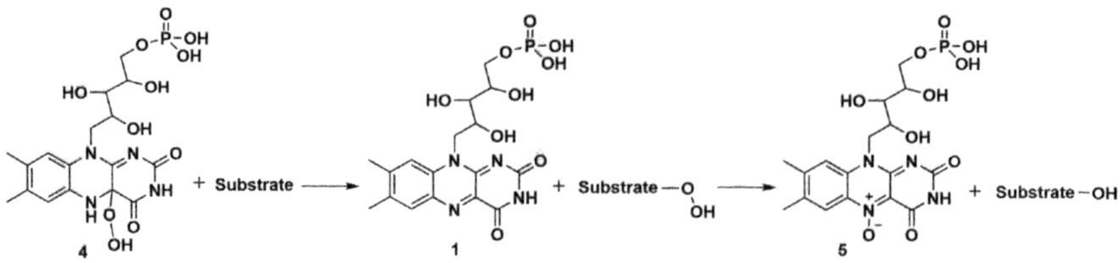

Fig. 8 Proposal for the use of flavin-N5-oxide intermediates by flavoenzymes catalyzing substrate hydroperoxide formation in the absence of an active site cysteine.

4. As of yet, protein amino acid sequence analysis has not generated a characteristic sequence motif that enables the identification of flavin-N5-oxide-forming enzymes. However, mechanistic analysis has proven useful in the identification of such enzymes. The analysis of the DszA reaction mechanism suggested that flavin-N5-oxides are likely intermediates in any flavoenzyme involving substrate hydroperoxide formation in the absence of an active site cysteine (Fig. 8) (Hall et al., 2009). This model led to the identification of RutA and HcbA1 as possible FMN-N5-oxide generating enzymes, but does not include the catalytic motif shown in the EncM-catalyzed reaction. It is highly likely that many additional examples of flavin-N5-oxide-utilizing-flavoenzymes remain to be discovered.

Funding

This research was supported by the Robert A. Welch Foundation Grant A-0034.

References

Adak, S., & Begley, T. P. (2016). Dibenzothiophene catabolism proceeds via a flavin-N5-oxide intermediate. *Journal of the American Chemical Society*, *138*(20), 6424–6426.

Adak, S., & Begley, T. P. (2017a). Flavin-N5-oxide: A new, catalytic motif in flavoenzymology. *Archives of Biochemistry and Biophysics*, *632*, 4–10.

Adak, S., & Begley, T. P. (2017b). RutA-catalyzed oxidative cleavage of the uracil amide involves formation of a flavin-N5-oxide. *Biochemistry*, *56*(29), 3708–3709.

Adak, S., & Begley, T. P. (2019). Hexachlorobenzene catabolism involves a nucleophilic aromatic substitution and flavin-N5-oxide formation. *Biochemistry*, *58*(9), 1181–1183.

Barber, J. L., Sweetman, A. J., van Wijk, D., & Jones, K. C. (2005). Hexachlorobenzene in the global environment: Emissions, levels, distribution, trends and processes. *Science of the Total Environment*, *349*(1–3), 1–44.

Barbosa, A. C. C., Neves, R. P. P., Sousa, S. F., Ramos, M. J., & Fernandes, P. A. (2018). Mechanistic studies of a flavin monooxygenase: Sulfur oxidation of dibenzothiophenes by DszC. *ACS Catalysis*, *8*(10), 9298–9311.

Denome, S. A., Oldfield, C., Nash, L. J., & Young, K. D. (1994). Characterization of the desulfurization genes from Rhodococcus sp. strain IGTS8. *Journal of Bacteriology*, *176*(21), 6707–6716.

Denome, S. A., Olson, E. S., & Young, K. D. (1993). Identification and cloning of genes involved in specific desulfurization of dibenzothiophene by Rhodococcus sp. strain IGTS8. *Applied and Environmental Microbiology*, *59*(9), 2837–2843.

Enns, K., & Burgess, W. H. (1965). The photochemical oxidation of ethylenediaminetetraacetic acid and methionine by riboflavine. *Journal of the American Chemical Society*, *87*(24), 5766–5770.

Gonzalez-Osorio, L., Luong, K., Jirde, S., Palfey, B. A., & Vey, J. L. (2016). Initial investigations of C4a-(hydro)peroxyflavin intermediate formation by dibenzothiophene monooxygenase. *Biochemical and Biophysical Research Communications*, *481*(1–2), 189–194.

Gray, K. A., Pogrebinsky, O. S., Mrachko, G. T., Xi, L., Monticello, D. J., & Squires, C. H. (1996). Molecular mechanisms of biocatalytic desulfurization of fossil fuels. *Nature Biotechnology*, *14*(13), 1705–1709.

Guan, L.-J., Lee, W. C., Wang, S., Ohshiro, T., Izumi, Y., Ohtsuka, J., et al. (2015). Crystal structures of apo-DszC and FMN-bound DszC from Rhodococcus erythropolis D-1. *The FEBS Journal*, *282*(16), 3126–3135.

Hall, A., Karplus, P. A., & Poole, L. B. (2009). Typical 2-Cys peroxiredoxins—Structures, mechanisms and functions. *The FEBS Journal*, *276*(9), 2469–2477.

Heelis, P. F. (1982). The photophysical and photochemical properties of flavins (isoalloxazines). *Chemical Society Reviews*, *11*(1), 15–39.

Hu, J., Chuenchor, W., & Rokita, S. E. (2015). A switch between one- and two-electron chemistry of the human flavoprotein iodotyrosine deiodinase is controlled by substrate. *The Journal of Biological Chemistry*, *290*(1), 590–600.

Ito, K., Takagi, K., Iwasaki, A., Tanaka, N., Kanesaki, Y., Martin-Laurent, F., et al. (2017). Identification of the hcb gene operon involved in catalyzing aerobic hexachlorobenzene dechlorination in Nocardioides sp. strain PD653. *Applied and Environmental Microbiology*, *83*(19), e00824-17/1-e00824-17/17.

Ito, K., Takagi, K., Matsushima, Y., Igimi, S., Ito, K., Takagi, K., et al. (2018). Identification of the novel hcbB operon catalyzing the dechlorination of pentachlorophenol in the Gram-positive bacterium Nocardioides sp. strain PD653. *Journal of Pesticide Science*, *43*(2), 124–131.

Lee, W. C., Ohshiro, T., Matsubara, T., Izumi, Y., & Tanokura, M. (2006). Crystal structure and desulfurization mechanism of 2'-hydroxybiphenyl-2-sulfinic acid desulfinase. *The Journal of Biological Chemistry*, *281*(43), 32534–32539.

Loh, K. D., Gyaneshwar, P., Papadimitriou, E. M., Fong, R., Kim, K.-S., Parales, R., et al. (2006). A previously undescribed pathway for pyrimidine catabolism. *Proceedings of the National Academy of Sciences of the United States of America*, *103*(13), 5114–5119.

Mansurova, M., Koay, M. S., & Gaertner, W. (2008). Synthesis and electrochemical properties of structurally modified flavin compounds. *European Journal of Organic Chemistry*, (32), 5401–5406.

McCormick, D. B., Koster, J. F., & Veeger, C. (1967). Mechanisms of photochemical reductions of FAD and FAD-dependent flavoproteins. *European Journal of Biochemistry*, *2*(4), 387–391.

Mukherjee, T., Zhang, Y., Abdelwahed, S., Ealick, S. E., & Begley, T. P. (2010). Catalysis of a flavoenzyme-mediated amide hydrolysis. *Journal of the American Chemical Society*, *132*(16), 5550–5551.

Phatarphekar, A., Buss, J. M., & Rokita, S. E. (2014). Iodotyrosine deiodinase: A unique flavoprotein present in organisms of diverse phyla. *Molecular BioSystems*, *10*(1), 86–92.

Piddington, C. S., Kovacevich, B. R., & Rambosek, J. (1995). Sequence and molecular characterization of a DNA region encoding the dibenzothiophene desulfurization operon on Rhodococcus sp. strain IGTS8. *Applied and Environmental Microbiology*, *61*(2), 468–475.

Saleem-Batcha, R., Stull, F., Sanders, J. N., Moore, B. S., Palfey, B. A., Houk, K. N., et al. (2018). Enzymatic control of dioxygen binding and functionalization of the flavin cofactor. *Proceedings of the National Academy of Sciences of the United States of America*, *115*(19), 4909–4914.

Soleimani, M., Bassi, A., & Margaritis, A. (2007). Biodesulfurization of refractory organic sulfur compounds in fossil fuels. *Biotechnology Advances*, *25*(6), 570–596.

Sun, Z., Su, Q., & Rokita, S. E. (2017). The distribution and mechanism of iodotyrosine deiodinase defied expectations. *Archives of Biochemistry and Biophysics*, *632*, 77–87.

Teufel, R. (2017). Flavin-catalyzed redox tailoring reactions in natural product biosynthesis. *Archives of Biochemistry and Biophysics*, *632*, 20–27.

Teufel, R. (2018). Preparation and characterization of the Favorskiiase flavoprotein EncM and its distinctive flavin-N5-oxide cofactor. *Methods in Enzymology*, *604*, 523–540.

Teufel, R., Agarwal, V., & Moore, B. S. (2016). Unusual flavoenzyme catalysis in marine bacteria. *Current Opinion in Chemical Biology*, *31*, 31–39.

Teufel, R., Miyanaga, A., Michaudel, Q., Stull, F., Louie, G., Noel, J. P., et al. (2013). Flavin-mediated dual oxidation controls an enzymatic Favorskii-type rearrangement. *Nature*, *503*(7477), 552–556.

Teufel, R., Stull, F., Meehan, M. J., Michaudel, Q., Dorrestein, P. C., Palfey, B., et al. (2015). Biochemical establishment and characterization of EncM's flavin-N5-oxide cofactor. *Journal of the American Chemical Society*, *137*(25), 8078–8085.

Thomas, S. R., McTamney, P. M., Adler, J. M., LaRonde-LeBlanc, N., & Rokita, S. E. (2009). Crystal structure of iodotyrosine deiodinase, a novel flavoprotein responsible for iodide salvage in thyroid glands. *The Journal of Biological Chemistry*, *284*(29), 19659–19667.

Further reading

Agarwal, V., Miles, Z. D., Winter, J. M., Eustaquio, A. S., El Gamal, A. A., & Moore, B. S. (2017). Enzymatic halogenation and dehalogenation reactions: Pervasive and mechanistically diverse. *Chemical Reviews*, *117*(8), 5619–5674.

Crawford, R. L., Jung, C. M., & Strap, J. L. (2007). The recent evolution of pentachlorophenol (PCP)-4-monooxygenase (PcpB) and associated pathways for bacterial degradation of PCP. *Biodegradation*, *18*(5), 525–539.

Hlouchova, K., Rudolph, J., Pietari, J. M. H., Behlen, L. S., & Copley, S. D. (2012). Pentachlorophenol hydroxylase, a poorly functioning enzyme required for degradation of pentachlorophenol by Sphingobium chlorophenolicum. *Biochemistry*, *51*(18), 3848–3860.

CHAPTER EIGHTEEN

Prenylated FMN: Biosynthesis, purification, and Fdc1 activation

Anna N. Khusnutdinova[a], Johnny Xiao[a], Po-Hsiang Wang[a], Khorcheska A. Batyrova[a], Robert Flick[a], Elizabeth A. Edwards[a], Alexander F. Yakunin[a,b,*]

[a]Department of Chemical Engineering and Applied Chemistry, University of Toronto, Toronto, ON, Canada
[b]Centre for Environmental Biotechnology, School of Natural Sciences, Bangor University, Bangor, United Kingdom
*Corresponding author: e-mail address: a.iakounine@utoronto.ca

Contents

Abstract

Prenylated flavin mononucleotide (prFMN) is a recently discovered flavin cofactor produced by the UbiX family of FMN prenyltransferases, and is required for the activity of UbiD-like reversible decarboxylases. The latter enzymes are known to be involved in ubiquinone biosynthesis and biotransformation of lignin, aromatic compounds, and unsaturated aliphatic acids. However, exploration of uncharacterized UbiD proteins

Methods in Enzymology, Volume 620
ISSN 0076-6879
https://doi.org/10.1016/bs.mie.2019.03.021

for biotechnological applications is hindered by our limited knowledge about the biochemistry of prFMN and prFMN-dependent enzymes. Here, we describe experimental protocols and considerations for the biosynthesis of prFMN *in vivo* and *in vitro*, in addition to cofactor extraction and application for activation of UbiD proteins.

Abbreviations

DMAP	dimethylallyl phosphate
DMAPP	dimethylallyl pyrophosphate
DMSO	dimethyl sulfoxide
FMN	flavin mononucleotide
IPTG	isopropyl-β-D-1-thiogalactopyranoside
LC-MS	liquid chromatography mass spectrometry
PMSF	phenylmethane sulfonyl fluoride
polyP	polyphosphate
prFMN	prenylated flavin mononucleotide
prRiboflavin	prenylated riboflavin
RT	room temperature

1. Introduction

Flavin-containing enzymes account for approximately 1% of cellular proteins and catalyze numerous redox reactions with organic substrates, as well as acid/base and nucleophilic/electrophilic reactions that involve the formation of covalent flavin-substrate intermediates (Leys & Scrutton, 2016; Piano, Palfey, & Mattevi, 2017; Sobrado, 2012; Walsh & Wencewicz, 2013). Flavins are derived from riboflavin (vitamin B2), which serves as the precursor for the two major flavin cofactors, flavin mononucleotide (FMN) and flavin adenine dinucleotide (FAD), as well as their derivatives (Massey, 2000; Teufel, Agarwal, & Moore, 2016). The recently discovered prenylated FMN (prFMN, Fig. 1) was shown to be produced by UbiX-like FMN prenyltransferases (EC 2.5.1.129) found in bacteria, archaea, and lower eukaryotes (fungi and unicellular algae) (IPR004507, InterPro database, over 13,000 sequences). The prFMN cofactor is required for the activity of the UbiD family of reversible decarboxylases (EC 4.1.1.98) involved in ubiquinone biosynthesis and biotransformation of lignin, aromatic, hetero-aromatic, and unsaturated aliphatic acids (IPR002830, InterPro database, over 17,000 sequences). UbiD enzymes are also wide spread in nature with a high diversity in sequences, subunit organization,

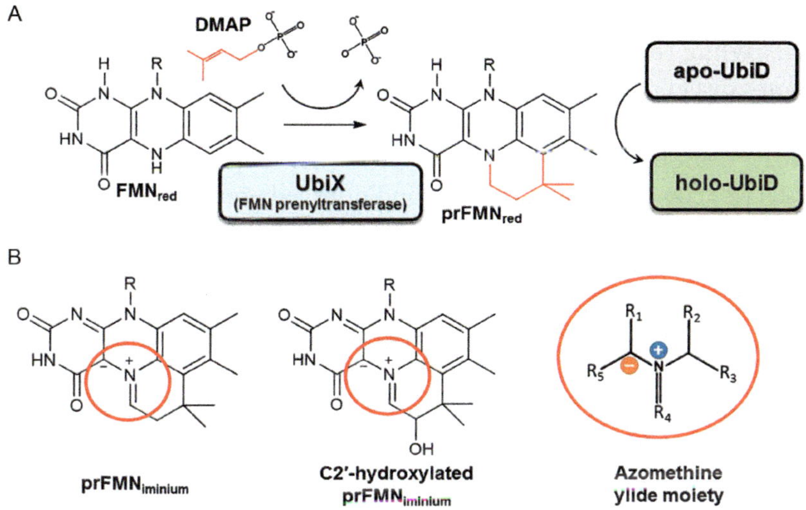

Fig. 1 FMN prenylation reaction catalyzed by UbiX and associated reaction products. (A), UbiX is a FMN prenyltransferase, which transfers a prenyl moiety from DMAP to reduced FMN (FMN$_{red}$). The product, prenylated FMN (prFMN), acts as a cofactor activating an inactive UbiD-like (de)carboxylase (apo-form) and producing an active enzyme (holo-UbiD) (Piano et al., 2017). (B), Two forms of prFMN showing the azomethine ylide structure (1,3 dipole surrounding a nitrogen atom), which is produced by the deprotonation of an iminium structure. This structure is necessary for the (de)carboxylation of aromatic substrates by UbiD (de)carboxylases *via* 1,3-dipolar addition (Payne et al., 2015). The two identified forms of protein-free prFMN with this structure are prFMN$_{iminium}$ (White et al., 2015) and C2′-hydroxylated prFMN$_{iminium}$ (Wang et al., 2018).

and substrate specificity (Marshall, Payne, & Leys, 2017). Although both carboxylation and decarboxylation reactions catalyzed by UbiDs have significant biotechnological potential, exploration of biochemical diversity of the uncharacterized members of this large family is restricted by our limited knowledge about the biosynthesis of prFMN and UbiD activation.

Recently, it has been shown that UbiX enzymes catalyze the transfer of the dimethylallyl moiety of dimethylallyl-monophosphate (DMAP) to the N5 and C6 atoms of reduced FMN, creating an additional, non-aromatic ring in the FMN isoalloxazine ring system (Payne et al., 2015; White et al., 2015) (Fig. 1). Crystal structures of UbiX proteins from *Escherichia coli*, *Pseudomonas aeruginosa* (PA4019), and *Colwellia psychrerythraea* revealed homooligomeric (dodecameric) complexes containing 12 individual monomers (22–23 kDa) with a Rossmann fold (a three-layered $\alpha/\beta/\alpha$ structure)

(Do et al., 2015; Kopec, Schnell, & Schneider, 2011; Rangarajan et al., 2004; White et al., 2015). Each dodecamer was found to contain 12 non-covalently bound FMN molecules attached at the interface between two and three subunits. Recent detailed biochemical characterization of PA4019 demonstrated the presence of DMAP-dependent FMN prenyltransferase activity under anaerobic conditions, whereas the structures of this protein in complex with DMAP + FMN, covalent intermediate (N5-C1′ adduct), and product (prFMN) revealed the molecular mechanisms of the catalyzed reactions (White et al., 2015). These reactions include breakage of the DMAP phosphate-C1′ bond (triggered by FMN reduction), creation of the FMN N5-C1′ bond followed by adduct reorientation and formation of the dimethylallyl C3′-C6 bond and $\text{prFMN}_{\text{reduced}}$, which rapidly oxidized under aerobic conditions producing a purple semiquinone-like species ($\text{prFMN}_{\text{radical}}$) (White et al., 2015). It is proposed that UbiX enzymes release prFMN, which is bound by apo-UbiD enzymes in a reduced form ($\text{prFMN}_{\text{reduced}}$) followed by cofactor oxidation and the formation of the catalytically relevant $\text{prFMN}_{\text{iminium}}$ form (White et al., 2015). It has been shown that ferulic acid decarboxylation activity of the *Aspergillus niger* UbiD-like protein Fdc1 can be reconstituted *in vitro* after anaerobic incubation with the reduced PA4019 + prFMN complex followed by aerobic oxidation resulting in the formation of the reactive N5=C iminium prFMN form with an azomethine ylide character (Payne et al., 2015; White et al., 2015) (Fig. 1). Structural studies with Fdc1 also revealed the presence of an inactive form of oxidized prFMN, $\text{prFMN}_{\text{ketimine}}$, which was produced by a light-dependent isomerization of $\text{prFMN}_{\text{iminium}}$ (Payne et al., 2015). Other known forms of prFMN cofactors include: protonated prFMN-C-$4\text{a}_{\text{radical}}$, C1′-hydroxylated prFMN, C1′-ene-prFMN, C2′-hydrolxylated prFMN, and C1′,C2′-dihydroxylated prFMN (Marshall et al., 2017; Payne et al., 2015; Wang et al., 2018; White et al., 2015). In addition, biochemical studies with the UbiX-UbiD system from *Saccharomyces cerevisiae* (PAD1 and FDC) demonstrated the lability of the cofactor produced by PAD1 that prevented its purification and identification (Lin, Ferguson, Boyer, Lin, & Marsh, 2015). Thus, the observed diversity and lability of prFMN forms hinders the studies on the biochemistry of prFMN and prFMN-dependent enzymes. In this chapter, we describe experimental protocols, reagents, and considerations for *in vivo* and *in vitro* FMN prenylation, prFMN extraction and quantitation, as well as for UbiD activation using both UbiX-bound and protein-free cofactors.

2. Biosynthesis of prFMN

Recently, we demonstrated that purified UbiX proteins from diverse organisms exhibit different affinities to prFMN, with some enzymes (*e.g.*, AF1214 from *Archaeoglobus fulgidus*) retaining the produced cofactor after the reaction and during affinity protein purification (Wang et al., 2018). This suggests that some UbiX proteins can be used as protein traps for the purification of prFMN and apo-UbiD activation, as well as for cofactor stabilization during purification and storage (in complex with UbiX). We also found that FMN prenylation can be stimulated by the addition of prenol and riboflavin to *E. coli* cultures expressing recombinant UbiX proteins, as well as by anaerobic incubation of *E. coli* cells before harvesting (Wang et al., 2018). Furthermore, we showed that the *E. coli* hydroxyethylthiazole kinase ThiM can phosphorylate prenol producing DMAP (Fig. 2) (Wang et al., 2018). Since commercial preparations of DMAP are very expensive (and usually contaminated by DMAPP and phosphate), purified ThiM can be

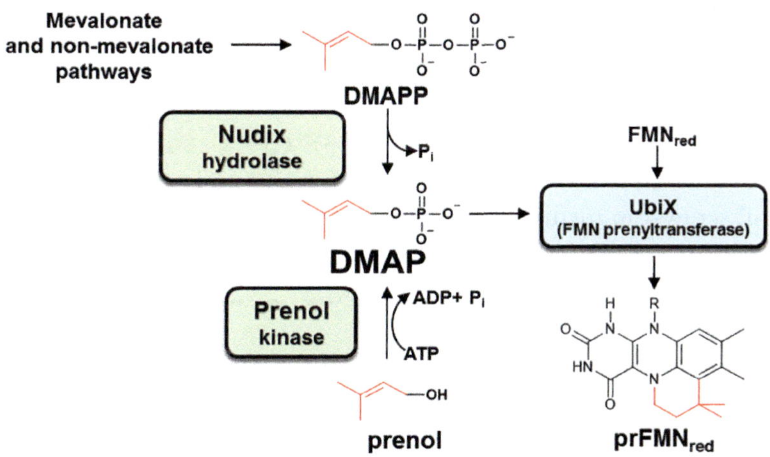

Fig. 2 Enzymes involved in the *in vivo* biosynthesis of DMAP and prFMN. Schematic diagram showing bacterial enzymes involved in the biosynthesis of DMAP and prFMN. DMAP can be produced by the dephosphorylation of DMAPP or by phosphorylation of prenol. In bacteria, the mevalonate (MEV) and non-mevalonate (MEP) pathways produce DMAPP, which can be dephosphorylated by Nudix hydrolases to DMAP. Alternatively, DMAP can be produced through the phosphorylation of prenol by the hydroxyethylthiazole kinase ThiM. Produced DMAP is then used by UbiX as an isoprenoid donor for prenylation of intracellular FMN.

used to generate DMAP from prenol for *in vitro* FMN prenylation reactions using purified UbiX enzymes. In this section, we describe two protocols for *in vivo* and *in vitro* biosynthesis of prFMN using *E. coli* cultures and purified enzymes, respectively.

2.1 *In vivo* FMN prenylation using *E. coli* cells expressing AF1214

Our recent work revealed that some UbiX proteins exhibit high affinity to prFMN, being capable of retaining this cofactor during protein expression in *E. coli* and subsequent protein purification using affinity chromatography (Wang et al., 2018). This work also demonstrated that FMN prenylation in *E. coli* cells appears to be limited by intracellular levels of DMAP and in some cases FMN, and can be improved by adding prenol and riboflavin to *E. coli* cultures during recombinant expression of UbiX proteins. Therefore, the proposed protocol for *in vivo* FMN prenylation is based on the recombinant expression of UbiX proteins with the addition of prenol and riboflavin, as well as using anaerobic conditions after IPTG induction (Fig. 3).

2.1.1 *Equipment*

1. UV–Vis spectrophotometer (*e.g.*, SpectraMax M2 from Molecular Devices)
2. Shaking incubator for culture tubes (37 °C)
3. Platform shaker for 2-L plastic flasks (37 °C)
4. 2-L baffled plastic flasks for *E. coli* cultures (autoclaved)
5. 1-L screw-cap media bottles with magnetic stir bars (autoclaved)

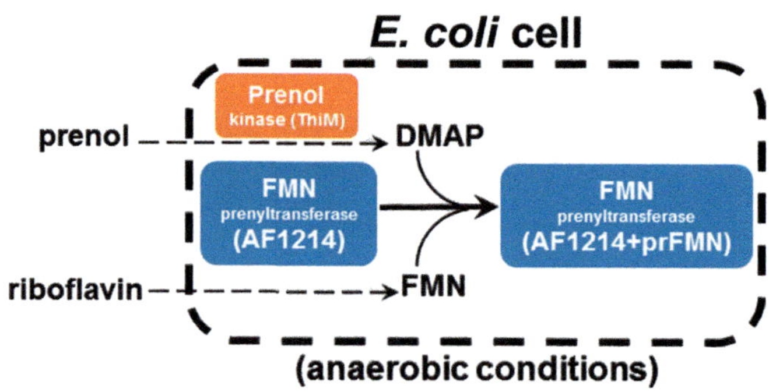

Fig. 3 Schematic diagram showing the *in vivo* biosynthesis of prFMN using *E. coli* cells over-expressing recombinant UbiX proteins (in this case AF1214). The addition of prenol and riboflavin to growth medium increases intracellular concentrations of FMN and DMAP in *E. coli* cells, whereas anaerobic incubation facilitates the reduction of FMN, which is necessary for FMN prenylation.

6. Preparative refrigerating centrifuges with rotors for *E. coli* cell harvesting and cell lysate clearing with 1-L centrifuge bottles and 50-mL centrifuge tubes
7. Sonicator (*e.g.*, QSonica Q700)

2.1.2 Materials

1. *E. coli* growth medium (Terrific Broth or equivalent)
2. Antibiotics for *E. coli* cultures (as appropriate)
3. Prenol (3-methyl-2-buten-1-ol, from Sigma-Aldrich, cat # W364703)
4. Riboflavin, stock solution in water (0.1 M)
5. Dimethyl sulfoxide (DMSO)
6. Isopropyl-β-D-1-thiogalactopyranoside (IPTG), 0.4 M stock solution in water
7. Phenylmethane sulfonyl fluoride (PMSF), 10 mM stock solution (in isobutanol)
8. *E. coli* strains (BL21(DE3)) expressing recombinant UbiX proteins from a T7 overexpression plasmid
9. Liquid nitrogen
10. Standard protein purification buffers for Ni-chelate affinity chromatography: binding buffer (BB: 50 mM HEPES-K, pH 7.5, 0.4 M NaCl, 5 mM imidazole, 5% glycerol), washing buffer (WB: BB with 30 mM imidazole), and elution buffer (EB: BB with 500 mM imidazole)
11. Centrifugal filters for protein concentration (Mw cut-off 3–5 kDa)
12. Glass or plastic columns for gravity flow protein purification

2.1.3 Procedure

1. Inoculate 10 mL of Terrific Broth (TB) medium containing the appropriate antibiotics with the *E. coli* strain (*e.g.*, BL21(DE3)) expressing a UbiX protein (PA4019 or AF1214) from a vector containing an IPTG-inducible promoter (*e.g.*, pET15b or p15TVL) and grow the starter culture overnight (12–16 h) at 37 °C (shaking at 200 rpm).
2. Use the starter *E. coli* culture to inoculate 1 L of TB medium containing the appropriate antibiotics in a 2-L flask and grow *E. coli* cultures aerobically at 37 °C to optical density OD_{600} of 0.6–0.8, (shaking at 200 rpm).
3. Transfer *E. coli* cultures to screw-capped 1-L bottles with a magnetic stir bar (leave at least 100 mL of head space), add 1% DMSO (final concentration, w/v), prenol (1 mM final concentration), riboflavin (0.1 mM final concentration), and IPTG (0.4 mM final concentration). Incubate overnight (12–16 h) at RT (stirring at 300–400 rpm).

4. Following overnight incubation, harvest cells (aerobically) by centrifugation at 8900 × *g* for 20 min at 4 °C.
5. Remove the supernatant and resuspend the pellet in a small volume (30–45 mL) of BB buffer. Freeze cell pellets at −80 °C or use them immediately for sonication and protein purification.
6. Before sonication, add PMSF (1 μM final concentration) to cell suspensions and sonicate on ice for 20 min (with 5 s on/off pulses) at 20 kHz.
7. Centrifuge sonicated samples at 40,000 × *g* for 20 min at 10 °C and transfer supernatants to 50-mL Falcon tubes.
8. Purify 6His-tagged proteins using Ni-chelate affinity chromatography and standard purification protocols (using washing and elution buffers containing 30 and 500 mM imidazole, respectively).
9. Concentrate purified proteins using centrifugal filters (Mw cut-off 3 kDa) to a final concentration 3–5 mg/mL (if necessary), flash freeze in liquid nitrogen and store at −80 °C.
10. Analyze protein purity using denaturing SDS-polyacrylamide gel electrophoresis (standard protocols).

2.1.4 Notes

1. Under these expression conditions, purified UbiX proteins with high affinity to prFMN (*e.g.*, AF1214 or PA4019) will have a purple color in solution, which can be easily followed during column purification. Yellow color of purified UbiX proteins would indicate that they contain mainly unmodified FMN. In this case, FMN prenylation can be performed using the *in vitro* FMN prenylation protocol with purified UbiX proteins described below (Section 2.2).
2. Riboflavin addition to *E. coli* cultures is not necessary for UbiX proteins with high affinity to FMN (*e.g.*, AF1214 or PA4019), but can be useful for other UbiX proteins.
3. Sometimes purified UbiX proteins (*e.g.*, AF1214) will precipitate on ice (or in the fridge at 4 °C). In this case, the protein solution should be maintained at room temperature before freezing in liquid nitrogen or during thawing.

2.2 *In vitro* biosynthesis of prFMN using purified UbiX proteins

The following section describes a protocol for *in vitro* biosynthesis of prFMN from FMN and prenol using purified UbiX proteins, as well as the *E. coli* hydroxyethylthiazole kinase ThiM and polyphosphate (polyP)

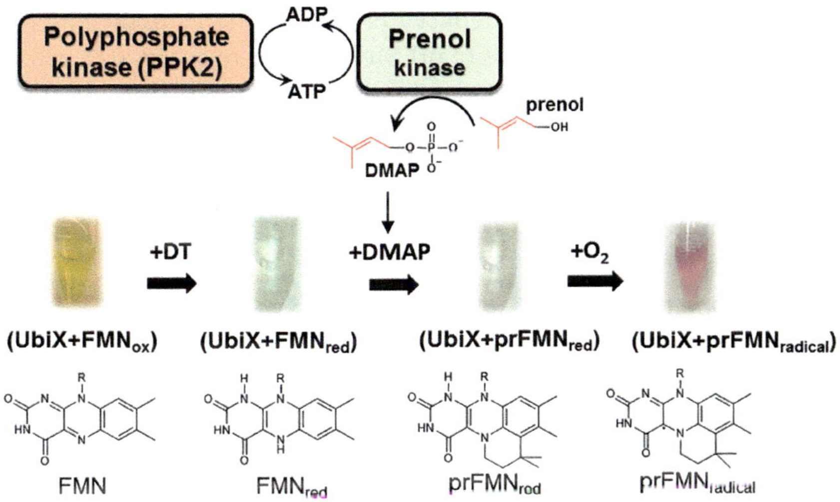

Fig. 4 *In vitro* synthesis of DMAP and prFMN. (A), Schematic representation of DMAP synthesis through *in vitro* phosphorylation of prenol using purified hydroxyethylthiazole kinase (ThiM). (B), Depiction of the color change and chemical structure associated with the enzyme-bound flavin cofactor as it is transformed from FMN to prFMN. The oxidized, enzyme-bound FMN is yellow, but turns colorless as it is reduced to FMNred during dithionite reduction. After incubation with DMAP, followed by further air oxidation, the FMN is transformed into $prFMN_{radical}$, which exhibits a purple color.

kinase CHU0107 from *Cytophaga hutchinsonii* (Fig. 4). This method is based on the saturation of UbiX proteins with FMN during recombinant protein expression in *E. coli* (by the addition of riboflavin) or after incubation of purified UbiX proteins with FMN in excess followed by FMN reduction by dithionite and prenylation in the presence of a synthetic DMAP or DMAP-generating system. Since synthetic DMAP preparations are expensive (and sometimes of low quality), we have developed an enzyme-based DMAP-generating system containing the *E. coli* hydroxyethylthiazole kinase ThiM and the polyP kinase CHU0107 from *C. hutchinsonii* (PPK2 family) (Fig. 4). ThiM, a salvage enzyme in thiamine biosynthesis, catalyzes the phosphorylation of 4-methyl-5-beta-hydroxyethylthiazole (Mizote & Nakayama, 1989), which is structurally similar to prenol. The polyP kinase family 2 (PPK2) protein CHU0107 has been shown to phosphorylate ADP using polyP as a phosphate donor (Khusnutdinova et al., 2017). In the DMAP-generating system, the first kinase (ThiM) phosphorylates prenol using ATP, producing DMAP and ADP, which is regenerated to ATP by the second kinase (CHU0107) using inorganic polyphosphate (polyP) as a phosphate donor. Generated DMAP is then used by purified UbiX proteins for the prenylation of reduced FMN (Fig. 4). Thus, the proposed protocol for *in vitro* biosynthesis of prFMN includes two steps: (1), aerobic expression and purification of the UbiX protein AF1214 with bound FMN, and (2), reduction of the AF1214-FMN complex with dithionite under anaerobic conditions followed by FMN prenylation in the presence of the DMAP-generating system and aerobic oxidation to produce the AF1214-prFMN$_{radical}$ complex.

2.2.1 Equipment

1. UV–Vis spectrophotometer
2. Shaking incubator for culture tubes (37 °C)
3. Bench-top centrifuge
4. Simple Schlenk line (a vacuum gas manifold including a vacuum pump and argon or nitrogen gas cylinder) for oxygen removal from reaction solutions (and inert gas saturation)
5. Standard equipment for affinity purification of recombinant proteins using Ni-NTA column purification (see Section 2.1.3)

2.2.2 Materials

1. Prenol (3-methyl-2-buten-1-ol) (Sigma-Aldrich, cat# W364703)
2. Reaction buffer for FMN prenylation: 100 mM HEPES (pH 7.5), 0.3 M NaCl, 10 mM $MgCl_2$, 5 mM ATP, 5 mM prenol, 2 mM polyphosphate (polyP), 0.5 M imidazole
3. Storage buffer (for protein bound prFMN): 50 mM HEPES (pH 7.5), 0.1 M NaCl, 0.5 M imidazole
4. Aerobically grown and purified AF1214
5. Purified hydroxyethylthiazole kinase ThiM from *E. coli*
6. Purified polyphosphate kinase CHU0107 from *C. hutchinsonii*
7. Centrifugal filters (Mw cut-off 3 kDa)
8. Anaerobic solution of dithionite (dithionite) (0.1 M, prepared on the day of use)
9. FMN (Sigma-Aldrich), 50 mM solution in water
10. 10 mL serum vials with rubber stoppers
11. 1 mL syringes and needles

2.2.3 Procedure

1. Using a Schlenk manifold, prepare ~1 mL of 0.1 M dithionite (keep under argon atmosphere).
2. Thaw purified AF1214 at RT and add an equimolar amount of FMN (to ensure complete saturation with FMN), keep at RT for 10–15 min.

3. Prepare the reaction mixture for FMN prenylation containing 0.3M NaCl, 10mM $MgCl_2$, 5mM ATP, 5mM prenol, and 2mM polyphosphate.
4. Thaw purified ThiM and CHU0107 proteins and add them to the reaction mixture with AF1214 in a serum vial (ThiM, 1mg/mL; CHU0107, 0.1mg/mL; AF1214, 2–3mg/mL).
5. Incubate the reaction mixture aerobically for 1–5h on a shaking incubator at 37°C (shaking at 200rpm) (this step produces DMAP from prenol and polyphosphate).
6. Close the glass vial with a rubber stopper and remove oxygen using a Schlenk manifold.
7. Using an argon-flushed syringe, add 10–20μL of anaerobic dithionite solution to the reaction mixture to reduce FMN (resulting in complete discoloration of the yellow reaction mixture)
8. Incubate the colorless reaction mixture for 1h at 37°C (with shaking at 200rpm) (this step produces the reduced form of AF1214+prFMN).
9. Expose the colorless reaction mixture to air (for ~10min) for the oxidation of the reduced AF1214+prFMN complex to the brownish colored AF1214+$prFMN_{radical}$. At this step, the reaction mixture has a brownish color as it also contains the yellow-colored unbound FMN.
10. Add acetonitrile to a final concentration of 70% (v/v) to the reaction mixture.
11. Centrifuge the reaction mixture at max speed (14,000rpm) and discard the supernatant containing unbound FMN.
12. Wash the protein pellet with 70% acetonitrile 3–5 times (steps 10–11) to remove unbound FMN.
13. Dissolve the protein pellet in a small volume (0.5–2mL) of Storage buffer (the protein color should be purple) and clear the preparation by centrifugation at 14,000rpm for 5min.
14. Using a UV–VIS spectrophotometer, determine the protein absorbance at λ_{460nm} and λ_{550nm} (samples with prFMN are expected to have $\lambda_{550nm}/\lambda_{460nm}=2$ or higher).
15. The protein sample (AF1214+$prFMN_{radical}$) can be frozen at −80°C, or it can be used immediately for UbiD activation or for extraction of protein-free prFMN (see Section 3.3).

2.2.4 Notes

1. Protein precipitation in AF1214 preparations can be reduced by keeping them at RT or by increasing imidazole concentrations in the reaction mixture or storage buffer.

2. The obtained preparation of AF1214 + $prFMN_{radical}$ can be used as a source of prFMN for activation of purified UbiD proteins.
3. The proposed reaction conditions for DMAP generation usually achieve a 25% conversion of prenol to DMAP.
4. If synthetic DMAP is available, it can be added directly to the FMN reduction mixture (step 2, omitting steps 3–6).
5. To decrease the UbiX sample contamination by unbound FMN during washing with 70% acetonitrile, FMN (1 mM) can be added at the cell lysis step during protein purification. In this case, FMN addition in step 2 is no longer needed as the UbiX protein (AF1214) will be mostly saturated with FMN.

3. Extraction and quantitation of protein-free prFMN

This section describes a protocol for the extraction and spectrophotometric quantitation of protein-free prFMN cofactor from the UbiX protein AF1214. This protocol is also applicable to other UbiX proteins with a high affinity to prFMN (including PA4019), which retain the produced prFMN cofactor after the reaction. Cofactor extraction from the purified UbiX-prFMN complexes is based on the use of low pH and acetonitrile for protein denaturation and precipitation resulting in the release of free cofactor into solution. Due to the relative hydrophobicity of prFMN, it will be partitioned mainly into the acetonitrile phase, which can be collected followed by cofactor concentration using evaporation. From several free prFMN forms (Fig. 5), the protonated $prFMN_{radical}$ ($prFMN_{radical}$-H) form was found to be the most stable when stored under anaerobic conditions. For this reason, the described extraction protocol is aimed at producing $prFMN_{radical}$-H, which can then be used to generate other prFMN forms (Fig. 5).

3.1 Equipment

1. Simple Schlenk line (a vacuum gas manifold; see the previous section)
2. Anaerobic chamber (glove-box) (*e.g.*, from Coy Lab Products)
3. Microcentrifuge
4. 2-mL screw-cap microtubes with O-ring caps
5. 10-mL serum vials with rubber stoppers
6. pH meter or probe
7. SpeedVac concentrator
8. UV–VIS spectrophotometer with glass cuvettes

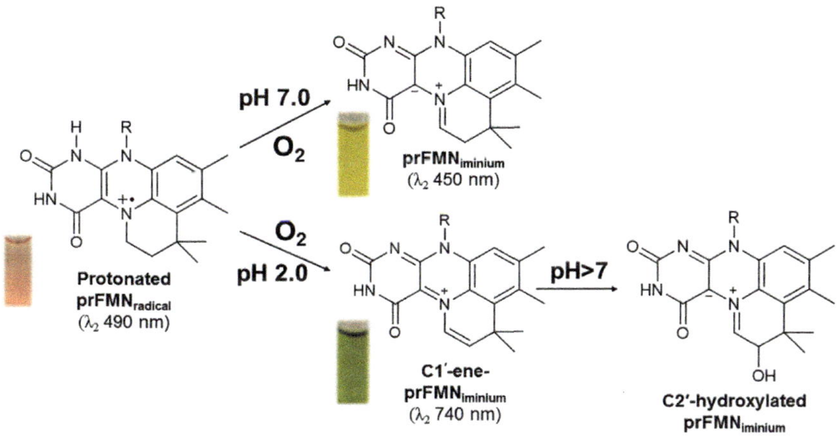

Fig. 5 Colorimetric and chemical structure depictions for transformation of protonated $prFMN_{radical}$ to other forms. Forms of prFMN and their relevant UV-VIS absorption spectra peaks (λ_2), where applicable. The standard form, protonated $prFMN_{radical}$, is obtained using the extraction protocol. This is the most stable form when stored anaerobically and is oxidized by air to form, depending on the pH, two different preparations which result in a higher concentration of one of two forms. At neutral pH (pH 7.0), an active, yellowish-colored $prFMN_{iminium}$ is formed. At a low pH, a greenish-colored C1′-ene-$prFMN_{iminium}$ is formed, which can then be readily hydroxylated under neutral pH into the active C2′-hydroxylated $prFMN_{iminium}$ form.

3.2 Materials

1. Anaerobic 6 N HCl solution
2. Anaerobic acetonitrile, LC-MS grade
3. Purified AF1214 + $prFMN_{radical}$ complex (freshly purified or from −80 °C freezer; see the previous sections for purification methods)
4. 5 mM ammonium acetate in water (pH 6.0) and in methanol (pH 6.0)

3.3 Procedure

1. Using a vacuum-gas manifold and serum vials closed with rubber stoppers, prepare anaerobic samples of purified UbiX proteins (in this case AF1214 + $prFMN_{radical}$).
2. Move the vials to an anaerobic glove-box and transfer protein aliquots to open screw-cap microtubes. Typically, all of the following steps can be performed anaerobically within an anaerobic chamber, but if necessary tightly closed microtubes with samples can be removed from the glove box (*e.g.*, for sample centrifugation, concentration, or storage at −80 °C).

3. Inside the glove-box, precipitate AF1214 + $prFMN_{radical}$ by adding acetonitrile to a final concentration of 70% (v/v) (a purple-colored protein precipitate will be formed).
4. Centrifuge the precipitated AF1214 + $prFMN_{radical}$ at 14,000 rpm (10 min) and wash protein pellets twice using 70% acetonitrile.
5. Discard supernatants and solubilize protein pellets in 100 mM HCl (the final pH should be around 2).
6. Precipitate the solubilized protein by adding acetonitrile (70%, v/v; higher acetonitrile volumes may also be used) and centrifuge the sample at 14,000 rpm. Transfer the supernatant containing the protein-free prFMN cofactor (in this case, protonated $prFMN_{radical}$ or $prFMN_{radical}$-H, Fig. 5) to a fresh microtube.
7. If necessary, concentrate diluted cofactor samples outside of the glove box using a SpeedVac concentrator (~200 μL aliquots, 15–20 min until full evaporation of acetonitrile) and return the samples back to an anaerobic glove box.
8. The concentration of prFMN preparations can be measured using UV–VIS spectrophotometry and serial dilutions (Fig. 6).
9. The cofactor solutions can be stored under inert atmosphere (argon or nitrogen) at −80 °C in tightly closed screw-cap microtubes with O-rings.
10. To produce $prFMN_{iminium}$ (Fig. 5), incubate the $prFMN_{radical}$-H sample (in 70% acetonitrile) for 6–8 h or overnight at 4 °C and pH 7.0 (titrate with NaOH). After this incubation, the $prFMN_{iminium}$ samples should be used immediately or stored at −80 °C under inert atmosphere (life time ~24 h).
11. To produce C2′-hydroxylated $prFMN_{iminium}$ (Fig. 5), incubate the $prFMN_{radical}$-H sample (in 70% acetonitrile) for 6–8 h or overnight at 4 °C and pH 2.0 (titrate with HCl). After the acid incubation, titrate the sample with NaOH to pH > 7 and use immediately or store at −80 °C under inert atmosphere (life time ~24 h).
12. For $prFMN_{radical}$-H quantitation (Fig. 6), dilute protein-free cofactor in 86% (v/v) acetonitrile +50 mM HCl (inside the glove box).
13. Transfer the samples to glass cuvettes and measure absorption at $\lambda = 490$ nm (the prFMN λ_2 maximum) (Fig. 6).
14. Using the prFMN extinction coefficient of $6.3\,mM^{-1}\,cm^{-1}$ (Fig. 6), calculate the concentration of the cofactor in your samples.
15. Verify the purity of prFMN preparations using appropriate analytical method (*e.g.*, LC-MS) (Fig. 6C).

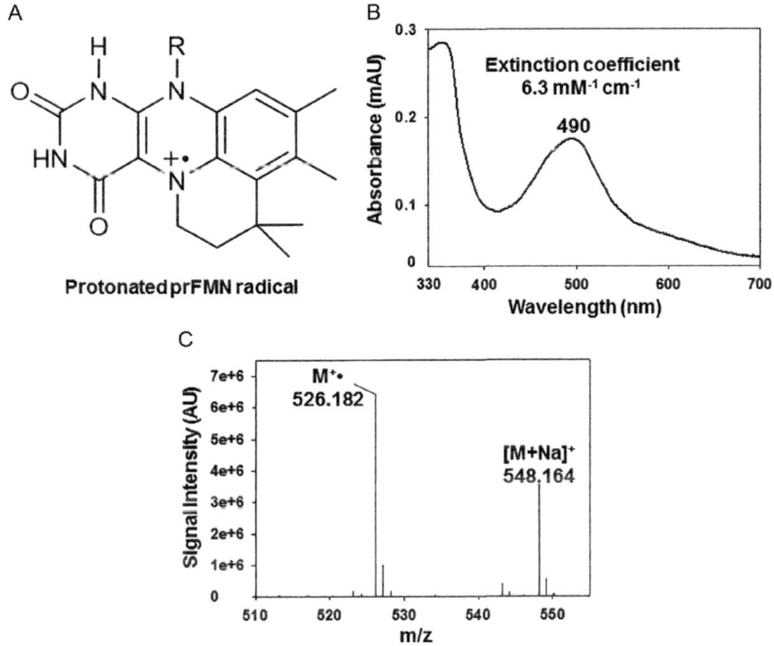

Fig. 6 Identification and quantitation of protonated $prFMN_{radical}$. (A), Chemical structure of protonated $prFMN_{radical}$ ($prFMN_{radical}$-H). (B), UV-VIS absorption spectrum of $prFMN_{radical}$-H. This cofactor has a distinctive λ_2 peak at 490 nm with an extinction coefficient of 6.3 mM^{-1} cm^{-1} (Wang et al., 2018). (C), MS spectrum of purified $prFMN_{radical}$-H. The main peak represents the protonated M^+ form ($m/z = 526.182$), whereas the second peak is the sodium adduct, $[M+Na]^+$ ($m/z = 548.164$).

3.4 Notes

1. As protein-bound prFMN appears to be sensitive to photooxidation (Bailey et al., 2018), it is recommended to avoid direct exposure to sunlight and reduce light exposure to a minimum (*e.g.*, using aluminum foil).
2. Centrifugation of protein samples during acetonitrile washing can be performed outside of the glove box using tightly closed screw-cap tubes with O-rings.
3. To verify the purity of cofactor samples, carry out LC-MS analysis using a Dionex Ultimate 3000 UHPLC system equipped with a photodiode array detector and a Q-Exactive MS equipped with a HESI II source. Perform runs at 40 °C on a Thermo Hypersil Gold C18 column (flow rate 0.2 mL/min; solvent A—5 mM ammonium acetate in water (pH 6.0), solvent B—5 mM ammonium acetate in methanol, pH 6.0).

Elution gradient: 2% B for 1 min; linear increase of B to 90% over 6 min; 90% B for 3.5 min; 2% B after 0.5 min; and 2% B for 7 min before starting next sample.

4. Since $prFMN_{radical}$-H is sensitive to oxygen, it is recommended to keep sample exposure to air at a minimum. However, it is quite stable during storage under anaerobic conditions at −80 °C.
5. During prolonged storage of $prFMN_{iminium}$ and $prGMN_{iminium}$-OH at −80 °C under inert atmosphere, they can be converted to inactive forms: prFMN-OH and prFMN-$(OH)_2$. Therefore, it is recommended to use the protein-free prFMN samples immediately after preparation.

4. Activation of the ferulic acid decarboxylase Fdc1 by protein-free and UbiX-bound prFMN cofactors

Most purified recombinant UbiD enzymes show negligible or low catalytic activity because there is not enough prFMN in *E. coli* cells for UbiD activation under standard conditions used during recombinant protein expression. This chapter describes a two-step activation protocol for UbiDs, which includes incubation of purified apo-UbiD with prFMN (UbiX-bound or protein-free cofactor) followed by the addition of components of the UbiD reaction mixture and enzyme assay. In this work, we used the UbiD enzyme, ferulic acid decarboxylase Fdc1 from *Aspergillus niger*, which is a well-known UbiD enzyme with available crystal structures and characterized interactions with prFMN (Bailey et al., 2018). It has been shown that purified apo-Fdc1 can be activated by *in vitro* anaerobic reconstitution with dithionite-reduced prFMN followed by aerobic oxidation using both UbiX-bound (PA4019 + $prFMN_{reduced}$ or AF1214 + $prFMN_{reduced}$) and protein-free ($prFMN_{iminium}$ and C2′-OH-$prFMN_{iminium}$) cofactors (Payne et al., 2015; Wang et al., 2018). In addition, we demonstrated that Fdc1 can be activated by UbiX-bound and protein-free prFMN cofactors under aerobic conditions (without reduction by dithionite) (Fig. 7), but the observed Fdc1 activity was 2–3 times lower than that after activation with dithionite reduction (Wang et al., 2018). Both protein-free prFMN forms can be prepared from purified $prFMN_{radical}$ by incubation at different pH (see Section 3.3, Fig. 5). Our work also revealed that apo-Fdc1 has high affinity to prFMN and can be saturated by micromolar concentrations of both protein-free cofactors (app. K_m 9–19 μM) (Wang et al., 2018). Thus, it appears that under aerobic conditions apo-Fdc1 quickly binds added protein-free prFMN cofactors, preventing their aerobic inactivation.

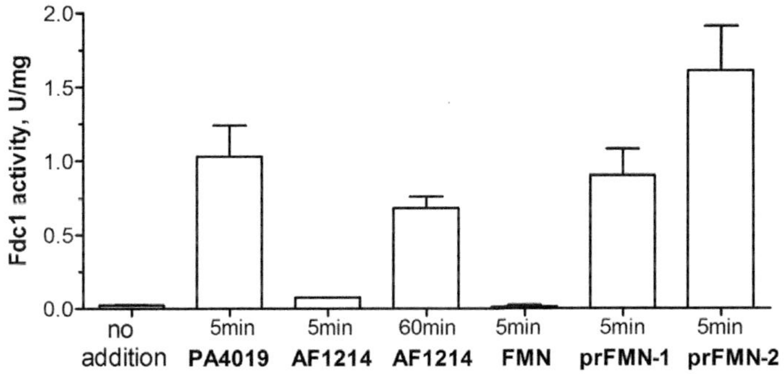

Fig. 7 Ferulic acid decarboxylation activity of Fdc1 activated by UbiX-bound or purified (protein-free) prFMN. (A), Fdc1 activity after 5 min or 60 min aerobic incubation with a 4-fold molar excess of UbiX-prFMN (PA4019 or AF1214). (B), Fdc1 activity after 5 min aerobic incubation with a fourfold molar excess of protein-free prFMN (prFMN-1: $prFMN_{iminium}$; prFMN-2: $prFMN_{iminium}$-C2′-OH).

4.1 Equipment

1. UV–Vis spectrophotometer
2. Centrifugal filters (Mw cut-off 3 kDa)
3. Microcentrifuge

4.2 Materials

1. Activation buffer for Fdc1 (100 mM K-phosphate buffer, pH 6.0)
2. Reaction buffer for Fdc1 (100 mM MES buffer (pH 6.0), 0.3 mM cinnamic acid, 200 mM NaCl, 0.2 mM $MnCl_2$, 5 mM $MgCl_2$)
3. Purified protein-free prFMN ($prFMN_{iminium}$, $prFMN_{iminium}$-OH, or protonated $prFMN_{radical}$) (0.05–2 mM)
4. Purified UbiX + $prFMN_{radical}$ (in this case, AF1214 + $prFMN_{radical}$ and PA4019 + $prFMN_{radical}$) (1–6 mg/mL stock)
5. Purified UbiD (in this case, Fdc1 from *A. niger* recombinantly expressed in *E. coli* and affinity purified) (7–14 mg/mL stock)

4.3 Procedure

1. Add 1–2 M equivalents of purified prFMN or 1–10 M equivalents of UbiX + prFMN to purified Fdc1 (0.25 nmol) in the activation buffer. Use aluminum foil to reduce light exposure.
2. Depending on the source of prFMN, incubate Fdc1 samples at RT for 5 min (for protein-free and PA4019-bound cofactors) or 60 min (for AF1214-bound cofactors).

3. Transfer small aliquots (0.1–10 μg) to a 96-well plate containing Fdc1 reaction buffer (200 μL reactions). Follow the reaction spectrophotometrically at 270 nm (with shaking every 30 s) at 30 °C (for 5–60 min).

4.4 Notes

1. Fdc1 activation can also be performed under anaerobic conditions with dithionite reduction followed by aerobic activation (Payne et al., 2015; White et al., 2015). This protocol usually produces holo-Fdc1 preparations, which are 2–3 times more active than those obtained under aerobic conditions.
2. Both apo-Fdc1 activation and holo-Fdc1 enzymatic reaction can be performed simultaneously with protein-free cofactors and some UbiX-bound cofactors (*e.g.*, PA4019), but Fdc1 activation in the presence of AF1214-bound prFMN is slower, resulting in a Fdc1 activity lag (Fig. 7).
3. Phosphate buffer is recommended for Fdc1 activation, since it was found that activation of these enzyme by UbiX-bound and protein-free prFMN cofactors was stimulated by phosphate addition (Wang et al., 2018).
4. For activation of other UbiD enzymes, the reaction conditions need to be optimized (pH, buffers, temperature, incubation time, substrates, metal cofactors).
5. Technically, activation of recombinant UbiD proteins can be performed *in vivo* using *E. coli* cells co-expressing both UbiD and UbiX proteins with the addition of prenol and riboflavin to growth medium.

5. Conclusions

In this chapter, we described a complete set of protocols starting from the preparation of various forms of prFMN using FMN prenylation reactions performed *in vivo* (in *E. coli* cells) or *in vitro* (using purified proteins), followed by prFMN extraction, quantitation, and then by their application for Fdc1 activation. In general, most of these experiments can be performed in a standard laboratory setting without using expensive chemicals or complex equipment. It is anticipated that future developments of these methods, including streamlining and optimization will make them more convenient and efficient. We believe that these protocols will contribute to our knowledge of prFMN biochemistry and facilitate functional exploration of uncharacterized enzymes from the large family of UbiD-like reversible decarboxylases eventually revealing novel enzymes for applications in biocatalysis and synthetic biology.

Acknowledgments

This work was supported by the NSERC Strategic Network grant Industrial Biocatalysis Network and Genome Canada (by the Government of Canada through Genome Canada and the Ontario Genomics Institute, 2009-OGI-ABC-1405).

References

Bailey, S. S., Payne, K. A. P., Fisher, K., Marshall, S. A., Cliff, M. J., Spiess, R., et al. (2018). The role of conserved residues in Fdc decarboxylase in prenylated flavin mononucleotide oxidative maturation, cofactor isomerization, and catalysis. *The Journal of Biological Chemistry*, *293*(7), 2272–2287. https://doi.org/10.1074/jbc.RA117.000881.

Do, H., Kim, S. J., Lee, C. W., Kim, H. W., Park, H. H., Kim, H. M., et al. (2015). Crystal structure of UbiX, an aromatic acid decarboxylase from the psychrophilic bacterium Colwellia psychrerythraea that undergoes FMN-induced conformational changes. *Scientific Reports*, *5*, 8196. https://doi.org/10.1038/srep08196.

Khusnutdinova, A. N., Flick, R., Popovic, A., Brown, G., Tchigvintsev, A., Nocek, B., et al. (2017). Exploring bacterial carboxylate reductases for the reduction of bifunctional carboxylic acids. *Biotechnology Journal*, *12*(11), 1700015. https://doi.org/10.1002/biot.201600751.

Kopec, J., Schnell, R., & Schneider, G. (2011). Structure of PA4019, a putative aromatic acid decarboxylase from Pseudomonas aeruginosa. *Acta Crystallographica. Section F, Structural Biology and Crystallization Communications*, *67*, 1184–1188. https://doi.org/10.1107/S174430911102923X. Pt. 10.

Leys, D., & Scrutton, N. S. (2016). Sweating the assets of flavin cofactors: new insight of chemical versatility from knowledge of structure and mechanism. *Current Opinion in Structural Biology*, *41*, 19–26. https://doi.org/10.1016/j.sbi.2016.05.014.

Lin, F., Ferguson, K. L., Boyer, D. R., Lin, X. N., & Marsh, E. N. (2015). Isofunctional enzymes PAD1 and UbiX catalyze formation of a novel cofactor required by ferulic acid decarboxylase and 4-hydroxy-3-polyprenylbenzoic acid decarboxylase. *ACS Chemical Biology*, *10*(4), 1137–1144. https://doi.org/10.1021/cb5008103.

Marshall, S. A., Fisher, K., Ní Cheallaigh, A., White, M. D., Payne, K. A. P., Parker, D. A., et al. (2017). Oxidative maturation and structural characterization of prenylated FMN binding by UbiD, a decarboxylase involved in bacterial ubiquinone biosynthesis. *Journal of Biological Chemistry*, *292*(11), 4623–4637. https://doi.org/10.1074/jbc.M116.762732.

Marshall, S. A., Payne, K. A. P., & Leys, D. (2017). The UbiX-UbiD system: The biosynthesis and use of prenylated flavin (prFMN). *Archives of Biochemistry and Biophysics*, *632*, 209–221. https://doi.org/10.1016/j.abb.2017.07.014.

Massey, V. (2000). The chemical and biological versatility of riboflavin. *Biochemical Society Transactions*, *28*(4), 283–296.

Mizote, T., & Nakayama, H. (1989). The thiM locus and its relation to phosphorylation of hydroxyethylthiazole in Escherichia coli. *Journal of Bacteriology*, *171*(6), 3228–3232.

Payne, K. A., White, M. D., Fisher, K., Khara, B., Bailey, S. S., Parker, D., et al. (2015). New cofactor supports alpha,beta-unsaturated acid decarboxylation via 1,3-dipolar cycloaddition. *Nature*, *522*(7557), 497–501. https://doi.org/10.1038/nature14560.

Piano, V., Palfey, B. A., & Mattevi, A. (2017). Flavins as covalent catalysts: New mechanisms emerge. *Trends in Biochemical Sciences*, *42*(6), 457–469. https://doi.org/10.1016/j.tibs.2017.02.005.

Rangarajan, E. S., Li, Y., Iannuzzi, P., Tocilj, A., Hung, L. W., Matte, A., et al. (2004). Crystal structure of a dodecameric FMN-dependent UbiX-like decarboxylase (Pad1) from Escherichia coli O157: H7. *Protein Science*, *13*(11), 3006–3016. https://doi.org/10.1110/ps.04953004.

Sobrado, P. (2012). Noncanonical reactions of flavoenzymes. *International Journal of Molecular Sciences*, *13*(11), 14219–14242. https://doi.org/10.3390/ijms131114219.

Teufel, R., Agarwal, V., & Moore, B. S. (2016). Unusual flavoenzyme catalysis in marine bacteria. *Current Opinion in Chemical Biology*, *31*, 31–39. https://doi.org/10.1016/j.cbpa.2016.01.001.

Walsh, C. T., & Wencewicz, T. A. (2013). Flavoenzymes: Versatile catalysts in biosynthetic pathways. *Natural Product Reports*, *30*(1), 175–200. https://doi.org/10.1039/c2np20069d.

Wang, P. H., Khusnutdinova, A. N., Luo, F., Xiao, J., Nemr, K., Flick, R., et al. (2018). Biosynthesis and activity of prenylated FMN cofactors. *Cell Chemical Biology*, *25*, P560–P570. https://doi.org/10.1016/j.chembiol.2018.02.007.

White, M. D., Payne, K. A., Fisher, K., Marshall, S. A., Parker, D., Rattray, N. J., et al. (2015). UbiX is a flavin prenyltransferase required for bacterial ubiquinone biosynthesis. *Nature*, *522*(7557), 502–506. https://doi.org/10.1038/nature14559.

CHAPTER NINETEEN

Heterologous production, reconstitution and EPR spectroscopic analysis of prFMN dependent enzymes

Stephen A. Marshall, Karl A.P. Payne, Karl Fisher, Deepankar Gahloth, Samuel S. Bailey, Arune Balaikaite, Annica Saaret, Irina Gostimskaya, Godwin Aleku, Huanming Huang, Stephen E.J. Rigby, David Procter, David Leys*

School of Chemistry, University of Manchester, Manchester, United Kingdom

*Corresponding author: e-mail address: david.leys@manchester.ac.uk

Contents

Methods in Enzymology, Volume 620
ISSN 0076-6879
https://doi.org/10.1016/bs.mie.2019.03.022

Abstract

The recent discovery of the prenylated FMN (prFMN) cofactor has led to a renewed interest in the prFMN-dependent UbiD family of enzymes. The latter catalyses the reversible decarboxylation of alpha-beta unsaturated carboxylic acids and features widely in microbial metabolism. The flavin prenyltransferase UbiX synthesizes prFMN from reduced FMN and phosphorylated dimethylallyl precursors. Oxidative maturation of the resulting prFMN^{reduced} species to the active prFMN^{iminium} form is required for UbiD activity. Heterologous production of active *holo*-UbiD requires co-expression of UbiX, but the levels of prFMN incorporation and oxidative maturation appear variable. Detailed protocols and strategies for *in vitro* reconstitution and oxidative maturation of UbiD are presented that can yield an alternative source of active *holo*-UbiD for biochemical studies.

1. Introduction

Flavoenzymes catalyse a remarkably wide range of reactions, courtesy of the broad chemistry repertoire of the flavin cofactor (Fraaije & Mattevi, 2000). In certain cases, the properties of the bound flavin are subtly altered by covalent modification, usually through linkage with the protein (Heuts, Scrutton, McIntire, & Fraaije, 2009). In contrast, the recently discovered UbiX-UbiD system depends on a heavily modified flavin, prenylated FMN, with substantially altered chemical properties as a direct consequence (Leys, 2018). The canonical UbiX-UbiD enzymes feature in the bacterial ubiquinone biosynthesis pathway and are associated with a key (de) carboxylation step (Marshall, Fisher, et al., 2017). The UbiX enzyme acts as a flavin prenyltransferase, converting reduced $FMNH_2$ to prFMN^{reduced} using the isoprenoid precursor DMAP(P) as the prenylating agent (White et al., 2015; Fig. 1, reaction **1**). The UbiD enzyme depends on an oxidized form of this cofactor, prFMN^{iminium} for activity. Oxidative maturation is suggested to occur within the UbiD enzyme, following binding of the prFMN^{reduced} by *apo*-UbiD (Fig. 1, reaction **2a**). The prFMN^{iminium} has azomethine ylide characteristics, leading to 1,3-dipolar cycloaddition with the substrate. The reversible (de)carboxylation step occurs following covalent substrate-prFMN^{iminium} adduct formation, and is followed by

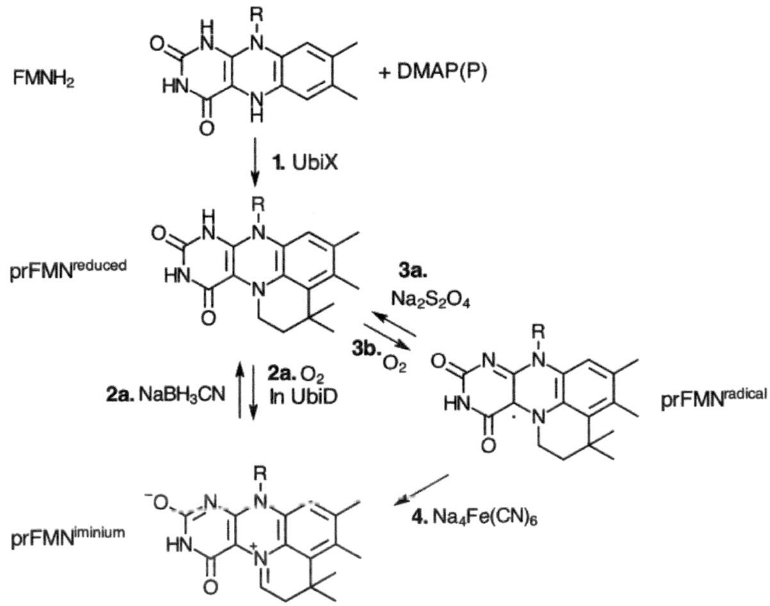

Fig. 1 Schematic overview of production and interconversion of key prFMN species.

cycloelimination to yield the prFMN^{iminium} and product (Payne et al., 2015). The UbiD family is widespread, and *ubiD* genes can be found in most microbes, associated with a wide range of metabolic pathways. UbiD substrates are thus remarkably diverse, but each contains an alpha-beta unsaturated carboxylic acid group. This acrylic acid moiety acts as the polarophile and is thus required for catalysis. Under ambient [CO_2] conditions, the equilibrium between an alpha-beta unsaturated acid and the corresponding alkene + CO_2 is poised toward the latter, and many UbiD enzymes act as decarboxylases. However, in some cases UbiD family members have been proposed to act as carboxylases (Marshall, Payne, & Leys, 2017).

Relatively few UbiD enzymes have been studied in detail Annaval et al., 2018; (Marshall, Fisher, et al., 2017; Mergelsberg et al., 2017; Payer et al., 2017; Payne et al., 2015) and it is unclear to what extent the properties of these enzymes are conserved throughout the wider family. The fungal UbiD homolog Fdc1 was the first to be characterized in detail and remains arguably the best characterized enzyme to date (Bailey et al., 2018; Payne et al., 2015). This enzyme acts to decarboxylate cinnamic acid and related compounds and binds prFMN in a Mn^{2+} and K^{+} dependent manner. Active *holo*-Fdc1 can be obtained *in vivo* through co-expression of the *fdc1* gene with a *ubiX* homolog in *E. coli*. When purified under ambient conditions, the enzyme is inactivated by light, which has been implicated in driving an unusual cofactor isomerisation (Bailey et al., 2018). The activity of other UbiD enzymes (but not Fdc1) has been found to be sensitive to oxygen, despite the fact an oxidative maturation step is required (Payer et al., 2017). A comparison of the various UbiD crystal structures reveals a common fold consisting of an N-terminal prFMN binding domain connecting to a C-terminal oligomerization domain *via* a central alpha helix. While Fdc1 is a dimer, other UbiD family members form higher order oligomers. The active site is located at the interface between both domains, and the relative position of the prFMN binding domain to the oligomerization domain appears to vary considerably between individual crystal structures. This is suggestive of domain motion, possibly associated with catalysis (Marshall, Fisher, et al., 2017; Payer et al., 2017). The Fdc1 crystal structures all represent a closed form of the enzyme, allowing substrate binding and catalysis. In contrast, other UbiD crystal structures appear to correspond to an open form, which is unable to bind substrate. Although the link between prFMN binding and domain motion has likewise not been firmly established, cofactor binding affinity appears weaker for those enzymes that are crystallized in the open form. As a consequence, heterologous production and purification of the latter often leads to *apo*-protein, necessitating the need for *in vitro* reconstitution. Here, we describe the heterologous production and purification of *holo*- or *apo*-UbiD, followed by various methods for *in vitro* reconstitution and oxidative maturation. This is followed by an overview of the use of EPR spectroscopy for detailed characterization of $\text{prFMN}^{\text{radical}}$ UbiX/UbiD complexes.

Specific forms of prFMN are indicated by the superscript used, while UbiD refers to a generic UbiD enzyme unless otherwise mentioned. Protocols are based on the production and/or reconstitution of Fdc1 or AroY UbiD-family members.

2. *In vivo* production of *holo*-UbiD/*apo*-UbiD

The levels of native UbiX activity in *E. coli* are insufficient to support the production of large quantities of *holo*-UbiD enzymes. Hence, co-expression with a highly active UbiX is required to support *holo*-UbiD production *in vivo*. Initial attempts carried out in our group focused on co-expression with the cognate UbiX homologs. However, we have subsequently found that co-expression of various UbiD homologs with either *E. coli* or *Pseudomonas aeruginosa* UbiX results in higher levels of prFMN incorporation (Payne et al., 2015).

In addition, prFMN production may also be boosted by increasing the levels of the UbiX substrate, DMAP. *E. coli* BL21(DE3) cell lysates have been shown to phosphorylate prenol to produce DMAP, possibly as a result of the activity of the hydroxyethylthiazole kinase, ThiM. Consistent with this, supplementation of *E. coli* cultures over-expressing recombinant UbiX with prenol was shown to stimulate prFMN production (Wang et al., 2018). In addition to supplementing the media with prenol, addition of a source of K^+ and Mn^{2+} ions during production and purification steps can also benefit *holo*-UbiD levels, as these are found to be required for prFMN binding.

2.1 Equipment

- Orbital shaking incubator
- Centrifuge and Rotor
- Cell disruptor
- Ni-NTA agarose column
- Econopac 10-DG column
- Anaerobic glove box

2.2 Buffers and reagents

- 1 L LB medium (autoclaved)
- 50 mg/mL ampicillin (sterile filtered through a 0.2 μm syringe filter)
- 50 mg/mL kanamycin (sterile filtered through a 0.2 μm syringe filter)
- 0.25 M IPTG (sterile filtered through a 0.2 μm syringe filter)
- 1 M $MnCl_2$ (sterile filtered through a 0.2 μm syringe filter)
- 1 L Ni-Binding buffer: 200 mM NaCl, 50 mM Tris-HCl, pH 7.5 in Milli-Q water
- 1 L Ni-Elution buffer: 200 mM NaCl, 50 mM Tris-HCl, 500 mM imidazole pH 7.5 in Milli-Q water
- 1 L 10-DG buffer: 100 mM NaCl, 20 mM Tris-HCl, pH 7.5 in Milli-Q water
- Leave buffers in cold room overnight to chill.

2.3 Procedures

A. niger Fdc1 was cloned into the *Nde*I/*Xho*I sites of pET30a (kanamycin resistant) to produce a protein with a C-terminal His-tag, while *E. coli* UbiX was cloned into the *Nde*I/*Xho*I sites of pET21b (ampicillin resistant) so as to yield untagged protein. Both plasmids were co-transformed into BL21 (DE3), with successful co-transformants selected for by resistance to both ampicillin and kanamycin. We have found the use of two pET vectors, both possessing the same origin of replication, to be effective. However, it may be preferable to either express Fdc1 and UbiX on two vectors with different origins of replication, or on a single plasmid with two multiple cloning sites.

2.3.1 Heterologous production

Set up an overnight pre-culture by inoculating 50 mL LB supplemented with 50 μg/mL ampicillin and 50 μg/mL kanamycin with a single colony from an agar plate of the co-expressing strain and grow at 37 °C overnight. Next day inoculate 1 L LB supplemented with 50 μg/mL ampicillin and 50 μg/mL kanamycin with 5 mL of the overnight pre-culture. Grow at 37 °C and 180 rpm until an OD_{600} of ~0.6–1 is reached and cool the culture to 17 °C. Supplement with 1 mM $MnCl_2$ (Mn^{2+} is involved in prFMN binding by Fdc1 and AroY, however the identity of the metal ions may be different in other UbiD homologs) and induce with 0.25 mM IPTG. The culture is then grown at 17 °C and 180 rpm overnight. Next day the cultures are harvested by centrifugation and the pellets stored at −20 °C until required.

2.3.2 **Holo-*UbiD purification***

Some members of the UbiD family have been shown to be sensitive to light (Bailey et al., 2018) and/or to oxygen (Payer et al., 2017), and it is desirable to take measures to mitigate the amount of light exposure during purification to a minimum (*e.g.*, wrapping the column/bottles in foil) and/or to perform protein purification within an anaerobic glove box. Providing a source of K^+ and Mn^{2+} in the purification or storage buffers can also boost prFMN retention levels.

All steps to be carried out with minimum light exposure, and in an anaerobic glove box if required. Thaw and resuspend cells in Ni-binding buffer. Supplement with EDTA-free protease inhibitors, lysozyme, DNase, and RNase. Allow to stir in cold room until homogeneous. Pass through the cell disruptor at 20,000 psi twice, prior to centrifugation at 100,000 × *g* at 4 °C for 1 h. Decant the supernatant and apply to a column loaded with 5 mL Ni^{2+}-NTA agarose resin equilibrated with Ni-Binding buffer and collect flow through. Wash the column with 4 column-volumes of Ni-binding buffer containing 10 mM imidazole supplemented with 1 mM $MnCl_2$ and collect flow through. Elute the protein with 2 column-volumes of buffer containing 250 mM imidazole supplemented with 1 mM $MnCl_2$, adding 1 mL at a time and collecting in individual tubes. Subject all fractions to

SDS-PAGE analysis to locate UbiD containing fractions, and pool those fractions containing pure protein. The imidazole is removed by exchanging the buffer of the protein into imidazole-free buffer using a 10-DG column. Equilibrate the column with 20 mL imidazole-free buffer supplemented with 1 mM $MnCl_2$ (in case of preparing samples for EPR, Mn^{2+} should be omitted). Reduce volume of the pooled fractions to <3 mL and apply to the column. Collect eluting protein in 0.5–1 mL fractions and use Bradford reagent to locate protein. In certain cases, the protein is colored ($prFMN^{radical}$ is purple, while $prFMN^{iminium}$ is faint yellow in color), which can allow simple visual or spectrophotometric inspection to identify fractions of interest. Fractions containing purified protein are pooled, aliquoted and snap-frozen in liquid nitrogen.

3. *In vitro* reconstitution of *apo*-UbiD using $prFMN^{reduced}$

We have found that many UbiD enzymes are purified in the *apo*-form, despite co-expression with UbiX. Furthermore, levels of active $prFMN^{iminium}$ can also vary in a batch-dependent manner. Hence, *in vitro* reconstitution of UbiD enzymes is sometimes necessary. Therefore, a source of $prFMN^{reduced}$ is required, and can be made in an anaerobic environment by the UbiX enzyme (Fig. 1, reaction **1**). We have found that using UbiX from *Pseudomonas aeruignosa* (*Pa*UbiX) is an efficient and effective way of producing prFMN from $FMNH_2$ and dimethylallyl monophosphate (DMAP; White et al., 2015). Although it has been shown that other UbiX homologs utilize the better known dimethylallyl pyrophosphate (DMAPP) (Arunrattanamook & Marsh, 2017), we continue to use *Pa*UbiX due to its stability in solution and the relative ease of synthesizing DMAP (see Section 5).

3.1 Equipment

- Anaerobic chamber
- Optional - UV–Vis spectrophotometer in anaerobic chamber
- Microcentrifuge

3.2 Buffers and reagents

- Anaerobic buffer
- FMN

- DMAP (for synthesis see Section 5)
- Purified *Pa*UbiX (for protocol see White et al., 2015)
- Purified *E. coli* Fre reductase and NADH or sodium dithionite as reductant

3.3 Procedures

3.3.1 prFMN^{reduced} production

All reagents should be dissolved in anaerobic buffer, *i.e.*, thoroughly purged with nitrogen, and left to equilibrate in the anaerobic chamber. Buffers used should be specific for the UbiD that is being reconstituted. A typical reaction consists of 50 μM UbiX incubated with 1 mM FMN in an anaerobic environment. FMN is reduced by addition of 10 μM Fre reductase and 2 mM NADH. Upon addition of NADH, the mixture should soon turn colorless. Alternatively, sodium dithionite can be used as the reductant, however, this may cause problems with further experiments. When dithionite is used, a stock of ~100 mM can be made and added in minimal volume until the solution is colorless. Once all FMN is reduced, DMAP is added to a final concentration of 5 mM. The reaction is then allowed to proceed at ambient temperature for at least an hour. If desired, this reaction can be left overnight, in which case NADH is increased to 10 mM to minimize reoxidation. Finally, the prFMN^{reduced} mixture is filtered using a centrifugal concentrator to remove UbiX and Fre. Although not strictly necessary, this step ensures no contamination with UbiX or Fre. It can be useful at this point to remove the filter and expose this to oxygen. A purple color on the filter developing confirms production of prFMN^{radical} species (Fig. 2), and by extension the presence of prFMN^{reduced} in the filtrate.

3.3.2 Reconstitution of UbiD with prFMN^{reduced}

The prFMN^{reduced} filtrate is added to UbiD, along with $MnCl_2$/KCl, in a molar ratio of 2:2:2:1 (prFMN^{reduced}:$MnCl_2$:KCl:UbiD). This mixture is then incubated for at least 10 min in the anaerobic chamber. To remove excess prFMN^{reduced} and NAD^+, the mixture is applied to a desalting column equilibrated with anaerobic buffer and the protein is eluted according to the manufacturer's protocol.

3.4 Notes

When using dithionite as the reductant, there is a risk of prFMN-sulfite adduct formation (Marshall, Fisher, et al., 2017). For this reason, we

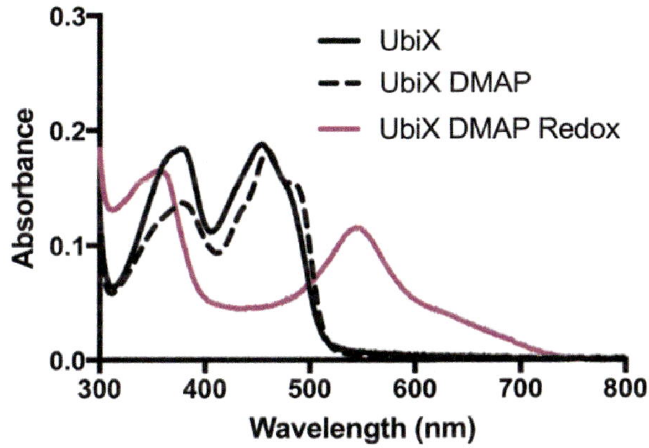

Fig. 2 UV–Vis spectra of purified *Pa*UbiX in complex with FMN, FMN and DMAP, and following reduction and oxidation (redox) respectively.

developed a protocol that utilizes the natural FMN reductase Fre (Shepard et al., 2015). Following removal of Fre by filtration, the prFMN^{reduced} is readily oxidized, leading to a purple prFMN^{radical} species. For this reason, the oxygen content of the anaerobic chamber should be kept to a minimum. To ensure the desalting resin contains no oxygen, wash with anaerobic buffer supplemented with dithionite. To remove dithionite the column should be washed thoroughly (>4 column-volumes) with anaerobic buffer.

4. In vitro oxidative maturation of prFMN containing UbiD

Exposure of prFMN^{reduced}:UbiD to oxygen leads to either direct formation of the active prFMN^{iminium} form, or formation of the inactive prFMN^{radical} form (Fig. 1, reactions **2a** and **3b**, respectively), dependent on the UbiD variant used. The level of activation also appears subject to considerable variability. We have discovered that in addition to molecular oxygen, a variety of alternative oxidants can be used to achieve oxidative maturation of prFMN^{reduced}:UbiD or prFMN^{radical}:UbiD (the latter is stable in presence of oxygen). The chemical oxidants potassium ferricyanide and hydrogen peroxide in particular can substitute for molecular oxygen to oxidize reduced prFMN species, giving rise to the iminium species. These oxidants are cheap, easy to prepare and can be accurately used in a quantitative manner.

4.1 Equipment

- Anaerobic chamber
- Optional—UV–Vis spectrophotometer in anaerobic chamber

4.2 Buffers and reagents

- Anaerobic buffer
- Hydrogen peroxide (30%)
- Potassium ferricyanide
- UbiD substrate, *i.e.*, cinnamic acid in case of Fdc1

4.3 Procedure

A typical oxidation reaction consists of 25 μM anaerobic reduced *holo*-UbiD and a range of desired oxidant concentrations (10–100 μM) with the final volume adjusted with the buffer of choice. Oxidation reactions should be incubated in sealed vials at ambient temperature for 10–30 min to allow the oxidation to occur before samples are taken for analysis. All chemical oxidants and substrates should be dissolved in buffer that has been thoroughly purged with nitrogen, and left to equilibrate in the anaerobic chamber overnight. In case molecular oxygen is to be used, the concentration of oxygen dissolved in distilled water can be determined from the solubility of oxygen in water of a given salinity, specific temperature and pressure. As an example, water at 20 °C typically contains about 280 μM oxygen. Oxygenated water samples can be taken into the anaerobic chamber in "Suba Seal" capped vials and aliquots removed using a gas tight Hamilton syringe can be added to UbiD in anoxic crimp sealed glass vials.

In the case of Fdc1, enzyme maturation can be monitored by determining the initial rate of decarboxylation through following the consumption of cinnamic acid by UV–vis spectroscopy at 270 nm. The rate of cinnamic acid decarboxylation is determined from its rate of consumption using the extinction coefficient $\varepsilon^{270} = 20{,}000\ M^{-1}\ cm^{-1}$. A typical 500 μL assay contains 80 μM cinnamic acid and 1 μM enzyme in 50 mM sodium phosphate pH 6, 50 mM KCl in a 10 mm path length cuvette at 25 °C. Fig. 3 shows the result of incubating $Fdc1^{prFMN}$ with 1–5 M equivalents of oxygen, H_2O_2 and potassium ferricyanide for 30 min prior to assaying 1 μM enzyme for decarboxylation of cinnamic acid. The initial rates of reaction used to determine k_{cat} values are expressed as a percentage of maximum activity for each

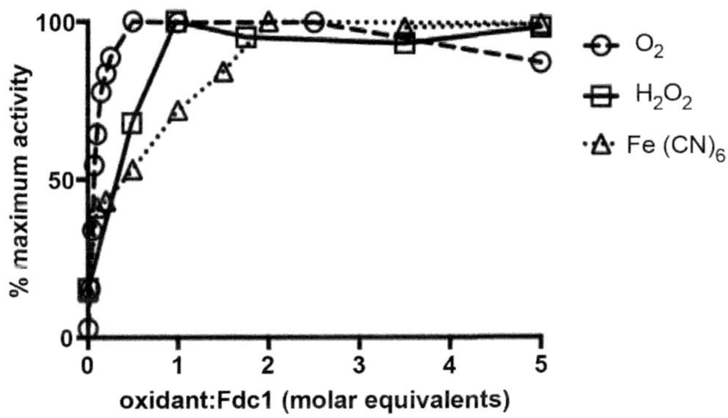

Fig. 3 Graph showing the change in enzyme activity upon addition of varying equivalents of oxidant to purified Fdc1 under anaerobic conditions. The measured activity has been normalized to 100% maximum activity determined for each set of conditions. Activity was determined by measuring the decarboxylation of cinnamic acid at 270 nm.

assay. Full maturation is achieved at a two-equivalent excess of oxidant over enzyme. An excess of oxidant can lead to a decrease in activity levels by an as yet unknown process.

4.4 Notes

It is crucial that any oxidant used is made fresh and kept under strict anaerobic conditions immediately prior to use. Potassium ferricyanide is an orange solid that can be flushed with nitrogen before being dissolved in anaerobic buffer in the anoxic chamber. We recommend making a concentrated stock solution and performing serial dilutions down to a working concentration to ensure it is fully anaerobic. Hydrogen peroxide can be purchased as a 30% or roughly 10 M stock solution but it is unstable and slowly decomposes in light. Stock solutions can be sparged with oxygen free nitrogen before being taken into the anaerobic chamber and then diluted down to a working concentration with anaerobic buffer before use.

5. Synthesis of DMAP

Although there are commercial sources of DMAP and DMAPP, these can be expensive and/or lack purity. We therefore synthesize DMAP using the following protocol.

5.1 Equipment

- Round-bottomed flask (25 mL) and Erlenmeyer flasks (25 mL)
- Disposable syringes (3 and 6 mL)
- Measuring cylinder, magnetic stirrer bar, rubber septa, needles, stirring plates, laboratory balance, rotary evaporator, vacuum pump, oven
- TLC plates (TLC silica gel 60, Merck, cat. no. 105553)
- Pasteur pipettes
- Glass chromatography column
- NMR tubes & NMR spectrophotometer

5.2 Reagents

- Concentrated phosphoric acid
- Triethylamine
- Acetonitrile
- 3-Methylbut-2-en-1-ol
- Trichloroacetonitrile
- Isopropanol
- 38% aqueous NH_4OH
- H_2O, D_2O
- Silica gel 60 Angstrom (Å) with 240–400 mesh

5.3 Procedures

A literature protocol was followed (Keller & Thompson, 1993).

5.3.1 Preparing the bis-triethylammonium phosphate (TEAP) solution

Solution A was prepared by adding concentrated phosphoric acid (1.25 mL) to acetonitrile (4.7 mL). Solution B was prepared by adding triethylamine (5.5 mL) to acetonitrile (5.0 mL). The TEAP solution was prepared by slowly adding solution A (4.7 mL) to solution B (5.0 mL) with stirring.

5.3.2 Preparing the dimethylallyl monophosphate (DMAP)

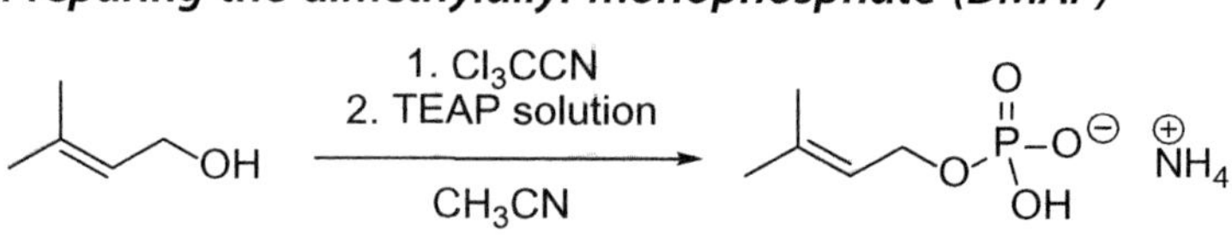

To the alcohol (2 mmol, 0.2 mL) in a 25 mL round-bottomed flask was added trichloroacetonitrile (5 mL), followed by dropwise addition of the TEAP solution (5 mL). The reaction mixture was stirred at 37 °C for

5 min. Additional TEAP solution (2 × 5 mL) was added with 5 min between each addition. The reaction mixture was then stirred overnight and the solvent was removed *in vacuo*. The crude product was purified by column chromatography on silica gel (isopropanol:38% aq. NH_4OH:H_2O, 6.5:2.5:0.5) to give the product (0.232 g, 1.27 mmol, 64%) as a pale yellow solid.

5.4 Notes

The purified compound can be characterized by NMR and MS analysis. The data should be consistent with those reported in the literature (Lira, Vasilev, Pilli, & Wessjohann, 2013). ^{1}H NMR (400 MHz, D_2O): δ ppm 5.48–5.41 (m, 1H, $(CH_3)_2C{=}CH$), 4.47 (t, J=7.3 Hz, 1H, 1H from $(CH_3)_2C{=}CHCH_2$), 4.36 (t, J=6.8 Hz, 1H, 1H from $(CH_3)_2C{=}CHCH_2$), 1.77 (d, J=3.1 Hz, 3H, CH_3), 1.72 (d, J=5.8 Hz, 3H, CH_3). ^{13}C NMR (126 MHz, D_2O): δ ppm 140.1 (C), 119.5 (d, J=7.6 Hz, $(CH_3)_2C{=}CH$), 62.1 (d, J=5.0 Hz, $(CH_3)_2C{=}CHCH_2$), 24.9 (CH_3), 17.1 (CH_3). MS (ESI$^-$) m/z (%): 165.0 (M-NH_4^+).

6. In vitro reconstitution of *apo*-UbiD using prFMN^{iminium}

As an alternative to the reconstitution with prFMN^{reduced}, which requires subsequent oxidative maturation within UbiD, the enzyme can be reconstituted with prFMN^{iminium} directly. The latter can be prepared by extraction of prFMN^{radical} from single turnover UbiX^{prFMN} (Wang et al., 2018), and converted to prFMN^{iminium} to be used to reconstitute *apo*-UbiD.

6.1 Equipment

- Anaerobic chamber (*e.g.*, Belle technologies glovebox)
- Microfuge (anaerobic)
- 85 °C water bath
- O-ring capped plastic tubes
- Freeze drier

6.2 Buffers and reagents

- Anaerobic buffer containing 1 mM $MnCl_2$
- Anaerobic water
- Anaerobic acetonitrile

6.3 Procedures

Anaerobic water and acetonitrile are prepared by sparging with nitrogen before being taken into the anaerobic chamber. Inside the anaerobic chamber set up a 1 mL UbiX single-turnover reaction containing 800 μM UbiX, 2 mM DMAP, 2 mM FMN and 2 mM sodium dithionite. After 1 h add an equal volume of ice-cold degassed 100% acetonitrile to the anaerobic solution, to give a final acetonitrile concentration of 50%. In order to remove the buffer and other components in the solution, precipitate the protein using a microfuge. Discard the supernatant and resuspend the pellet with anaerobic 70% acetonitrile, made by diluting the stock acetonitrile solution with cold anaerobic water. Centrifuge again and discard the supernatant to remove any remaining contaminants.

Resuspend the pellet with anaerobic 70% acetonitrile and transfer the suspension to a suitable O-ring capped plastic tube (for example, CryoPure O-ring plastic cryo-tubes, Sarstedt). Remove the sealed tube from the box and heat in a 85 °C water bath for 5 min with intermittent vortexing to release the cofactor from the protein. Inside the anaerobic chamber, centrifuge the suspension once again. If you have a white pellet and purple supernatant then you have successfully separated the prFMN^{radical}. Transfer the supernatant to a new tube and add one equivalent of anaerobic potassium ferricyanide to the prFMN^{radical} extract, to oxidize the prFMN^{radical} to the iminium form. After vortexing, the color of the solution changes from purple to light yellow, indicating conversion to the iminium form.

The extracted prFMN^{iminium} cofactor can be frozen in liquid nitrogen and freeze-dried for future use. When dried, it is stable and can be stored in the anaerobic chamber for future use for reconstitution of *apo*-UbiD. Alternatively one can freeze-dry the radical form of the prFMN to remove the acetonitrile and oxidize it after reconstitution.

6.4 Notes

A similar extraction procedure can also be applied to heterologously produced UbiD enzymes that contain prFMN^{radical} or prFMN^{iminium}, bypassing the need for *in vitro* production of prFMN using UbiX. *Apo*-UbiD enzymes can be reconstituted using prFMN^{iminium} extracted in the manner outlined in Section 4.

7. EPR spectroscopy as a tool to study prFMN^{radical} species

The one electron reduced flavin semiquinone radical state, frequently formed using one electron-equivalent of sodium dithionite or *via* photoreduction in the presence of EDTA (Massey & Palmer, 1966), can be studied using electron paramagnetic resonance (EPR; aka electron spin resonance, ESR) spectroscopy that measures the absorption of electromagnetic radiation (usually microwave radiation in the GHz range) by an unpaired electron that has been placed in a magnetic field (Hagen, 2009). This absorption is considered "resonant" as it occurs only for certain frequency and field combinations, most commonly (for both historical and technical reasons) at "X-band," that is frequencies close to 9 GHz and magnetic fields close to 0.3T. EPR spectroscopy of protein-bound flavosemiquinones is typically carried out at low temperatures using frozen solutions of the protein under study. Under such conditions a single line EPR spectrum with few distinguishing features such as shown in Fig. 4A is produced. (EPR presents spectra as first derivatives of the absorption.) The single line is 2.2–1.8 mT wide (peak to trough) for neutral (blue) flavin semiquinones and 1.7–1.5 mT wide for anionic (red) flavin semiquinones, both having g values (the spectrometer-independent measure of resonance condition) between 2.0038 and 2.0032 (Edmondson, 1978; Palmer, Müller, & Massey, 1971). The shape and width of such EPR spectra is the result of inhomogeneous broadening of the spectrum through interaction (hyperfine interaction) of the unpaired electron with the magnetic moments of certain atomic nuclei, for example, ^{1}H or ^{14}N, *i.e.*, those nuclei that can be studied using the "sister" magnetic resonance technique nuclear magnetic resonance, NMR. The unpaired electron is delocalised over the three rings of the flavin semiquinone isoalloxazine structure in the singly occupied molecular π orbital (SOMO) and interacts with the ^{1}H and ^{14}N atoms of the flavin structure through the hyperfine interaction to produce the observed EPR spectrum. The major hyperfine interactions involve the 8-methyl group, the protons at C6 and N5 (in neutral radicals) and the ribityl methylene group at N10 (Kay et al., 1999; Kurreck, Kirste, & Lubitz, 1988). The result of these circumstances therefore renders the shape of the EPR spectrum sensitive to any changes in the number and/or location of ^{1}H and ^{14}N atoms attached to the three isoalloxazine rings. With this in mind, the X-band

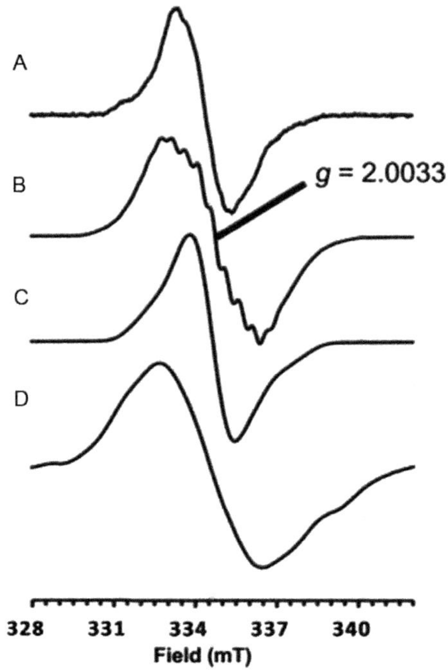

Fig. 4 X-band EPR spectra of frozen flavoprotein solutions obtained at 20 K showing a typical neutral (blue) flavin semiquinone (A), UbiX + DMAP reduced and reoxidized (B), UbiX + DMAP deuterated at C1 (C), and sodium cyanoborohydride-reduced Fdc1 (D).

EPR spectrum of UbiX reduced and reoxidized in the presence of excess DMAP, Fig. 4B, having a line width of 3.55 mT at $g=2.0033$ and some partially resolved fine structure, clearly indicates a change in the number and location of ^{1}H and/or ^{14}N atoms compared to a "normal" flavin semiquinone. This EPR spectrum is only observed under the stated conditions. Addition of DMAP only and reduction without DMAP do not give rise to an observable EPR spectrum (White et al., 2015). Repeating the reduction/reoxidation experiment but this time using DMAP deuterated at C1 gives rise to the spectrum of Fig. 4C, which is comparable to Fig. 4A. Deuterium, ^{2}H, has a much smaller magnetic moment than ^{1}H and so the effect of deuterium on EPR spectra is almost negligible. Therefore the contributions to the EPR spectrum from the DMAP C1 protons must be the difference between Fig. 4B and C. This shows that the DMAP molecule has been incorporated into the flavin structure in UbiX in such a way that the unpaired electron in the SOMO can interact with the protons that were formerly the C1 protons of the DMAP molecule. Detailed measurement of the hyperfine coupling to the DMAP-derived protons using the electron nuclear double resonance (ENDOR) spectroscopy technique (Marshall, Fisher, et al., 2017), further discussion of which is beyond the scope of this article, show that the additional protons must be attached to the flavin structure at N5, fully consistent with the location of the DMAP C1-derived prFMN C1′-protons.

The X-band EPR spectrum of sodium cyanoborohydride ($NaBH_3CN$) reduced and reoxidized Fdc1 (Fig. 1, reactions **2a** and **3b**) purified from a culture co-expressing UbiX and supplemented with Mn^{2+} is shown in Fig. 4D. The line width, 2.83 mT, and g value, 2.0033, resemble that of the UbiX spectrum shown in Fig. 4B, *i.e.*, prFMN semiquinone. However, the line shape differs in lacking the fine structure observed in Fig. 4B. The weak line at approximately 340 mT and further studies using the ENDOR technique (Bailey et al., 2018; Marshall, Fisher, et al., 2017) show this spectrum to be the result of a magnetic interaction between the unpaired electron of the prFMN semiquinone and the five unpaired electrons of the protein-bound Mn^{2+} ion that is essential for the activity of Fdc1. A much wider EPR spectrum covering 150 mT and centered around $g = 2.00$ reveals the six line spectrum associated with that Mn^{2+} ion (Payne et al., 2015), which arises from interaction between the unpaired electrons of the paramagnetic ion and the ^{55}Mn nucleus (Mn^{2+} is a $S = 5/2$ ion with 5 unpaired d electrons and the nucleus has nuclear spin $I = 5/2$). The Mn^{2+} spectrum is observable before and after sodium cyanoborohydride reduction and while superficially similar to the spectrum of free Mn^{2+} ions in solution, the Fdc1 spectrum is distinct in its detail.

8. Summary and conclusion

The discovery of the prFMN cofactor is a recent event, which has sparked interest in the study and application of prFMN dependent enzymes. As of yet, only the UbiD family of enzymes has been found to use prFMN, although the widespread presence of the prFMN producing UbiX gene might suggest other prFMN dependent enzymes exist. The prenylation of FMN leads to the transformation of the chemical properties of this molecule, as exemplified by the diverse prFMN species that have been described. Central to UbiD functionality is the oxidized $prFMN^{iminium}$ species, which is obtained through oxidation of the $prFMN^{reduced}$ UbiX product. While this process can occur *in vivo*, leading to the heterologous production of *holo*-UbiD as outlined in Section 2, prFMN incorporation levels can be unsatisfactory. Furthermore, the fact $prFMN^{reduced}$ oxidation can also result in $prFMN^{radical}$ species, and the observation that prFMN^{iminium} appears light sensitive, can give rise to inactive prFMN:UbiD complexes. *In vitro* reconstitution of *apo*-UbiD can provide an alternative route to production of active *holo*-UbiD, and either makes use of *in vitro* produced prFMN species (necessitating the synthesis of the DMAP precursor) or extracted prFMN from *in vivo* produced prFMN containing proteins. Oxidation to the level of prFMN^{iminium} can occur either within UbiD or prior to UbiD complex formation, and both molecular oxygen as well as alternative oxidants can be used. In this chapter, we provide protocols for the various strategies (Fig. 5), based on studies carried out with Fdc1 and AroY enzymes. Which route leads to the most efficient production of the active *holo*-form for a specific UbiD enzyme is at present difficult to predict. From the limited study of the various UbiD enzymes at present, even single point mutations appear to affect prFMN incorporation and maturation (Bailey et al., 2018). Hence, both *in vivo* and *in vitro* strategies might need to be trialed when attempting production and purification of a novel UbiD-enzyme or variant.

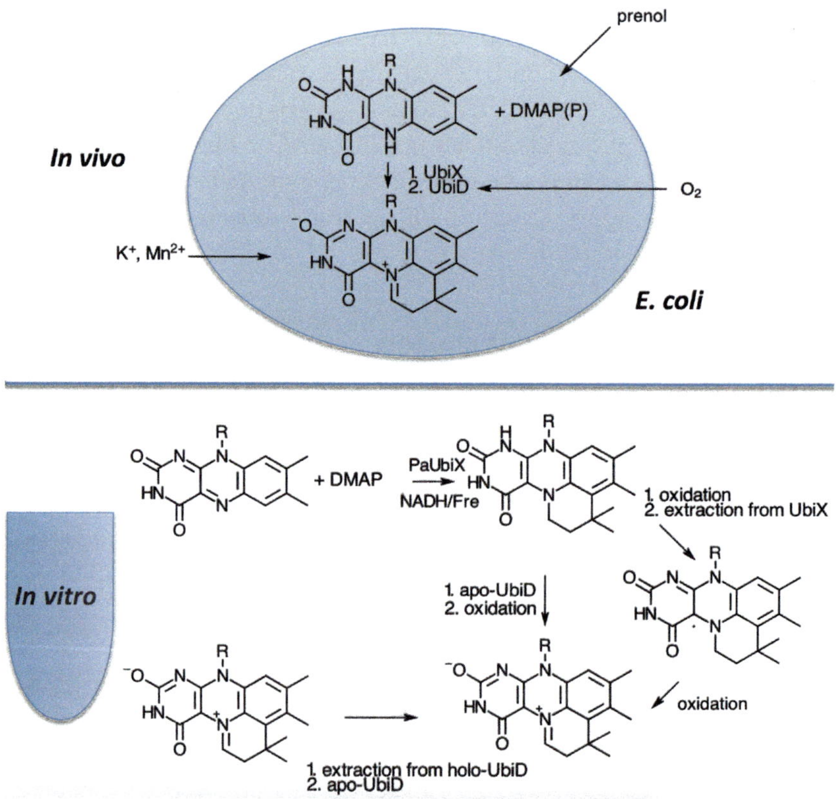

Fig. 5 A schematic overview of the various strategies for *holo*-UbiD production.

Acknowledgments

This works was supported by the BBSRC Grant BB/K017802/1 and the ERC Grant 695013:pre-FAB. David Leys is a Royal Society Wolfson Merit Award holder.

References

Annaval, T., et al. (2018). Biochemical and structural characterisation of TtnD, a prenylated FMN-dependent decarboxylase from the tautomycetin biosynthetic pathway. *ACS Chemical Biology*, *13*, 2728–2738.

Arunrattanamook, N., & Marsh, E. N. G. (2017). Kinetic characterization of prenyl-flavin synthase from *Saccharomyces cerevisiae*. *Biochemistry*, *57*, 696–700.

Bailey, S. S., et al. (2018). The role of conserved residues in Fdc decarboxylase in prenylated flavin mononucleotide oxidative maturation, cofactor isomerization, and catalysis. *The Journal of Biological Chemistry*, *293*, 2272–2287.

Edmondson, D. E. (1978). ESR of free radicals in enzymatic systems. *Biological Magnetic Resonance*, *1*, 205–237.

Fraaije, M. W., & Mattevi, A. (2000). Flavoenzymes: Diverse catalysts with recurrent features. *Trends in Biochemical Sciences*, *25*, 126–132.

Hagen, W. R. (2009). *Biomolecular EPR spectroscopy*. Boca Raton: CRC Press.

Heuts, D. P., Scrutton, N. S., McIntire, W. S., & Fraaije, M. W. (2009). What's in a covalent bond? *The FEBS Journal*, *276*, 3405–3427.

Kay, C. M. W., Feicht, R., Schulz, K., Sadewater, P., Sancar, A., Bacher, A., et al. (1999). EPR, ENDOR and TRIPLE resonance spectroscopy on the neutral flavin radical in Escherichia coli DNA photolyase. *Biochemistry*, *38*, 16740–16748.

Keller, R. K., & Thompson, R. (1993). Rapid synthesis of isoprenoid diphosphates and their isolation in one step using either thin layer or flash chromatography. *Journal of Chromatography*, *A645*, 161–167.

Kurreck, H., Kirste, B., & Lubitz, W. (1988). *Electron nuclear double resonance of radicals in solution: Application to organic and biological chemistry* (pp. 308–327). Weinheim: VCH.

Leys, D. (2018). Flavin metamorphosis: Cofactor transformation through prenylation. *Current Opinion in Chemical Biology*, *47*, 117–125.

Lira, L. M., Vasilev, D., Pilli, R. A., & Wessjohann, L. A. (2013). *Tetrahedron Letters*, *54*, 1690–1692.

Marshall, S. A., Fisher, K., Ni Cheallaigh, A., White, M. D., Payne, K. A. P., Parker, D. A., et al. (2017). Oxidative maturation and structural characterization of prenylated-FMN binding by UbiD, a decarboxylase involved in bacterial ubiquinone biosynthesis. *The Journal of Biological Chemistry*, 292, 4623–4637.

Marshall, S. A., Payne, K. A. P., & Leys, D. (2017). The UbiX-UbiD system: The biosynthesis and use of prenylated flavin. *Archives of Biochemistry and Biophysics*, *632*, 209–221.

Massey, V., & Palmer, G. (1966). On the existence of spectrally distinct classes of flavoprotein semiquinones. A new method for the quantitative production of flavoprotein semiquinones. *Biochemistry*, *5*, 3181–3189.

Mergelsberg, M., et al. (2017). Phthaloyl-coenzyme a decarboxylase from *Thauera chlorobenzoica*: The prenylated flavin-, K- and Fe-dependent key enzyme of anaerobic phthalate degradation. *Environmental Microbiology*, *19*, 3734–3744.

Palmer, G., Müller, F., & Massey, V. (1971). Electron paramagnetic resonance studies on flavoprotein radicals. In H. Kamin (Ed.), *Third international symposium. Flavins and Flavoproteins* (pp. 123–137). Baltimore: University Park Press.

Payer, S. E., et al. (2017). Regioselective Para-carboxylation of catechols with a prenylated falvin dependent decarboxylase. *Angewandte Chemie (International Ed. in English)*, *56*, 13893–13897.

Payne, K. A. P., White, M. D., Fisher, K., Khara, B., Bailey, S. S., Parker, D., et al. (2015). New cofactor supports α,β-unsaturated acid decarboxylation via 1,3-dipolar cycloaddition. *Nature*, *522*, 497–501.

Shepard, S. A., Karthikeyan, C., Latham, J., Struck, A.-W., Thomson, M. L., Menon, B., et al. (2015). Extending the biocatalytic scope of regiocomplementarity flavin-dependent halogenase enzymes. *Chemical Science*, *6*, 3454–3460.

Wang, P. H., et al. (2018). Biosynthesis and activity of prenylated FMN cofactors. *Cell Chemical Biology*, *25*, 560–570.

White, M. D., Payne, K. A. P., Fisher, K., Marshall, S. A., Parker, D., Rattray, N. J. W., et al. (2015). UbiX is a flavin prenyltransferase required for bacterial ubiquinone. *Nature*, *522*, 502–506.

CHAPTER TWENTY

Physical methods for studying flavoprotein photoreceptors

Estella F. Yee, Siddarth Chandrasekaran, Changfan Lin, Brian R. Crane*

Department of Chemistry and Chemical Biology, Cornell University, Ithaca, NY, United States
*Corresponding author: e-mail address: bc69@cornell.edu

Contents

Abstract

Molecular mechanisms of dark-to-light state transitions in flavoprotein photoreceptors have been the subject of intense investigation. Blue-light sensing flavoproteins fall into three general classes that share aspects of their activation processes: LOV domains, BLUF proteins, and cryptochromes. In all cases, light-induced changes in flavin redox, protonation, and bonding states result in hydrogen-bond and conformational rearrangements important for regulation of downstream targets. Physical characterization of these flavoprotein states can provide valuable insights into biological function, but clear

Methods in Enzymology, Volume 620
ISSN 0076-6879
https://doi.org/10.1016/bs.mie.2019.03.023

conclusions are often challenging to draw owing to complexities of data collection and interpretation. In this chapter, we briefly review the three classes of flavoprotein photoreceptors and provide methods for their recombinant production, reconstitution with flavin cofactor, and characterization. We then relate best practices and special considerations for the application of several types of spectroscopies, redox potential measurements, and X-ray scattering experiments to photosensitive flavoproteins. The methods presented are generally accessible to most laboratories.

1. Introduction

Blue-light-sensing flavoproteins comprise a family of photoreceptors that are widely recognized for diverse physiological roles and an increasing number of applications in the realm of optogenetics. For example, blue-light regulatory responses in plants, such as entrainment of the circadian clock, phototropism, chloroplast movement, stomatal opening, flowering, and growth and repair, rely upon flavoprotein photoreceptors (Christie et al., 1998; Conrad, Manahan, & Crane, 2014; Crosson, Rajagopal, & Moffat, 2003; Losi & Gärtner, 2012, 2017). In addition, archaea, prokaryotes, bacteria, fungi, animals, and protists utilize flavoprotein light sensors for regulation of phototaxis, DNA repair, stress response, and metabolic processes, among other functions (Becker, Zhu, & Moxley, 2011; Losi & Gärtner, 2012; Losi, Polverini, Quest, & Gärtner, 2002; Singh, Doerks, Letunic, Raes, & Bork, 2009). Despite this breadth of physiological activity, the primary light-sensing mechanisms of the flavin-containing domains fall into a few general categories. Excitation of the flavin cofactor can induce hydrogen-bond rearrangements, redox chemistry, or covalent bond formation (Becker et al., 2011; Conrad et al., 2014; Losi & Gärtner, 2012) that lead to conformational and structural changes in the local protein environment. The signals eventually propagate toward the protein surface and elicit alterations in protein-protein interactions to affect downstream signaling and ultimately biological behavior.

Although flavoprotein light sensors share a chromophore, the isoalloxazine ring, differences in photochemical responses are tuned by the protein environment, which makes studying the specific molecular mechanism of signal activation, conversion, and transduction critical for not only understanding biological function but also enabling the rational design of responses in optogenetic applications. Of particular challenge is the characterization of transient intermediary states of the chromophore and protein conformation during photoconversion. Continued investigation of these properties will necessitate the use of robust spectroscopic methods and redox measurements. Such techniques often utilize the flavin cofactor as a spectroscopic handle to follow the unique properties of each redox state. Herein we discuss several biophysical methods useful for understanding how the flavin cofactor influences protein conformational state. As with other chromophoric systems, an armamentarium of spectroscopic techniques have been applied to flavoprotein photoreceptors; we do not attempt to be comprehensive in our coverage and instead focus on methods used extensively in our laboratory. Notable omissions are ultra-fast and nuclear magnetic resonance (NMR) spectroscopies, for which we refer the reader elsewhere (Kennis & Groot, 2007; Li, Wang, & Zhong, 2013; Zoltowski & Gardner, 2011).

1.1 Classes of flavoprotein photoreceptors

Blue-light photoreceptors are grouped into three major classes based on the overall protein fold and mechanism of flavin activation: light-oxygen-voltage (LOV) domains, cryptochrome/photolyases, and blue light sensing using FAD (BLUF) domains.

1.1.1 Light-oxygen-voltage domains

LOV domains, typically found in plants and fungi and predicted to exist in animals (Krauss et al., 2009), are sensors for blue-light responses. These domains are a subset of the Per-Arnt-Sim superfamily and non-covalently bind to a molecule of either flavin mononucleotide (FMN) or flavin adenine dinucleotide (FAD). LOV domains were first discovered in the plant phototropins (phot-1 and phot-2) in *Arabidopsis* and were determined to be important for phototropism in *Arabidopsis* (Christie et al., 1998; Christie, Swartz, Bogomolni, & Briggs, 2002; Crosson et al., 2003). Since their discovery, these domains have garnered steady interest, most recently in regard to engineering light-mediated signal transduction processes and optogenetics through the coupling of LOV domains to effector domains (Ziegler & Möglich, 2015) and designing photoresponsive molecules (Seifert & Brakmann, 2018).

The canonical LOV photocycle forms a covalent cysteinyl-thioether adduct bond between a strictly conserved cysteine on the Eα helix and the C4a carbon of the flavin isoalloxazine ring. Whereas it is generally accepted that flavin excitation results in the singlet S1 state, which intersystem crosses to the triplet T* state, the subsequent steps to form the adduct state remain somewhat unclear. It was first proposed that the triplet state becomes protonated to form an ionic, charge-separated species between the C4a carbon and N5 nitrogen that then facilitates nucleophilic attack by the negatively-charged cysteinyl thiolate (Swartz et al., 2001). However, more recent electron spin resonance (ESR) and transient spectroscopy studies, provide evidence for a transient radical pair intermediate (Bauer, Rabl, Heberle, & Kottke, 2011; Kutta, Magerl, Kensy, & Dick, 2015; Schleicher et al., 2004; Zhu et al., 2016). Regardless, subsequent cysteinyl-thioether bond formation and N5 nitrogen protonation cause conformational flipping of the conserved glutamine residue and hydrogen-bond reorganization that induces the release and unfolding of adjacent helices (Christie, Blackwood, Petersen, & Sullivan, 2015; Ganguly, Thiel, & Crane, 2017; Jones, Feeney, Kelly, & Christie, 2007; Peter, Dick, & Baeurle, 2012; Yee et al., 2015).

The mechanism of signal transduction encompasses a broad range of conformational changes within the protein (Herrou & Crosson, 2011). In phototropins, adduct formation is followed by release of the Jα helix and activates the serine/threonine kinase effector domain. In Vivid (VVD), a circadian clock fungal protein in *Neurospora crassa*, the adduct state causes rearrangement of the N-cap and formation of a rapidly exchanging dimer (Zoltowski et al., 2007; Zoltowski & Crane, 2008) that then represses the primary blue-light photoreceptor White Collar-1 (Chen, DeMay, Gladfelter, Dunlap, & Loros, 2010; Hunt, Thompson, Elvin, & Heintzen, 2010). Still other LOV-LOV interactions can result in the rotation or unfolding of the Jα helix to change interactions between effector domains, as in the stress response protein YtvA in *Bacillus subtilis* (Herrou & Crosson, 2011).

1.1.2 Blue light sensing using FAD domains

BLUF domains are primarily found in bacteria and play important roles in nucleotide metabolism, phototaxis, photosystem synthesis, and biofilm formation (Gomelsky & Klug, 2002; Jung et al., 2005; Kanazawa et al., 2010; Masuda, 2013; Masuda & Bauer, 2002). Unlike LOV domains, spectral changes between the dark and the light-adapted states are minimal, with blue light excitation producing a reversible red-shift of $\sim$10 nm; both dark and light states retain the fully oxidized FAD (Masuda & Bauer, 2002). Crystal and solution state structures indicate that the domain has a ferrodoxin-like fold (Anderson et al., 2005; Grinstead et al., 2006; Jung et al., 2005; Jung, Reinstein, Domratcheva, Shoeman, & Schlichting, 2006), with the cofactor lying in between two parallel alpha helical chains and surrounded by a hydrogen-bonding network with four conserved, neighboring residues: Tyr, Gln, Trp, and Met, all of which have been established by mutagenesis to be essential for function (Dragnea, Waegele, Balascuta, Bauer, &

Dragnea, 2005; Laan et al., 2006; Masuda, Hasegawa, & Ono, 2005b; Masuda, Hasegawa, Ohta, & Ono, 2008; Unno, Masuda, Ono, & Yamauchi, 2006).

Time-resolved spectroscopic studies have led to many proposed photochemical mechanisms that have been complicated by conflicting interpretations of spectroscopic and structural data. Spectroscopic experiments have suggested rapid formation of a transient flavin radical via proton-coupled electron transfer from an adjacent, conserved Tyr on the ultrafast timescale (Bonetti et al., 2008; Gauden et al., 2006), and other studies have demonstrated that the rate of electron transfer can be modulated by tuning the redox potential of flavin and Tyr (Mathes, Van Stokkum, Stierl, & Kennis, 2012). However, not all photoactive BLUF domain proteins generate this charge-separated species (Lukacs et al., 2014), and the role of the biradical state is still unclear. Additional Fourier-transform infrared spectroscopy (FTIR) studies observed a downshift of the C4 carbonyl stretch by $\sim$20 cm^{-1} after photoexcitation that points to changes in the hydrogen bond at the oxygen (Masuda, Hasegawa, Ishii, & Ono, 2004; Masuda, Hasegawa, & Ono, 2005a; Mathes & Götze, 2015) and is suspected to correspond to the red-shifted UV-Vis spectrum (Hasegawa, Masuda, & Ono, 2005). Rotation of the conserved Gln and keto-enol tautomerism to an imidic acid results in hydrogen-bond rearrangements that instigate Trp and Met conformational changes, which have been proposed to propagate the signal through a β-strand toward the attached effector region (Anderson et al., 2005; Domratcheva, Grigorenko, Schlichting, & Nemukhin, 2008; Masuda et al., 2005b).

1.1.3 Cryptochromes/photolyases

Cryptochromes (CRYs) and photolyases (PLs) are structurally homologous, both consisting of an N-terminal nucleotide-interacting Rossmann-fold domain and a C-terminal α-helical FAD-binding domain (Sancar, 2000); however, their functionalities vary distinctly. PLs have an additional light-harvesting antenna cofactor (typically a pteridine or deazaflavin moiety) that transfers light energy to the two-electron reduced flavohydroquinone ($FADH^-$) during DNA repair (Ganguly et al., 2016; Mees et al., 2004; Sancar, 1994; Weber, 2005). The excited state of the $FADH^-$ then donates electrons to break the UV-generated cross-linked pyrimidine dimers via a cyclic electron-transfer reaction (Kao, Saxena, Wang, Sancar, & Zhong, 2005; Liu et al., 2011).

CRYs, on the other hand, are central signaling components in the circadian clock of animals and plants and exhibit different functionalities in other kingdoms (Cashmore, 2003; Cashmore, Jarillo, Wu, & Liu, 1999).

Excitation of flavin cofactor is generally followed by electron transfer from the surface of the protein via a tryptophan chain or tetrad (Lin, Top, Manahan, Young, & Crane, 2018; Müller, Yamamoto, Martin, Iwai, & Brettel, 2015), reducing the cofactor to either the anionic semiquinone (ASQ) or following proton transfer from an adjacent Asp, Glu, or Asn (Hense, Herman, Oldemeyer, & Kottke, 2015; Iwata, Zhang, Hitomi, Getzoff, & Kandori, 2010; Kottke, Batschauer, Ahmad, & Heberle, 2006), to the neutral semiquinone (NSQ). Flavin reduction ultimately causes the C-terminal extension or tail (CTT) to undergo a conformational change.

CRYs and PLs are grouped into several subclasses: cyclobutane pyrimidine dimer (CPD) PLs; 6-4 PLs; CRY-DASH; plant CRYs; and animal CRYs (Lin & Todo, 2005). In *Drosophila* CRY (dCRY), light excitation of the flavin cofactor results in the anionic semiquinone (ASQ) that is purported to alter the protonation state of a nearby His and induces the release of the CTT (Ganguly et al., 2016). dCRY then binds to the circadian clock co-repressor Timeless (TIM) and recruits Jetlag (JET), an E3-ubiquitin ligase, causing degradation of TIM and dCRY, which photoentrains and resets the circadian clock (Crane & Young, 2014). Presently though, it is uncertain which redox state (oxidized or ASQ) of animal-type CRY exists in vivo and contributes to downstream signaling (Öztürk, Song, Selby, & Sancar, 2008), although recent data show a strong correlation between flavin photoreduction and biological activity (Lin et al., 2018). Moreover, CRY has garnered ongoing interest owing to it being implicated in magnetoreception via generation of a radical pair upon photoactivation (Biskup et al., 2009; Foley, Gegear, & Reppert, 2011; Gegear, Foley, Casselman, & Reppert, 2010; Liedvogel & Mouritsen, 2010; Rodgers & Hore, 2009; Solov'yov, Chandler, & Schulten, 2007) and its possible applications in optogenetics for mammalian systems (Müller & Weber, 2013; Pathak, Vrana, & Tucker, 2013; Tischer & Weiner, 2014).

2. Recombinant expression and purification of flavoproteins

Obtaining properly folded protein samples is often not trivial. We have outlined several methods to improve the yield of soluble, flavin-incorporated proteins, including expression with CmpX13 cells (which are engineered to express the riboflavin transporter RibM), reconstitution with excess flavin, insect cell expression, and refolding from inclusion bodies.

2.1 Overexpression of proteins in *Escherichia coli*

Materials

BL21(DE3) cells (New England Biolabs) or CmpX13 cells (C41(DE3) manX::ribM) (Mehlhorn et al., 2013)
8 L of LB-Miller media (Terrific Broth when using CmpX13 cells)
100 mM stock of isopropyl β-D-1-thiogalactopyranoside (IPTG)
Riboflavin
DNA constructs cloned into a T7 expression vector
Antibiotics to select for cells transformed with plasmid

1. Transform competent *E. coli* cells with the plasmid of choice and grow a 150-mL inoculation culture overnight (~16 h) in LB with the selective antibiotic at 37 °C.
2. Dilute an equivalent amount of inoculation culture into four flasks with 2 L of autoclaved LB-Miller media and antibiotic. Add 50–100 mg of riboflavin directly to each 2 L of media. Grow to an optical density at 600 nm ($O.D._{600}$) of 0.6–0.8 before reducing the temperature of the culture if necessary. Induce with 2 mL of 100 mM IPTG stock per flask. Express for the desired period of time. Note that the temperature and expression duration can be optimized through small-scale growths and visualization by SDS-PAGE. These conditions are highly dependent on the construct; for instance, VVD and dCRY express at 17 °C for 16 h, while BAT-LOV* and WC-1 (356–507) express at 37 °C for 3 h.
3. Collect cells by centrifugation at 5000 × *g* for 10 m and remove spent media. Store at −20 °C until ready for next steps.

2.2 Isolation and purification of soluble proteins

Materials

Gel filtration buffer (GFB), pH buffered and passed through a 0.22-μm filter
Affinity column; Ni-NTA resin for His_6 tags.
1 M stock of imidazole, pH 7

1. Resuspend cells in GFB. Lyse the cells on ice using a sonic dismembrator (2 s pulse at 30% power followed by 2 s pause, 4 m total time per cycle; 3 cycles in total). In the case of dCRY the cells were sonicated in 50 mM HEPES (pH 8), 400 mM NaCl, 10% glycerol (v/v), 10 mM imidazole, 0.5% Triton X-100 (v/v), 0.5 mM TCEP (fresh), and protease inhibitor. 10% glycerol acts as a stabilizing agent for some proteins and can be important to use throughout the purifications.

2. Centrifuge the mixture at 20,000 RPM for 1 h at 4 °C to remove cell detritus.
3. Flow supernatant through an affinity column. For a Ni-NTA column, wash the resin with 5 mL of GFB + 20 mM imidazole to remove non-specifically bound proteins and then elute with 10 mL of GFB + 200 mM imidazole. Concentrate down the elution using centrifugal filter units.
4. Further purify the protein using an isocratic elution through a size-exclusion column (SEC) 26/60, collecting 6-mL fractions in the region of interest.
5. If the purity is still poor, the protein is likely highly unstable and changes to the buffer may help (glycerol, sodium chloride, adjusting the pH, etc.). An additional polishing step can be applied by using an anion or cation exchange column (Hi-Prep Q XL 16/10 or HiPrep CMFF 16/10, respectively). Elution with an ionic strength gradient (0–100%) can separate charged species.
6. Check for flavin incorporation by UV-Vis. Some constructs, such as BAT-LOV*, can bind to free flavin. For these proteins, add approximately 1 molar equivalent of free FAD or FMN to the solution with protein and incubate overnight at 4 °C. Exchange the buffer using a centrifugal filter unit until the flow-through is colorless and free of flavin.

2.3 Isolation and purification of insoluble proteins

Materials

Lysis buffer (20 mM Tris, 150 mM NaCl, pH 7.5, 10% glycerol)
Denaturation buffer (20 mM Tris, 150 mM NaCl, pH 7.5, 10% glycerol, 6 M guanidine hydrochloride, 1 mM EDTA, 5 mM dithiothreitol (DTT))
Refolding buffer (50 mM HEPES, 150–500 mM NaCl, 5% glycerol, pH 7.5)

1. After lysing the cells in lysis buffer (see Step 1 in Section 2.2) and centrifugation at 12,000 × *g* for 20 m, white inclusion bodies will be visible at the bottom of the centrifugal tube. Wash the inclusion bodies by resuspending the pellet in lysis buffer + 1% (w/v) Triton-X and centrifuging at 12,000 × *g* for 10 m to remove the spent buffer. Remove all traces of the wash buffer by resuspending the pellet in lysis buffer and centrifuging again for 10 m. Resuspend the resultant pellet in denaturation buffer with dithiothreitol (DTT) to keep the thiol moieties reduced. Centrifuge the solution briefly to remove undissolved particulates.

2. Further purification can be accomplished by running the solubilized, denatured protein through a Ni-NTA column, washing and eluting the bound protein with denaturation buffer + imidazole.
3. Completely dissolve 1–3 mM of free FAD or FMN into the protein solution and slowly drip into 250 mL of refolding buffer with constant stirring at 4 °C. The dripping can be facilitated and controlled by placing the protein solution into a separatory funnel, anchoring it to a ring stand above the beaker, and gradually opening the stopcock until a slow, continuous drip is attained (1 drop per 5–10 s).
4. After the entire sample of denatured protein is added, stir at 4 °C for ~2–16 h. Remove precipitated protein by gravity filtration. Collect soluble protein either by affinity resin or by concentrating with a centrifugal filter. Remove excess flavin by buffer-exchange or dialysis. Check for flavin binding by looking for the three distinct vibronic peaks centered at 450 nm or for features of the reduced state using UV-Vis spectroscopy and test for photoreduction activity.

Note: Many variables go into the successful reconstitution of protein. Inclusion bodies must be unfolded to a degree that allows free flavin to interact and bind with the domain. However, if proteins are completely denatured, refolding to its native state is nigh impossible. As such, some studies recommend monitoring the unfolding behavior by circular dichroism and fluorescence. Other factors, including the protein concentration, refolding buffer, and temperature, can play critical roles on success and yield (Hefti, Vervoort, & Van Berkel, 2003). Additives, like arginine, in the refolding buffer have also been determined to aid in the refolding of flavoproteins from inclusion bodies (Cafaro et al., 2002; Leartsakulpanich, Antonkine, & Ferry, 2000). Additionally, refolding flavoproteins in phosphate buffers may be problematic because the interaction of the backbone phosphate moieties in FAD or FMN directs binding of the isoalloxazine ring to the protein; free phosphate molecules may interfere with the binding kinetics, as in the case of apoflavodoxin (Hefti et al., 2003; Murray, Foster, & Swenson, 2003).

2.4 Insect cell expression

Cryptochrome proteins are often best expressed from insect cells. For *Drosophila* cryptochrome, the following protocols were followed:

1. Clone gene of interest into vector pFastBac-HTa (Invitrogen). The resultant plasmid provides for protein expression, with an N-terminal His_6 tag and a TEV protease cleavage site under the control of a baculovirus-specific promoter.

2. Amplify the baculovirus by infecting Sf9 cells at a multiplicity of infection (MOI) of 0.02 and incubation for 96 h at 27 °C using serum-free culture medium F900-II (Life Technologies Corp., Carlsbad, CA, USA) and a cell density at infection of 10^6 cells per mL. Collect culture supernatant containing the high-titer virus after centrifugation at 3000 × *g* for 10 m. at 4 °C and filter sterilized using a 0.2-μm sterile filter unit. The virus can be stored at 4 °C protected from exposure to direct light. The virus titer can be determined using a standard plaque enumeration assay by plating dilutions of the virus onto monolayers of Sf9 cells. By applying an agarose overlay, the plaques can be read after a 7 day incubation at 27 °C. Generally cells are grown for 72 h post infection. Cell pellets from 10 L productions were aliquoted and stored at −80 °C.
3. Pellets from 2 L dCRY cultures are sonicated briefly in ca. 50 mL lysis buffer (50 mM HEPES, 5 mM imidazole, 300 mM NaCl and 10% (v/v) glycerol, pH 8.0). Sonication is conducted using a Fisher 550 sonic dismembrator at power level 6 for 1 m (2 s on, 2 s off). Centrifuge the lysate for 1 h at 58,000 × *g* and incubate the supernatant for 1 h with 2 mL of Ni-NTA resin (Qiagen) with gentle rocking on ice. Transfer resin to a column, wash with lysis buffer and elute the protein eluted in lysis buffer supplemented with 100 mM imidazole, a lower concentration of NaCl (150 mM) and 2 mM DTT. Cleave the His_6 tag by treating at 4 °C overnight with TEV protease. Cleaved protein can then be further purified by a Superdex 200 prep grade (SEC 26/60; GE Healthcare Life Sciences) in 50 mM HEPES, 150 mM NaCl, 10% glycerol, pH 8.0. The yield is typically 40 mg of purified protein from 2 L of culture. Protein preparation should be carried out under red light or in the dark so as to minimize blue light exposure.

3. Photoreduction/photoactivation of flavoprotein photoreceptors

The flavin chromophore can be reduced or converted to one of several states (Fig. 1). These conversions are readily observed by irradiating free flavin in solution using blue light under anaerobic conditions with 10 mM ethylenediaminetetraacetic acid (EDTA) as an extrinsic reductant (Frisell, Chung, & Mackenzie, 1958; Massey, Stankovich, & Hemmerich, 1978). The lifetime of these states can be extended by stabilizing the chromophore in a protein scaffold to regulate solvent accessibility and the rate of quenching.

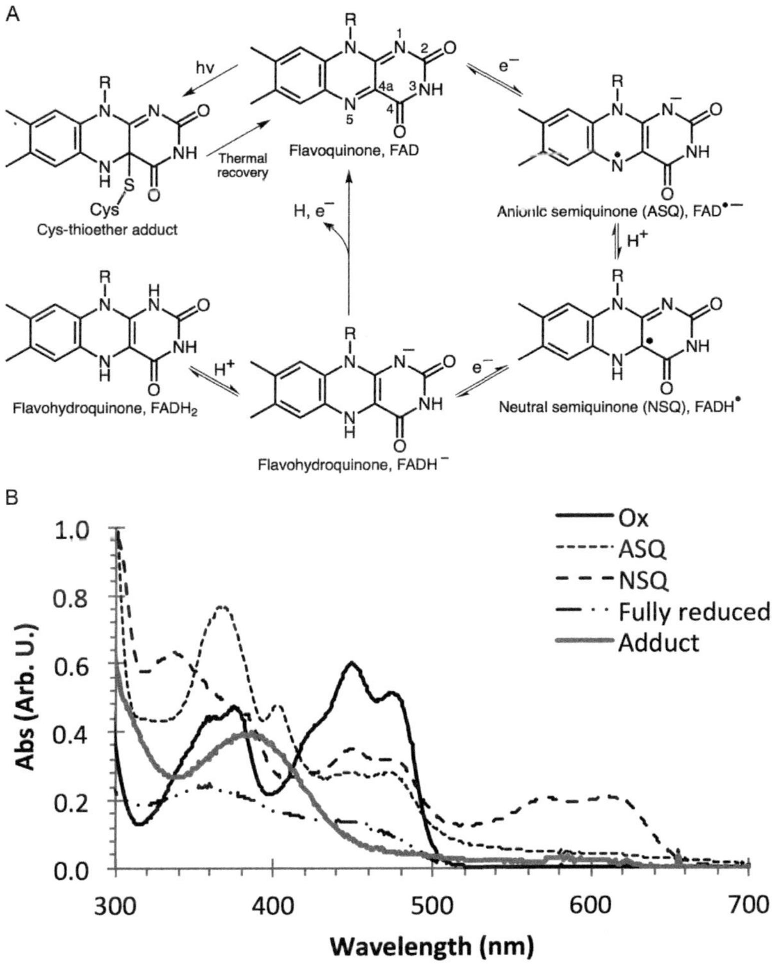

Fig. 1 (A) Redox states of flavin. Numbering convention is indicated in the oxidized flavoquinone structure. (B) Representative UV-Vis spectra of each flavin state.

Each redox state exhibits a unique, spectrophotometric handle. In general, excitation by UVA or blue light leads to an excited triplet state that then generates one of three reduced states. Recovery of the oxidized state requires the removal of H^+ and $2e^-$ and can be catalyzed by the presence of base, as in the case of LOV domain proteins (Alexandre, Arents, Van Grondelle, Hellingwerf, & Kennis, 2007; Zoltowski, Vaccaro, & Crane, 2009).

The decay of the 450-nm peak intensities can be followed to observe the kinetics of photoproduct formation. However, photobleaching is a pervasive concern with prolonged, intense light irradiation. Thus, it is recommended to also observe the development of product species and compare the rates of formation to the decay at 450 nm. Notably, some flavoproteins exhibit marked fluorescence during photoreduction, causing a broad, negative UV-Vis peak from ~475 nm to 600 nm that can interfere with absorption measurements, and so observation at these wavelengths should be avoided. Also worth noting, in the case of LOV-domain proteins, the apparent adduct-state lifetimes, which can be monitored through the return of the oxidized flavin spectrum from that of the adduct 390-nm species by UV/Vis spectroscopy, are strongly dependent on the frequency of data acquisition. Rapid sampling and monitoring of LOV spectra with broad spectrum light causes net photo-induced repopulation of the adduct state. It is advisable to report lifetimes for the longest separation between sampling times after which no change in lifetime is observed (Zoltowski & Crane, 2008). Rates and lifetimes can be obtained by fitting the time traces to exponential curves using curve-fitting programs, such as MatLab (MathWorks) and MagicPlot (MagicPlot Systems, LLC).

The applications of photochemical studies in the flavoprotein field are ubiquitous and diverse. We have found UV-Vis spectroscopy to be useful in comparing the photoactivity of different variant proteins. For instance, in the protein BAT-LOV*, we implemented a single Trp → Phe mutation that increased the quantum yield and rate of photoreduction to the NSQ state. Under similar light excitation conditions, an analogous active-Cys-less VVD variant was found to photoreduce at a slower rate (Yee et al., 2015).

3.1 Kinetics experiments by UV-Vis spectroscopy

Materials

UV-Vis spectrophotometer (no lid) and temperature control (HP Agilent 8453 UV-Visible Spectrophotometer & Peltier Controller Module)

Microcuvette (Quartz, Starna cells, path-length of 0.2–1 cm only requires 10–80 μL of sample)

Laser source (World Star Tech laser; 30 mW; $\lambda = 448$ nm)

Neutral density filter

Ring stand and clamps

1. Using a ring stand for support, clamp the laser orthogonal to the UV-Vis observation path, aiming the center of the laser beam directly above the sample well of the microcuvette such that the entire sample is ideally fully illuminated to minimize diffusion effects from the lack of stirring. If the light does not irradiate the entire sample, the beam size can be broadened using an intervening diffuser or diverging lens (Fig. 2).

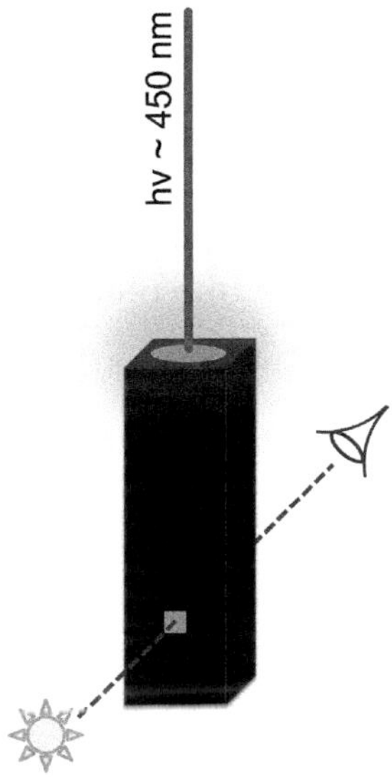

Fig. 2 Collection of photoreduction kinetics by UV-Vis spectrophotometer. A small laser diode is mounted orthogonal to the detector observation path.

2. Set temperature controller to room temperature or lower depending on protein stability or if studying thermal effects.
3. Collect kinetics data, observing at 450–460 nm and at the spectroscopic features of the photoreduced species (e.g., 365 nm and 403 nm for ASQ). Optimize the cycle time so that fast photoreduction events are collected and unwanted chromophore photoreduction by the probe light is minimized. Collect recovery data similarly.

For example, in the case of dCRY 10 μL of protein (~15 mg/mL) in buffer containing 50 mM HEPES (pH 8), 150 mM NaCl, 10% glycerol (v/v) and 2 mM TCEP as solvent reductant is measured in quartz cuvettes with 2-mm path-length, which gives an absorbance around 0.35 at 450 nm. Kinetics of photoreduction are measured every 2 s at 365 nm, 403 nm, and 450 nm under blue laser. Neutral density filters (Thorlabs) can be used to adjust the intensity of light source.

3.2 Transient absorption measurements

To observe transient spectrophotometric features, experimental setups necessitate rapid detection speeds. For this reason, spectroscopic systems are often homebuilt and utilize photomultiplier tubes (PMTs), charge-coupled device (CCD) sensors, or streak cameras for detection. Pump-probe systems rely on a rapid laser excitation pulse followed by irradiation with broad-spectrum white light to collect the absorption spectrum. Time resolution is obtained by implementing a time delay between the excitation and collection events.

Materials

Laser capable of emitting at 450 nm (Nd:YAG; Opotek; 1–20 Hz, 5 mJ, 8 ns pulse width).
Optical table (Thorlabs).
Mirrors, lenses, neutral density filters.
Broad-spectrum arc lamp (75–300-W xenon and/or mercury source; Oriel).
Photonic multichannel analyzer (PMA; Hamamatsu).
Digital Delay/Pulse generator (DG535).
Computer with PMA interface software and Glotaran analysis program.
Cuvette (either microcuvette (0.2-cm or 1-cm path-length; Starna) or a demountable O-Shaped Circular Dichroism Cuvette; FireflySci).

Setup the system as depicted in Fig. 3 for unidirectional microcuvettes and flat circular dichroism cuvettes. The probe light is directed at a shallow angle relative to the laser excitation axis through the use of two concave mirrors with a center hole (Thorlabs). The laser travels unperturbed through the hole in the mirrors to excite the sample and the diffuse probe light is then focused on the sample along the same axis to be collected by the photonic multichannel analyzer (PMA). Alternatively, if using a fluorescence cuvette

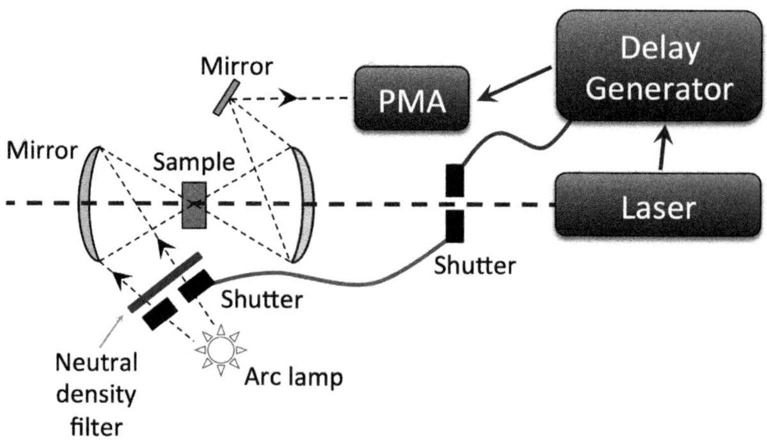

Fig. 3 Schematic of a home-built transient absorption setup. Laser light and probe lights are arranged either parallel and opposite to each other or at a shallow angle, with the focused beams crossing and overlapping at the center of the sample.

(glass or quartz on all sides), the arc lamp can be directed perpendicular to the excitation axis. The latter geometry is generally preferable to minimize stray laser light from overloading the PMA; however, this step may require greater sample volume.

The Q-switch of the laser is used as an external trigger to synchronize the PMA and light shutters. Time delays can be implemented through the PMA software and the delay generator. As the trigger and t_0 are not necessarily the same times, an initial delay needs to be set to account for instrument response time and omit any intense fluorescence/scattering signals that may overwhelm the detector. A helpful, slow-decaying sample such as zinc protoporphyrin (10^{-5} M anaerobic solution in 50% ethanol and 50% phosphate buffer, 10 mM, pH 7; Feitelson & Barboy, 1986) can be ideal for this purpose, as the triplet state only develops after laser excitation. Each spectrum is collected at time $t_0 + x$, where x is the user input delay time, with a time resolution determined by the selected gate width. For every time point, a reference spectrum of the un-excited sample is collected with the probe light and averaged to improve signal-to-noise. Subsequently, a similar excitation measurement is obtained with both laser and probe light. Difference spectra are calculated as –log[*data*/*ref*]. Unfortunately, repetitive irradiation pulses without recovery in the dark will cause samples to photobleach over time and the difference spectra to lose signal intensity and detail. As flavoproteins are particularly susceptible to bleaching, fresh samples are often exchanged in between time point collections. PMA software programs may have a pause option, which closes the physical light shutters for a period of time to allow for sample replacement.

Small cuvette volumes allow fresh samples to be manually exchanged with a micropipettor in between measurements without requiring a large quantity of sample. Some less laborious methods employ flow-cells coupled to a peristaltic pump that slowly circulates fresh sample through the excitation path to swap out photobleached sample (Bauer et al., 2011; Lewis & Kliger, 1993); however, this technique can require milliliters of concentrated protein. Other options include using a larger, mL-cuvette that can accommodate a stir bar. A pulsed stirring mechanism can then dissipate bleached protein through the rest of the non-irradiated cuvette in between measurements. Bleaching can also be reduced by adjusting the intensity of the probe light through the neutral density filter in front of the shutter (Fig. 3) and setting the repetition frequency of the laser to 1 Hz or less to increase recovery time in between spectra acquisitions. The disadvantage of lowering the probe light intensity is a reduced signal for fast time resolution.

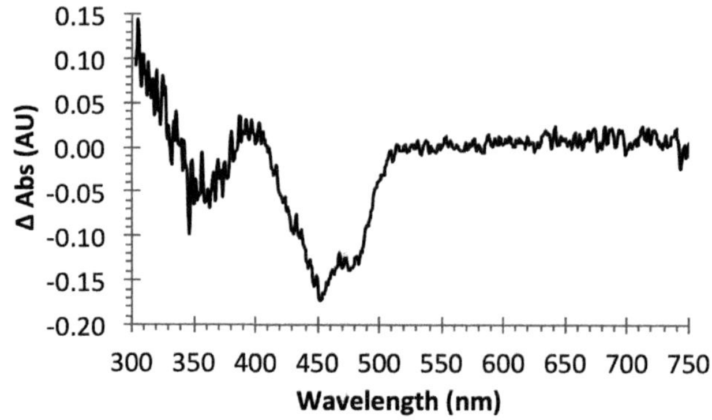

Fig. 4 Transient absorption spectrum of the LOV protein VVD (~400 μM) 10 μs after excitation at 450 nm, with a time resolution of 5 μs. Adduct formation is observed at 400 nm and concurs with spectra collected on LOV2 proteins (Swartz et al., 2001).

With a 75-W Xe lamp, a time resolution of 1 μs is the minimum that can be used to detect clear, spectral detail for most samples. Despite this limitation, we can easily discern a prominent bleach of the oxidized flavoquinone and adduct formation in a VVD sample at 10 μs after laser excitation (Fig. 4).

Analysis of the set of time resolved data is accomplished using a global and target analysis software (Glotaran (Snellenburg, Laptenok, Seger, Mullen, & van Stokkum, 2012)), which applies singular value decomposition (SVD) methods on time-resolved data sets in order to model the spectral components and respective concentrations. Thus, the rate of growth and decay of a particular species of interest can be monitored as a function of time. Glotaran also provides a clear graphical interface and visual modeling, which can facilitate careful inspection of fits and residuals.

4. Electron paramagnetic resonance spectroscopy

Electron paramagnetic resonance (EPR), also known as electron spin resonance (ESR) spectroscopy, and related techniques such as electron nuclear double resonance (ENDOR) are powerful magnetic resonance techniques to probe unpaired electrons spins, such as the semiquinone radical found in flavoproteins. Generally, in the absence of an external magnetic field, the unpaired electron spin states are degenerate, whereas in the presence of a magnetic field (B_0), this spin state degeneracy is removed, and the energy levels split, i.e., Zeeman splitting. This difference in energy levels between spin states directly depends on the magnetic field applied and is given by the following expression for a free electron, $\Delta E = h\nu = g_e \beta_e B_0$, where g_e is the g-factor of the electron, β_e is the Bohr magneton, and B_0 is the magnitude of the external magnetic field applied. Additionally, the presence of nuclei carrying a non-zero magnetic moment around the electron spin can lead to further splitting of the electronic spin states; this splitting due to the coupling of the nuclear to electronic spin states is referred to as the hyperfine (hf) coupling. The hf coupling constant (A) provides information about the strength of the interaction between the electron and nuclear spin, which relates to the delocalization of the electron spin over the radical (Weil, Bolton, & Wertz, 1994).

An electron with spin of 1/2 is split into two energy levels in the presence of a magnetic field (Fig. 5). These two electronic spin states are further split into three energy levels each due to hf coupling with a nuclear spin of 1. The differences in these spin states can be probed by irradiating the sample in the EPR spectrometer with microwave (mw) radiation, and the frequencies of microwaves absorbed can reveal information about the environment of the electron spin such as the hf interactions. The linewidths of EPR peaks can also reveal information about the local dynamics and the electronic structure at the site of the electronic spin.

Most molecules found in nature are diamagnetic and are therefore not EPR active; this is advantageous as EPR spectra have little background signal. Further, EPR can be used to study sub-micromolar spin concentrations. These low spin concentrations are particularly important when studying flavoproteins that are difficult to express and/or prone to light degradation. This sensitivity is also advantageous when only a small amount of FAD

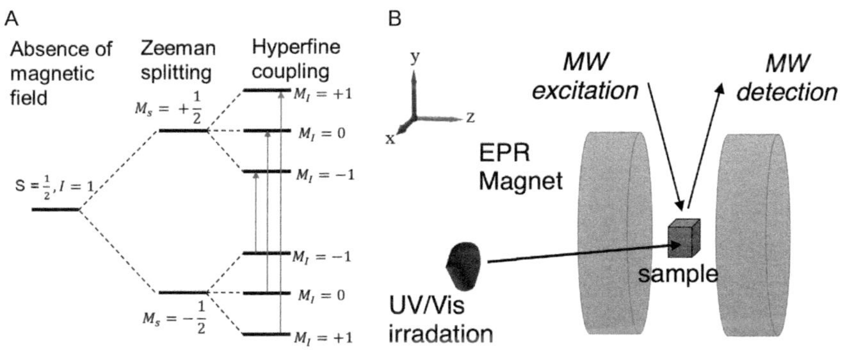

Fig. 5 (A) The electron spin energy level diagram for an electron spin ($S = 1/2$) coupled to a nuclear spin ($I = 1$). (B) Setup of a standard EPR experiment. The main magnetic field (B_0), the microwave radiation (B_1), and the UV/Vis radiation are applied along the z-, y-, x-directions respectively.

radicals are formed due to low quantum yield. Of the three redox states found of FAD, only the semiquinone state is EPR active, while the quinone and hydroquinone states are EPR silent. This makes EPR-based methods suited to study light induced changes to the redox states of FAD. Furthermore, EPR experiments are carried out using microwave radiation, and this provides two major advantages: (a) UV/vis radiation does not affect the EPR measurements and therefore can be used to irradiate samples directly in the EPR cavity and (b) mw radiation is not scattered by cells, opening up the possibility to study the flavoproteins both in vitro as well as in vivo (Hoang et al., 2008; Yee et al., 2015). In this chapter, we will address three important EPR based methods (a) continuous wave (cw)-EPR, (b) pulsed dipolar spectroscopy (PDS), and (c) ENDOR.

4.1 Continuous-wave EPR

cw-EPR has been extensively used to probe FAD and other related flavin based radicals in a variety of systems (Hoang et al., 2008; Kay et al., 1999; Müller, Hemmerich, Ehrenberg, Palmer, & Massey, 1970; Schleicher, Bittl, & Weber, 2009; Yee et al., 2015). In cw-EPR experiments, a continuous microwave field of constant frequency is applied on the sample while the magnetic field (B_0) is swept until the resonance condition is met. The cw-EPR spectrum is recorded by observing the reflected microwave radiation (see Fig. 5) as a function of the magnetic field. An amplitude modulation typically with a frequency of 100 kHz of the magnetic field is carried out to obtain better signal to noise. This additional modulation leads to the characteristic first derivative line shapes observed in cw-EPR. There are three parameters that can be readily obtained upon analysis of cw-EPR spectra: (a) field position, (b) hyperfine interactions, and (c) linewidth. The field positions can be used to calculate the g-factor values for the electron spin, which reveals the nature of the radical species formed and can be considered analogous of the chemical shift in NMR spectroscopy. In the case of flavoproteins, two kinds of EPR active FAD radicals are possible—the anion semiquinone (ASQ) or the neutral semiquinone form (NSQ). Both these FAD radical forms have a g-factor of $\sim$2.003 (Fuchs et al., 2002) indicative of a semiquinone radical. Further, the hf interactions reveal information about the localization of the electron spin on the FAD moiety. However, for FAD radicals the hyperfine (hf) interactions are relatively small (with the strongest interaction being $\sim$8 G) (Barquera et al., 2003; Kay et al., 1999) and are typically hidden by the broad linewidth

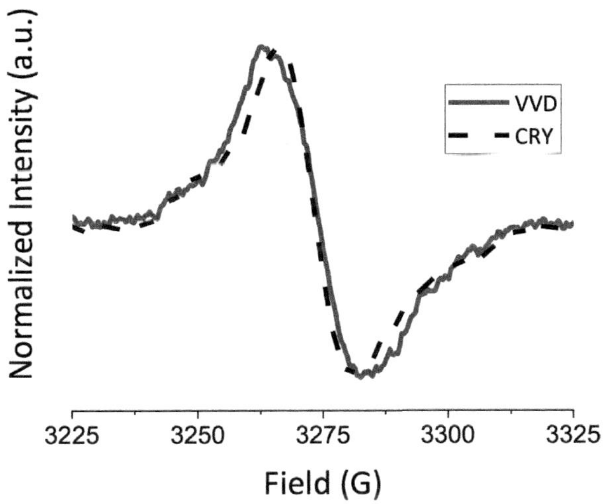

Fig. 6 The photoinduced radical formation in two flavoproteins—CRY and VVD. The experiments were carried out at ~9.4GHz (X-band) at 5°C with a modulation amplitude of 2G.

of these radicals. These hf couplings can however be resolved by ENDOR as discussed below. Finally, the peak-to-peak linewidth (ΔB_{pp}, measured as the field value difference between the crest and the trough of the peak) also indicates key properties of the radical. Generally, the NSQ radicals (ΔB_{pp} ~ 16–21 G) are broader than the ASQ radicals (ΔB_{pp} ~ 12–16 G), due to the additional broadening provided by the extra hydrogen atom on the isoalloxazine ring of FAD. Fig. 6 shows the FAD radical formed by two flavoproteins (CRY and VVD) upon photoexcitation by blue light, as can be seen clearly the NSQ in VVD (ΔB_{pp} ~ 20.2 G) has broader line shapes than the ASQ in CRY (ΔB_{pp} ~ 14.7 G). Additional cw-EPR measurements also reveal that ASQ radical of CRY is formed instantaneously upon photoexcitation while NSQ radical in VVD forms roughly after ~2 m of photoexcitation (data not shown here).

4.2 Electron nuclear double resonance

The ENDOR technique involves simultaneously irradiating the sample with radiofrequency (rf) and microwave (mw) fields. The result is observation of the nuclear spin transitions (like NMR), but through its coupling to the electron spin, i.e., via the hf interaction. The details of the ENDOR experiment have been extensively reviewed elsewhere (Hoffman, 1991; Kulik et al., 2009; Kwiram, 1971). ENDOR is extremely sensitive to small hyperfine coupling constants that are usually broadened out in the cw-EPR and thereby allows detection of small changes to the environment around a free radical. In FAD, the strongest hf-couplings are from the nitrogen and hydrogen atoms on the isoalloxazine ring of the FAD moiety; therefore ENDOR can be used to distinguish between the NSQ and the ASQ forms of the FAD radical. Additionally, the presence of solvent molecules, as well as atoms on the protein backbone and sidechains that are coupled to the FAD radical, also contribute to the ENDOR signal. ENDOR has been used to study the flavin radical formed by FAD upon photooxidation in the bacterial photolyase system (Kay et al., 1999) as well as the circadian clock protein CRY (Hoang et al., 2008).

4.3 Pulsed dipolar ESR spectroscopy

Pulsed dipolar ESR spectroscopy (PDS) has become a powerful tool for the measurement of nanometer range distances in biological systems (Banham, Timmel, Abbott, Lea, & Jeschke, 2006; Borbat & Freed, 2007; Park et al., 2006; Schiemann & Prisner, 2007). This technique is analogous to fluorescence-based methods such as Förster resonance energy transfer (FRET). Currently, PDS has been used to study 1–16 nm distances in a range of biological macromolecules (e.g., membrane proteins, RNA, DNA). PDS utilizes the weak dipolar interactions between two spin probes (S_A; S_B) attached to a biomolecule to identify distance distributions between the biomolecules (Borbat, Costa-Filho, Earle, Moscicki, & Freed, 2001). The frequency of the coupling between the two electron spins is given by the $\omega_d = (\gamma_e^2 \hbar)/r^3$ where $\hbar$ is the reduced Planck's constant, γ_e is the gyromagnetic ratio of the electronic spin, r is the distance between the dipoles. The dipolar frequency between the two spins is detected by observing the effect of flipping S_B on a series of appropriately spaced spin echoes of S_A. The details of the PDS experiment are beyond the scope of this discussion; however interested readers are pointed to the following reviews (Borbat & Freed, 2007, 2013; Casey & Fanucci, 2015; Jeschke, 2012; Misra & Freed, 2011). PDS has been used to study the distance between two flavin (ASQ) radicals in the homodimer formed by the bacterial aerotaxis protein Aer (Samanta, Widom, Borbat, Freed, & Crane, 2016). The distance between the two flavin moieties was determined to be ~4.2 nm, which agrees well with homology models for this protein. PDS was also used to study the light-induced dimerization of circadian clock LOV-protein VVD from *Neurospora*. In VVD (Fig. 7), the distance between the flavin radicals (NSQ) was found to be ~37 Å, which agrees well with the prediction for the light adapted dimer (Yee et al., 2015).

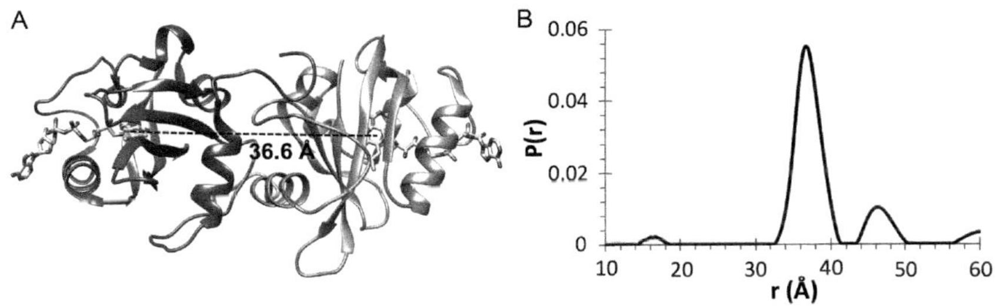

Fig. 7 (A) The structure of the light state dimer of VVD showing that the inter-flavin distance is ~3.7 nm. (B) The PDS distance distribution obtained for two flavin radicals on light irradiated VVD from *Neurospora* (Yee et al., 2015).

4.4 Transient EPR

Transient EPR (trEPR) (Bittl & Weber, 2005) is an infrequently used, but potentially impactful tool where an EPR spectrum is recorded within nanoseconds of light irradiation of samples with a pulsed laser of a defined wavelength. The EPR spectra can be recorded in either pulse or cw-EPR mode; however, pulse transient EPR has gained popularity due to its higher spectral and temporal resolution. In transient EPR, radicals formed by laser excitations will initially have non-Boltzmann population distributions; this leads to a phenomenon called chemically induced dynamic electron polarization (CIDEP). CIDEP is developed in trEPR mainly by the triplet (TM) or radical pair (RPM) mechanisms. CIDEP can manifest itself in the form of enhanced absorption (A) and/or stimulated emission (E) peaks in the EPR spectra (Van Willigen, Levstein, & Ebersole, 1993). The E/A shape of the EPR peaks depends not only on the radical formation mechanism (TM vs RPM) but also on the time between the laser pulse and the EPR measurement. Given sufficient time after the laser pulse, the developed radicals (if not decayed) will relax to the Boltzmann population states and any EPR spectra measured after this time will have steady-state EPR line shapes. Therefore, the time dependence of all the spectral parameters: frequencies, amplitudes, line widths, and phases (E/A) of all the peaks, is carried out to identify critical intermediates and the kinetics of the radical formation.

TrEPR has been applied to the study of radical propagation in flavoproteins, particularly with respect to electron transfer reactions between the flavin ring and amino acid radicals of tryptophan or tyrosine. For example, Weber et al. have shown that the electron donation to FAD in the *Xenopus laevis* (6-4) photolyase occurs from a tyrosine site rather than a tryptophan site as initially hypothesized (Weber et al., 2002). These measurements would not have been possible with conventional cw-EPR methods because the electron hole (formed by electron transfer from the protein sidechain to flavin moiety) would have reacted further in the time scale of a cw experiment.

4.5 Some practical considerations for EPR

1. Typical ESR concentrations are in the range of ~10–50 μM of protein (~5 mg/mL) and usually only 20–100 μL of sample is required for obtaining spectra with a good signal-to-noise ratio.
2. The flavin radicals can be observed both in aqueous (5 °C or room temperature) and in frozen state (100 K). For samples in frozen state, it is important to have a cryoprotectant included in the buffer such as ~20% (w/v) of glycerol.
3. Both PDS and pulse ENDOR measurements are carried out in the frozen state (~60 K). Deuterated buffers (including deuterated glycerol, i.e., d^8-glycerol) are recommended for pulse measurements as they reduce spin-spin relaxation rates allowing longer accumulation times. Deoxygenating sample buffers before use will also help with reducing spin relaxation in addition to increasing the lifetime of the flavin radical.
4. For cw-EPR measurements use a modulation amplitude of ~4 G (the recommended modulation amplitude ~1/3 of the peak to peak linewidth of the flavin signal).
5. For ENDOR measurements, there are two pulse experiments available: (a) Mims—for small hyperfine coupling constants and (b) Davies—for large hyperfine coupling constants. Strongly coupled nuclei may not be visible in a Mims experiment.
6. The radical lifetime is limited in aqueous solution, so measurements need to be carried out immediately after photoexcitation and/or in the presence of photoexcitation. However, if photoexcitation is carried out in the cavity, it can lead to sample heating and loss of sample tuning. Therefore, when photoexcitation is performed, it is also important to maintain the temperature in the EPR cavity using a variable temperature unit.
7. To carry out measurements on FAD photoexcitation in the frozen state, it is important to photoreduce the sample in aqueous state before flash freezing the sample in liquid nitrogen. The formation of the semiquinone radical is greatly reduced when the photoexcitation is performed on frozen samples.

5. Measurement of flavoprotein redox potentials

We have applied a spectroscopic method (Massey, 1991) to measure the redox potential of flavoprotein photoreceptors by equilibrating them with a reference dye that has a similar redox potential to the flavoprotein but gives an absorbance change different from that of the flavin cofactor. Below is the procedure used for *Drosophila* cryptochrome dCRY (Lin et al., 2018).

1. 100 μL dCRY (4 mg/mL) was prepared in 50 mM potassium phosphate buffer (pH 7) with 5% glycerol and added to quartz cuvettes with 1-cm path-length ($A_{450} \sim 0.4$).
2. dCRY solution was equilibrated with safranin T ($A_{498} \sim 0.5$), 30 μM methyl viologen and 300 μM xanthine.
3. 10 mM glucose, 25 μg/mL glucose oxidase, and 50 μg/mL catalase were prepared fresh and added to protein solution. The cuvette with protein solution was placed in glove box for 15 min to remove oxygen from the protein solution.
4. The cuvette was sealed with a septum to maintain low oxygen concentration and taken out of the glove box to be measured with UV-Vis spectrum.
5. A syringe with a sharp needle was used to inject 100–200 nM xanthine oxidase through the septum to initiate the chemical reduction. The absorbance changes at 450 nm for FAD and 498 nm for safranin T were measured every 10 s (Fig. 8A). The change at 450 nm caused by the reduction of safranin T was calculated according to its relation with the absolute change at 498 nm (Fig. 8B and C). The slow and simultaneous reduction of dCRY and the reference dye over 1 h was achieved by adjusting the concentration of xanthine oxidase and methyl viologen.
6. When no more major absorbance changes were observed, 5 mM sodium dithionite was added to complete the reduction.

As described (Lin et al., 2018), if the reduction process is given enough time so that equilibrium conditions are met, the protein redox potential can be calculated from the reference dye. The potential will be related to the standard redox potential of $FAD_{ox}/FAD^{\bullet -}$, as in Eq. (1), where R is the gas constant (in $J \times K^{-1}\ mol^{-1}$), T is the absolute temperature (293 K during the experiment), F is the Faraday constant (in $C \times mol^{-1}$), n is the number of moles of electrons transferred during the reaction, E° is the standard reduction potential, $E^{\varnothing}$ is the reduction potential at nonstandard conditions, and D and P indicate the dye and protein, respectively.

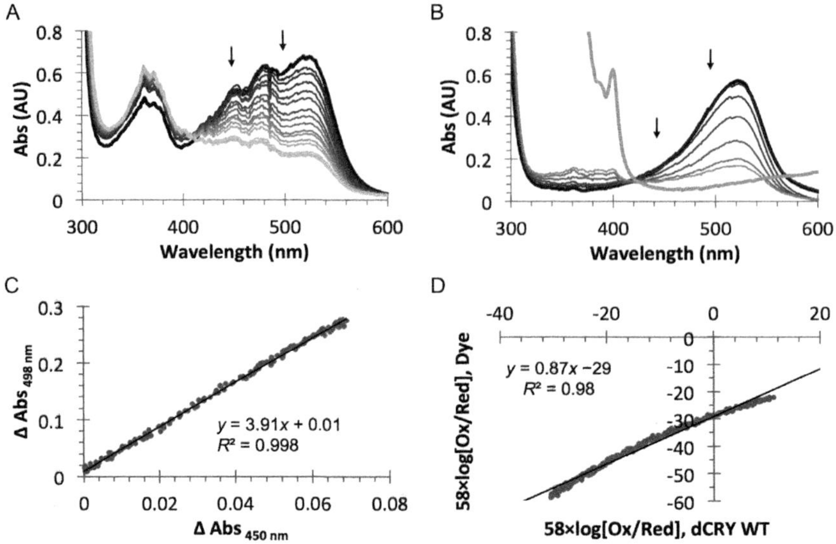

Fig. 8 Electrochemistry of dCRY. (A) The solution of dCRY and reference dye safranin T ($E^{o\prime} = -289$ mV, pH 7 vs SHE) are reduced by xanthine/xanthine oxidase simultaneously. The absorbance at wavelength 450 and 498 nm decreases as reduction proceeds. UV/Vis spectra are shown for every 2 m over 1 h. (B) Safranin T is reduced by xanthine/xanthine oxidase with absorbance decrease at both 450 and 498 nm. UV/Vis spectra are shown for every 2 m, and the last spectrum is collected after adding sodium dithionite. (C) Absorbance change at wavelength 450 and 498 nm is related for safranin T. (D) Nernst plot of safranin T ($E^{o\prime} = -289$ mV vs SHE) vs dCRY WT.

$$E_{equilibrated} = E_D^{\emptyset} + \frac{RT}{n_D F} \ln\left(\frac{[D_{ox}]}{[D_{red}]}\right) = E_P^{\emptyset} + \frac{RT}{n_P F} \ln\left(\frac{[P_{ox}]}{[P_{red}]}\right) \quad (1)$$

Thus,

$$\frac{58}{n_D} \log\left(\frac{[D_{ox}]}{[D_{red}]}\right) = \frac{58}{n_P} \log\left(\frac{[P_{ox}]}{[P_{red}]}\right) + \left(E_P^{o} - E_D^{o}\right) \quad (2)$$

where

$$\frac{[\mathrm{Ox}]}{[\mathrm{Red}]} = \frac{\Delta A_{\mathrm{total}}}{\Delta A} - 1$$

and ΔA means the change of absorbance.

Since $n_D = 1$ for safranin T and $n_P = 1$ for dCRY, according to Eq. (2), $58 \times \log$ ([oxidized]/[reduced]) of safranin T is plotted as a function of $58 \times \log$([oxidized]/[reduced]) for dCRY. The y-intercept indicates the difference of standard redox potential at pH 7 between dCRY and safranin

T (Fig. 8D), and the slope 0.87 is close to 1, as expected for a one-electron process. Thus, the standard redox potential of dCRY is −316 mV vs SHE. Note that a dye must be chosen with a potential in range of that of the protein of interest (±60 mV), and any means of reduction (including photochemical) can be applied to drive the reaction.

6. Protein crystallography

Protein crystallography is a mature method that has been applied with great impact to all classes of flavin-containing photoreceptors. Special considerations arise in attempts to characterize the structures of reactive states and/or photocycle intermediates. A good example is structural characterization of the light-induced adduct state of LOV proteins, which has limited lifetimes and is sensitive to photoinduced reduction, as are the other redox states of protein-bound flavins. In characterizing such states, one must first achieve a sufficient population of the light state in the crystal, and second, the state must be protected from beam damage. To accumulate the light-state, several strategies have been employed. When light-activated states are long lived, they can be crystallized (Circolone et al., 2012; Vaidya, Chen, Dunlap, Loros, & Crane, 2011). In the case of the clock protein VVD, residue substitutions were made to raise the flavin potential and thereby stabilize the cysteine-adduct (which can be considered a reduced form of the flavin ring) for days (Vaidya et al., 2011). Continual or periodic irradiation during crystallization is also effective when light states are sufficiently long-lived and quantum yields are relatively high (Crosson & Moffat, 2002; Heintz & Schlichting, 2016; Pudasaini et al., 2017). The advantage of crystallizing proteins in light-activated states is that light-activation is often accompanied by large-scale conformational changes that would not be tolerated within a crystal lattice (Heintz & Schlichting, 2016; Vaidya et al., 2011). Nonetheless, irradiation of crystals followed by rapid cryo-cooling can trap photocycle intermediates with coupled conformational changes that are relatively small in amplitude and local to the flavin-binding pocket (Crosson & Moffat, 2002; Fedorov et al., 2003; Möglich & Moffat, 2007; Zoltowski et al., 2007). Once formed, a high occupancy light state is sensitive to beam reduction. X-ray beams liberate secondary electrons that are mobile even at cryogenetic temperatures (Ravelli & Garman, 2006). The radiation required to generate such photoelectrons can be an order of magnitude less than what would cause noticeable deterioration of crystal diffraction (Yano et al., 2005). Radiation damage is most often mitigated by spreading a complete diffraction data set over multiple crystals, or more ideally, several positions on the same crystal (Fedorov et al., 2003; Heintz & Schlichting, 2016; Zoltowski et al., 2007). Modern, sensitive, fast-framing detectors coupled to efficient detection strategies can also limit X-ray dose (Wojdyla et al., 2018). In general, damage depends on the total radiation exposure (Murray et al., 2005; Ravelli & Garman, 2006). Procedures have been developed to predict diffraction decay, thereby minimizing damage and optimizing data collection (Bourenkov & Popov, 2010; Heintz & Schlichting, 2016; Wojdyla et al., 2018). Furthermore, extrapolation methods based on the progressive collection of redundant diffraction spots can be used to reconstruct the recorded intensities to values at time zero (Diederichs, McSweeney, & Ravelli, 2003). Advances in synchrotron-based serial crystallography (Huang et al., 2015), where thousands of microcrystals are used to compile a complete data set at room temperature, offer great promise for photosensitive and photoactivatable samples (Huang et al., 2015; Selikhanov, Fando, Dontsova, & Gabdulkhakov, 2018). Although the principles of such data collection were pioneered at X-ray free-electron lasers (XFELS) in so-called "diffract and destroy" experiments (Selikhanov et al., 2018; Spence, Weierstall, & Chapman, 2012), similar procedures are now operative at third-generation synchrotron sources (Owen, Juanhuix, & Fuchs, 2016; Zander et al., 2015). These experiments offer the advantages of room-temperature data collection and minimal beam exposure, but also rely on crystal homogeneity. Finally, when possible, it is advisable to confirm the flavin redox in the crystal by use of microspectrophotometric measurements on the crystals, a capability that is available at many synchrotron beam lines (Ellis, Buffey, Hough, & Hasnain, 2008).

7. Small-angle X-ray scattering

Small-angle X-ray scattering is a synchrotron radiation technique that provides information on molecular mass, shape, and flexibility for samples in solution at ambient temperatures (Mertens & Svergun, 2010; Putnam, Hammel, Hura, & Tainer, 2007; Skou, Gillilan, & Ando, 2014). The method is particularly useful in determining oligomeric states, characterizing conformational transitions in response to stimuli (e.g., light absorption), and testing the arrangement of domain constituents in full-length proteins and higher-order complexes (Mertens & Svergun, 2010; Putnam et al., 2007; Skou et al., 2014). For the case of photoactive proteins, SAXS provides a means to compare the overall structural changes of light vs dark states and even follow such perturbations in a time-resolved manner (Conrad, Bilwes, & Crane, 2013; Lamb, Zoltowski, Pabit, Crane, & Pollack, 2008; Lamb et al., 2009; Vaidya et al., 2013). Size-exclusion chromatography coupled to SAXS (SEC-SAXS) effectively separates complex mixtures (Skou et al., 2014), which can be extremely useful if light activation causes alterations in oligomeric state (Conrad et al., 2013; Lamb et al., 2008, 2009). In some cases, light activation only perturbs equilibria among different association states and even with SEC separation, the resulting species can be challenging to delineate (Zoltowski & Crane, 2008). Computational methods invoking evolving factor analysis show promise to deconvolute closely related components of such mixtures (Meisburger et al., 2016). Several comprehensive software packages have been developed to analyze SAXS data and extract information on molecular size, shape, and flexibility (Franke et al., 2017; Hopkins, Gillilan, & Skou, 2017).

With flavin-containing photoreceptors, SAXS is often used to determine differences between light and dark states (Conrad et al., 2013; Heintz & Schlichting, 2016; Lamb et al., 2008, 2009; Purcell, McDonald, Palfey, & Crosson, 2010; Rollen, Granzin, Batra-Safferling, & Stadler, 2018; Vaidya et al., 2013). As with crystallography, photobleaching, reduction, and beam damage are considerations to heed. At the Cornell Synchrotron's BioSAXS line, samples are typically robotically loaded into a 2-mm path-length capillary tube from a 96-well PCR plate (Conrad et al., 2013; Vaidya et al., 2013). To reduce radiation damage during SAXS data collection, the liquid sample plug is oscillated in the X-ray beam using a computer-controlled syringe pump. Six or more sequential exposures are used to assess possible radiation damage. Light samples are irradiated both before and after loading into the sample capillary, with continuous irradiation during data collection also possible. For each sample, a protein-absent buffer background image is usually acquired first, followed by an image of the photoreceptor in buffer in the "dark" state. Samples are then appropriately irradiated and images of the photoreceptor in the "light" state acquired immediately after exposure. Finally, a second buffer background image is taken. Scattering profiles for the photoreceptor can then be obtained by subtracting the scattering of the protein-absent or background profiles from the protein-present profiles. Changes in scattering between dark and light photoreceptors are often small, and thus replicate experiments with different samples are recommended.

Another advantage of SAXS is the ability to carry out time-resolved experiments. With the VVD protein, a flow cell format was used to follow its light-induced dimerization reaction (Lamb et al., 2008, 2009). A continuous flow cell made of a 1-mm polyester tube through which both an excitation laser and the probe X-ray beam are passed to allow for relatively long exposures on samples that are being continually replenished. The X-ray beam was focused with a glass capillary for better position definition. The laser focal point was then moved against the direction of fluid flow to introduce time delay. Time delays are limited by beam size, positioning, and flow rate, which can all be altered to produce different intervals. In the case of VVD, time delays in the range of 20 ms to 10 s were recorded (Lamb et al., 2008, 2009). Importantly, a PIN diode mounted onto the X-ray beam-stop measures beam intensity so that the diffraction patterns can be normalized for beam fluctuations. More recent advances in beam-line technology have extended time-resolved SAXS on photoreceptors to the microsecond timescale (Cho et al., 2010; Takala et al., 2014).

Acknowledgments

The EPR measurements for this manuscript were collected at ACERT (Cornell University) supported by the NIH/NIGMS Grant P41GM103521. This work supported by NSF grant MCB 1715233 and NIH grant R35GM066775 to B.R.C.

References

Alexandre, M. T. A., Arents, J. C., Van Grondelle, R., Hellingwerf, K. J., & Kennis, J. T. M. (2007). A base-catalyzed mechanism for dark state recovery in the *Avena sativa* phototropin-1 LOV2 domain. *Biochemistry*, *46*(11), 3129–3137.

Anderson, S., Dragnea, V., Masuda, S., Ybe, J., Moffat, K., & Bauer, C. (2005). Structure of a novel photoreceptor, the BLUF domain of AppA from *Rhodobacter sphaeroides*. *Biochemistry*, *44*(22), 7998–8005.

Banham, J. E., Timmel, C. R., Abbott, R. J. M., Lea, S. M., & Jeschke, G. (2006). The characterization of weak protein-protein interactions: Evidence from DEER for the trimerization of a von Willebrand factor A domain in solution. *Angewandte Chemie—International Edition*, *45*(7), 1058–1061.

Barquera, B., Morgan, J. E., Lukoyanov, D., Scholes, C. P., Gennis, R. B., & Nilges, M. J. (2003). X- and W-band EPR and Q-band ENDOR studies of the flavin radical in the Na+-translocating NADH:quinone oxidoreductase from *Vibrio cholerae*. *Journal of the American Chemical Society*, *125*(1), 265–275.

Bauer, C., Rabl, C. R., Heberle, J., & Kottke, T. (2011). Indication for a radical intermediate preceding the signaling state in the LOV domain photocycle. *Photochemistry and Photobiology*, *87*(3), 548–553.

Becker, D. F., Zhu, W., & Moxley, M. A. (2011). Flavin redox switching of protein functions. *Antioxidants and Redox Signaling*, *14*(6), 1079–1091.

Biskup, T., Schleicher, E., Okafuji, A., Link, G., Hitomi, K., Getzoff, E. D., et al. (2009). Direct observation of a photoinduced radical pair in a cryptochrome blue-light photoreceptor. *Angewandte Chemie—International Edition*, *48*(2), 404–407.

Bittl, R., & Weber, S. (2005). Transient radical pairs studied by time-resolved EPR. *Biochimica et Biophysica Acta—Bioenergetics*, *1707*(1), 117–126.

Bonetti, C., Mathes, T., Van Stokkum, I. H. M., Mullen, K. M., Groot, M. L., Van Grondelle, R., et al. (2008). Hydrogen bond switching among flavin and amino acid side chains in the BLUF photoreceptor observed by ultrafast infrared spectroscopy. *Biophysical Journal*, *95*(10), 4790–4802.

Borbat, P. P., Costa-Filho, A. J., Earle, K. A., Moscicki, J. K., & Freed, J. H. (2001). Electron spin resonance in studies of membranes and proteins. *Science*, *291*(5502), 266–269.

Borbat, P. P., & Freed, J. H. (2007). Measuring distances by pulsed dipolar ESR spectroscopy: Spin-labeled histidine kinases. *Methods in Enzymology*, *423*, 52–116.

Borbat, P. P., & Freed, J. H. (2013). Pulse dipolar electron spin resonance: Distance measurements. In C. R. Timmel & J. R. Harmer (Eds.), *Vol. 152. Structure and bonding* (pp. 1–82). Berlin, Heidelberg, New York Springer.

Bourenkov, G. P., & Popov, A. N. (2010). Optimization of data collection taking radiation damage into account. *Acta Crystallographica Section D: Biological Crystallography*, *66*, 409–419.

Cafaro, V., Scognamiglio, R., Viggiani, A., Izzo, V., Passaro, I., Notomista, E., et al. (2002). Expression and purification of the recombinant subunits of toluene/o-xylene monooxygenase and reconstitution of the active complex. *European Journal of Biochemistry*, *269*(22), 5689–5699.

Casey, T. M., & Fanucci, G. E. (2015). Spin labeling and double electron-electron resonance (DEER) to deconstruct conformational ensembles of HIV protease. In P. Z. Qin & K. Warncke (Eds.), *Vol. 564. Methods in enzymology* (pp. 153–187). London,; San Diego, CA; Waltham, MA; Kidlington, OxfordAcademic Press.

Cashmore, A. R. (2003). Cryptochromes: Enabling plants and animals to determine circadian time. *Cell*, *114*(5), 537–543.

Cashmore, A. R., Jarillo, J. A., Wu, Y.-J., & Liu, D. (1999). Cryptochromes: Blue-light receptors for plants and animals. *Science*, *284*(5415), 760–765.

Chen, C.-H., DeMay, B. S., Gladfelter, A. S., Dunlap, J. C., & Loros, J. J. (2010). Physical interaction between VIVID and white collar complex regulates photoadaptation in *Neurospora*. *Proceedings of the National Academy of Sciences of the United States of America*, *107*(38), 16715–16720.

Cho, H. S., Dashdorj, N., Schotte, F., Graber, T., Henning, R., & Anfinrud, P. (2010). Protein structural dynamics in solution unveiled via 100-ps time-resolved x-ray scattering. *Proceedings of the National Academy of Sciences of the United States of America*, *107*(16), 7281–7286.

Christie, J. M., Blackwood, L., Petersen, J., & Sullivan, S. (2015). Plant flavoprotein photoreceptors. *Plant and Cell Physiology*, *56*(3), 401–413.

Christie, J. M., Reymond, P., Powell, G. K., Bernasconi, P., Raibekas, A. A., Liscum, E., et al. (1998). Arabidopsis NPH1: A flavoprotein with the properties of a photoreceptor for phototropism. *Science*, *282*(5394), 1698–1701.

Christie, J. M., Swartz, T. E., Bogomolni, R. A., & Briggs, W. R. (2002). Phototropin LOV domains exhibit distinct roles in regulating photoreceptor function. *The Plant Journal*, *32*(2), 205–219.

Circolone, F., Granzin, J., Jentzsch, K., Drepper, T., Jaeger, K. E., Willbold, D., et al. (2012). Structural basis for the slow dark recovery of a full-length LOV protein from *Pseudomonas putida*. *Journal of Molecular Biology*, *417*(4), 362–374.

Conrad, K. S., Bilwes, A. M., & Crane, B. R. (2013). Light-induced subunit dissociation by a light-oxygen-voltage domain photoreceptor from *Rhodobacter sphaeroides*. *Biochemistry*, *52*, 378–391.

Conrad, K. S., Manahan, C. C., & Crane, B. R. (2014). Photochemistry of flavoprotein light sensors. *Nature Chemical Biology*, *10*(10), 801–809.

Crane, B. R., & Young, M. W. (2014). Interactive features of proteins composing eukaryotic circadian clocks. *Annual Review of Biochemistry*, *83*(1), 191–219.

Crosson, S., & Moffat, K. (2002). Photoexcited structure of a plant photoreceptor domain reveals a light-driven molecular switch. *Plant Cell*, *14*, 1067–1075.

Crosson, S., Rajagopal, S., & Moffat, K. (2003). The LOV domain family: Photoresponsive signaling modules coupled to diverse output domains. *Biochemistry*, *42*(1), 2–10.

Diederichs, K., McSweeney, S., & Ravelli, R. B. G. (2003). Zero-dose extrapolation as part of macromolecular synchrotron data reduction. *Acta Crystallographica Section D: Biological Crystallography*, *59*, 903–909.

Domratcheva, T., Grigorenko, B. L., Schlichting, I., & Nemukhin, A. V. (2008). Molecular models predict light-induced glutamine tautomerization in BLUF photoreceptors. *Biophysical Journal*, *94*(10), 3872–3879.

Dragnea, V., Waegele, M., Balascuta, S., Bauer, C., & Dragnea, B. (2005). Time-resolved spectroscopic studies of the AppA blue-light receptor BLUF domain from *Rhodobacter sphaeroides*. *Biochemistry*, *44*(49), 15978–15985.

Ellis, M. J., Buffey, S. G., Hough, M. A., & Hasnain, S. S. (2008). On-line optical and X-ray spectroscopies with crystallography: An integrated approach for determining metalloprotein structures in functionally well defined states. *Journal of Synchrotron Radiation*, *15*, 433–439.

Fedorov, R., Schlichting, I., Hartmann, E., Domratcheva, T., Fuhrmann, M., & Hegemann, P. (2003). Crystal structures and molecular mechanism of a light induced signaling switch: The Phot-LOV1 domain from *Chlamydomonas reinhardtii*. *Biophysical Journal*, *84*(4), 2474–2482.

Feitelson, J., & Barboy, N. (1986). Triplet-state reactions of zinc protoporphyrins. *Journal of Physical Chemistry*, *90*(2), 271–274.

Foley, L. E., Gegear, R. J., & Reppert, S. M. (2011). Human cryptochrome exhibits light-dependent magnetosensitivity. *Nature Communications*, *2*, 356.

Franke, D., Petoukhov, M. V., Konarev, P. V., Panjkovich, A., Tuukkanen, A., Mertens, H. D. T., et al. (2017). ATSAS 2.8: A comprehensive data analysis suite for small-angle scattering from macromolecular solutions. *Journal of Applied Crystallography*, *50*, 1212–1225.

Frisell, W. R., Chung, C. W., & Mackenzie, C. G. (1958). Catalysis of oxidation coenzymes of nitrogen compounds in the presence of light. *Journal of Biological Chemistry*, *234*(5), 1297–1302.

Fuchs, M. R., Schleicher, E., Schnegg, A., Kay, C. W. M., Törring, J. T., Bittl, R., et al. (2002). G-tensor of the neutral flavin radical cofactor of DNA photolyase revealed by 360-GHz electron paramagnetic resonance spectroscopy. *The Journal of Physical Chemistry B*, *106*(34), 8885–8890.

Ganguly, A., Manahan, C. C., Top, D., Yee, E. F., Lin, C., Young, M. W., et al. (2016). Changes in active site histidine hydrogen bonding trigger cryptochrome activation. *Proceedings of the National Academy of Sciences of the United States of America*, *113*(36), 10073–10078.

Ganguly, A., Thiel, W., & Crane, B. R. (2017). Glutamine amide flip elicits long distance allosteric responses in the LOV protein vivid. *Journal of the American Chemical Society*, *139*(8), 2972–2980.

Gauden, M., van Stokkum, I. H., Key, J. M., Luhrs, D. C., van Grondelle, R., Hegemann, P., et al. (2006). Hydrogen-bond switching through a radical pair mechanism in a flavin-binding photoreceptor. *Proceedings of the National Academy of Sciences of the United States of America*, *103*(29), 10895–10900.

Gegear, R. J., Foley, L. E., Casselman, A., & Reppert, S. M. (2010). Animal cryptochromes mediate magnetoreception by an unconventional photochemical mechanism. *Nature*, *463*(7282), 804–807.

Gomelsky, M., & Klug, G. (2002). BLUF: A novel FAD-binding domain involved in sensory transduction in microorganisms. *Trends in Biochemical Sciences*, *27*(10), 497–500.

Grinstead, J. S., Hsu, S. T. D., Laan, W., Bonvin, A. M. J. J., Hellingwerf, K. J., Boelens, R., et al. (2006). The solution structure of the AppA BLUF domain: Insight into the mechanism of light-induced signaling. *Chembiochem*, 7(1), 187–193.

Hasegawa, K., Masuda, S., & Ono, T. A. (2005). Spectroscopic analysis of the dark relaxation process of a photocycle in a sensor of blue light using FAD (BLUF) protein Slr1694 of the cyanobacterium *Synechocystis* sp. PCC6803. *Plant and Cell Physiology*, *46*(1), 136–146.

Hefti, M. H., Vervoort, J., & Van Berkel, W. J. H. (2003). Deflavination and reconstitution of flavoproteins: Tackling fold and function. *European Journal of Biochemistry*, *270*(21), 4227–4242.

Heintz, U., & Schlichting, I. (2016). Blue light-induced LOV domain dimerization enhances the affinity of Aureochrome 1a for its target DNA sequence. *eLife*, *5* e11860.

Hense, A., Herman, E., Oldemeyer, S., & Kottke, T. (2015). Proton transfer to flavin stabilizes the signaling state of the blue light receptor plant cryptochrome. *Journal of Biological Chemistry*, *290*(3), 1743–1751.

Herrou, J., & Crosson, S. (2011). Function, structure and mechanism of bacterial photosensory LOV proteins. *Nature Reviews Microbiology*, *9*(10), 713–723.

Hoang, N., Schleicher, E., Kacprzak, S., Bouly, J. P., Picot, M., Wu, W., et al. (2008). Human and Drosophila cryptochromes are light activated by flavin photoreduction in living cells. *PLoS Biology*, *6*(7), 1559–1569.

Hoffman, B. M. (1991). Electron nuclear double resonance (ENDOR) of metalloenzymes. *Accounts of Chemical Research*, *24*(6), 164–170.

Hopkins, J. B., Gillilan, R. E., & Skou, S. (2017). BioXTAS RAW: Improvements to a free open-source program for small-angle X-ray scattering data reduction and analysis. *Journal of Applied Crystallography*, *50*, 1545–1553.

Huang, C. Y., Olieric, V., Ma, P. K., Panepucci, E., Diederichs, K., Wang, M. T., et al. (2015). In meso in situ serial X-ray crystallography of soluble and membrane proteins. *Acta Crystallographica Section D, Structural Biology*, *71*, 1238–1256.

Hunt, S. M., Thompson, S., Elvin, M., & Heintzen, C. (2010). VIVID interacts with the WHITE COLLAR complex and FREQUENCY-interacting RNA helicase to alter light and clock responses in *Neurospora*. *Proceedings of the National Academy of Sciences of the United States of America*, *107*(38), 16709–16714.

Iwata, T., Zhang, Y., Hitomi, K., Getzoff, E. D., & Kandori, H. (2010). Key dynamics of conserved asparagine in a cryptochrome/photolyase family protein by Fourier transform infrared spectroscopy. *Biochemistry*, *49*(41), 8882–8891.

Jeschke, G. (2012). DEER distance measurements on proteins. *Annual Review of Physical Chemistry*, *63*(1), 419–446.

Jones, M. A., Feeney, K. A., Kelly, S. M., & Christie, J. M. (2007). Mutational analysis of phototropin 1 provides insights into the mechanism underlying LOV2 signal transmission. *Journal of Biological Chemistry*, *282*(9), 6405–6414.

Jung, A., Domratcheva, T., Tarutina, M., Wu, Q., Ko, W.-H., Shoeman, R. L., et al. (2005). Structure of a bacterial BLUF photoreceptor: Insights into blue light-mediated signal transduction. *Proceedings of the National Academy of Sciences of the United States of America*, *102*(35), 12350–12355.

Jung, A., Reinstein, J., Domratcheva, T., Shoeman, R. L., & Schlichting, I. (2006). Crystal structures of the AppA BLUF domain photoreceptor provide insights into blue light-mediated signal transduction. *Journal of Molecular Biology*, *362*(4), 717–732.

Kanazawa, T., Ren, S., Maekawa, M., Hasegawa, K., Arisaka, F., Hyodo, M., et al. (2010). Biochemical and physiological characterization of a BLUF protein-EAL protein complex involved in blue light-dependent degradation of cyclic diguanylate in the purple bacterium *Rhodopseudomonas palustris*. *Biochemistry*, *49*(50), 10647–10655.

Kao, Y.-T., Saxena, C., Wang, L., Sancar, A., & Zhong, D. (2005). Direct observation of thymine dimer repair in DNA by photolyase. *Proceedings of the National Academy of Sciences of the United States of America*, *102*(45), 16128–16132.

Kay, C. W. M., Feicht, R., Schulz, K., Sadewater, P., Sancar, A., Bacher, A., et al. (1999). EPR, ENDOR, and TRIPLE resonance spectroscopy on the neutral flavin radical in *Escherichia coli* DNA photolyase. *Biochemistry*, *38*(51), 16740–16748.

Kennis, J. T. M., & Groot, M. L. (2007). Ultrafast spectroscopy of biological photoreceptors. *Current Opinion in Structural Biology*, *17*(5), 623–630.

Kottke, T., Batschauer, A., Ahmad, M., & Heberle, J. (2006). Blue-light-induced changes in Arabidopsis cryptochrome 1 probed by FTIR difference spectroscopy. *Biochemistry*, *45*(8), 2472–2479.

Krauss, U., Minh, B. Q., Losi, A., Gärtner, W., Eggert, T., Von Haeseler, A., et al. (2009). Distribution and phylogeny of light-oxygen-voltage-blue-light-signaling proteins in the three kingdoms of life. *Journal of Bacteriology*, *191*(23), 7234–7242.

Kulik, L. V., Rapatsky, L. L., Pivtsov, A. V., Surovtsev, N. V., Adichtchev, S. V., Grigor'ev, I. A., et al. (2009). Electron-nuclear double resonance study of molecular librations of nitroxides in molecular glasses: Quantum effects at low temperatures, comparison with low-frequency Raman scattering. *The Journal of Chemical Physics*, *131*(6) 064505.

Kutta, R. J., Magerl, K., Kensy, U., & Dick, B. (2015). A search for radical intermediates in the photocycle of LOV domains. *Photochemical & Photobiological Sciences*, *14*(2), 288–299.

Kwiram, A. L. (1971). Electron nuclear double resonance. *Annual Review of Physical Chemistry*, *22*(1), 133–170.

Laan, W., Gauden, M., Yeremenko, S., Van Grondelle, R., Kennis, J. T. M., & Hellingwerf, K. J. (2006). On the mechanism of activation of the BLUF domain of AppA. *Biochemistry*, *45*(1), 51–60.

Lamb, J. S., Zoltowski, B. D., Pabit, S. A., Crane, B. R., & Pollack, L. (2008). Time-resolved dimerization of a PAS-LOV protein measured with photocoupled small angle X-ray scattering. *Journal of the American Chemical Society*, *130*(37), 12226–12227.

Lamb, J. S., Zoltowski, B. D., Pabit, S. A., Li, L., Crane, B. R., & Pollack, L. (2009). Illuminating solution responses of a LOV domain protein with photocoupled small-angle X-ray scattering. *Journal of Molecular Biology*, *393*(4), 909–919.

Leartsakulpanich, U., Antonkine, M. L., & Ferry, J. G. (2000). Site-specific mutational analysis of a novel cysteine motif proposed to ligate the 4Fe-4s cluster in the iron-sulfur flavoprotein of the thermophilic methanoarchaeon Methanosarcina thermophila. *Journal of Bacteriology*, *182*(19), 5309–5316.

Lewis, J. W., & Kliger, D. S. (1993). Microliter flow cell for measurement of irreversible optical absorbance transients. *Review of Scientific Instruments*, *64*(10), 2828–2833.

Li, J., Wang, L. J., & Zhong, D. P. (2013). Ultrafast dynamics of flavins and flavoproteins. In R. Hille, S. Miller, & B. Palfey (Eds.), *Vol. 2. Handbook of flavoproteins: Complex flavoproteins, dehydrogenases and physical methods* (pp. 393–428). Berlin: De Gruyter.

Liedvogel, M., & Mouritsen, H. (2010). Cryptochromes—A potential magnetoreceptor: What do we know and what do we want to know? *Journal of the Royal Society Interface*, 7, S147–S162.

Lin, C., & Todo, T. (2005). The cryptochromes. *Genome Biology*, *6*(5), 220.

Lin, C., Top, D., Manahan, C. C., Young, M. W., & Crane, B. R. (2018). Circadian clock activity of cryptochrome relies on tryptophan-mediated photoreduction. *Proceedings of the National Academy of Sciences of the United States of America*, *115*(15), 3822–3827.

Liu, Z., Tan, C., Guo, X., Kao, Y.-T., Li, J., Wang, L., et al. (2011). Dynamics and mechanism of cyclobutane pyrimidine dimer repair by DNA photolyase. *Proceedings of the National Academy of Sciences of the United States of America*, *108*(36), 14831–14836.

Losi, A., & Gärtner, W. (2012). The evolution of flavin-binding photoreceptors: An ancient chromophore serving trendy blue-light sensors. *Annual Review of Plant Biology*, *63*(1), 49–72.

Losi, A., & Gärtner, W. (2017). Solving blue light riddles: New lessons from flavin-binding LOV photoreceptors. *Photochemistry and Photobiology*, *93*(1), 141–158.

Losi, A., Polverini, E., Quest, B., & Gärtner, W. (2002). First evidence for phototropin-related blue-light receptors in prokaryotes. *Biophysical Journal*, *82*(5), 2627–2634.

Lukacs, A., Brust, R., Haigney, A., Laptenok, S. P., Addison, K., Gil, A., et al. (2014). BLUF domain function does not require a metastable radical intermediate state. *Journal of the American Chemical Society*, *136*(12), 4605–4615.

Massey, V. (1991). A simple method for the determination of redox potentials. In B. Curti, S. Ronchi, & G. Zanetti (Eds.), *Flavins and flavoproteins* (pp. 59–66). Berlin: De Gruyter.

Massey, V., Stankovich, M., & Hemmerich, P. (1978). Light-mediated reduction of flavoproteins with flavins as catalysts. *Biochemistry*, *17*, 1–8.

Masuda, S. (2013). Light detection and signal transduction in the BLUF photoreceptors. *Plant and Cell Physiology*, *54*(2), 171–179.

Masuda, S., & Bauer, C. E. (2002). AppA is a blue light photoreceptor that antirepresses photosynthesis gene expression in *Rhodobacter sphaeroides*. *Cell*, *110*(5), 613–623.

Masuda, S., Hasegawa, K., Ishii, A., & Ono, T. A. (2004). Light-induced structural changes in a putative blue-light receptor with a novel FAD binding fold sensor of blue-light using FAD (BLUF); Slr1694 of *Synechocystis* sp. PCC6803. *Biochemistry*, *43*(18), 5304–5313.

Masuda, S., Hasegawa, K., Ohta, H., & Ono, T. A. (2008). Crucial role in light signal transduction for the conserved Met93 of the BLUF protein PixD/Slr1694. *Plant and Cell Physiology*, *49*(10), 1600–1606.

Masuda, S., Hasegawa, K., & Ono, T. A. (2005a). Light-induced structural changes of apoprotein and chromophore in the sensor of blue light using FAD (BLUF) domain of AppA for a signaling state. *Biochemistry*, *44*(4), 1215–1224.

Masuda, S., Hasegawa, K., & Ono, T. A. (2005b). Tryptophan at position 104 is involved in transforming light signal into changes of β-sheet structure for the signaling state in the BLUF domain of AppA. *Plant and Cell Physiology*, *46*(12), 1894–1901.

Mathes, T., & Götze, J. P. (2015). A proposal for a dipole-generated BLUF domain mechanism. *Frontiers in Molecular Biosciences*, *2*, 62.

Mathes, T., Van Stokkum, I. H. M., Stierl, M., & Kennis, J. T. M. (2012). Redox modulation of flavin and tyrosine determines photoinduced proton-coupled electron transfer and photoactivation of BLUF photoreceptors. *Journal of Biological Chemistry*, *287*(38), 31725–31738.

Mees, A., Klar, T., Gnau, P., Hennecke, U., Eker, A. P. M., Carell, T., et al. (2004). Crystal structure of a photolyase bound to a CPD-like DNA lesion after in situ repair. *Science*, *306*(5702), 1789–1793.

Mehlhorn, J., Steinocher, H., Beck, S., Kennis, J. T. M., Hegemann, P., & Mathes, T. (2013). A set of engineered *Escherichia coli* expression strains for selective isotope and reactivity labeling of amino acid side chains and flavin cofactors. *PLoS One*, *8*(11), e79006.

Meisburger, S. P., Taylor, A. B., Khan, C. A., Zhang, S. N., Fitzpatrick, P. F., & Ando, N. (2016). Domain movements upon activation of phenylalanine hydroxylase characterized by crystallography and chromatography-coupled small-angle X-ray scattering. *Journal of the American Chemical Society*, *138*(20), 6506–6516.

Mertens, H. D. T., & Svergun, D. I. (2010). Structural characterization of proteins and complexes using small-angle X-ray solution scattering. *Journal of Structural Biology*, *172*(1), 128–141.

Misra, S. K., & Freed, J. H. (2011). Distance measurements: Continuous-wave (CW)- and pulsed dipolar EPR. In S. K. Misra (Ed.), *Multifrequency electron paramagnetic resonance* (pp. 545–588). Weinheim: Wiley-VCH.

Möglich, A., & Moffat, K. (2007). Structural basis for light-dependent signaling in the dimeric LOV domain of the photosensor YtvA. *Journal of Molecular Biology*, *373*(1), 112–126.

Müller, F., Hemmerich, P., Ehrenberg, A., Palmer, G., & Massey, V. (1970). The chemical and electronic structure of the neutral flavin radical as revealed by electron spin resonance spectroscopy of chemically and isotopically substituted derivatives. *European Journal of Biochemistry*, *14*(1), 185–196.

Müller, K., & Weber, W. (2013). Optogenetic tools for mammalian systems. *Molecular BioSystems*, *9*(4), 596–608.

Müller, P., Yamamoto, J., Martin, R., Iwai, S., & Brettel, K. (2015). Discovery and functional analysis of a 4th electron-transferring tryptophan conserved exclusively in animal cryptochromes and (6-4) photolyases. *Chemical Communications*, *51*(85), 15502–15505.

Murray, T. A., Foster, M. P., & Swenson, R. P. (2003). Mechanism of flavin mononucleotide cofactor binding to the *Desulfovibrio vulgaris* flavodoxin. 2. Evidence for cooperative conformational changes involving tryptophan 60 in the interaction between the phosphate- and ring-binding subsites. *Biochemistry*, *42*(8), 2317–2327.

Murray, J. W., Rudino-Pinera, E., Owen, R. L., Grininger, M., Ravelli, R. B. G., & Garman, E. F. (2005). Parameters affecting the X-ray dose absorbed by macromolecular crystals. *Journal of Synchrotron Radiation*, *12*, 268–275.

Owen, R. L., Juanhuix, J., & Fuchs, M. (2016). Current advances in synchrotron radiation instrumentation for MX experiments. *Archives of Biochemistry and Biophysics*, *602*, 21–31.

Öztürk, N., Song, S. H., Selby, C. P., & Sancar, A. (2008). Animal type 1 cryptochromes: Analysis of the redox state of the flavin cofactor by site-directed mutagenesis. *Journal of Biological Chemistry*, *283*(6), 3256–3263.

Park, S.-Y., Borbat, P. P., Gonzalez-Bonet, G., Bhatnagar, J., Pollard, A. M., Freed, J. H., et al. (2006). Reconstruction of the chemotaxis receptor-kinase assembly. *Nature Structural & Molecular Biology*, *13*(5), 400–407.

Pathak, G. P., Vrana, J. D., & Tucker, C. L. (2013). Optogenetic control of cell function using engineered photoreceptors. *Biology of the Cell*, *105*(2), 59–72.

Peter, E., Dick, B., & Baeurle, S. A. (2012). Illuminating the early signaling pathway of a fungal light-oxygen-voltage photoreceptor. *Proteins: Structure, Function, and Bioinformatics*, *80*(2), 471–481.

Pudasaini, A., Shim, J. S., Song, Y. H., Shi, H., Kiba, T., Somers, D. E., et al. (2017). Kinetics of the LOV domain of ZEITLUPE determine its circadian function in *Arabidopsis*. *eLife*, *6*, e21646.

Purcell, E. B., McDonald, C. A., Palfey, B. A., & Crosson, S. (2010). An analysis of the solution structure and signaling mechanism of LovK, a sensor histidine kinase integrating light and redox signals. *Biochemistry*, *49*, 6761–6770.

Putnam, C. D., Hammel, M., Hura, G. L., & Tainer, J. A. (2007). X-ray solution scattering (SAXS) combined with crystallography and computation: Defining accurate macromolecular structures, conformations and assemblies in solution. *Quarterly Reviews of Biophysics*, *40*(3), 191–285.

Ravelli, R. B. G., & Garman, E. F. (2006). Radiation damage in macromolecular cryocrystallography. *Current Opinion in Structural Biology*, *16*(5), 624–629.

Rodgers, C. T., & Hore, P. J. (2009). Chemical magnetoreception in birds: The radical pair mechanism. *Proceedings of the National Academy of Sciences of the United States of America*, *106*(2), 353–360.

Rollen, K., Granzin, J., Batra-Safferling, R., & Stadler, A. M. (2018). Small-angle X-ray scattering study of the kinetics of light-dark transition in a LOV protein. *PLoS One*, *13*(7), e0200746.

Samanta, D., Widom, J., Borbat, P. P., Freed, J. H., & Crane, B. R. (2016). Bacterial energy sensor Aer modulates the activity of the chemotaxis kinase CheA based on the redox state of the flavin cofactor. *Journal of Biological Chemistry*, *291*(50), 25809–25814.

Sancar, A. (1994). Structure and function of DNA photolyase. *Biochemistry*, *33*(1), 2–9.

Sancar, A. (2000). Cryptochrome: The second photoactive pigment in the eye and its role in circadian photoreception. *Annual Review of Biochemistry*, *69*(1), 31–67.

Schiemann, O., & Prisner, T. F. (2007). Long-range distance determinations in biomacromolecules by EPR spectroscopy. *Quarterly Reviews of Biophysics*, *40*(1), 1–53.

Schleicher, E., Bittl, R., & Weber, S. (2009). New roles of flavoproteins in molecular cell biology: Blue-light active flavoproteins studied by electron paramagnetic resonance. *FEBS Journal*, *276*(16), 4290–4303.

Schleicher, E., Kowalczyk, R. M., Kay, C. W. M., Hegemann, P., Bacher, A., Fischer, M., et al. (2004). On the reaction mechanism of adduct formation in LOV domains of the plant blue-light receptor phototropin. *Journal of the American Chemical Society*, *126*(35), 11067–11076.

Seifert, S., & Brakmann, S. (2018). LOV domains in the design of photoresponsive enzymes. *ACS Chemical Biology*, *13*(8), 1914–1920.

Selikhanov, G. K., Fando, M. S., Dontsova, M. V., & Gabdulkhakov, A. G. (2018). Investigations of photosensitive proteins by serial crystallography. *Biochemistry (Moscow)*, *83*, S163–S175.

Singh, A. H., Doerks, T., Letunic, I., Raes, J., & Bork, P. (2009). Discovering functional novelty in metagenomes: Examples from light-mediated processes. *Journal of Bacteriology*, *91*(1), 32–41.

Skou, S., Gillilan, R. E., & Ando, N. (2014). Synchrotron-based small-angle X-ray scattering of proteins in solution. *Nature Protocols*, *9*(7), 1727–1739.

Snellenburg, J. J., Laptenok, S. P., Seger, R., Mullen, K. M., & van Stokkum, I. H. M. (2012). Glotaran: A Java-based graphical user interface for the R package TIMP. *Journal of Statistical Software*, *49*(3), 1–22.

Solov'yov, I. A., Chandler, D. E., & Schulten, K. (2007). Magnetic field effects in *Arabidopsis thaliana* cryptochrome-1. *Biophysical Journal*, *92*(8), 2711–2726.

Spence, J. C. H., Weierstall, U., & Chapman, H. N. (2012). X-ray lasers for structural and dynamic biology. *Reports on Progress in Physics*, *75*(10), 102601.

Swartz, T. E., Corchnoy, S. B., Christie, J. M., Lewis, J. W., Szundi, I., Briggs, W. R., et al. (2001). The photocycle of a flavin-binding domain of the blue light photoreceptor phototropin. *Journal of Biological Chemistry*, *276*(39), 36493–36500.

Takala, H., Bjorling, A., Berntsson, O., Lehtivuori, H., Niebling, S., Hoernke, M., et al. (2014). Signal amplification and transduction in phytochrome photosensors. *Nature*, *509*(7499), 245.

Tischer, D., & Weiner, O. D. (2014). Illuminating cell signalling with optogenetic tools. *Nature Reviews Molecular Cell Biology*, *15*(8), 551–558.

Unno, M., Masuda, S., Ono, T. A., & Yamauchi, S. (2006). Orientation of a key glutamine residue in the BLUF domain from AppA revealed by mutagenesis, spectroscopy, and quantum chemical calculations. *Journal of the American Chemical Society*, *128*(17), 5638–5639.

Vaidya, A. T., Chen, C. H., Dunlap, J. C., Loros, J. J., & Crane, B. R. (2011). Structure of a light-activated LOV protein dimer that regulates transcription. *Science Signaling*, *4*(184), ra50.

Vaidya, A. T., Top, D., Manahan, C. C., Tokuda, J. M., Zhang, S., Pollack, L., et al. (2013). Flavin reduction activates Drosophila cryptochrome. *Proceedings of the National Academy of Sciences of the United States of America*, *110*(51), 20455–20460.

Van Willigen, H., Levstein, P. R., & Ebersole, M. H. (1993). Application of Fourier transform electron paramagnetic resonance in the study of photochemical reactions. *Chemical Reviews*, *93*(1), 173–197.

Weber, S. (2005). Light-driven enzymatic catalysis of DNA repair: A review of recent biophysical studies on photolyase. *Biochimica et Biophysica Acta—Bioenergetics*, *1707*(1), 1–23.

Weber, S., Kay, C. W. M., Mogling, H., Mobius, K., Hitomi, K., & Todo, T. (2002). Photoactivation of the flavin cofactor in *Xenopus laevis* (6-4) photolyase: Observation of a transient tyrosyl radical by time-resolved electron paramagnetic resonance. *Proceedings of the National Academy of Sciences of the United States of America*, *99*(3), 1319–1322.

Weil, J. A., Bolton, J. R., & Wertz, J. E. (1994). *Electron paramagnetic resonance: Elementary theory and practical applications*. New York: Wiley.

Wojdyla, J. A., Kaminski, J. W., Panepucci, E., Ebner, S., Wang, X. Q., Gabadinho, J., et al. (2018). DA plus data acquisition and analysis software at the Swiss light source macromolecular crystallography beamlines. *Journal of Synchrotron Radiation*, *25*, 293–303.

Yano, J., Kern, J., Irrgang, K. D., Latimer, M. J., Bergmann, U., Glatzel, P., et al. (2005). X-ray damage to the Mn4Ca complex in single crystals of photosystem II: A case study for metalloprotein crystallography. *Proceedings of the National Academy of Sciences of the United States of America*, *102*(34), 12047–12052.

Yee, E. F., Diensthuber, R. P., Vaidya, A. T., Borbat, P. P., Engelhard, C., Freed, J. H., et al. (2015). Signal transduction in light-oxygen-voltage receptors lacking the adduct-forming cysteine residue. *Nature Communications*, *6*(1), 10079.

Zander, U., Bourenkov, G., Popov, A. N., de Sanctis, D., Svensson, O., McCarthy, A. A., et al. (2015). MeshAndCollect: An automated multi-crystal data-collection workflow for synchrotron macromolecular crystallography beamlines. *Acta Crystallographica Section D, Structural Biology*, *71*, 2328–2343.

Zhu, J., Mathes, T., Hontani, Y., Alexandre, M. T. A., Toh, K. C., Hegemann, P., et al. (2016). Photoadduct formation from the FMN singlet excited state in the LOV2 domain of *Chlamydomonas reinhardtii* phototropin. *Journal of Physical Chemistry Letters*, 7(21), 4380–4384.

Ziegler, T., & Möglich, A. (2015). Photoreceptor engineering. *Frontiers in Molecular Biosciences*, *2*, 30.

Zoltowski, B. D., & Crane, B. R. (2008). Light activation of the LOV protein vivid generates a rapidly exchanging dimer. *Biochemistry*, *47*(27), 7012–7019.

Zoltowski, B. D., & Gardner, K. H. (2011). Tripping the light fantastic: Blue-light photoreceptors as examples of environmentally modulated protein-protein interactions. *Biochemistry*, *50*(1), 4–16.

Zoltowski, B. D., Schwerdtfeger, C., Widom, J., Loros, J. J., Bilwes, A. M., Dunlap, J. C., et al. (2007). Conformational switching in the fungal light sensor Vivid. *Science*, *316*(5827), 1054–1057.

Zoltowski, B. D., Vaccaro, B., & Crane, B. R. (2009). Mechanism-based tuning of a LOV domain photoreceptor. *Nature Chemical Biology*, *5*(11), 827–834.

9780128168295

9780128168295